Information Technology Project Management

Third Edition

Kathy Schwalbe, Ph.D., PMP

Augsburg College

THOMSON

COURSE TECHNOLOGY

Australia • Canada • Mexico • Singapore • Spain • United Kingdom • United States

THOMSON

COURSE TECHNOLOGY

Information Technology Project Management, Third Edition
is published by Course Technology.

Senior Vice President, Publisher
Kristen Duerr

Executive Editor
Jennifer Locke

Product Manager
Janet Aras

Developmental Editor
Karen Lyons

Production Editor
Aimee Poirier

Senior Marketing Manager
Jason Sakos

Text Designer
GEX Publishing Services

Cover Designer
Betsy Young

Cover Art
Rakefet Kenaan

For Dan, Anne, Bobby, and Scott

Preface

The future of many organizations depends on their ability to harness the power of information technology, and good project managers are in high demand. Colleges are responding to this need by establishing courses in project management and making them part of the information technology, management, or engineering curriculum. Corporations are investing in continuing education to help develop information technology project managers and effective project teams. This text provides a much-needed framework for teaching courses in information technology project management. The first and second editions of this text were extremely well received by people in academia and the workplace. The third edition builds on the strengths of the first two editions and adds new, important information and features.

It's impossible to read a newspaper, magazine, or Web page without hearing about the impact of information technology on our society. Information is traveling faster and being shared by more individuals than ever before. You can do your banking, order your groceries, read personalized news, or vote for your favorite entertainer on the World Wide Web. Companies have linked their many systems together to help them fill orders on time and better serve their customers. Software companies are coming out with new products every day to help us streamline our work and get better results. Technology is changing almost everything about the way we live and work today. When this technology works well, it is almost invisible. But did it ever occur to you to ask, "Who makes these complex technologies and systems happen?"

Because you're reading this text, you must have an interest in the "behind-the-scenes" aspects of technology. If I've done my job correctly, as you read you'll begin to see the many innovations society is currently experiencing as the result of thousands of successful information technology projects. In this text, you'll read about projects that went well, like Bank Of America's interstate banking project, Motorola's Six Sigma project, The Boots Company's systems infrastructure project, and Kodak's Advantix Advanced Photo System project. Of course, not all projects are successful; factors such as time, money, and unrealistic expectations, among many others, can sabotage a promising effort if it is not properly managed. In this text, you'll also learn from the mistakes people made on many projects that were not successful. I have written this book in an effort to educate you, tomorrow's information technology project managers, about what will help make a project succeed—and what can make it fail.

Although project management has been an established field for many years, managing information technology projects requires ideas and information that go beyond standard project management. For example, many information technology projects fail because of a lack of user input, incomplete and changing requirements, and a lack of executive support. This book includes suggestions on dealing with these issues. New technologies can also aid in managing information technology projects, and examples of using software to assist in project management are included throughout the book.

Information Technology Project Management, Third Edition is still the only textbook to apply all nine project management knowledge areas—project integration, scope, time, cost, quality, human resource, communications, risk, and procurement management—and all five process groups—initiating, planning, executing, controlling, and closing—to information technology projects. This text builds on the PMBOK® Guide 2000, an American National Standard, to provide a solid framework and context for managing information technology projects. It also includes an appendix, *Guide to Using Microsoft Project 2002*, and a 120-day trial version of Microsoft Project 2002 software. A second appendix provides advice on earning and maintaining Project Management Professional (PMP) certification from the Project Management Institute (PMI) as well as information on other certification programs, such as CompTIA's IT Project+ certification. The third edition features two additional appendices to help you practice and apply the skills learned in this text by providing running cases and templates.

Information Technology Project Management, Third Edition provides practical lessons in project management for students and practitioners alike. By weaving together theory and practice, this text presents an understandable, integrated view of the many concepts, skills, tools, and techniques involved in information technology project management. The comprehensive design of the text provides a strong foundation for students and practitioners in information technology project management.

NEW TO THE THIRD EDITION

Building on the success of the first and second editions, *Information Technology Project Management, Third Edition* introduces a uniquely effective combination of features. The main changes made to the third edition include the following:

■ A new, fictional case study follows a project through the initiating, planning, executing, controlling, and closing project management process groups—all in one chapter. This new case study, JWD Consulting's Project Management Intranet Site Project, appears in Chapter 3, The Project Management Process Groups: A Case Study. The new case study replaces the second edition's five-chapter ResNet case study, which is still available in its entirety on the new companion Web site for this text. The third edition now includes twelve chapters instead of sixteen.

■ A free, companion Web site for the third edition (*www.course.com/mis/ schwalbe*) provides you with access to informative links from the Suggested Readings and footnotes, lecture notes, chapter summaries, project management templates, the entire ResNet case, additional running cases, and many other items to enhance your learning.

- Appendix C, Running Case Studies, is brand new and provides additional, intensive project ideas that you can use throughout the semester to practice your project management skills. Instructors can assign all or parts of the running cases as individual or group assignments. Appendix C includes four running cases, and several more are available on the companion Web site. Each running case includes five parts—initiating, planning, executing, controlling, and closing—with scenario-based information and tasks for you to complete. You will use project management templates, which are provided in the second new appendix, as described next. The instructor manual for this text provides suggested solutions and solution files for all of the running cases.

- Appendix D, Templates, is also new and provides hard copies of more than thirty templates used in this text as well as information on where to find hundreds of additional, free templates. Research shows that organizations that excel in managing projects have a strong toolkit, including many standard templates. The companion Web site for this text includes links to the templates (in Word, Excel, PowerPoint, and Project 2002) as well as links to many other Web sites that provide additional templates. The templates on the companion Web site are organized by knowledge areas and by process groups.

- Updated examples are provided throughout the text. You'll notice several new examples in the third edition to reflect recent events in managing real information technology projects. Several of the What Went Right and What Went Wrong examples have been updated, and many new Suggested Readings have been added to keep you up-to-date. Results of new studies are also included throughout the text.

- User feedback is incorporated. Based on student and faculty feedback on the second edition of this text, you'll see several additional changes, mostly additions, to help clarify information. For example, Chapter 2, The Project Management and Information Technology Context, includes a new section describing the unique aspects of information technology projects. Chapter 8, Project Quality Management, includes detailed information on Six Sigma projects. Chapter 10, Project Communications Management, includes a new section on using e-mail effectively. Learning objectives and discussion questions also map more closely to the major headings in each chapter.

APPROACH

Many people have been practicing some form of project management with little or no formal study in this area. New books and articles are being written each year as we discover more about the field of project management, and project management software continues to advance. Because the project management field and the technology industry are changing rapidly, you cannot assume that what worked twenty years ago is still the best approach today. This text provides up-to-date information on how good project management and effective use of project management software can help you manage information technology projects. Five distinct features of this text include its relationship to the Project Management Body of Knowledge, its bundling with Microsoft Project 2002, its value in preparing for Project Management Professional and other certification exams, its addition of running case studies and templates, and its free, companion Web site.

Based on the PMBOK® Guide 2000

The Project Management Institute (PMI) created the Guide to the Project Management Body of Knowledge (the PMBOK® Guide 2000) as a framework and starting point for understanding project management. It includes an introduction to project management, brief descriptions of all nine project management knowledge areas, and a glossary of terms. The PMBOK® Guide 2000 is, however, just that, a guide. This text uses the PMBOK® Guide 2000 as a foundation, but goes beyond it by providing more details, highlighting additional topics, and providing a real-world context for project management. *Information Technology Project Management, Third Edition* explains project management specifically as it applies to managing information technology projects in the 21st century. It includes several unique features to bring you the excitement of this dynamic field (for more information on features, see the section entitled "Pedagogical Features").

Bundled with Microsoft Project 2002 and a Guide on How to Use It

Software has advanced tremendously in recent years, and it is important for project managers and their teams to use software to help manage information technology projects. Each copy of *Information Technology Project Management, Third Edition* includes a 120-day trial version of the leading project management software on the market—Microsoft Project 2002. Examples using Project 2002 and other software tools are integrated throughout the text, not as an afterthought. Appendix A, *Guide to Using Microsoft Project 2002*, teaches you, step-by-step, to use this powerful software to help in project scope, time, cost, human resource, and communications management. You do not need to buy a separate book to learn how to use Project 2002 effectively, but suggested resources are provided if you want to learn even more about Project 2002.

Resource for PMP and Other Certification Exams

Professional certification is an important factor in recognizing and ensuring quality in a profession. PMI provides certification as a Project Management Professional (PMP), and this text is an excellent resource for studying for the certification exam. This text will also help you pass other certification exams, such as PMI's Certificate of Added Qualification (CAQ) in systems development and CompTIA's IT Project+ exam. Having experience working on projects does not mean you can easily pass the PMP or other certification exams.

I like to tell my students a story about taking a driver's license test after moving to Minnesota. I had been driving very safely and without accidents for over sixteen years, so I thought I could just walk in and take the test. I was impressed by the sophisticated computer system used to administer the test. The questions were displayed on a large touch-screen monitor, often along with an image or video to illustrate different traffic signs or driving situations. I became concerned when I found I had no idea how to answer several questions, and I was perplexed when the test seemed to stop and a message displayed saying, "Please see the person at the service counter." This was a polite way of saying I had failed the test! After controlling my embarrassment, I picked up one of the Minnesota driving test brochures, studied it for an hour or two that night, and successfully passed the test the next day.

The point of this story is that it is important to study information from the organization that creates the test and not be over-confident that your experience is enough. Because this text is based on PMI's PMBOK® Guide 2000, it provides a valuable reference for studying for PMP certification. It is also an excellent reference for CompTIA's IT Project+ exam. I have earned that CompTIA IT Project+ certification and written a book specifically for that exam as well.

Provides Running Cases and Templates

Based on feedback from faculty and students, the third edition now includes two new appendices. Appendix C, Running Cases, provides four cases that you can use to develop your project management skills individually or in a team setting. Each running case provides scenarios and tasks for you to complete that apply your skills in initiating, planning, executing, controlling, and closing projects. Appendix D, Templates, provides more than thirty templates used in this text and links to hundreds more that you can use to help manage your projects.

Includes a Free, Companion Web Site

A free, companion Web site provides you with a one-stop location to access informative links and tools to enhance your learning. Similar to other companion

Web sites provided by Course Technology, this site will be a valuable resource as you download templates, student files, important articles, and references.

Organization and Content

Information Technology Project Management, Third Edition is organized into three main sections to provide a framework for project management, a detailed description of each project management knowledge area, and four appendices to provide practical information for applying project management. The first three chapters form the first section that introduces the project management framework and sets the stage for the remaining chapters. Chapter 1 introduces the critical need for better project management in the information technology field. It provides an overview of the field of project management, including definitions of fundamental terms, the relationship between project management and other disciplines, and a brief history of the field and how it has changed. It also contains an introduction to the project management profession, including information on careers, recent research, certification, ethics, and project management software. Chapter 2 provides a context for project management in general and information technology projects in particular. It focuses on the need to understand how projects fit into an entire organizational system and applies common project management terms to concepts used in the information technology field. It describes the importance of using a systems view in managing all projects, the effects of different organizational structures on project management, the importance of top management support for projects, the difference between the project and product life cycles, the unique attributes and diverse nature of information technology projects, and important skills of good project managers. Chapter 3 describes the project management process groups—initiating, planning, executing, controlling, and closing. Chapter 3 also provides a matrix that relates the process groups to each knowledge areas. It describes how to develop an information technology project management methodology based on the PMBOK® Guide 2000 and unique organizational needs. It also provides a comprehensive case study of an organization applying the project management process groups to manage an intranet project.

Chapters 4 through 12 form the second section of the text that describes each of the project management knowledge areas—project integration, scope, time, cost, quality, human resource, communications, risk, and procurement management—in the context of information technology projects. An entire chapter is dedicated to each knowledge area. Each knowledge area chapter includes sections that map to their major processes as described in the PMBOK® Guide 2000. For example, the chapter on project integration management includes sections on project plan development, project plan execution, and integrated change control. Additional sections highlight other important concepts related to each knowledge area. Each chapter also includes detailed examples of key project management tools and techniques as applied to information technology projects. For example, the chapter on project scope management includes samples of a project charter, net present value analysis, and work breakdown structures for several information technology projects.

Appendices A through D form the third section of the text that provides practical information to help you apply project management skills on real or practice projects. You will learn how to use Project 2002 by following the detailed, step-by-step guide in Appendix A, which includes more than sixty screen shots to help you check your work. Appendix B summarizes what you need to know to earn PMP or other certifications related to project management. Appendix C provides four running cases to help you practice your new skills, and Appendix D provides useful templates you can start using immediately. You can download the files for the templates from the companion Web site for this text, along with many other useful resources.

Pedagogical Features

Several pedagogical features are included in this text to enhance the presentation of the materials so that you can more easily understand the concepts and apply them. Throughout the text, emphasis is placed on applying concepts to up-to-date, real-world information technology project management.

Learning Objectives, Chapter Summaries, Discussion Questions, and Exercises

Learning Objectives, Chapter Summaries, Discussion Questions, and Exercises are designed to function as integrated study tools. Learning Objectives reflect what you should be able to accomplish after completing each chapter. Chapter Summaries highlight key concepts you should master. The Discussion Questions help guide critical thinking about those key concepts. Exercises provide opportunities to practice important techniques.

Opening Case and Case Wrap Up

To set the stage, each chapter begins with an opening case related to the materials in that chapter. These scenarios (most based on the author's experiences) spark your interest and introduce important concepts in a real-world context. As project management concepts and techniques are discussed, they are applied to the opening case and other similar scenarios. Each chapter then closes with a case wrap up—with some ending successfully and some, realistically, failing—to further illustrate the real world of project management.

What Went Right and What Went Wrong

Failures, as much as successes, can be valuable learning experiences. Each chapter of the text includes one or more examples of real information technology projects that went right as well as examples of projects that went wrong. These examples further illustrate the importance of mastering key concepts in each chapter.

Minicases

Minicases provide challenging assignments to help you apply some of the concepts, tools, and techniques discussed in each chapter. Each Minicase has at least two parts based on the same business scenario. Instructors can assign Minicases for homework or use them as in-class exercises to help you apply what you are learning.

Running Cases

The new running cases provided in Appendix C make excellent individual or group project assignments that you can use progressively throughout the course. Four different running cases are included in the third edition, and each case includes five parts—initiating, planning, executing, controlling, and closing—with scenario-based information and several questions to answer or tasks to complete. Several of the questions or tasks involve using the templates provided in Appendix D or from the companion Web site for this text.

Suggested Readings

Every chapter ends with a list of Suggested Readings with annotations that offer opportunities for further research. Several exercises require you to review these readings or research related topics. The World Wide Web is widely referenced throughout the text as a research tool, and many of the Suggested Readings can be found on the Web.

Key Terms

The fields of information technology and project management both include many unique terms that are key to creating a workable language when the two fields are combined. Key terms are displayed in bold face and are defined the first time they appear. Definitions of key terms are provided in alphabetical order at the end of each chapter and in a glossary at the end of the text.

Application Software

Learning becomes much more dynamic with hands-on practice using the top project management software tool in the industry, Microsoft Project 2002, as well as other tools, such as spreadsheet software and Internet browsers. Each chapter offers you many opportunities to get hands-on and build new software skills. This text is written from the point of view that reading about something only gets you so far; in order to understand project management, you have to do it for yourself. In addition to the exercises and Minicases found at the end of each chapter and the running cases in Appendix C, several challenging projects are provided at the end of Appendix A, Guide to Using Microsoft Project 2002.

SUPPLEMENTS

The following supplemental materials are available when this text is used in a classroom setting. All of the teaching tools available with this text are provided to the instructor on a single CD-ROM.

Electronic Instructor's Manual The Instructor's Manual that accompanies this textbook includes additional instructional material to assist in class preparation, including suggestions for lecture topics and additional discussion questions.

ExamView® This textbook is accompanied by ExamView, a powerful testing software package that allows instructors to create and administer printed, computer (LAN-based), and Internet exams. ExamView includes hundreds of questions that correspond to the topics covered in this text, enabling students to generate detailed study guides that include page references for further review. The computer-based and Internet testing components allow students to take exams at their computers, and also save the instructor time by grading each exam automatically.

PowerPoint Presentations This text comes with Microsoft PowerPoint slides for each chapter. These are included as a teaching aid for classroom presentation, to make available to students on the network for chapter review, or to be printed for classroom distribution. Instructors can add their own slides for additional topics they introduce to the class.

Solution Files Solutions to end-of-chapter questions can be found on the Instructor Resource CD-ROM and may also be found on the Course Technology Web site at *www.course.com.* The solutions are password protected.

Distance Learning Course Technology is proud to present online testbanks in WebCT and Blackboard, to provide the most complete and dynamic learning experience possible. We hope you will make the most of your course, both online and offline. For more information on how to bring distance learning to your course, contact your local Course Technology sales representative.

Student Online Companion Visit the companion Web site for this text at *www.course.com/mis/schwalbe* for links from the Suggested Readings and footnotes, lecture notes, chapter summaries, project management templates, Project 2002 student files used in Appendix A, the entire ResNet case, additional running cases, and other helpful resources.

ACKNOWLEDGEMENTS

I never would have taken on this project—writing this book, the first, second, or third edition—without the help of many people. I would like to thank the staff at Course Technology for their dedication and hard work in helping me produce this book and in doing such an excellent job of marketing it. Jennifer Locke was a great project manager and editor in coordinating all of the people involved in producing this book. I owe tremendous gratitude to Marilyn Freedman, my development editor for both the first and second edition, and Karen Lyons, my development editor for the third edition. It was valuable having someone read every word that I wrote and offer helpful suggestions on how to make the book even better. I learned a lot about collaborative writing from Marilyn and Karen. Janet Aras did an excellent job as the product manager and Aimee Poirier was an excellent production editor. It's amazing how many little details are involved in writing a good book, and Janet and the other editors and artists supporting Course Technology did a great job in producing a high quality text.

I thank my many colleagues who contributed information to this book. I again used materials in the project risk management chapter provided by David Jones of Lockheed Martin and an adjunct professor at Augsburg College and the University of St. Thomas; cost management information provided by Jodi Curtis, a member of my PMP study group; and procurement management information provided by Rita Mulcahy, who ran my PMP review course. I also thank Karen Boucher of the Standish Group, Bill Munroe of Blue Cross Blue Shield, and Tess Galati of Practical Communications, Inc. for providing excellent materials for the third edition of this book.

Special thanks go to Rachel Hollstadt, CEO of Hollstadt & Associates, for providing materials on testing for Chapter 8, Project Quality Management, and also for suggesting that I contact people at Northwest Airlines to document their highly successful ResNet projects. I thank Bob Borlik, Vice President of Information Services at Northwest Airlines in the late 1990s, and Lori Manacke, Director of Business Results, for letting me access information and spend time interviewing Arvid Lee and Kathy Christenson. Arvid and Kathy were wonderful collaborators. Their enthusiasm for the ResNet projects was obvious, as were their intelligence, diligence, and fun-loving natures. We had several marathon sessions discussing and documenting the initiating, planning, executing, controlling, and closing of the ResNet projects. Many thanks also go to Peeter Kivestu, the project manager for the ResNet projects. Peeter used many creative techniques to make the ResNet projects so successful.

I also want to thank my students at Augsburg College and the University of Minnesota for providing feedback on the first and second edition of this book. I received many valuable comments from them on ways to improve the text and structure of my courses. I am also grateful for the articles students gave me, which provided additional examples of information technology projects, material for the What Went Right and What Went Wrong features, and additional references. Student papers and discussions on Six Sigma also helped me to further understand that topic and write about it in this edition.

Faculty and staff at Augsburg College were very supportive of my work. I am grateful for summer research money provided by the faculty development office, which encouraged me to begin writing this book. I also thank Lee Clarke at Augsburg College for agreeing to adjust his teaching schedule so I could have more time to write the second and third editions.

Three faculty reviewers provided excellent feedback for me in writing this third edition. Tom Norris of SUNY Cobleskill, James Wetterau of Wake Technical Community College, and Patti Wolf of the University of Maryland University College provided outstanding suggestions for improving the text in this third edition. I also wish to thank the reviewers of the first edition, Barbara Denison, Wright State University; Barbara Grabowski, Benedictine University; Bruce Hartman, Lucent Corporation; and Tom Logan, National American University, as well as reviewers from the second edition, Sherry Thatcher of the University of Arizona, Ann Digilio of The University of Findlay, and Steven White of Anne Arundel Community College. I also thank the many other faculties who have contacted me with praise as well as suggestions for improving this text. I really appreciate the feedback and do my best to incorporate as much as I can.

I thank the people involved in Minnesota's PMI chapter. Cliff Sprague and Michael Branch gave me the opportunity to solicit information for the book at a dinner meeting, and I made several valuable contacts there. I continue to be actively involved with the Minnesota PMI chapter and enjoy working with all of the people involved.

Most of all, I am grateful to my family. Without their support, I never could have written this book. My wonderful husband, Dan, was very patient and supportive of me working like crazy to write the first edition of this book. The second and third editions weren't quite as hectic, but I still needed support from Dan. Dan has written textbooks himself, so he provided some great advice. He even referred to parts of my book to help him in his job as a lead architect for software development with Com Squared Systems, Inc. and helped me in updating some of the sections on new software development programs and techniques. Our three children, Anne, Bobby, and Scott, continue to be very supportive of their mom's work. Our oldest child, Anne, started college in the fall of 2001, and she actually thinks it's cool that her mom wrote a textbook. Our children all understand the main reason why I write—I have a passion for educating future leaders of the world, including them.

As always, I am eager to receive your feedback on this book. Please send all feedback to Course Technology at mis@course.com. My editors will make sure it gets to me!

Kathy Schwalbe, Ph.D., PMP
Augsburg College

ABOUT THE AUTHOR

Kathy Schwalbe is an Associate Professor in the Department of Business Administration at Augsburg College in Minneapolis, where she teaches courses in project management, problem solving for business, systems analysis and design, information systems projects, and electronic commerce. She is the area coordinator for the management information systems major and program manager for Augsburg's Information Technology Certificate program. Kathy is also an adjunct faculty member at the University of Minnesota, where she teaches a graduate-level course in project management. Kathy worked for ten years in industry before entering academia in 1991. She was an Air Force officer, systems analyst, project manager, senior engineer, and information technology consultant. Kathy is an active member of PMI, serving as the Student Chapter Liaison for the Minnesota chapter of PMI, past VP of Education for the Minnesota chapter, past Editor of the ISSIG Review, and member of PMI's test-writing team. Kathy earned her Ph.D. in Higher Education at the University of Minnesota, her MBA at Northeastern University's High Technology MBA program, and her B.S. in mathematics at the University of Notre Dame.

Brief Contents

Contents

Chapter 8 Project Quality Management

Chapter 9 Project Human Resource Management

1

Introduction to Project Management

Objectives

After reading this chapter, you will be able to:

1. *Understand the growing need for better project management, especially for information technology projects*
2. *Explain what a project is and provide examples of information technology projects*
3. *Describe what project management is and discuss key elements of the project management framework*
4. *Discuss how project management relates to other disciplines*
5. *Understand the history of project management*
6. *Describe the project management profession, including recent trends in project management research, certification, and software products*

OPENING CASE

Anne Roberts, the new Chief Project Officer (CPO) for a large retail chain, stood in front of five hundred people in the large corporate auditorium to explain the company's new strategies. She was also broadcasting to thousands of other employees, suppliers, and stockholders throughout the world via the Internet. The company had come a long way in implementing new information systems to improve inventory control, sell products using the Web, and streamline the sales and distribution processes. However, the stock price was down, the nation's economy was weak, and people were anxious to hear about the company's new strategies.

Anne began to address the audience, "Good morning. As many of you know, our CEO promoted me to a new position as Chief Project Officer. Most of what we do is manage projects, and my role in this new position is to turn this company around by helping us effectively select and manage those projects. Our challenge is to develop a culture in which we all work together to provide high-quality goods and services to our consumers while earning a profit in this difficult market. To meet this challenge, we must collaborate to focus on finding solutions to complex problems. We must decide what projects will most benefit the company, how we can continue to leverage the power of information technology to support our business, and how we can exploit our human capital to successfully plan and execute those projects. If we succeed, we'll become a world-class corporation."

"And if we fail?" someone asked from the audience.

"Let's just say that failure is not an option," Anne replied.

INTRODUCTION

Many people and organizations today have a new or renewed interest in project management. Until the 1980s, project management primarily focused on providing schedule and resource data to top management in the military and construction industries. This tracking of a few key project parameters is still an important element, but today's project management involves much more, and people in every industry manage projects. Today, new technologies have become a significant factor in many businesses. Computer hardware, software, networks, and the use of interdisciplinary and global work teams have radically changed the work environment. The statistics below demonstrate the significance of project management in today's society:

- The U.S. spends $2.3 trillion on projects every year, an amount equal to one-quarter of the nation's gross domestic product.
- The world as a whole spends nearly $10 trillion of its $40.7 trillion gross product on projects of all kinds.
- More than sixteen million people regard project management as their profession.
- On average, a project manager earns more than $82,000 per year.[1]
- More than half a million new information technology (IT) application development projects were initiated during 2001, up from 300,000 in 2000.[2]
- Famous business authors and consultants are stressing the importance of project management. As Tom Peters writes in his book, *Reinventing Work: the Project 50*, "To win today you must master the art of the project!"

[1] Project Management Institute (PMI), *The PMI Project Management Fact Book, Second Edition*, 2001.
[2] The Standish Group, "CHAOS 2001: A Recipe for Success" (*www.standishgroup.com*) (2001).

Today's corporations, governments, and non-profit organizations are recognizing that to be successful, they need to be conversant with and use modern project management techniques.

What Went Wrong?

In 1995, the Standish Group published an often-quoted study entitled "CHAOS."[3] This prestigious consulting firm surveyed 365 information technology executive managers in the United States who managed more than 8380 information technology application projects. As the title of the study suggests, the projects were in a state of chaos. United States companies spent more than $250 billion each year in the early 1990s on approximately 175,000 Information technology application development projects. Examples of these projects included creating a new database for a state department of motor vehicles, developing a new system for car rental and hotel reservations, and implementing a client–server architecture for the banking industry. The survey found that the average cost of an information technology application development project for a large company was more than $2.3 million; for a medium company, it was more than $1.3 million; and for a small company, it was more than $434,000. Their study reported that the overall success rate of information technology projects was *only* 16.2 percent. The surveyors defined success as meeting project goals on time and on budget.

The study also found that more than 31 percent of information technology projects were canceled before completion, often due to poor project management. These project cancellations cost U.S. companies and government agencies more than $81 billion. The authors of this study were adamant about the need for better project management in the information technology industry. They explained, "Software development projects are in chaos, and we can no longer imitate the three monkeys—hear no failures, see no failures, speak no failures."[3]

Many organizations claim that using project management provides advantages, such as:

- Better control of financial, physical, and human resources
- Improved customer relations
- Shorter development times
- Lower costs
- Higher quality and increased reliability
- Higher profit margins
- Improved productivity
- Better internal coordination
- Higher worker morale

[3] The Standish Group, "CHAOS" (*www.standishgroup.com*) (1995). See also Johnson, Jim, "CHAOS: The Dollar Drain of IT Project Failures," *Application Development Trends* (January 1995).

This chapter introduces projects and project management, compares project management to other disciplines, gives a brief history of project management, and provides information on this growing profession. Although project management applies to many different industries and types of projects, this textbook focuses on applying project management to information technology projects.

WHAT IS A PROJECT?

To discuss project management, it is important to understand the concept of a project. A **project** is "a temporary endeavor undertaken to create a unique product or service."[4] Projects normally involve several people performing interrelated activities, and the main customer for the project is often interested in the effective use of resources to complete the project in an efficient and timely manner. The following attributes help define a project further:

- *A project has a unique purpose.* Every project should have a well-defined objective. For example, Anne Roberts, the Chief Project Officer in the opening case, might sponsor an information technology collaboration project to develop a list and initial analysis of potential information technology projects that might improve operations for the company. The unique purpose of this project would be to create a collaborative report with ideas from people throughout the company. The results would provide the basis for further discussions and projects. As in this example, projects result in a unique product, service, or outcome.

- *A project is temporary.* A project has a definite beginning and a definite end. In the information technology collaboration project, Anne might form a team of people to immediately work on the project, and then expect a report and an executive presentation of the results in one month.

- *A project requires resources, often from various areas.* Resources include people, hardware, software, or other assets. Many projects cross departmental or other boundaries to achieve their unique purposes. For the information technology collaboration project, people from information technology, marketing, sales, distribution, and other areas of the company would need to work together to develop ideas. The company might also hire outside consultants to provide input. Once the project team has selected key projects, they will probably require additional hardware, software, and network resources. People from other companies—product suppliers and consulting companies—will become resources for meeting project objectives. Resources, however, are limited. They must be used effectively to meet project and other corporate goals.

---

[4] Project Management Institute, Inc., *A Guide to the Project Management Body of Knowledge (PMBOK® Guide)* (2000), p. 4.

■ *A project should have a primary customer or sponsor.* Most projects have many interested parties or stakeholders, but someone must take the primary role of sponsorship. The **project sponsor** usually provides the direction and funding for the project. In this case, Anne Roberts would be the sponsor for the information technology collaboration project. Once further information technology projects are selected, however, the sponsors for those projects would be senior managers in charge of the main parts of the company affected by the projects. For example, if the vice president of sales initiates a project to improve direct product sales using the Internet, he or she might be the project sponsor. If several projects related to Internet technologies were undertaken, the organization might form a program. A **program** is "a group of projects managed in a coordinated way to obtain benefits not available from managing them individually."[5] A program manager provides leadership for the project managers directing those projects, and the sponsors might come from several business areas.

■ *A project involves uncertainty.* Because every project is unique, it is sometimes difficult to clearly define the project's objectives, estimate how long it will take to complete, or how much it will cost. This uncertainty is one of the main reasons project management is so challenging, especially on projects involving new technologies.

A good **project manager** is crucial to a project's success. Project managers work with the project sponsors, the project team, and the other people involved in a project to meet project goals.

Every project is constrained in different ways by its scope, time, and cost goals. These limitations are sometimes referred to in project management as the **triple constraint**. To create a successful project, a project manager must consider scope, time, and cost and balance these three often-competing goals. He or she must consider the following:

■ *Scope:* What is the project trying to accomplish? What unique product or service does the customer or sponsor expect from the project?

■ *Time:* How long should it take to complete the project? What is the project's schedule?

■ *Cost:* What should it cost to complete the project?

Figure 1-1 illustrates the three dimensions of the triple constraint. Each area—scope, time, and cost—has a target at the beginning of the project. For example, the information technology collaboration project might have an initial scope of producing a forty to fifty-page report and a one-hour presentation on thirty potential information technology projects. The project manager might further define project scope by providing a description of each potential project, an investigation of what other companies have implemented for similar projects, a

---

[5] Ibid, p. 10

rough time and cost estimate, and assessments of the risk and potential payoff as high, medium, or low. The initial time estimate for this project might be one month, and the cost estimate might be $50,000. These expectations provide the targets for the scope, time, and cost dimensions of the project.

Successful project management means meeting all three goals (scope, time, and cost)—and satisfying the project's sponsor!

Target

Scope goal

Cost goal

Time goal

Figure 1-1. The Triple Constraint of Project Management

Managing the triple constraint involves making trade-offs between scope, time, and cost goals for a project. For example, you might need to increase the budget for a project in order to meet scope and time goals. Alternatively, you might have to reduce the scope of a project in order to meet time and cost goals. Because of the uncertain nature of projects and competition for resources, it is rare to complete many projects according to the exact scope, time, and cost plans originally predicted. The project's sponsor, team members, or other stakeholders might have different views of the project as time progresses. For example, to generate project ideas, suppose the project manager for the information technology collaboration project began by sending an e-mail survey to all employees. In addition, to coordinate survey responses, suppose the project manager assigned one key contact in each department of the company to the project for one week. Now, suppose the e-mail survey generated only a few good project ideas, and several department contacts were uncooperative or unqualified to work on the project. What would happen if the project team did not have the in-house expertise to analyze the survey data? Should the project manager ask for more money to hire an outside consultant? Should he or she ask for more time? On the other hand, should the project manager renegotiate the scope of the project? As another example, suppose the CEO heard about the project and thought the project team should come up with at least forty potential projects instead of thirty. Should the project team try to accomplish this increase in scope without changing the cost and schedule goals? It is the project manager's job to negotiate with the project team and project sponsor in order to make qualified decisions about scope, time, and cost goals.

Although the triple constraint describes how the basic elements of a project— scope, time, and cost—interrelate, other elements can also play significant roles. Quality is often a key factor in projects, as is customer or sponsor satisfaction. Some people, in fact, refer to the "quadruple" constraint of project management, including quality along with scope, time, and cost. Others believe that quality considerations, including customer satisfaction, must be inherent in setting the scope, time, and cost goals of a project. A project team may meet scope, time, and cost goals but fail to meet quality standards or satisfy their sponsor, if they have not adequately addressed these concerns. For example, Anne Roberts may receive a fifty-page report describing thirty potential information technology projects and hear a presentation on the findings of the report. The project team may have completed the work on time and within the cost constraint, but the quality may have been unacceptable. Anne's view of a project description may be very different from the project team's view.

How can you avoid the problems that occur when you meet scope, time, and cost goals, but lose sight of quality or customer satisfaction? The answer is *good project management, which includes more than meeting the triple constraint.*

WHAT IS PROJECT MANAGEMENT?

Project management is "the application of knowledge, skills, tools, and techniques to project activities in order to meet project requirements."[6] Project managers must not only strive to meet specific scope, time, cost, and quality goals of projects, they must also facilitate the entire process to meet the needs and expectations of the people involved in or affected by project activities.

Figure 1-2 provides a framework for beginning to understand project management. Key elements of this framework include the project stakeholders, project management knowledge areas, project management tools and techniques, and the project's contribution to the enterprise.

Figure 1-2. Project Management Framework

Stakeholders are the people involved in or affected by project activities and include the project sponsor, project team, support staff, customers, users, suppliers, and even opponents to the project. These stakeholders often have very different needs and expectations. For example, building a new house is a well-known example of a project. There are several stakeholders involved in a home construction project.

- The project sponsors would be the potential new homeowners. They would be the people paying for the house who also want to live in it. They might be on a very tight budget, so they would expect the contractor to provide accurate estimates of the costs involved in building the house. They would also need a realistic idea of when they could move in and what type of home they could afford given their budget constraints. The new homeowners would have to make important decisions to keep the

---

[6] Ibid, p. 6

costs of the house within their budget. Can they afford to finish the basement right away? If they can afford to finish the basement, will it affect the projected move-in date? In this example, the project sponsors are also the customers and users for the product, which is the house.

■ The project manager in this example would normally be the general contractor responsible for building the house. He or she needs to work with all the project stakeholders to meet their needs and expectations.

■ The project team for building the house would include several construction workers, electricians, carpenters, and so on. These stakeholders would need to know exactly what work they must do and when they need to do it. They would need to know if the required materials and equipment will be at the construction site or if they are expected to provide the materials and equipment. Their work would need to be coordinated since there are many interrelated factors involved. For example, the carpenter cannot put in kitchen cabinets until the walls are completed.

■ Support staff might include the buyers' employers, the general contractor's secretary, and other people who support other stakeholders. The buyers' employers might expect their employees to still complete their work but allow some flexibility so they can visit the building site or take phone calls related to building the house. The contractor's secretary would support the project by coordinating meetings between the buyers, the contractor, suppliers, and so on.

■ Building a house requires many suppliers. The suppliers would provide the wood, windows, flooring materials, appliances, and so on. Suppliers would expect exact details on what items they need to provide, where and when to deliver those items, and so on.

■ There may or may not be opponents to a project. In this example, there might be a neighbor who opposes the project because the workers are making so much noise that she cannot concentrate on her work at home, or the noise might wake her sleeping children. She might interrupt the workers to voice her complaints or even file a formal complaint. Or the neighborhood might have association rules concerning new home design and construction. If the homeowners did not follow these rules, they might have to halt construction due to legal issues.

As you can see from this example, there are many different stakeholders on projects, and they all have different needs and expectations. People's needs and expectations are important in the beginning and throughout the life of a project. Successful project managers develop good relationships with project stakeholders to understand and meet their needs and expectations.

Project management knowledge areas describe the key competencies that project managers must develop. The center of Figure 1-2 shows the nine knowledge areas of project management. The four core knowledge areas of project management include project scope, time, cost, and quality management.

These are core knowledge areas because they lead to specific project objectives. Brief descriptions of each core knowledge area are below:

- Project scope management involves defining and managing all the work required to complete the project successfully.
- Project time management includes estimating how long it will take to complete the work, developing an acceptable project schedule, and ensuring timely completion of the project.
- Project cost management consists of preparing and managing the budget for the project.
- Project quality management ensures that the project will satisfy the stated or implied needs for which it was undertaken.

The four facilitating knowledge areas of project management are human resources, communications, risk, and procurement management. These are called facilitating areas because they are the processes through which the project objectives are achieved. Brief descriptions of each facilitating knowledge area are below:

- Project human resource management is concerned with making effective use of the people involved with the project.
- Project communications management involves generating, collecting, disseminating, and storing project information.
- Project risk management includes identifying, analyzing, and responding to risks related to the project.
- Project procurement management involves acquiring or procuring goods and services for a project from outside the performing organization.

Project integration management, the ninth knowledge area, is an overarching function that affects and is affected by all of the other knowledge areas. Project managers must have knowledge and skills in all nine of these areas.

Project management tools and techniques assist project managers and their teams in carrying out scope, time, cost, and quality management. Additional tools help project managers and teams carry out human resource, communications, risk, procurement, and integration management. For example, some popular time-management tools and techniques include Gantt charts, network diagrams, and critical path analysis. Table 1-1 lists some commonly used tools and techniques by knowledge area. You will learn more about these and other tools and techniques throughout this text.

Project managers must work with key stakeholders to define what constitutes success for a particular project and strive to complete their projects successfully by applying appropriate tools and techniques. In many organizations, project managers also support an emerging business strategy of **project portfolio management**, in which organizations group and manage projects as a portfolio of investments that contribute to the entire enterprise's success. You will learn more about project portfolio management in Chapter 7, Project Cost Management.

Table 1-1: Common Project Management Tools and Techniques by Knowledge Area

KNOWLEDGE AREA/CATEGORY	TOOLS AND TECHNIQUES
Integration management	Stakeholder analysis
	Project plans
	Project management software
	Change control boards
	Configuration management
	Project review meetings
	Work authorization systems
	Project leadership
	Executive sponsorship
Scope management	Net present value, return on investment, payback
	Weighted scoring models
	Business cases
	Project charters
	Scope statements
	Work breakdown structures
	Statements of work
	Requirements analysis
	Scope change control
Time management	Gantt charts
	Network diagrams
	Critical path analysis
	Program evaluation review technique
	Critical chain scheduling
	Crashing
	Fast tracking
	Milestone reviews
Cost management	Earned value management
	Project portfolio management
	Cost estimates
	Cost management plan
	Financial software
Quality management	Six sigma
	Quality control charts
	Pareto diagrams
	Fishbone or Ishikawa diagrams
	Quality audits
	Maturity models
	Statistical methods

Table 1-1: Common Project Management Tools and Techniques by Knowledge Area (continued)

KNOWLEDGE AREA/CATEGORY	TOOLS AND TECHNIQUES
Human resource management	Motivation techniques
	Empathic listening
	Team contracts
	Responsibility assignment matrices
	Resource histograms
	Resource loading
	Resource leveling
	Team-building exercises
Communications management	Communications management plan
	Conflict management
	Communications media selection
	Communications infrastructure
	Status reports
	Meetings
	Virtual communications
	Templates
	Project Web sites
Procurement management	Make-or-buy analysis
	Contracts
	Requests for Proposals or Quotes
	Source selection
	Negotiating
	E-procurement
Risk management	Risk management plan
	Probability/impact matrix
	Risk ranking
	Monte Carlo simulation
	Top-Ten Risk Item Tracking

What Went Right?

A follow-up study done by the Standish Group showed some improvement in the statistics for information technology projects. The cost of failed projects went down from $81 billion in 1995 to an estimated $75 billion in 1998. There was a major decrease in cost overruns from the $59 billion spent in 1995 to the estimated $22 billion in 1998. The study also showed that 26 percent of information technology projects succeeded in meeting scope, time, and cost goals. In spite of these improvements, 46 percent of information technology projects were challenged (completed over budget and past the original deadline) and 28 percent failed.[7]

[7] The Standish Group, "CHAOS: A Recipe for Success" (1999).

The 2001 Standish Group report showed decided improvement in information technology project management compared to the 1995 study.

■ Time overruns significantly decreased to 163%, compared to 222%.

■ Cost overruns were down to 145%, compared to 189%.

■ Required features and functions were up to 67%, compared to 61%.

■ 78,000 U.S. projects were successful, compared to 28,000.

■ 28% of information technology projects succeeded, compared to 16%.[8]

Even though there have been significant improvements in managing information technology projects, there is still much room for improvement. The best news is that project managers are learning how to succeed more often. "The reasons for the increase in successful projects vary. First, the average cost of a project has been more than cut in half. Better tools have been created to monitor and control progress and better skilled project managers with better management processes are being used. The fact that there are processes is significant in itself."[9]

Despite its advantages, project management is not a silver bullet that guarantees success on all projects. Project management is a very broad, often complex discipline. What works on one project may not work on another, so it is essential for project managers to continue to develop their knowledge and skills in managing projects. The unique nature of projects, their stakeholders, their organizations, and their business environments provide many challenges that excite many people about working in project management.

HOW PROJECT MANAGEMENT RELATES TO OTHER DISCIPLINES

Much of the knowledge needed to manage projects is unique to the discipline of project management. However, project managers must also have knowledge and experience in general management and must understand the application area of the project in order to work effectively with specific industry groups and technologies. For example, project managers must understand general management concepts such as organizational behavior, financial analysis, and planning techniques, to name a few. If a project involves sales force automation, the project manager needs to understand the sales process, sales automation software, and mobile computing.

Although being a project manager requires some knowledge of and practice in general management areas, the role of a project manager is different from the

[8] The Standish Group, "CHAOS 2001: A Recipe for Success" (2001).
[9] Ibid.

role of a corporate manager or executive. The nature of projects distinguishes project management from general or operations management. Since projects are unique, temporary, and involve various resources, project managers must focus on integrating all the various activities required to complete the project successfully. In contrast, most of the tasks performed by a general manager or operations manager are repetitive, ongoing, and done as day-to-day activities.

General or operations managers also focus on a particular discipline or functional area. For example, a manager of an accounting department focuses on the discipline of accounting. If a project manager is hired to manage an information technology project for the accounting department, then he or she would need to know some things about accounting as well as information technology. However, the project manager's responsibility would be to manage the project, not to perform accounting or information technology functions.

Project management also requires some knowledge of the particular industry or knowledge area of the project itself. For example, this textbook focuses on information technology projects—projects that involve computer hardware, computer software, and telecommunications technology. It would be very difficult for someone with little or no background in information technology to become the project manager for a large information technology project. Not only would it be difficult to work with other managers and suppliers, it would also be difficult to earn the respect of the project team. However, project managers for large information technology projects do not have to be experts in the field of information technology. They should have some working knowledge of various technologies, but more important, they should understand how the project they are managing would enhance the business. Many companies have found that a good business manager can be a very good information technology project manager because both roles focus on meeting business needs and rely on key project members to handle the technical details. Although information technology project managers need to draw on their information technology expertise or the expertise of key team members, they must spend more time becoming better project managers and less time becoming information technology experts in order to successfully lead their project teams.

Lively debates continue on the differences between managing information technology projects and managing other types of projects. There are several differences, but there are even more similarities. Several articles and speakers joke about the differences between construction projects and software development projects. No, you cannot blow up an old information system like you can an old building and start from scratch. No, there are often no specific engineering principles and building codes that everyone knows and follows. Nevertheless, information technology project managers, like all project managers, still have the responsibility for working with their sponsors, project teams, and other stakeholders to achieve specific project and organizational goals. All project managers should continue to develop their knowledge and experience in project management, general management, and the industries they support.

HISTORY OF PROJECT MANAGEMENT

Some people might argue that building the Egyptian pyramids was a project. Others might argue that building the Great Wall of China was also a project. Most people agree, however, that the modern concept of project management began with the Manhattan Project, which the U.S. military led to develop the atomic bomb. The Manhattan Project involved many people with different skills at several different locations. It also clearly separated the overall management of the project's mission, schedule, and budget under General Leslie R. Groves and the technical management of the project under Dr. Robert Oppenheimer. The Manhattan Project lasted about three years and cost almost $2 billion in 1946.

In developing the project, the military realized that scientists and other technical specialists often did not have the desire or the necessary skills to manage large projects. For example, after being asked several times for each team member's responsibilities at the new Los Alamos laboratory in 1943, Dr. Oppenheimer threw a piece of paper with an organization chart on it at his director and said, "Here's your damn organization chart."[10]

The military was the key industry behind the development of several project management techniques. In 1917, Henry Gantt developed the famous Gantt chart as a tool for scheduling work in factories. A **Gantt chart** is a standard format for displaying project schedule information by listing project activities and their corresponding start and finish dates in a calendar format. Managers drew Gantt charts by hand to show project tasks and schedule information, and this tool provided a standard format for planning and reviewing all the work on early military projects.

Today's project managers still use the Gantt chart as the primary tool to communicate project schedule information, but with the aid of computers, it is no longer necessary to draw the charts by hand. Figure 1-3 displays a Gantt chart for developing an intranet. This version of the chart was created with Microsoft Project, the most widely used project management software today. You will learn more about using Project 2002 in Appendix A.

[10] The Regents of the University of California, Manhattan Project History. "Here's Your Damned Organization Chart" (*www.childrenofthemanhattanproject.org*) (1998–2001).

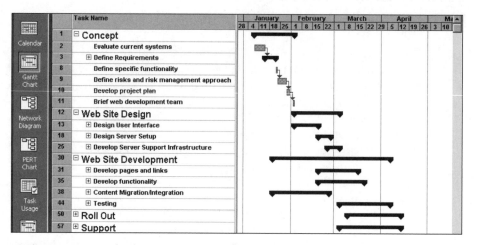

Figure 1-3. Sample Gantt Chart in Microsoft Project 2002

Members of the Navy Polaris missile/submarine project first used network diagrams in 1958. These diagrams helped managers model the relationships among project tasks, which allowed them to create schedules that were more realistic. Figure 1-4 displays a network diagram created using Microsoft Project 2002. Note that the diagram includes arrows that show which tasks are related and the sequence in which team members must perform the tasks. The concept of determining relationships among tasks is essential in helping to improve project scheduling. This concept allows you to find and monitor the **critical path**—the longest path through a network diagram that determines the earliest completion of a project. You will learn more about Gantt charts, network diagrams, critical path analysis, and other time management concepts in Chapter 6, Project Time Management.

Figure 1-4. Sample Network Diagram in Microsoft Project 2002

By the 1970s, the military had begun to use software to help manage large projects. Early project management software products were very expensive and ran on mainframe computers. For example, Artemis was an early project management software product that helped managers analyze complex schedules for designing aircraft. A full-time employee was often required to run the complicated software, and expensive pen plotters were used to draw network diagrams and Gantt charts. As computer hardware became smaller and more affordable and software became graphical and easy to use, project management software became less expensive and more widely used. Today, many industries use project management software on all types and sizes of projects. New software makes basic tools, such as Gantt charts and network diagrams, inexpensive, easy to create, and available for anyone to update. See the section in this chapter on project management software for more information.

In the latter part of the twentieth century, people in virtually every industry began to investigate and apply different aspects of project management to their projects. The sophistication and effectiveness with which project management tools are being applied and used today is influencing the way companies do business, use resources, and respond to market requirements with speed and accuracy. Web-based project management tools are now available to help organizations manage individual projects as well as portfolios of projects. Figure 1-5 provides a sample chart that a project manager could use for an enterprise project management solution. **Enterprise project management software** integrates information from multiple projects to show the status of active, approved, and future projects across an entire organization and provides links to more detailed information. Summary charts often show status as green to indicate things are going well, yellow to indicate that there are some problems, and red to indicate major problems.

Company ABC Project Portfolio				
Project Name	Scope	Schedule	Budget	Links
Active Projects				
Project 1	○	◉	◉	
Project 2	●	●	◉	
Project 3	○	○	○	
Project 4	○	◉	●	
Approved Projects				
Project 10	○	○	○	
Project 11	○	○	○	
Project 12	○	○	○	
Project 13	○	○	○	
Project 14	○	○	○	
Opportunities				
Project 100				
Project 200				
○	White = going well			
◉	Gray = some problems			
●	Black = major problems			

Figure 1-5. Sample Enterprise Project Management Tool

Generally, a project manager is someone leading the construction of a new sports arena, planning a fund-raiser for a charitable organization, or managing development of an electronic commerce application. Being an information technology project manager involves understanding project management, the information technology function, and the business environment of the project. Many organizations, including the one described in the opening case, are also moving toward enterprise and portfolio project management to organize and maximize their investments in projects. All of these examples illustrate one key point: no matter what industry, you need to understand the problems at hand if you are to manage projects successfully. The project manager's challenge is to understand the concepts of project management and determine what tools and techniques should be applied on specific projects and in specific organizations.

THE PROJECT MANAGEMENT PROFESSION

Like it or not, change is a constant in this world, especially in information technology. Year-2000 issues dominated the work of many information technology departments in the late 1990s. The Year-2000 focus produced unmet demand for other types of information technology projects—projects meant to exploit

Web-based applications, take advantage of faster telecommunications technologies, establish data warehousing capabilities, and so on. Having experienced the backlog resulting from this demand, many companies needed to adopt a far more rigorous approach to project management to make up for lost time while improving the success rate of information technology projects. In the early 2000s, organizations had to respond to societal and economic changes to remain competitive. To meet these demands, project managers are needed in all disciplines, especially information technology, and researchers and practitioners need to continue to develop the project management profession.

Project Management Careers

Paul Ziv, a recruitment strategist at ComputerJobs.com, described why project management made the top-ten list of information technology skills in demand in late 2002. He explained that information technology project managers are expected to do much more today. They have to understand the field and possess an executive skill set in order to lead teams in developing products and services that improve the bottom line. Table 1-2 summarizes the top ten information technology skills and average salaries based on job postings in 2002.

Table 1-2: Top Ten Most In Demand Information Technology Skills

RANK	IT SKILL/JOB	AVERAGE ANNUAL SALARY
1	SQL Database Analyst	$80,664
2	Oracle Database Analyst	$87,144
3	C/C++ Programmer	$95,829
4	Visual Basic Programmer	$76,903
5	E-Commerce/Java Developer	$89,163
6	Windows NT/2000 Expert	$80,639
7	Windows/Java Developer	$93,785
8	Security Architect	$86,881
9	Project Manager	$95,719
10	Network Engineer	$82,906

Paul Ziv, "The Top 10 IT Skills in Demand," Global Knowledge Webcast (*www.globalknowledge.com*) (11/20/2002).

Many people recognize the need to fill these other information technology careers, but most are unaware of the need for project managers or do not really understand what project management is all about. Each of the positions listed above is carried out in conjunction with the role of project management as the specific technologies are deployed. For example, an SQL database analyst would be a team member on a project that requires development or support of SQL databases.

Most people in information technology work on projects in some capacity. Some people become project managers, at least on a part-time basis, very early in their careers. Many people think you need many years of experience and vast technical knowledge to be a project manager. Although this need for experience and knowledge might be the case for very large, complex, and expensive projects, many information technology projects can be, and are, led by project managers who are just starting their careers. Project managers need some general management and technical knowledge, but they primarily need the skills and desire to be project managers.

The project management profession is growing at a very rapid pace. Although many professional societies are suffering from declining memberships, the **Project Management Institute (PMI)**, an international professional society for project managers, has continued to attract and retain members, reporting more than 100,000 members worldwide in early 2003. A large percentage of PMI members work in the information technology field and more than 15,000 pay additional dues to join the Information Systems Specific Interest Group.[11] Because there are so many people working on projects in various industries, PMI has created Specific Interest Groups (SIGs) that enable people to share ideas about project management in their particular application areas, such as information systems. PMI also has SIGs for aerospace/defense, financial services, healthcare, hospitality management, manufacturing, new product development, retail, and urban development, to name a few. Consult PMI's Web site (*www.pmi.org*) for more information.

In the 1990s, many companies began creating project management offices to help them handle the increasing number and complexity of projects. A **Project Management Office (PMO)** is an organizational group responsible for coordinating the project management function throughout an organization. There are different ways to structure a PMO, and they can have various roles and responsibilities. Below are possible goals of a PMO:

- Collect, organize, and integrate project data for the entire organization
- Develop and maintain templates for project documents
- Develop or coordinate training in various project management topics
- Develop and provide a formal career path for project managers
- Provide project management consulting services
- Provide a structure to house project managers while they are acting in those roles or between projects

---

[11] Project Management Institute, "PMI Today," (January 2003)

Many colleges, universities, and firms now offer courses related to various aspects of project management. The complexities of managing projects, the publicity surrounding project management, and the belief that project management really can make a difference continue to contribute to the growth of this field.

Project Management Research

Several individuals and organizations have made significant strides in advancing the study of topics important to the field of project management. There are several degree programs available, and an increasing number of people are earning Master's degrees and doctorates in project management. Even more common are the research studies funded by corporations, governments, and professional societies to help understand and advance the field of project management.

As already noted, groups like the Project Management Institute (PMI) and the Standish Group are actively studying various aspects of project management. PMI held its second international research conference on project management in July 2002. More than 100 academics and 200 practitioners exchanged ideas about important issues related to the profession. The Standish Group continues to hold its CHAOS University to update and analyze information related to information technology projects. *The International Journal of Project Management* and other publications provide a means for disseminating new project management research.

This text, *Information Technology Project Management, Third Edition*, summarizes the results of several important research studies throughout, and the companion Web site (available at *www.course.com/mis/schwalbe*) provides links to several articles and sites about project management.

Project Management Certification

Professional certification is an important factor in recognizing and ensuring quality in a profession. PMI provides certification as a **Project Management Professional (PMP)**—someone who has documented sufficient project experience and education, agreed to follow the PMI code of professional conduct, and demonstrated knowledge of the field of project management by passing a comprehensive examination. Appendix B provides more information on PMP certification as well as other certification programs, such as CompTIA's IT Project+ certification.

The number of people earning PMP certification continues to increase. In 1993, there were about 1,000 certified project management professionals. By the end of 2002, there were 52,443 certified project management professionals. Figure 1-6 shows the rapid growth in the number of people earning project management professional certification from 1993 to 2002.

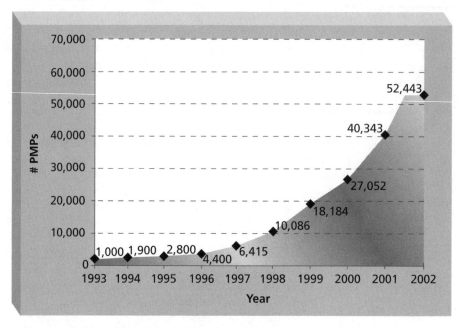

Figure 1-6. Growth in PMP Certification, 1993–2002

Several studies show that organizations supporting technical certification programs tend to operate in more complex information technology environments and are more efficient than companies that do not support certification. Likewise, organizations that support PMP certification recognize the value of investing in programs to improve their employees' knowledge of project management. Many employers today require specific certification to ensure their workers have current skills, and job seekers find that they often have an advantage when they earn and maintain marketable certifications.

As information technology projects become more complex and global in nature, the need for people with demonstrated knowledge and skills in project management will continue. Just as passing the CPA exam is a standard for accountants, passing the PMP exam is becoming a standard for project managers. Some companies require that all project managers be PMP certified. Project management certification is also enabling professionals in the field to share a common base of knowledge. For example, any person with PMP certification can list, describe, and implement the nine project management knowledge areas. Sharing a common base of knowledge is important because it helps advance the theory and practice of project management.

Ethics in Project Management

Ethics is an important part of all professions. Project managers often face ethical dilemmas. For example, several projects involve different payment methods, such as a lump sum payment for a specific job or a specified fee per hour for services. If a project manager can make more money on a job that pays an hourly fee by taking more time to do the work, should he or she take more time than needed? If the project manager reports taking one hundred hours to do tasks that only took eighty hours, will the customer be able to find out? If a project manager is personally opposed to the development of nuclear weapons, should he or she refuse to manage a project that helps produce them? It is important for project managers to make decisions in an ethical manner and act as a role model for their project team members, suppliers, and other project stakeholders.

PMI developed a PMP code of professional conduct that all applicants must agree to in order to become certified project management professionals (PMPs). PMI states that it is vital for all PMPs to conduct their work in an ethical manner. Conducting work in an ethical manner helps the profession earn the confidence of the public, employers, employees, and project team members. The PMP code of professional conduct lists responsibilities to the profession, such as compliance with all organizational rules and policies, professional practice, and advancement of the profession. It also lists responsibilities to customers and the public, such as qualifications, experience, and performance of professional services as well as conflict-of-interest situations. For example, one of the responsibilities to customers and the public includes refraining "from offering or accepting inappropriate payments, gifts or other forms of compensation for personal gain, unless in conformity with applicable laws or customs of the country where project management services are being provided."[12] PMI even added a new series of question to the PMP certification exam in March 2002 to emphasize the importance of professional responsibility. See Appendix B for more information.

Project Management Software

Unlike the cobbler neglecting to make shoes for his own children, the project management and software development communities have definitely responded to the need to provide more software to assist in managing projects. PMI published a Project Management Software Survey in 1999 that describes, compares, and contrasts more than 200 project management software tools. The Project Management Center, a Web site for people involved in project management, provides an alphabetical listing of and links to hundreds of products that help manage projects.[13] This site and others demonstrate the growth in available project management software products, especially Web-based tools. Deciding which project management software to use has become a project in itself. This section provides a summary of the basic types of project management software available and references for finding more information. In

[12] Project Management Institute, Project Management Institute Certification Handbook, (January 2003), p. 22.
[13] The Project Management Center (*www.infogoal.com/pmc/pmcswr.htm*) (January 2003).

Appendix A, you will learn how to use Microsoft Project 2002, the most widely used project management software tool today.

Many people still use basic productivity software, such as Microsoft Word or Excel, to perform many project management functions, such as determining project scope, time, and cost, assigning resources, preparing project documentation, and so on. People often use productivity software instead of specialized project management software because they already have it and know how to use it. However, there are hundreds of project management software tools that provide specific functionality for managing projects. These project management software tools can be divided into three general categories based on functionality and price.

- *Low-end tools:* These tools provide basic project management features and generally cost less than $200 per user. They are often recommended for small projects and single users. Most of these tools allow users to create Gantt charts, which cannot be done using current productivity software. For example, Milestones Simplicity by KIDASA Software, Inc., has a Schedule Setup Wizard that walks users through simple steps to produce a Gantt chart. For $49 per user, this tool also includes a large assortment of symbols, flexible formatting, an outlining utility, and an Internet Publishing Wizard. Another product, called "How's it going?" was written in Microsoft Access 97, 2000, and XP by LogicAbility. For $145 per user, this tool includes an online guide and templates for many project management deliverables, such as reports for project tracking, status reporting, budgeting, time reporting and resource management features, and scheduling features for creating Gantt charts and performing critical path analysis.

- *Midrange tools:* A step up from low-end tools, midrange tools are designed to handle larger projects, multiple users, and multiple projects. All of these tools can produce Gantt charts and network diagrams, and can assist in critical path analysis, resource allocation, project tracking, status reporting, and so on. Prices range from about $200 to $500 per user, and several tools require additional server software for using workgroup features. Microsoft Project is still the most widely used project management software used today, and Project 2002 includes an enterprise version, as described briefly below and in more detail in Appendix A. Other companies that sell midrange project management tools include Artemis, PlanView, Primavera, and Welcom, to name just a few.

- *High-end tools:* Another category of project management software is high-end tools, sometimes referred to as enterprise project management software. These tools provide robust capabilities to handle very large projects, dispersed workgroups, and enterprise functions that summarize and combine individual project information to provide an enterprise view of all projects. These products are generally licensed on a per-user basis, integrate with enterprise database management software, and are accessible via the Internet. For example, Pacific Edge provides several products to help organizations manage portfolios of projects across the enterprise. Bell System uses Niku's Workbench, formerly known as Project Workbench, to manage many of their projects. In mid 2002, Microsoft introduced their Enterprise Project Management version of Project 2002, which requires SQL server and several other components. Several companies that provide midrange tools are also starting to offer enterprise versions of their software. See the Project Management Center Web site or similar sites for links to many companies that provide project management software.

As mentioned earlier, there are many reasons to study project management, particularly as it relates to information technology projects. The number of information technology projects continues to grow, the complexity of these projects continues to increase, and the profession of project management continues to expand and mature. As more people study and work in this important field, the success rate of information technology projects should continue to improve.

CASE WRAP-UP

Anne Roberts worked with the other VPs and the CEO to begin reorganizing several parts of the company to support their new emphasis on projects. They formed a project team to implement an enterprise project management software tool across the organization. They formed another team to develop project-based reward systems for all employees. They also authorized funds for a project to educate all employees in project management and to develop a mentoring program. Anne had successfully convinced everyone that effectively managing projects was crucial to their company's future.

CHAPTER SUMMARY

A project is a temporary endeavor undertaken to create a unique product or service. Projects are unique, temporary, require resources, have a sponsor, and involve uncertainty.

The triple constraint of project management refers to managing the scope, time, and cost dimensions of a project.

Project management is the application of knowledge, skills, tools, and techniques to project activities in order to meet project requirements. Stakeholders are the people involved in or affected by project activities.

A framework for project management includes the project stakeholders, project management knowledge areas, and project management tools and techniques. The nine knowledge areas are project integration management, scope, time, cost, quality, human resources, communications, risk, and procurement management.

Project portfolio management involves organizing and managing projects as a portfolio of investments that contribute to the entire enterprise's success.

The discipline of information technology is focusing on improving project management methods to increase the success rate of information technology projects. A 1995 study found that only 16 percent of information technology projects met scope, time, and cost goals. This performance improved to 28 percent in a 2001 follow-up study, but there is still great room for improvement.

Project management overlaps somewhat with general management and application area knowledge. Project managers, however, use unique project management skills to focus on integrating all of the knowledge areas required to successfully complete a project.

The Manhattan Project was the first project that used modern project management techniques. The Gantt chart, network diagram, and critical path analysis are project-scheduling tools that were developed in the early 1900s. Today, project management is used in some form in virtually all organizations and disciplines, and hundreds of project management software products are available to assist people in managing projects.

The need for more project managers, especially in information technology, is growing. The Project Management Institute (PMI) is an international professional society that provides certification as a Project Management Professional (PMP). PMI publishes many documents, including comparisons of the hundreds of project management software products available. These products fall into three basic categories: low-end tools, midrange tools, and high-end or enterprise tools. Microsoft Project is the most widely used project management software tool today.

DISCUSSION QUESTIONS

1. Why is there a new or renewed interest in the field of project management?
2. What is a project? How is it different from what most people do in their day-to-day jobs?
3. What is project management? Briefly describe the project management framework.
4. How does project management relate to other disciplines?
5. Briefly describe some key events in the history of project management. What project was the first to use "modern" project management?

6. What is the career outlook for project managers in information technology? Describe recent trends in project management certification and software tools.

EXERCISES

1. Visit the Standish Group's Web site (*www.standishgroup.com*). Read one of the CHAOS articles or a similar report on information technology project management on their site or a similar site. Write a one-page summary of the report, its key conclusions, and your opinion of the report.

2. Find someone who works as a project manager or is a member of a project team. Prepare several interview questions, and then ask that person your questions in person, via the phone, or via the Internet. Write a one- to two-page summary of your findings.

3. Search the Internet for the terms *project management*, *project management careers*, *project portfolio management*, and *information technology project management*. Write down the number of hits that you received for each of these phrases. Find at least three Web sites that provide interesting information on one of the topics. In a one- to two-page paper, summarize key information about these three Web sites, as well as the Project Management Institute's Web site (*www.pmi.org*).

4. Scan information technology industry journals or Web sites such as *Information Week*, *Computer World*, and *Information World*, and find one good article about information technology project management. Write a one- to two-page paper describing the article.

5. Skim through Appendix A on Microsoft Project 2002. Review information about Project 2002 from Microsoft's Web site (*www.microsoft.com/project*). Write a one- to two-page paper answering the following questions:

    a. What functions does this software provide that cannot be done easily using other tools such as a spreadsheet or database?

    b. Does the software appear easy to learn?

    c. What are key inputs and outputs of the software?

    d. What is the main difference between the enterprise version of Microsoft Project 2002 and Project Standard 2002?

SUGGESTED READINGS

1. Ibbs, C. William and Young H. Kwak. "Calculating Project Management's Return on Investment." *Project Management Journal* (*www.ce.berkeley.edu/pmroi/calculating-PMROI.pdf*) (June 2000).

    Many companies want proof that project management is a good investment. This paper describes a procedure to measure the return on investment for project management. Information is based on 38 companies and government agencies in four different industries.

2. Peters, Tom. *The Project 50 (Reinventing Work): Fifty Ways to Transform Every "Task" into a Project That Matters!* Alfred A. Knopf, Inc. (1999).

> *Tom Peters is a well-known author and speaker in business. This short book stresses the importance of projects and in following them from start to finish. Peters breaks the WOW or "Way Cool" project into four stages (create, sell, implement, and exit) and provides 50 lists to follow under these stages.*

3. Project Management Institute Web site (*www.pmi.org*).

> *PMI's Web site changes often, but it provides useful information on project management as well as links to other sites, several articles and publications, and their online bookstore. Students can join PMI at reduced rates and receive copies of* PM Network *magazine monthly and the* Project Management Journal *quarterly.*

4. Project Management Institute. Proceedings of PMI Research Conference 2002 (2002) and *Project Management Research at the Turn of the Millennium* (2000).

> *PMI sponsored its first international research conference in Paris, France in June 2000, and a second research conference in Seattle in July 2002. These documents contain almost one hundred invited papers written by some of the best researchers in the field of project management.*

5. The Standish Group. CHAOS (1995). (*www.standishgroup.com*).

> *This article is often quoted for its survey results showing the poor track record for information technology projects. It provides statistics and good examples of successful and unsuccessful information technology projects. This report is based on data collected in 1994 (published in 1995). Their report entitled "CHAOS: A Recipe for Success," published in 1999, provides similar statistics based on 1998 data. These older reports are available to the public from www.standishgroup.com. The Standish Group continues to provide updates to this study.*

6. Yourdon, Ed. "Surviving a Death March Project." *Software Development* (July 1997).

> *Ed Yourdon is famous for his books on software development, including a 1999 text titled* Death March. *Yourdon offers practical advice on how to handle software projects that are plagued with problems from the start. He describes the importance of a project manager's ability to negotiate schedules, budgets, and other aspects of a project with users, managers, and other stakeholders.*

KEY TERMS

- **critical path** — the longest path through a network diagram that determines the earliest completion of a project
- **enterprise project management software** — software that integrates information from multiple projects to show the status of active, approved, and future projects across an entire organization
- **Gantt chart** —a standard format for displaying project schedule information by listing project activities and their corresponding start and finish dates in a calendar format
- **program** — a group of projects managed in a coordinated way
- **project** — a temporary endeavor undertaken to create a unique product or service
- **project management** — the application of knowledge, skills, tools, and techniques to project activities in order to meet project requirements
- **Project Management Institute (PMI)** — international professional society for project managers
- **project management knowledge areas** — project integration management, scope, time, cost, quality, human resource, communications, risk, and procurement management
- **Project Management Office (PMO)** — an organizational group responsible for coordinating the project management functions throughout an organization
- **Project Management Professional (PMP)** — certification provided by PMI that requires documenting project experience and education, agreeing to follow the PMI code of ethics, and passing a comprehensive examination
- **project management tools and techniques** — methods available to assist project managers and their teams.
- **project manager** — the person responsible for working with the project sponsor, the project team, and the other people involved in a project to meet project goals
- **project portfolio management** — when organizations group and manage projects as a portfolio of investments that contribute to the entire enterprise's success
- **project sponsor** — the person who provides the direction and funding for a project
- **stakeholders** — people involved in or affected by project activities
- **triple constraint** — balancing scope, time, and cost goals

2

The Project Management and Information Technology Context

Objectives

After reading this chapter, you will be able to:

1. *Understand the systems view of project management and how it applies to information technology projects*
2. *Analyze a formal organization using the structural, human resources, political, and symbolic organizational frames*
3. *Describe the differences among functional, matrix, and project organizational structures*
4. *Explain why stakeholder management and top management commitment are critical for a project's success*
5. *Understand the concept, development, implementation, and close-out phases of the project life cycle*
6. *Distinguish between project development and product development*
7. *Discuss the unique attributes and diverse nature of information technology projects*
8. *List the skills and attributes of a good project manager in general and in the information technology field*

OPENING CASE

Tom Walters recently accepted a new position at his college as the Director of Information Technology. Tom had been a respected faculty member at the college for the past fifteen years. The college—a small, private college in the Southwest—offered a variety of programs in the liberal arts and professional areas. Enrollment included 1,500 full-time traditional students and about 1,000 working-adult students attending an evening program. Many instructors supplemented their courses with information on the Internet and course Web sites, but they did not offer any distance-learning programs. The college's niche was serving students in that region who liked the setting of a small liberal arts college. Like most colleges, its use of information technology had grown tremendously in the past five years. There were a few classrooms on campus with computers for the instructors and students, and a few more with just instructor stations and projection systems. Tom knew that several colleges throughout the country had begun to require that all students lease laptops and that these colleges incorporated technology components into most courses. This idea fascinated him. He and two other members of the Information Technology Department visited a local college that had required all students to lease laptops for the past three years, and they were very impressed with what they saw and heard. Tom and his staff developed plans to start requiring students to lease laptops at their college the next year.

Tom sent an e-mail to all faculty and staff in September, which briefly described this and other plans. He did not get much response, however, until the February faculty meeting when, as he described some of the details of his plan, the chairs of the History, English, Philosophy, and Economics departments all voiced their opposition to the idea. They eloquently stated that the college was not a technical training school, and that they thought the idea was ludicrous. Members of the Computer Science Department voiced their concern that all of their students already had state-of-the art computers and would not want to pay a mandatory fee to lease less-powerful laptops. The director of the adult education program expressed her concern that many adult-education students would balk at an increase in fees. Tom was in shock to hear his colleagues' responses, especially after he and his staff had spent a lot of time planning details of how to implement the use of laptops at their campus. Now what should he do?

Many of the theories and concepts of project management are not difficult to understand. What *is* difficult is implementing them in various environments. Project managers must consider many different issues when managing projects. Just as each project is unique, so is its environment. This chapter discusses some of the components involved in understanding the project environment, such as using a systems approach, understanding organizations, managing stakeholders, matching product life cycles to the project environment, the context of information technology projects, and developing important skills for successful project management.

A SYSTEMS VIEW OF PROJECT MANAGEMENT

Even though projects are temporary and intended to provide a unique product or service, organizations cannot run projects in isolation. If project managers lead projects in isolation, it is unlikely that those projects will ever truly serve the needs of the organization. Therefore, projects must operate in a broad organizational environment, and project managers need to consider projects within the greater organizational context. To handle complex situations effectively, project managers need to take a holistic view of a project and understand how it relates to the larger organization. **Systems thinking** describes this holistic view of carrying out projects within the context of the organization.

The term **systems approach** emerged in the 1950s to describe a holistic and analytical approach to solving complex problems that includes using a systems philosophy, systems analysis, and systems management. A **systems philosophy** is an overall model for thinking about things as systems. **Systems** are sets of interacting components working within an environment to fulfill some purpose. For example, the human body is a system composed of many subsystems—the brain, the skeletal system, the circulatory system, the digestive system, and so on. **Systems analysis** is a problem-solving approach that requires defining the scope of the system, dividing it into its components, and then identifying and evaluating its problems, opportunities, constraints, and needs. The analyst then examines alternative solutions for improving the current situation, identifies an optimum, or at least satisfactory, solution or action plan, and examines that plan against the entire system. **Systems management** addresses the business, technological, and organizational issues associated with making a change to a system.

Using a systems approach is critical to successful project management. Top managers and project managers must follow a systems philosophy to understand how projects relate to the whole organization. They must use systems analysis to address needs with a problem-solving approach. They must use systems management to identify key business, technological, and organizational issues related to each project in order to identify and satisfy key stakeholders and do what is best for the entire organization.

In the opening case, when Tom Walters planned the laptop project, he did not use a systems approach. Members of the Information Technology Department did all of the planning. Even though Tom sent an e-mail describing the laptop project to all faculty and staff, he did not address many of the organizational issues involved in such a complex project. Most faculty and staff are very busy at the beginning of fall term and many may not have read the entire message. Others may have been too busy to communicate their concerns to the Information Technology Department. Tom was unaware of the effects the laptop project would have on other parts of the college. He did not clearly define the business, technological, and organizational issues associated with the project. Tom and the Information Technology Department began work on the laptop project in isolation. If they had taken a systems approach, considering other dimensions of the project, and involving key stakeholders, they could have identified and addressed many of the issues raised at the February faculty meeting *before* the meeting.

Figure 2-1 provides a sample of some of the business, organizational, and technological issues that could be factors in the laptop project. In this case, technological issues, though not simple by any means, are probably the least difficult to identify and resolve. However, projects must address issues in all three spheres of the systems management model—business, organizational, and technological. Although it is easier to focus on the immediate and sometimes narrow concerns of a particular project, project managers and other staff must keep in mind the effects of any project on the interests and needs of the entire system or organization.

Figure 2-1. Three-Sphere Model for Systems Management

Many information technology professionals become captivated with the technology and day-to-day problem solving involved in working with information systems. They tend to become frustrated with many of the "people problems" or politics involved in most organizations. In addition, many information technology professionals ignore important business issues. Does it make financial sense to pursue this new technology? Should the company develop this software in-house or purchase it off-the-shelf? Using a more holistic approach helps project managers integrate business and organizational issues into their planning. It also helps them look at projects as a series of interrelated phases. When you integrate business and organizational issues into project planning and look at projects as a series of interrelated phases, you do a better job of ensuring project success.

UNDERSTANDING ORGANIZATIONS

The systems approach requires that project managers always view their projects in the context of the larger organization. Organizational issues are often the most difficult part of working on and managing projects. For example, many people believe that most projects fail because of politics. Project managers often do not spend enough time identifying all the stakeholders involved in projects, especially the people opposed to the projects. Similarly, they often do not consider the political context of a project or the culture of the organization. In order to improve the success rate of information technology projects, it is important for project managers to develop a better understanding of people as well as organizations.

Organizations can be viewed as having four different frames: structural, human resources, political, and symbolic.[1]

- The **structural frame** deals with how the organization is structured (usually depicted in an organizational chart) and focuses on different groups' roles and responsibilities in order to meet the goals and policies set by top management. This frame is very rational and focuses on coordination and control. For example, within the structural frame, a key information technology issue is whether a company should centralize the information technology personnel in one department or decentralize across several departments. You will learn more about organizational structures in the next section.

- The **human resources frame** focuses on producing harmony between the needs of the organization and the needs of the people. It recognizes that there are often mismatches between the needs of the organization and the needs of individuals and groups and works to resolve any potential

[1] Bolman, Lee G. and Deal, Terrence E., *Reframing Organizations*, Jossey-Bass Publishers, 1991.

problems. For example, many projects might be more efficient for the organization if personnel worked 80 or more hours a week for several months. This work schedule would probably conflict with the personal lives of those people. Important issues in information technology related to the human resources frame are the shortage of skilled information technology workers within the organization and unrealistic schedules imposed on many projects.

■ The **political frame** addresses organizational and personal politics. **Politics** in organizations take the form of competition among groups or individuals for power and leadership. The political frame assumes that organizations are coalitions composed of varied individuals and interest groups. Often, important decisions need to be made based on the allocation of scarce resources. Competition for scarce resources makes conflict a central issue in organizations, and power improves the ability to obtain scarce resources. Project managers must pay attention to politics and power if they are to be effective. It is important to know who opposes your projects as well as who supports them. Important issues in information technology related to the political frame are the power shifts from central functions to operating units or from functional managers to project managers.

■ The **symbolic frame** focuses on symbols and meanings. What is most important about any event in an organization is not what actually happened, but what it means. Was it a good sign that the CEO came to a kick-off meeting for a project, or was it a threat? The symbolic frame also relates to the company's culture. How do people dress? How many hours do they work? How do they run meetings? Many information technology projects are international and include stakeholders from various cultures. Understanding those cultures is also a crucial part of the symbolic frame.

What Went Wrong?

Several large organizations have installed or tried to install enterprise resource planning (ERP) systems to integrate business functions such as ordering, inventory, delivery, accounting, and human resource management. They understand the potential benefits of an ERP system and can analyze its various technical issues, but many companies do not realize how important the organizational issues are to ERP implementations.

For example, in early 2001, Sobey's, Canada's second largest grocery store chain with 1400 stores, abandoned its two-year, $90-million investment in an ERP system. The system was developed by SAP, the largest enterprise software company and the third-largest software supplier. Unfortunately, the system did not work properly due to several organizational challenges. Every department has to work together to implement an ERP system, and it is often difficult to get departments to communicate their needs. As Dalhousie University Associate Professor Sunny Marche states, "The problem of building an integrated system that can accommodate different people is a very serious challenge. You can't divorce technology from the sociocultural issues. They have an equal role." Sobey's ERP system shut down for five days and employees were scrambling to stock potentially empty shelves in several stores for weeks. The system failure cost Sobey's more than $90 million and caused shareholders to take an 82-cent after-tax hit per share.[2]

Project managers must learn to work within all four of these frames in order to function well in organizations. Chapter 9, Project Human Resource Management, and Chapter 10, Communications Management, further develop some of the organizational issues. The following sections on organizational structures, stakeholder management, and the need for top management commitment provide additional information related to the structural and political frames.

Organizational Structures

Many discussions of organizations focus on organizational structure. Three general classifications of organizational structures are functional, project, and matrix. Figure 2-2 portrays these three organizational structures. A **functional organizational structure** is the hierarchy most people think of when picturing an organizational chart. Functional managers or vice presidents in specialties such as engineering, manufacturing, information technology (IT), and human resources (HR) report to the chief executive officer (CEO). Their staffs have specialized skills in their respective disciplines. For example, most colleges and universities have very strong functional organizations. Only faculty in the Business Department teach business courses; faculty in the History Department teach history; faculty in the Art Department teach art, and so on.

A **project organizational structure** also has a hierarchical structure, but instead of functional managers or vice presidents reporting to the CEO, program managers report to the CEO. Their staffs have a variety of skills needed to complete the projects within their programs. Many large defense organizations use project structures. For example, major aircraft corporations usually have vice presidents in charge of each aircraft program. Many consulting firms also follow a project organization and hire people specifically to work on particular projects.

A **matrix organizational structure** represents the middle ground between functional and project structures. Personnel often report to both a functional manager and one or more project managers. For example, information technology personnel at many companies often split their time between two or more projects, but they report to their manager in the Information Technology Department. Project managers in matrix organizations have staff from various functional areas working on their projects, as shown in Figure 2-2. Matrix

2 Hoare, Eva. "Software hardships," *The Herald*, Halifax, Nova Scotia, Canada (2001)

structures can be strong, weak, or balanced, based on the amount of control exerted by the project managers.

Functional

		CEO	
VP Engineering	VP Manufacturing	VP IT	VP HR
Staff	Staff	Staff	Staff

Project

	CEO	
Program Manager A	Program Manager B	Program Manager C
Staff	Staff	Staff

Matrix

		CEO		
Program Managers	VP Engineering	VP Manufacturing	VP IT	VP HR
Staff	Staff	Staff	Staff	Staff
Project Manager A:	2 engineering	1 manufacturing	3½ IT	½ HR
Project Manager B:	5 engineering	3 manufacturing	10 IT	1 HR
Project Manager C:	1 engineering	0 manufacturing	4 IT	1/10 HR

Figure 2-2. Functional, Project, and Matrix Organizational Structures

Table 2-1 summarizes how organizational structure influences projects and project managers. Project managers have the most authority in a pure project organization and the least amount of authority in a pure functional organization. It is important that project managers understand the current structure under which they are working. For example, if someone in a functional organization is asked to lead a project that requires strong support from several different functional areas, he or she should ask for senior management sponsorship. This sponsor should solicit support from all relevant functional managers to ensure that they cooperate on the project and that qualified people are available to work as needed. The project manager might also ask for a separate budget to pay for project-related trips, meetings, and training or to provide financial incentives to the people supporting the project.

Table 2-1: Organizational Structure Influences on Projects

	FUNCTIONAL	MATRIX			PROJECT
		WEAK MATRIX	BALANCED MATRIX	STRONG MATRIX	
Project manager's authority	Little or none	Limited	Low to Moderate	Moderate to high	High to almost total
Percent of performing organization's personnel assigned full-time to project work	Virtually none	0–25%	15–60%	50–95%	85–100%
Project manager's role	Part-time	Part-time	Full-time	Full-time	Full-time
Common title for project manager's role	Project Coordinator/ Project Leader	Project Coordinator/ Project Leader	Project Manager/ Project Officer	Project Manager/ Program Manager	Project Manager/ Program Manager
Project management administrative staff	Part-time	Part-time	Part-time	Full-time	Full-time

PMBOK® Guide 2000, 19.

Even though project managers have the most authority in the project organizational structure, this type of organization is often inefficient for the company as a whole. Assigning staff full-time to the project often creates underutilization and/or misallocation of staff resources. For example, if a technical writer is assigned full-time to a project but there is no work for him or her on a particular day, the organization is wasting money by paying that person a full-time wage. Project organizations may also miss economies of scale available through the pooling of requests for materials with other projects. Disadvantages such as these illustrate the benefit of using a systems approach to managing projects. For example, the project manager might suggest hiring an independent contractor to do the technical writing work instead of using a full-time employee. This approach would save the organization money while still meeting the needs of the project. When project managers use a systems approach, they are better able to make decisions that address the needs of the entire organization.

STAKEHOLDER MANAGEMENT

Recall from Chapter 1 that project stakeholders are the people involved in or affected by project activities. Stakeholders can be internal to the organization, external to the organization, directly involved in the project, or simply affected by the project. Internal project stakeholders generally include the project sponsor, project team, support staff, and internal customers for the project. Other internal stakeholders include top management, other functional managers, and other project managers. Since organizations do not have unlimited resources, projects affect top management, other functional managers, and other project managers by using some of the organization's limited resources. Thus, while additional internal stakeholders may not be directly involved in the project, they are still stakeholders because the project affects them in some way. External project stakeholders include the project's customers (if they are external to the organization), competitors, suppliers, and other external groups potentially involved in or affected by the project, such as government officials or concerned citizens. Since the purpose of project management is to meet project requirements and satisfy stakeholders, it is critical that project managers take adequate time to identify, understand, and manage relationships with all project stakeholders. The expectations of all project stakeholders is most accurately met using the four frames (structural, human resources, political, and symbolic) of organizations.

Consider again the laptop project from the opening case. Tom Walters seemed to focus on just a few internal project stakeholders. He viewed only part of the structural frame of the college. Since his department would do most of the work in administering the laptop project, he focused on those stakeholders. Tom did not even involve the main customers for this project—the students at the college. Even though Tom sent an e-mail to faculty and staff, he did not hold meetings with senior administration or faculty at the college. Tom's view of who the stakeholders were for the laptop project was very limited.

During the faculty meeting, it became evident that the laptop project had many stakeholders in addition to the Information Technology Department and students. If Tom had expanded his view of the structural frame of his organization by reviewing an organizational chart for the entire college, he could have identified other key stakeholders. He would have been able to see that the laptop project would affect academic department heads and members of different administrative areas. If Tom had focused on the human resources frame, he would have been able to tap his knowledge of the college and identify individuals who would most support or oppose requiring laptops. By using the political frame, Tom could have considered the main interest groups that would be most affected by this project's outcome. Had he used the symbolic frame, Tom could have tried to address what moving to a laptop environment would really mean for the college. He then could have anticipated some of the opposition from people who were not in favor of increasing the use of technology on campus. He also could have solicited a strong endorsement from the college president or dean before talking at the faculty meeting.

Tom Walters, like many new project managers, learned the hard way that his technical and analytical skills were not enough to guarantee success in project management. To be more effective, he had to identify and address the needs of different stakeholders and understand how his project related to the entire organization.

The Importance of Top Management Commitment

People in top management positions, of course, are key stakeholders in projects. A very important factor in helping project managers successfully lead projects is the level of commitment and support they receive from top management. In fact, without top management commitment, many projects will fail. As described earlier, projects are part of the larger organizational environment, and many factors that might affect a project are out of the project manager's control. Several studies cite executive support as one of the key factors associated with the success of virtually all projects.

Table 2-2 summarizes the results of the 2001 Standish Group study describing, in order of importance, what factors contribute most to the success of information technology projects. Note that the study lists executive support as the most important factor, overtaking user involvement, which was most important in the 1995 CHAOS study. Executive support can also help projects succeed when top managers encourage users to actively participate in projects, assign experienced project managers, provide clear business objectives, and so on.

Table 2-2: What Helps Projects Succeed?

1. Executive support

2. User involvement

3. Experienced project manager

4. Clear business objectives

5. Minimized scope

6. Standard software infrastructure

7. Firm basic requirements

8. Formal methodology

9. Reliable estimates

The Standish Group, "CHAOS 2001: A Recipe for Success" (2001).

Top management commitment is crucial to project managers for the following reasons:

- Project managers need adequate resources. The best way to kill a project is to withhold the required money, people, resources, and visibility for the project. If project managers have top management commitment, they will also have adequate resources and not be distracted by events that do not affect their specific projects.

- Project managers often require approval for unique project needs in a timely manner. For example, on large information technology projects, top management must understand that unexpected problems may result from the nature of the products being produced and the specific skills of the people on the project team. For example, the team might need additional hardware and software halfway through the project for proper testing, or the project manager might need to offer special pay and benefits to attract and retain key project personnel. With top management commitment, project managers can meet these specific needs in a timely manner.

- Project managers must have cooperation from people in other parts of the organization. Since most information technology projects cut across functional areas, top management must help project managers deal with the political issues that often arise in these types of situations. If certain functional managers are not responding to project managers' requests for necessary information, top management must step in to encourage functional managers to cooperate.

- Project managers often need someone to mentor and coach them on leadership issues. Many information technology project managers come from technical positions and often are inexperienced as managers. Senior managers should take the time to pass on leadership advice and encourage new project managers to take classes to develop leadership skills and allocate the time and funds for them to do so.

Information technology project managers work best in an environment in which top management values information technology. Working in an organization that values good project management and sets standards for its use also helps project managers succeed.

The Need for Organizational Commitment to Information Technology

Another factor affecting the success of information technology projects is the organization's commitment to information technology in general. It is very difficult for a large information technology project (or a small one, for that matter) to be successful if the organization itself does not value information technology. Many companies have realized that information technology is integral to their business and have created a vice president or equivalent-level position for the head of information technology, often called the Chief Information Officer (CIO). Some companies assign people from non-information technology areas to work on large projects full-time to increase involvement from end users of the systems. Some CEOs even take a strong leadership role in promoting the use of information technology in their organizations.

The Gartner Group, Inc., a well-respected information technology consulting firm, awarded Boston's State Street Bank and Trust Company's CEO, Marshall Carter, the 1998 Excellence in Technology Award. Carter provided the vision and leadership for his organization to implement new information technology that successfully expanded the bank's business. They had to gather, coordinate, and analyze vast amounts of data from around the globe to provide new asset management services to their customers. It took six years to transform State Street Bank and Trust into a company providing state-of-the-art tools and services to its customers. The bank's revenues, profits, and earnings per share more than doubled during Carter's first five years as CEO. One key to Carter's success was his vision that technology was an integral part of the business and not just a means of automating old banking services. Carter used a highly personal style to keep his people motivated, and he often showed up at project review meetings to support his managers on information technology projects.[3]

The Need for Organizational Standards

Another problem in most organizations is not having standards or guidelines to follow that could help in performing project management. These standards or guidelines might be as simple as providing standard forms or templates for common project documents, examples of good project plans, or guidelines on how the project manager should provide status information to top management. The content of a project plan and how to provide status information might seem like common sense to senior managers, but many new information technology project managers have never created plans or given a nontechnical status report. Top management must support the development of these standards and guidelines and encourage, or even enforce, their use. For example, an organization might require all potential project information in a standard format to make project portfolio management decisions. If a project manager does not submit a potential project in the proper format, it could be rejected.

Some organizations invest heavily in project management by creating a project management office or center of excellence, as described in Chapter 1. A project management office or center of excellence is an organizational entity created to assist project managers in achieving project goals. Rachel Hollstadt, founder and CEO of a project management consulting firm, suggests that organizations consider adding a new position, a Chief Project Officer (CPO), the title Anne

---

[3] Melymuka, Kathleen, "Old Bank, New Ideas," *ComputerWorld* (February 15, 1999).

Roberts held in the opening case for Chapter 1, to elevate the importance of project management. Some organizations develop career paths for project managers. Some require that all project managers have project management professional (PMP) certification and that all employees have some type of project management training. The implementation of all of these standards demonstrates an organization's commitment to project management.

PROJECT PHASES AND THE PROJECT LIFE CYCLE

Since projects operate as part of a system and involve uncertainty, it is good practice to divide projects into several phases. A **project life cycle** is a collection of project phases. Project phases vary by project or industry, but some general phases include concept, development, implementation, and close-out. The first two phases (concept and development) focus on planning and are often referred to as **project feasibility**. The last two phases (implementation and close-out) focus on delivering the actual work and are often referred to as **project acquisition**. A project must successfully complete each phase before moving on to the next. This project life cycle approach provides better management control and appropriate links to the ongoing operations of the organization.

Figure 2-3 provides a summary framework for the general phases of the project life cycle. In the concept phase of a project, managers usually briefly describe the project—they develop a very high-level or summary plan for the project, which describes the need for the project and basic underlying concepts. A preliminary or rough cost estimate is developed in this first phase, and an overview of the work involved is created. A work breakdown structure (WBS) defines project work and is a deliverable-oriented document that defines the total scope of the project. (You will learn more about the work breakdown structure in Chapter 5, Project Scope Management.) For example, if Tom Walters (from the opening case) had followed the project life cycle instead of moving full-steam ahead with the laptop project, he could have created a committee of faculty and staff to study the concept of increasing the use of technology on campus. This committee might have developed a management plan that included an initial, smaller project to investigate alternative ways of increasing the use of technology. They might have estimated that it would take six months and $20,000 to conduct a detailed technology study. The WBS at this phase of the study might have three levels and partition the work to include a competitive analysis of what five similar campuses were doing; a survey of local students, staff, and faculty; and a rough assessment of how using more technology would affect costs and enrollments. At the end of the concept phase, the committee would be able to deliver a report and presentation on its findings. The report and presentation would be an example of a **deliverable**—a product produced as part of a project.

Project Feasibility		Project Acquisition	
Concept →	Development →	Implementation →	Close-out

Sample deliverables for each phase:

Management plan	Project plan	Last work package	Completed work
Preliminary cost estimate	Budgetary cost estimate	Definitive cost estimate	Lessons learned
3-level WBS	6+-level WBS	Performance reports	Customer acceptance

Figure 2-3. Phases of the Project Life Cycle

After the concept phase is completed, the next project phase of development begins. In the development phase, the project team creates a more detailed project plan, a more accurate cost estimate, and a more thorough WBS. In the example under discussion, suppose the concept phase report suggested that requiring students to have laptops was one means of increasing the use of technology on campus. The project team could then further expand this idea in the development phase. They would have to decide if students would purchase or lease the laptops, what type of hardware and software the laptops would require, how much to charge students, how to handle training and maintenance, how to integrate the use of the new technology with current courses, and so on. If, however, the concept phase report showed that the laptop idea was not a good idea for the college, then the project team would no longer consider increasing the use of technology by requiring laptops in the development phase. This phased approach minimizes the time and money spent developing inappropriate projects. A project idea must pass the concept phase before evolving into the development phase.

The third phase of the project life cycle is implementation. In this phase, the project team delivers the required work, creates a definitive or very accurate cost estimate, and provides performance reports to stakeholders. Suppose Tom Walter's college took the idea of requiring students to have laptops through the development phase. During the implementation phase, the project team would need to obtain the required hardware and software, install the necessary network equipment, deliver the laptops to the students, create a process for collecting fees, provide training to students, faculty, and staff, and so on. Other people on campus would also be involved in the implementation phase. Faculty would need to consider how best to take advantage of the new technology. The recruiting staff would have to update their materials to reflect this new feature of the college. Security would need to address new problems that might result from having students carry around expensive equipment. The project team usually spends the bulk of their efforts and money during the implementation phase of projects.

The last phase of the project life cycle is close-out. In the close-out phase, all of the work is completed, and there should be some sort of consumer acceptance of the entire project. The project team should document their experiences on the project in a lessons-learned report. If the laptop idea made it all the way through the implementation phase and all students received laptops, the project team would then complete the project by closing out any related activities. They might administer a survey to students, faculty, and staff in order to gather opinions on how the project fared. They would ensure that any contracts with suppliers were completed and appropriate payments made. They would transition future work related to the laptop project to other parts of the organization. The project team could also share their lessons learned with other college campuses considering implementing a similar program.

Just as a *project* has a life cycle, so does a *product*. Information technology projects help produce products such as new software, hardware, networks, research reports, and training on new systems. Understanding the product life cycle is just as important to good project management as understanding the phases of the project life cycle.

Product Life Cycles

All products follow some type of life cycle—cars, buildings, even amusement parks. The Disney corporation, for example, follows a rigorous process to design, build, and test new products. They assign project managers to oversee the development of all new products, such as rides, parks, and cruise lines. Likewise, major automotive companies follow product life cycles to produce new cars, trucks, and other products. Most information technology professionals are familiar with the concept of a product life cycle, especially for developing software.

Software development projects are one subset of information technology projects. In general, information technology projects involve researching, analyzing, and then purchasing and installing new hardware and software with little or no actual software development required. However, some projects involve minor software modifications to enhance existing software or to integrate one application with another. Other projects involve a major amount of software development. Many argue that developing software requires project managers to modify traditional project management methods, depending on a particular product's life cycle.

A **systems development life cycle (SDLC)** is a framework for describing the phases involved in developing information systems. Some popular models of a SDLC include the waterfall model, the spiral model, the incremental build model, the prototyping model, and the Rapid Application Development (RAD) model. These life cycle models are examples of a **predictive life cycle**, meaning the scope of the project can be clearly articulated and the schedule and cost can be accurately predicted. The project team spends a large portion of the project effort attempting to clarify the requirements of the entire system and then producing a design. Developers need to take time to understand requirements and create design documents before they can write any code for users to review. Users often are unable to see any tangible results in terms of working software for an extended period. Below are brief descriptions of several predictive systems development life cycle models: [4]

- The waterfall life cycle model has well-defined, linear stages of systems development and support. This life cycle model assumes that requirements will remain stable after they are defined.

- The spiral life cycle model was developed based on experience with various refinements of the waterfall model as applied to large government software projects. It recognizes the fact that most software is developed using an iterative or spiral approach rather than a linear approach.

- The incremental build life cycle model provides for progressive development of operational software, with each release providing added capabilities.

- The prototyping life cycle model is used for developing software prototypes to clarify user requirements for operational software. It requires heavy user involvement, and developers use a model to generate functional requirements and physical design specifications simultaneously. Developers can throw away or keep prototypes, depending on the project.

- The Rapid Application Development (RAD) life cycle model uses an approach in which developers work with an evolving prototype. This life cycle model also requires heavy user involvement and helps produce systems quickly without sacrificing quality. Developers use RAD tools such as CASE (Computer Aided Software Engineering), JRP (Joint Requirements Planning), and JAD (Joint Application Design) to facilitate rapid prototyping and code generation.

In contrast to the predictive life cycle, the **Adaptive Software Development (ASD)** life cycle model assumes that software development follows an adaptive approach. Important attributes of this approach are that the projects are mission driven and component based, using time-based cycles to meet target

---

[4] Desaulniers, Douglas H. and Anderson, Robert J. "Matching Software Development Life Cycles to the Project Environment," Proceedings of the Project Management Institute Annual Seminars & Symposium, (November 1–10, 2001).

dates. Requirements are developed using an iterative approach, and development is risk-driven and change tolerant to address and incorporate rather than mitigate risks. More recently, the term agile software development has become popular to describe new approaches for managing software development projects. Two popular ASD life cycle models are extreme programming and Scrum.

- **Extreme Programming (XP)** has become a popular buzzword. This life cycle model meets the needs of people developing software in rapidly changing environments. An XP development team includes software developers, managers, and user representatives. One unique feature of the XP life cycle model is that developers program in pairs to promote synergy and increase productivity. Another unique feature of XP is that software developers must write the tests for their own code. XP development teams emphasize user satisfaction by producing tangible results in a short time while promoting teamwork and collective code ownership. Some disadvantages of the XP life cycle model are that it does not work for large projects (more than twenty people), it requires full-time user representatives, it assumes developers are highly disciplined, and it will not work if the team must predict the cost and schedule at the project's onset and then try to meet those commitments. The term extreme project management loosely describes how to manage these types of projects.

- **Scrum** is similar to other life cycle models that use iterative development to address changing requirements, but in the Scrum life cycle model, the repetitions are referred to as sprints, which normally last thirty days. Each day the entire team meets for a short meeting, called a scrum, where they decide what to accomplish that day. Team members identify any obstacles they might have in accomplishing their goals, and the project manager must resolve and remove those obstacles. Project managers can use Scrum on large development projects because each team focuses on a discrete piece of work or object, with clear interfaces and behaviors. Scrum works best for projects using object-oriented technology and requires strong leadership to coordinate the work of each smaller team.

These life cycle models are all examples of software development product life cycles. Many Web sites and introductory management information systems texts describe each of them in detail. The type of software and complexity of the information system in development determines which life cycle model to use. It is important to understand the product life cycle to meet the needs of the project environment.

Do not confuse the project life cycle with the product life cycle. The project life cycle applies to all projects, regardless of the products being produced. On the other hand, product life cycle models vary considerably based on the nature of the product.

Most large information technology products are developed as a series of projects. For example, the systems planning phase for a new information system can include a project to hire an outside consulting firm to help identify and evaluate potential strategies for developing a particular business application, such as a new order processing system or general ledger system. It can also include a project to develop, administer, and evaluate a survey of users to get their opinions on the current information systems used for performing that business function in the organization. The systems analysis phase might include a project to create process models for certain business functions in the organization. It can also include a project to create data models of existing databases in the company related to the business function and application. The implementation phase might include a project to hire contract programmers to code a part of the system. The close-out phase might include a project to develop and run several training sessions for users of the new application. All of these examples show that developing large information technology products are usually composed of several smaller projects.

Because some aspects of project management need to occur during each phase of the product life cycle, it is critical for information technology professionals to understand and practice good project management.

The Importance of Project Phases and Management Reviews

Due to the complexity and importance of many information technology projects and their resulting products, it is important to take time to review the status of a project at each phase. A project should successfully pass through each of the main project phases before continuing to the next. Since the organization usually commits more money as a project continues, a management review should occur after each phase to evaluate progress, potential success, and continued compatibility with organizational goals. These management reviews, called **phase exits** or **kill points**, are very important for keeping projects on track and determining if they should be continued, redirected, or terminated. Recall that projects are just one part of the entire system of an organization. Changes in other parts of the organization might affect a project's status, and a project's status might likewise affect what is happening in other parts of the organization. By breaking projects into phases, top management can make sure that the projects are still compatible with the needs of the rest of the company.

Let's take another look at the opening case. Suppose Tom Walters' college did a study on increasing the use of technology that was sponsored by the college president. At the end of the concept phase, the project team could have presented information to the faculty, president, and other staff members that described different options for increasing the use of technology, an analysis of what competing colleges were doing, and results of a survey of local stakeholders' opinions on the subject. This presentation at the end of the concept phase represents one form of a management review. Suppose the study reported that 90 percent of students, faculty, and staff surveyed strongly opposed the idea of requiring all students to have laptops and that many adult students said they would attend other colleges if they were required to pay for the additional technology. The college would probably decide not to pursue this idea any further. Had Tom taken a phased approach, he and his staff would not have wasted the time and money it took to develop detailed plans.

In addition to formal management reviews, it is important to have top management involvement throughout the life cycle of most projects. It is unwise to wait for the end of project phases to have management inputs. Many projects are reviewed by management on a regular basis, such as weekly or even daily, to make sure they are progressing well. Everyone wants to be successful in accomplishing goals at work, and having management involvement ensures that they are on track in accomplishing both project and organizational goals.

What Went Right?

Having specific deliverables and kill points at the end of project phases helps managers make better decisions about whether to proceed, redefine, or kill a project. Improvement in information technology project success rates reported by the Standish Group has been due, in part, to an increased ability to know when to cancel failing projects. Standish Group Chairman Jim Johnson made the following observation:

"The real improvement that I saw was in our ability to—in the words of Thomas Edison—know when to stop beating a dead horse . . . Edison's key to success was that he failed fairly often; but as he said, he could recognize a dead horse before it started to smell . . . as a result he had 14,000 patents and was very successful. . . . In information technology we ride dead horses—failing projects—a long time before we give up. But what we are seeing now is that we are able to get off them; able to reduce cost overrun and time overrun. That's where the major impact came on the success rate."[5]

Another example of the power of management oversight comes from Huntington Bancshares, Inc. This company, like many others, had an **executive steering committee**, a group of senior executives from various parts of the organization, who regularly reviewed important corporate projects and issues.

[5] Cabanis, Jeannette, "A Major Import: The Standish Group's Jim Johnson On Project Management and IT Project Success," *PM Network*, PMI, (September 1998), 7.

This Ohio-based, $26 billion bank holding company completed a year-long Web site redesign effort using XML technology to give its online customers access to real-time account information as well as other banking services. The CIO, Joe Gottron, said there were "four or five very intense moments" when the whole project was almost stopped due to its complexity. An executive steering committee met weekly to review the project's progress and discuss work planned for the following week. Gottron said the meetings ensured that "if we were missing a beat on the project, no matter which company [was responsible], we were on top of it and adding additional resources to make up for it." [6]

THE CONTEXT OF INFORMATION TECHNOLOGY PROJECTS

As described earlier, software development projects can follow several different product life cycles based on the project context. There are several other issues related to information technology projects. In fact, several groups have recognized the additional knowledge that information technology project managers must posses by creating separate certification programs specifically for them. PMI launched a new Certificate of Added Qualification (CAQ) exam in 2002 that PMPs can take to demonstrate their knowledge of managing projects in information technology. CompTIA's IT Project+ exam includes project scenarios and information specific to working in the information technology industry. Appendix B includes more information on these exams. This section highlights some of the issues unique to the information technology industry that affect project management, including the nature of projects, the characteristics of project team members, and the diverse nature of technologies involved.

The Nature of Information Technology Projects

Unlike projects in many other industries, projects labeled as information technology projects can be very diverse. Some involve a small number of people installing off-the-shelf hardware and associated software. Others involve hundreds of people analyzing several organizations' business processes and then developing new software in a collaborative effort with users to meet business needs. Even for small hardware-oriented projects, there is a wide range of types of hardware that could be involved, such as personal computers, mainframe computers, network equipment, kiosks, or small mobile devices, to name a few. The network equipment might be wireless, phone-based, cable-based, or require satellite connection. The nature of software development projects is even more

---

[6] Mearian, Lucas, "Bank Hones Project Management Skills With Redesign," *ComputerWorld* (April 29, 2002).

diverse than hardware-oriented projects. A software development project might include developing a simple, stand-alone Excel or Access application or a sophisticated, global e-commerce system using state-of-the art programming languages.

Information technology projects also support every possible industry and business function. Managing an information technology project for a film company's animation department would require different knowledge and skills of the project manager and team members than a project to improve a federal tax collection system or install a communication infrastructure in a third-world country. Because of the diversity of information technology projects and the newness of the field, people are still struggling to find and develop the best processes (or sometimes any processes) to follow in managing them. Many information technology project managers feel as though they are often striking new ground with every project.

Characteristics of Information Technology Team Members

Due to the nature of information technology projects, the people involved come from very diverse backgrounds and possess different skill sets. Most trade schools, colleges, and universities did not start offering degrees in computer technology, computer science, management information systems, or other information technology areas until the 1980s. Therefore, many people in the field do not have a common educational background. Many companies purposely hire graduates with degrees in other fields such as business, mathematics, or the liberal arts to provide different perspectives on information technology projects. Even with these different educational backgrounds, there are some common job titles for people working on most information technology projects, such as business analyst, programmer, network specialist, database analyst, quality assurance expert, technical writer, security specialist, hardware engineer, software engineer, and system architect. Within the category of programmer, there are several other job titles used to describe the specific technologies the programmer uses, such as Java programmer, XML programmer, C/C++ programmer, and so on.

Some information technology projects require the skills of people in just a few of these job functions, but many require inputs from many or all of them. Occasionally, information technology professionals move around between these job functions, but more often people become technical experts in one area or they decide to move into a management position. It is also rare for technical specialists or project managers to remain with the same company for a long time, and in fact, many information technology projects include a large number of contract workers. Working with this "army of free agents," as Rob Thomsett, author and consultant for the Cutter Consortium, calls them, creates special challenges. (See the Suggested Readings for an article by Thomsett on this topic.)

Diverse Technologies

Many of the job titles for information technology professionals reflect the different technologies required to hold that position. Unfortunately, hardware specialists often do not understand the language of database analysts, and vise versa. Security specialists have a hard time communicating with business analysts. It is also unfortunate that people within the same information technology job function often do not understand each other because each uses different technology. For example, someone with the title of programmer can often use several different programming languages. However, a COBOL programmer cannot be of much help on a Java project, and vise versa. These highly specialized positions also make it difficult for project managers to form and lead project teams.

Another problem with diverse technologies is that many of them change rapidly. A project team might be close to finishing a project when they discover a new technology that can greatly enhance the project and better meet long-term business needs. New technologies have also shortened the timeframe many businesses have to develop, produce, and distribute new products and services. This fast-paced environment requires equally fast-paced processes to manage and produce information technology projects and products.

SUGGESTED SKILLS FOR A PROJECT MANAGER

As you can imagine, good project managers should possess many skills. They must be comfortable with change, since most projects introduce changes in organizations, as well as involving changes within the projects themselves. Project managers need to understand the organizations they work in and how products are developed and services are provided. Achieving high performance on projects requires soft skills or people skills, such as strong management, communication, leadership, and political skills. These soft skills are often instrumental in achieving customer satisfaction for project results. Project managers also need skills in organization, teamwork, coping, and making effective use of technology.

Why do project managers need good soft skills? One reason is that to understand, navigate, and meet stakeholders' needs and expectations, project managers need to lead, communicate, negotiate, problem solve, and influence the organization at large. They must be able to actively listen to what others are saying, help develop new approaches for solving problems, and persuade others to work toward achieving project goals. Project managers must lead their project teams by providing vision, delegating work, creating an energetic and positive environment, and setting an example of appropriate and effective behavior.

Project managers must also have strong organizational skills to be able to plan, analyze, set, and achieve project goals. Project managers must focus on teamwork skills in order to use their people effectively. They need to be able to motivate different types of people and develop *esprit de corps* within the project team and with other project stakeholders. Since most projects involve changes and trade-offs between competing goals, it is important for project managers to have strong coping skills as well. Project managers are better able to maintain their sanity and reduce their stress levels if they are able to cope with criticism and constant change. Project managers must be flexible, creative, and sometimes patient in working toward project goals; they must also be persistent in illustrating project needs. Lastly, project managers must be able to make effective use of technology as it relates to the specific project, often including having special product knowledge or experience with a particular industry. Project managers must make many decisions and deal with people in a wide variety of disciplines, so it helps tremendously to have a project manager who is confident in using the special tools or technologies that are the most effective in particular settings.

The above skills are important for all project managers. What additional skills do information technology project managers need? People in industry and academia often debate the answer to this question. Some people believe that it is important for information technology project managers to understand the technologies they use on the projects they are managing. They do not have to be expert users of any specific technology, but they have to know enough to build a strong team and ask the right questions to keep things on track. Others believe it is more important for information technology project managers to have strong business and behavioral skills so they can lead a project team and deliver a solution that will meet business needs.

In an interview with two CIOs, John Oliver of True North Communications, Inc. and George Nassef of Hotjobs.com, both men provided very different responses when asked what skills good information technology project managers need. They could not agree on which skills were more important, such as can-do optimism versus assume-the-worst realism, or detail-oriented versus visionary. However, both agreed that the most important skills depend on the uniqueness of the project and the people involved.[7] Project managers must possess a wide variety of skills and be able to decide which particular skills are more important in different situations.

Unfortunately, many people in information technology do not want to develop anything but their technical skills. They do not regard soft skills or business skills as essential to improving their product delivery. Top managers disagree with that mindset because they recognize the need to improve communications between information technology professionals and their consumers. Business people are now more savvy with information technology, but few information technology professionals have spent the time developing their

---

7 Brandel, Mary, "The Perfect Project Manager," *ComputerWorld* (August 6, 2001).

business savvy.[8] Organizations train information technology project managers by starting them off as members of a project team. However, they must be willing to develop more than just their technical skills to be more productive team members and potential project managers.

Several studies compare project management job responsibilities with the characteristics associated with success. The National Science Foundation's 1999 study, "Building a Foundation for Tomorrow: Skills Standards for Information Technology, Millennium Edition" documents the results of a collaborative effort between industry and academia to improve the education of the information technology workforce. This study found that project management is a skill needed in every major information technology field, from database administrator to network specialist to technical writer. Table 2-3 lists fifteen project management job functions that are essential for good project management.

Table 2-3: Fifteen Project Management Job Functions

1. Define scope of project

2. Identify stakeholders, decision-makers, and escalation procedures

3. Develop detailed task list (work breakdown structures)

4. Estimate time requirements

5. Develop initial project management flow chart

6. Identify required resources and budget

7. Evaluate project requirements

8. Identify and evaluate risks

9. Prepare contingency plan

10. Identify interdependencies

11. Identify and track critical milestones

12. Participate in project phase review

13. Secure needed resources

14. Manage the change control process

15. Report project status

"Building a Foundation for Tomorrow: Skills Standards for Information Technology, Millennium Edition," Northwest Center for Emerging Technologies (NWCET), Belleview, WA, 1999.

Each of the job functions listed in Table 2-3 requires different performance criteria, technical knowledge, foundation skills, and personal qualities. In a recent study, 100 project managers listed the characteristics they believed were critical for effective project management and the characteristics that made

[8] Thomsen-Moore, Lauren, "No 'soft skills' for us, we're techies," *Computerworld Today* (December 16, 2002).

project managers ineffective. Table 2-4 lists the results. The study found that effective project managers provide leadership by example, are visionary, technically competent, decisive, good communicators, good motivators, stand up to top management when necessary, support team members, and encourage new ideas. The study also found that respondents believed positive leadership is the strongest contributing factor to project success. The most important characteristics and behaviors of positive leaders include being a team builder and communicator, having high self-esteem, focusing on results, demonstrating trust and respect, and setting goals.

Table 2-4: Most Significant Characteristics of Effective and Ineffective Project Managers

Effective Project Managers	Ineffective Project Managers
Lead by example	Set bad examples
Are visionaries	Are not self-assured
Are technically competent	Lack technical expertise
Are decisive	Are poor communicators
Are good communicators	Are poor motivators
Are good motivators	
Stand up to top management when necessary	
Support team members	
Encourage new ideas	

Zimmerer, Thomas W. and Mahmoud M. Yasin, "A Leadership Profile of American Project Managers," *Project Management* (March 1998), 31–38.

Several organizations, including PMI, have developed project management competency models to help people identify what skills they need to develop to improve their performance as project managers. Project success is more likely when project managers develop these skills and organizations promote the use of good project management.

CASE WRAP-UP

After several people voiced concerns about the laptop idea at the faculty meeting, the president of the college directed that a committee be formed to formally review the concept of requiring students to have laptops in the near future. Because the college was dealing with several other important enrollment-related issues, the president named the vice president of enrollment to head the committee. Other people soon volunteered or were assigned to the committee, including Tom Walters as head of Information Technology, the director of the adult education program, the chair of the Computer Science Department, and the chair of the History Department. The president also insisted that the committee include at least two members of the student body. The president knew everyone was busy, and he questioned whether the laptop idea was a high-priority issue for the college. He directed the committee to present a proposal at the next month's faculty meeting, either to recommend the creation of a formal project team (of which these committee members would commit to be a part) to fully investigate requiring laptops, or to recommend terminating the concept. At the next facility meeting, few people were surprised to hear the recommendation to terminate the concept. Tom Walters learned that he had to pay much more attention to the needs of the entire college before proceeding with detailed information technology plans.

CHAPTER SUMMARY

Projects operate in an environment broader than the project itself. Project managers need to take a systems approach when working on projects; they need to consider projects within the greater organizational context.

Organizations have four different frames: structural, human resources, political, and symbolic. Project managers need to understand all of these aspects of organizations to be successful. The structural frame focuses on different groups' roles and responsibilities to meet the goals and policies set by top management. The human resources frame focuses on producing harmony between the needs of the organization and the needs of people. The political frame addresses organizational and personal politics. The symbolic frame focuses on symbols and meanings.

The structure of an organization has strong implications for project managers, especially in terms of the amount of authority the project manager has. The three basic organizational structures include functional, matrix, and project. Project managers have the most authority in a pure project organization, an intermediate amount of authority in the matrix organization, and the least amount of authority in a pure functional organization.

Project stakeholders are individuals and organizations actively involved in the project or whose interests may be positively or negatively affected because of project execution or successful project completion. Project managers must identify and understand the different needs of all stakeholders on their projects.

Top management commitment is crucial for project success. Since projects often affect many areas in an organization, top management must assist project managers if the project is to be successfully integrated. Organizational commitment to information technology is also important to the success of information technology projects. Development of standards and guidelines assist most organizations in managing projects.

A project life cycle is a collection of project phases. Most projects include concept, development, implementation, and close-out phases. Projects often produce products, which follow product life cycles. Examples of product life cycles for software development include the waterfall, spiral, incremental build, prototyping, RAD, extreme programming, and Scrum models. Project managers must understand the specific life cycle of the products they are producing as well as the general project life cycle model.

A project must successfully pass through each of the project phases in order to continue to the next phase. A management review should occur at the end of each project phase, and more frequent management inputs are often needed. These management reviews and inputs are important for keeping projects on track and determining if projects should be continued, redirected, or terminated.

Project managers need to consider several factors due to the unique context of information technology projects. The diverse nature of information technology projects and wide range of business areas and technologies involved make information technology projects especially challenging to manage. Project team members with a wide variety of specialized skills and the rapidly changing technologies are also important considerations.

Project managers perform a variety of job functions, and therefore must have many skills and personal characteristics to do their jobs well. Project managers need strong management skills, as well as particularly strong communication, leadership, and political skills. Project managers also need skills in organization, teamwork, coping, and making effective use of technology.

DISCUSSION QUESTIONS

1. What does it mean to take a systems view of a project? How does taking a systems view of a project apply to project management?

2. Explain the four frames of organizations. How can they help project managers understand the organizational context for their projects?

3. Briefly explain the differences between functional, matrix, and project organizations. Describe how each structure affects the management of the project.

4. Discuss the importance of top management commitment and the development of standards for successful project management. Provide examples to illustrate the importance of these items based on your experience on any type of project.

5. How does a project life cycle differ from a product life cycle? Why does a project manager need to understand both?

6. Which skills do you think are most important for an information technology project manager? Can a project manager learn all of these skills, or are some innate?

7. Do you think project managers of large information technology projects need strong technical skills? Why or why not? What about project managers for small information technology projects?

EXERCISES

1. Apply the information on the four frames of organizations to an information technology project with which you are familiar. If you cannot think of a good information technology project, use your personal experience in deciding where to attend college to apply this framework. Write a one- to two-page paper describing key issues related to the structural, human resources, political, and symbolic frames. Which frame seemed to be the most important and why? For example, did you decide where to attend college primarily because of the curriculum and structure of the program? Did you follow your friends? Did your parents have a lot of influence in your decision? Did you like the culture of the campus?

2. Read one of the suggested readings by Desaulniers and Anderson, McConnell, and/or Thomsett or search the Internet for an interesting article about software development life cycles, including agile software development. Also, find two informative Web sites related to this subject. What do these sources say about project management? Write a one- to two-page summary of your findings.

3. Read one of the suggested readings by Crawford, NWCET, or Posner about project management skills or find another reference about the skills required for a good project manager. Write a one- to two-page paper describing the article and your opinion of the skills it suggests are important.

4. Search the Internet and scan information technology industry magazines or Web sites to find an example of an information technology project that had problems. Write a one- to two-page paper summarizing who the key stakeholders were for the project and how they influenced the outcome.

5. Write a one- to two-page summary of an article about the importance of top management support for project success.

MINICASE

You have been part of your company's Information Technology Department for three years. You have learned a lot about the company and about many new technologies in your latest assignment—developing applications for your corporate intranet. Since you are an avid recreational athlete, you have spent a fair amount of time thinking about how you would write a sophisticated application to help people learn about the many corporate athletic teams, register on the intranet, determine team schedules, post team statistics, and so on. You have heard some rumors about profits not being as high as expected in the past year, and you know that mostly the junior employees participate on the athletic teams.

Part 1. Using Figure 2-1 as a guide, use the three-sphere model of systems management to identify potential issues that could be factors in deciding whether you should proceed with your idea to develop an application for recreational athletics at your company. Include at least three questions for each sphere.

Part 2: Your immediate supervisor likes your idea of developing a recreational sports application on the intranet, but he has to convince his supervisor that this project is valuable to the company. Prepare a short presentation with five to ten slides and speaker notes to convince top management to approve the recreational athletics application project. Be sure to list benefits of the project and suggest a phased approach. For example, the first phase might involve just posting information about various sports teams on the intranet. The second phase might include on-line registration, and so on.

RUNNING CASES

Appendix C provides several running cases that use a particular project scenario, such as the recreational athletics intranet described in the minicase above. Questions are provided to apply several project management concepts and tools to each of these scenarios. These questions progress from initiating the projects to planning, executing, controlling, and closing them. Appendix D provides several templates to use for answering some of these questions. Review these cases to supplement your understanding and practice of project management.

SUGGESTED READINGS

1. Bolman, Lee G. and Terrence E. Deal. *Reframing Organizations*. Jossey-Bass Publishers (1991).

   This book describes the four frames of organizations: structural, human resources, political, and symbolic. It explains how to look at situations from more than one vantage point to help bring order out of confusion.

2. Crawford, Lynn. "Profiling the Competent Project Manager." *Project Management Research at the Turn of the Millennium.* Project Management Institute (2000).

> *The competence of the project manager is a key factor in project success. In this article, Crawford presents a review and analysis of research-based literature concerning the knowledge, skills, and personal attributes of project managers.*

3. Desaulniers, Douglas H. and Robert J. Anderson. "Matching Software Development Life Cycles to the Project Environment," Proceedings of the Project Management Institute Annual Seminars & Symposium, (November 1–10, 2001).

> *Software development projects require special attention to the product life cycle. This article provides an excellent summary of several software development life cycles and the importance of using the correct one for your project environment.*

4. McConnell, Steve. "10 Myths of Rapid Software Development," Project Management Institute Information Systems Specific Interest Group (PMI-ISSIG) Webinar (*www.construx.com/docs/open/10MythsOfRapidDevelopment-Keynote.pdf*) March 2002.

> *Steve McConnell has written several books and articles on rapid software development. McConnell developed this presentation to highlight common myths about the subject. Visit Construx's Web site (www.construx.com) for further information and access to free resources to advance the field of software engineering.*

5. Northwest Center for Emerging Technologies (NWCET). *Building a Foundation for Tomorrow: Skills Standards for Information Technology, Millennium Edition.* Belleview, WA (1999). (Copies may be ordered from *www.nwcet.org*, including the latest edition published in March 2003.)

> *The National Science Foundation provided the vision and major funding for this collaborative study done by industry and academia. The report documents suggested skill standards for information technology career clusters. It provides detailed information on the performance criteria, technical knowledge, foundation skills, and personal qualities for each of fifteen major job functions of project management.*

6. Posner, Barry Z. "What It Takes to Be a Good Project Manager." *Project Management Institute* (March 1987).

> *This article summarizes key attributes of successful project managers and lists common causes for problems in managing projects based on a survey of 287 project managers. Posner also co-authored a book called* The Leadership Challenge in 2002 *to share more of his insight into what makes successful leaders.*

7. Thomsett, Rob. "Extreme Project Management," *Cutter Consortium* (*www.cutter.com*) (2002).

> *Rob Thomsett, an author and senior consultant for the Cutter Consortium, suggests the need for a new paradigm on project management called extreme project management. In this article, he discusses how three major forces—a power shift, the free-agent army, and the global e-economy—have changed the way projects should be managed. Thomsett also wrote a book in 2002 called* Radical Project Management.

KEY TERMS

Adaptive Software Development (ASD) — a SDLC model that assumes that systems development follows an adaptive or non-prescriptive approach because the scope of the project cannot be clearly articulated and the schedule and cost cannot be accurately predicted

deliverable — a product, such as a report or segment of software code, produced as part of a project

executive steering committee — a group of senior executives from various parts of the organization who regularly review important corporate projects and issues

Extreme Programming (XP) — an approach to programming for developing software in rapidly changing environments

functional organizational structure — an organizational structure that groups people by functional areas such as information technology, manufacturing, engineering, and accounting

human resources frame — focuses on producing harmony between the needs of the organization and the needs of people

matrix organizational structure — an organizational structure in which employees are assigned to both functional and project managers

phase exit or **kill point** — management review that should occur after each project phase to determine if projects should be continued, redirected, or terminated

political frame — addresses organizational and personal politics

politics — competition between groups or individuals for power and leadership

predictive life cycle — a SDLC model in which the scope of the project can be clearly articulated and the schedule and cost can be accurately predicted

project acquisition — the last two phases of a project life cycle, implementation and close-out

project feasibility — the first two phases of a project life cycle, concept and development

project life cycle — the collection of project phases—concept, development, implementation, and close-out

project organizational structure — an organizational structure that groups people by major projects, such as specific aircraft programs

Scrum — an iterative development process for software where the iterations are referred to as sprints, which normally last thirty days

structural frame — deals with how the organization is structured (usually depicted in an organizational chart) and focuses on different groups' roles and responsibilities to meet the goals and policies set by top management

symbolic frame — focuses on the symbols, meanings, and culture of an organization

systems — sets of interacting components working within an environment to fulfill some purpose

systems analysis — a problem-solving approach that requires defining the scope of the system to be studied, and then dividing it into its component parts for identifying and evaluating its problems, opportunities, constraints, and needs

systems approach — a holistic and analytical approach to solving complex problems that includes using a systems philosophy, systems analysis, and systems management

systems development life cycle (SDLC) — a framework for describing the phases involved in developing and maintaining information systems

systems management — addressing the business, technological, and organizational issues associated with making a change to a system

systems philosophy — an overall model for thinking about things as systems

systems thinking — taking a holistic view of an organization to effectively handle complex situations

3

The Project Management Process Groups: A Case Study

Objectives

After reading this chapter, you will be able to:

1. *Describe the five project management process groups, the typical level of activity for each, and the interactions among them*
2. *Understand how the project management process groups relate to the project management knowledge areas*
3. *Discuss how organizations develop information technology project management methodologies to meet their needs*
4. *Review a case study of an organization applying the project management process groups to manage an information technology project*
5. *Understand the contribution that effective project initiation, project planning, project execution, project control, and project closing make to project success*

OPENING CASE

Erica Bell was in charge of the Project Management Office (PMO) for her consulting firm. The firm, JWD (Job Well Done) Consulting, had grown to include more than two hundred full-time consultants and even more part-time consultants. JWD Consulting provides a variety of consulting services to assist organizations in selecting and managing information technology projects. The firm focuses on finding and managing high-payoff projects and developing strong metrics to measure project performance and estimated benefits to the organization after the project is implemented. The firm's emphasis on metrics and working collaboratively with its customers gives it an edge over many competitors.

Joe Fleming, the CEO, wanted his company to continue to grow and become a world-class consulting organization. Since the core of the business was helping other organizations with project management, he felt it was crucial for JWD Consulting to have an exemplary process for managing its own projects. He asked Erica to work with her team and other consultants in the firm to develop an intranet where they could share their project management knowledge. He also thought it would make sense to make some of the information available to the firm's clients. For example, the firm could provide project management templates, tools, articles, links to other sites, and an "Ask the Expert" feature to help build relationships with current and future clients. Since JWD Consulting emphasizes the importance of high-payoff projects, Joe also wanted to see a business case for this project before proceeding.

Recall from Chapter 1 that project management consists of nine knowledge areas: integration, scope, time, cost, quality, human resource, communications, risk, and procurement management. You will learn more about each of these knowledge areas and how they apply to managing information technology projects in Chapters 4 through 12. Another important concept is that projects involve five project management process groups: initiating, planning, executing, controlling, and closing. This chapter describes each project management process group in detail through a simulated case study based on JWD Consulting. Appendix C provides additional case studies and questions you can use to practice applying the concepts, tools, and techniques you are learning throughout this and subsequent chapters. This chapter also uses several templates for various project management documents. Copies of these templates are available in Appendix D and on the companion Web site for this text (*www.course.com/mis/schwalbe*).

PROJECT MANAGEMENT PROCESS GROUPS

Project management is an integrative endeavor; decisions and actions taken in one knowledge area at a certain time usually affect other knowledge areas. Managing these interactions often requires making trade-offs among the project's scope, time, and cost—the triple constraint of project management described in Chapter 1. A project manager may also need to make trade-offs between other knowledge areas, such as between risk and human resources. Consequently, you can view project management as a number of related processes. A **process** is a series of actions directed toward a particular result. **Project management process groups** progress from initiation activities to planning activities, executing activities, controlling activities, and closing activities.

- **Initiating processes** include actions to begin or end projects and project phases. To initiate a project in the concept phase, someone must define the business need for the project, someone must sponsor the project, and someone must take on the role of project manager. Initiating processes take place during each phase of a project. For example, project managers and teams should reexamine the business need for the project during every phase of the project life cycle to determine if the project is worth continuing. Initiating processes are also required to end a project. Someone must initiate activities to ensure the project team completes all the work, documents lessons learned, reassigns project resources, and that the customer accepts the work.

- **Planning processes** include devising and maintaining a workable scheme to ensure the project addresses the company's needs. Project plans define each knowledge area as it relates to the project at that point in time. For example, a project team must develop a project plan to define the scope of the project, to estimate costs, to decide what resources to procure, to define the schedule activities and who will perform them, and so on. To account for changing conditions on the project and in the organization, project teams often revise project plans during each phase of the project life cycle.

- **Executing processes** include coordinating people and other resources to carry out the project plans and produce the deliverables of the project or phase. Examples of executing processes include developing the project team, providing leadership, assuring project quality, disseminating information, procuring necessary resources, and delivering the actual work.

- **Controlling processes** ensure that the project team meets the project objectives. The project manager and staff monitor and measure progress against the project plan and take corrective action when necessary. A common controlling process is a performance review during which any necessary changes are identified, and a staff member is selected to analyze and manage those changes.

- **Closing processes** include formalizing acceptance of the phase or project and ending it efficiently. Administrative activities are often involved in this process group, such as archiving project files, documenting lessons learned, and receiving formal acceptance of the delivered work as part of the phase or project.

Figure 3-1 shows the project management process groups and how they relate to each other in terms of typical level of activity, time frame, and overlap. Notice that the process groups are not isolated events. They occur at varying levels of intensity throughout each phase of a project. The level of activity and length of each process group varies for every project. Normally, the executing processes require the most resources and time, followed by the planning processes. The initiating and closing processes are usually the shortest and require the least amount of resources and time. However, every project is unique, so there can be exceptions. Also, notice that this figure is called "overlap of process groups in a phase," which means you can apply the process groups for each major phase of a project—concept, development, implementation, and close-out, as described in Chapter 2—or you can apply the process groups to an entire project, as the JWD Consulting case study does in this chapter.

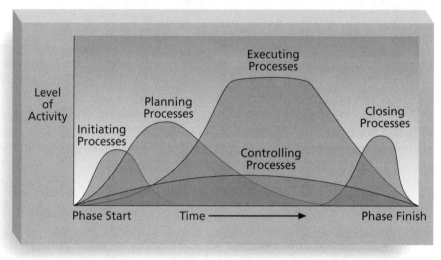

PMBOK® Guide, 2000, 31.

Figure 3-1. Overlap of Process Groups in a Phase

Each of the five project management process groups is characterized by the completion of certain tasks. During initiating processes for a new project, the organization recognizes that a new project exists. Often, completing a business case and project charter accomplishes this recognition (see Chapter 5, Project Scope Management). These documents identify the main stakeholders for a project, justify the project, and specify the high-level scope, time, and cost goals of the project. Usually, the project manager and key team members are also selected during the initiating process group.

Outcomes of the planning process group include completing the work breakdown structure and scope statement, the project schedule, and the project budget (see Chapters 5, 6, and 7, respectively). Planning processes are especially important for information technology projects. Everyone who has ever worked on a large information technology project that involves new technology knows the saying, "A dollar spent up front in planning is worth one hundred dollars spent after the system is implemented." Planning is crucial in information technology projects because once a project team implements a new system it takes a considerable amount of effort to change the system. Research suggests that companies working to implement best practices should spend at least 20 percent of project time in initiating and planning activities.[1]

The executing process group involves taking the actions necessary to complete the work described in the planning activities. The main outcome of this process group is the delivery of the actual work of the project. For example, if an information technology project involves providing new hardware, software, and training, the executing processes would include leading the project team and other stakeholders to purchase the hardware, develop and test the software, and deliver and participate in the training. The executing process group should overlap the other process groups and generally requires the most resources.

Controlling processes measure progress toward the project objectives, monitor deviation from the plan, and take corrective action to match progress with the plan. The ideal outcome of the controlling process group is to complete a project successfully by delivering the agreed-upon project scope within time, cost, and quality constraints. If changes to project objectives or plans are required, controlling processes ensure that these changes are made efficiently and effectively to meet stakeholder needs and expectations. Controlling processes overlap all of the other project management process groups because changes can occur at any time.

During the closing processes, the project team works to gain acceptance of the end product and bring the phase or project to an orderly end. Key outcomes of this process group are formal acceptance of the work and creation of closing documents, such as a final project report and lessons-learned report.

You can map the main activities of each project management process group, which apply to an entire project or a phase of a project, into the nine project management knowledge areas. Table 3-1 provides a big-picture view of the relationships among the thirty-nine project management activities, the time in which they are typically completed, and the knowledge areas into which they fit. The processes listed in the table are the main processes for each knowledge area listed in the PMBOK® Guide 2000. Many organizations use PMI's information as a foundation for developing their own project management methodologies, as described later in this chapter. For example, many project managers believe that knowledge areas other than scope management are important in initiating. Notice in the table that the majority of project management processes occur as part of the planning process group. Since each project is unique, project teams are always trying to do something that has not been done before. To succeed at unique and new activities, project teams must do a fair amount of planning. Recall, however, that the most time and money should be spent on executing.

---

[1] PCI Group, "PM Best Practices Report," (October 2001).

Table 3-1: Relationships Among Project Management Process Groups and Knowledge Areas

KNOWLEDGE AREA	PROJECT MANAGEMENT PROCESS GROUPS				
	INITIATING	PLANNING	EXECUTING	CONTROLLING	CLOSING
Integration		Project plan development	Project plan execution	Integrated change control	
Scope	Initiating	Scope planning Scope definition		Scope verification Scope change control	
Time		Activity definition Activity sequencing Activity duration estimating Schedule development		Schedule control	
Cost		Resource planning Cost estimating Cost budgeting		Cost control	
Quality		Quality planning	Quality assurance	Quality control	
Human Resource		Organizational planning Staff acquisition	Team development		
Communications		Communications planning	Information distribution	Performance reporting	Administrative closure
Risk		Risk management planning Risk identification Qualitative risk analysis Quantitative risk analysis Risk response planning		Risk monitoring and control	
Procurement		Procurement planning Solicitation planning	Solicitation Source Selection Contract administration		Contract closeout

PMBOK® Guide 2000, 38.

DEVELOPING AN INFORMATION TECHNOLOGY PROJECT MANAGEMENT METHODOLOGY

Many organizations spend a great deal of time and money on training efforts for project management, but after the training, a project manager may still not know how to tailor their project management skills to the organization's particular needs. Because of this problem, some organizations develop their own internal information technology project management methodologies.

For example, after implementing a systems development life cycle (SDLC) at Blue Cross Blue Shield of Michigan, the methods department became aware that developers and project managers were often working on information technology projects in different ways. Deliverables were often missing or looked different from project to project. There was a general lack of consistency and a need for standards to guide both new and experienced project managers. Top management decided to authorize funds to develop a methodology for project managers that could also become the basis for information technology project management training within the organization. It was also part of an overall effort to help raise the company's Software Capability Maturity Model level. (See Chapter 8, Project Quality Management, for more information on maturity models.)

Blue Cross Blue Shield of Michigan launched a three-month project to develop its own project management methodology. Some of the project team members had already passed the PMP certification examination, so they decided to base their methodology on the PMBOK® Guide 2000, making adjustments as needed to best describe how their organization managed information technology projects. Figure 3-2 shows a picture of the resulting information technology project management processes and the information flow among them.[2]

The project team realized that they would have to drop or de-emphasize some processes in the PMBOK® Guide 2000 to fit the organization's needs. For example, in contrast to other industries, the overriding financial investment in software development is in salaries, not in materials. In addition, Blue Cross Blue Shield of Michigan did not negotiate with contracting firms in the information technology area. Therefore, most of the procurement functions were absorbed into other processes, such as Scope Planning and Definition, and Resource Planning.

The project team also added additional processes. For example, to keep track of the large amount of documentation necessary on an information technology project, the team decided to develop a process for maintaining and updating a project workbook that would serve as an information resource for team members and a hard-copy record of project activities. Therefore, the project team added Project Book Records as a separate process under Information Distribution. They added another new process, Issue Control, because of its

[2] Munroe, William, "Developing and Implementing an IT Project Management Process," *ISSIG Review* (First Quarter 2001). A copy of the article is available on the companion Web site for this text.

importance to information technology projects. Issues are prevalent in information technology projects, in part because of the inherent complexity of information systems and rapidly changing technology.

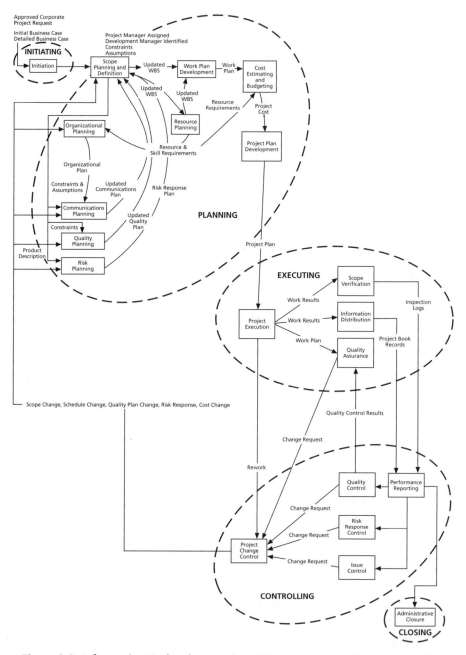

Figure 3-2. Information Technology Project Management Methodology at Blue Cross Blue Shield of Michigan

The team also decided to combine the PMBOK® Guide 2000's processes of Activity Sequencing, Activity Definition, Activity Duration Estimating, and Schedule Development into one process they called Work Plan Development. In addition, to enhance usability and simplify the overall process, the Integrated Change Control, Scope Change Control, Schedule Control, and Cost Control processes were combined into one process called Project Change Control.

Blue Cross Blue Shield of Michigan wanted its information technology project management methodology to work with any of the SDLC models. This forced team members to separate their thinking on what is involved in creating product-related deliverables such as design documents, code, or test plans from the process-related deliverables such as work breakdown structures, resource requirements, or change requests. Understanding the difference between the SDLC and the information technology project management methodology amounted to a paradigm shift for some members of the team accustomed to filling a dual role of developer and project manager. Today, Blue Cross Blue Shield of Michigan's information technology project management methodology is the basis for its training programs, and the company uses it to develop and implement its information technology projects.

The following section examines an example of applying the project management process groups to a project at JWD Consulting. It uses some of the ideas from the PMBOK® Guide 2000, some ideas from the Blue Cross Blue Shield of Michigan methodology, and some new ideas to meet unique project needs. This example also includes the use of Microsoft Project to demonstrate how project management software can assist in several aspects of managing a project. Several templates will illustrate how project teams prepare various project management documents. Details on creating many of the documents shown are provided in later chapters, so do not worry if you do not understand everything right now. You might want to read this section again to enhance your learning. The following case provides an example of the elements involved in managing a project from start to finish.

CASE STUDY: JWD CONSULTING'S PROJECT MANAGEMENT INTRANET SITE PROJECT

Project Initiation

In project management, initiating is the process of recognizing and starting a new project or project phase. An organization should put considerable thought into project selection to ensure that it initiates the right kinds of projects for the right reasons. It is better to have a moderate or even small amount of success on an important project than huge success on one that is unimportant. The selection of projects for initiation, therefore, is crucial, as is the selection of project managers. You will learn more about project selection in Chapter 5.

Project Scope Management

It is important to remember that strategic planning should serve as the foundation for deciding which projects to pursue. The organization's strategic plan expresses the vision, mission, goals, objectives, and strategies of the organization. It also provides the basis for information technology project planning. Information technology is usually a support function in an organization, so it is critical that the people initiating information technology projects understand how those projects relate to current and future needs of the organization. For example, JWD Consulting's main business is providing consulting services to other organizations, not developing its own intranet. Information systems, therefore, must support the firm's business goals, such as providing consulting services more effectively and efficiently.

An organization may initiate information technology projects for several reasons, but the most important reason is to support business objectives. Providing a good return on investment at a reasonable level of risk is also important, especially in tough economic times. As mentioned in the opening case, JWD Consulting wants to follow an exemplary process for managing its projects since its core business is helping other organizations manage projects. Developing an intranet to share its project management knowledge could help JWD Consulting reduce internal costs by working more effectively, and by allowing existing and potential customers to access some of the firm's information. JWD Consulting could also increase revenues by bringing in more business as people saw the value of their services.

The PMBOK® Guide 2000 lists project scope management as the only knowledge area involved in initiation, and it identifies only four outputs—the project charter, assignment of the project manager, clarification of constraints, and a list of assumptions. JWD Consulting also believes it is important to identify key project stakeholders and to write a business case for the project during project initiation. Some organizations, such as Blue Cross Blue Shield of Michigan, require an approved corporate project request and an initial and detailed business case before project initiation begins (see Figure 3-1). Some organizations require these items in a phase called pre-project initiation.

To simplify JWD Consulting's project, Erica decides to use the five project management process groups, including the pre-project initiation tasks of identifying stakeholders and developing a business case. Since the firm does not require formal paperwork for internal project requests, she decides not to include that item. The firm's business case includes a clarification of constraints and a list of assumptions, so the firm does not include those items as separate outputs of initiation. Table 3-2 summarizes what JWD Consulting considers typical outputs of project initiation. The project charter includes the preliminary budget and schedule information for the project.

Table 3-2: Outputs of Project Initiation

Project manager assigned
Key stakeholders identified
Business case completed
Project charter completed and signed

As described in the opening case, Joe Fleming, the CEO at JWD Consulting, has assigned Erica Bell as the project manager for the intranet project, so that particular initiating task is finished. Erica has managed several projects in the past, and as the head of the firm's Project Management Office, she had expertise that would be instrumental to this project's success.

Erica met with Joe Fleming, the project's sponsor, to help identify other key stakeholders for this project. They decided to contact one full-time consultant with an outstanding record, Michael Chen, one-part-time consultant, Jessie Faue, who was new to the company and supported the Project Management Office, and two members of the information technology department who supported the current intranet, Kevin Dodge and Cindy Dawson. They also knew that client inputs would be important for this project, so Joe agreed to call the CEOs of two of the firm's largest clients to see if they would be willing to provide representatives to work on this project at their own expense. After Joe and Erica made the preliminary contacts, Erica documented the stakeholders' roles, names, organizations, and contact information. All of the internal staff Joe and Erica recommended agreed to work on the project, and the two client representatives would be Kim Phuong and Page Miller. Erica would have these stakeholders review and sign the project charter.

Erica drafted a business case for the project, getting input and feedback from Joe, one of her senior staff members in the Project Management Office, and a member of the finance department. Table 3-3 provides the business case. (Note that this example and others are abbreviated examples. See the companion Web site for this text to download a business case template and other templates. You will learn how to perform the financial calculations in Chapter 5, Project Scope Management.) Notice the following information included in the business case:

- introduction/background
- business objective
- current situation and problem/opportunity statement
- critical assumptions and constraints
- analysis of options and recommendation
- preliminary project requirements
- budget estimate and financial analysis

- schedule estimate
- potential risks
- exhibits

Since this project is relatively small and is for an internal sponsor, the business case is not as long as many other business cases. Erica reviewed the business case with Joe before proceeding to develop the project charter. Joe agreed that the project was worth pursuing, and he told Erica to proceed with developing the project charter to officially recognize the existence of this project.

Table 3-3: JWD Consulting's Business Case

1.0 INTRODUCTION/ BACKGROUND

JWD Consulting's core business goal is to provide world-class project management consulting services to various organizations. The CEO, Joe Fleming, believes it can streamline operations and increase business by providing information related to project management on its intranet site, making some information and services accessible to current and potential clients.

2.0 BUSINESS OBJECTIVE

JWD Consulting's strategic goals include continuing growth and profitability. The Project Management Intranet Site project will support these goals by increasing visibility of the firm's expertise to current and potential clients by allowing client and public access to some sections of the intranet. It will also improve profitability by reducing internal costs by providing standard tools, techniques, templates, and project management knowledge to all internal consultants. Since JWD Consulting focuses on identifying profitable projects and measuring their value after completion, this project must meet those criteria.

3.0 CURRENT SITUATION AND PROBLEM/OPPORTUNITY STATEMENT

JWD Consulting has a corporate Web site as well as an intranet. The firm currently uses the Web site for marketing information. The primary use of the intranet is for human resource information, such as where consultants enter their hours on various projects, change and view their benefits information, access an online directory and Web-based e-mail system, and so on. The firm also uses an enterprise-wide project management system to track all project information, focusing on the status of deliverables and meeting scope, time, and cost goals. There is an opportunity to provide a new section on the intranet dedicated to sharing consultants' project management knowledge across the organization. JWD Consulting only hires experienced consultants and gives them freedom to manage projects as they see fit. However, as the business grows and projects become more complex, even experienced project managers are looking for suggestions on how to work more effectively.

4.0 CRITICAL ASSUMPTION AND CONSTRAINTS

The proposed intranet site must be a valuable asset for JWD Consulting. Current consultants and clients must actively support the project, and it must pay for itself within one year by reducing internal operating costs and generating new business. The Project Management Office manager must lead the effort, and the project team must include participants from several parts of the company, as well as current client organizations. The new system must run on existing hardware and software, and it should require minimal technical support. It must be easily accessible by clients and the public yet secure from unauthorized users.

Table 3-3: JWD Consulting's Business Case (continued)

5.0 ANALYSIS OF OPTIONS AND RECOMMENDATION

There are three options for addressing this opportunity:

1. Do nothing. The business is doing well, and we can continue to operate without this new project.
2. Purchase access to specialized software to support this new capability with little in-house development.
3. Design and implement the new intranet capabilities in-house using mostly existing hardware and software.

Based on discussions with stakeholders, we believe that option 3 is the best option.

6.0 PRELIMINARY PROJECT REQUIREMENTS

The main features of the project management intranet site include the following:

1. Access to several project management templates and tools. Users must be able to search for templates and tools, read instructions on using these templates and tools, and see examples of how to apply them to real projects. Users must also be able to submit new templates and tools, which should be first screened or edited by the Project Management Office before being added to the intranet site.
2. Access to relevant project management articles. Many consultants and clients feel as though there is an information overload when they research project management information. They often waste time they should be spending with their clients. The new intranet should include access to several important articles on various project management topics, which are searchable by topic, and allow users to request the Project Management Office staff to find additional articles to meet their needs.
3. Links to other, up-to-date Web sites, with brief descriptions of the main features of the external site.
4. An "Ask the Expert" feature to help build relationships with current and future clients and share knowledge with internal consultants.
5. Appropriate security to make the entire intranet site accessible to internal consultants and certain sections accessible to others.
6. Ability to charge money for access to some information. Some of the information and features of the intranet site should prompt external users to pay for the information or service. Payment options should include a credit card option or similar online payment transactions. After the system verifies payment, the user should be able to access or download the desired information.
7. Other features suggested by users, if they add value to the business.

7.0 BUDGET ESTIMATE AND FINANCIAL ANALYSIS

A preliminary estimate of the cost for the entire project is $140,000. This estimate is based on the project manager working about 20 hours per week for six months and other internal staff working a total of about 60 hours per week for six months. The customer representatives would not be paid for their assistance. A staff project manager would earn $50 per hour. The hourly rate for the other project team members would be $70 per hour, since some hours normally billed to clients may be needed for this project. The initial cost estimate also includes $10,000 for purchasing software and services from suppliers. After the project is completed, maintenance costs of $40,000 are included for each year, primarily to update the information and coordinate the "Ask the Expert" and article retrieval features.

Projected benefits are based on a reduction in hours consultants spend researching project management information, appropriate tools and templates, and so on. Projected benefits are also based on a small increase in profits due to new business generated by this project. If each of more than 400 consultants saved just 40 hours each year (less than one hour per week) and could bill that time to other projects that generate $10 per hour

Table 3-3: JWD Consulting's Business Case (continued)

in profits, then the projected benefit would be $160,000 per year. If the new intranet increased business by just 1 percent, using past profit information, increased profits due to new business would be at least $40,000 each year. Total projected benefits, therefore, are about $200,000 per year.

Exhibit A summarizes the projected costs and benefits and shows the estimated net present value (NPV), return on investment (ROI), and year in which payback occurs. It also lists assumptions made in performing this preliminary financial analysis. All of the financial estimates are very encouraging. The estimated payback is within one year, as requested by the sponsor. The NPV is $272,800, and the discounted ROI based on a three-year system life is excellent at 112 percent.

8.0 SCHEDULE ESTIMATE

The sponsor would like to see the project completed within six months, but there is some flexibility in the schedule. We also assume that the new system will have a useful life of at least three years.

9.0 POTENTIAL RISKS

There are several risks involved with this project. The foremost risk is a lack of interest in the new system by our internal consultants and external clients. User inputs are crucial for populating information into this system and realizing the potential benefits from using the system. There are some technical risks in choosing the type of software used to search the system, check security, process payments, and so on, but the features of this system all use proven technologies. The main business risk is investing the time and money into this project and not realizing the projected benefits.

10.0 EXHIBITS

Exhibit A: Financial Analysis for Project Management Intranet Site Project

Discount rate		8%				
			Year			
Assume project done in about 6 mos.		0	1	2	3	Total
Costs		140,000	40,000	40,000	40,000	
Discount factor		1	0.93	0.86	0.79	
Discounted costs		140,000	37,200	34,400	31,600	243,200
Benefits			200,000	200,000	200,000	
Discount factor			0.93	0.86	0.79	
Discounted benefits			186,000	172,000	158,000	516,000
Discounted benefits - costs		(140,000)	148,800	137,600	126,400	
Cumulative benefits - costs		(140,000)	8,800	146,400	272,800	<---NPV
		Payback in year 1				
Discounted life cycle ROI----------->		112%				
Assumptions						
Costs		# hours	$/hour			
PM		500	50	25,000		
Staff		1500	70	105,000		
Outsourced software and services		10,000				
Total projected costs				140,000		
Benefits						
# consultants		400				
Hours saved		40				
$/hour profit		10				
Benefits from saving time		160,000				
Benefits from 1% increase in profits		40,000				
Total annual projected benefits		200,000				

Erica drafted a project charter and had the project team members review it before showing it to Joe. Joe made a few minor changes, which Erica incorporated, and then all the key stakeholders signed the project charter. Table 3-4 shows the final project charter (see Chapter 5 for more information on project charters). Note the items included on the project charter and its short length. JWD Consulting believed that project charters should preferably be one or two pages long, and they could refer to other documents, such as a business case, as needed. Erica felt the most important parts of the project charter were the signatures of key stakeholders and their individual comments. It is hard to get stakeholders to agree on even a one-page project charter, so everyone has a chance to make their concerns known in the comments section. Note that Michael Chen, the senior consultant asked to work on the project, was concerned about working on this project when he felt that his other assignments with external clients might have a higher priority. He offered to have an assistant help as needed. The information technology staff members mentioned their concerns about testing and security issues. Erica knew that she would have to consider these concerns when managing the project.

Table 3-4: Project Charter

Project Title: Project Management Intranet Site Project

Project Start Date: May 2, 2005   **Projected Finish Date:** November 4, 2005

Budget Information: The firm has allocated $140,000 for this project. The majority of costs for this project will be internal labor. An initial estimate provides a total of 80 hours per week.

Project Manager: Erica Bell, (310) 889-5896, erica_bell@jwdconsulting.com

Project Objectives: Develop a new capability accessible on JWD Consulting's intranet site to help internal consultants and external customers manage projects more effectively. The intranet site will include several templates and tools that users can download, examples of completed templates and related project management documents used on real projects, important articles related to recent project management topics, article retrieval service, links to other sites with useful information, and an "Ask the Expert" feature, where users can post questions they have about their projects and receive advice from experts in the field. Some parts of the intranet site will be accessible free to the public, other parts will only be accessible to current customers and/or internal consultants, and other parts of the intranet site will be accessible for a fee.

Approach:

- Develop a survey to determine critical features of the new intranet site and solicit input from consultants and customers.

- Review internal and external templates and examples of project management documents.

- Research software to provide features such as security, managing user inputs, and the article retrieval and "Ask the Expert" capability.

- Develop the intranet using an iterative approach, soliciting a great deal of user feedback.

- Determine a way to measure the value of the intranet site in terms of reduced costs and new revenues, both during the project and one year after project completion.

Table 3-4: Project Charter (continued)

ROLES AND RESPONSIBILITIES:			
NAME	ROLE	POSITION	CONTACT INFORMATION
Joe Fleming	Sponsor	JWD Consulting, CEO	joe_fleming@jwdconsulting.com
Erica Bell	Project Manager	JWD Consulting, manager	erica_bell@jwdconsulting.com
Michael Chen	Team Member	JWD Consulting, senior consultant	michael_chen@jwdconsulting.com
Jessie Faue	Team Member	JWD Consulting consultant	jessie_faue@jwdconsulting.com
Kevin Dodge	Team Member	JWD Consulting, IT department	kevin_dodge@jwdconsulting.com
Cindy Dawson	Team Member	JWD Consulting, IT department	cindy_dawson@jwdconsulting.com
Kim Phuong	Advisor	Client representative	kim_phuong@client1.com
Page Miller	Advisor	Client representative	page_miller@client2.com

Comments: (Handwritten or typed comments from above stakeholders, if applicable)
"I will support this project as time allows, but I believe my client projects take priority. I will have one of my assistants support the project as needed." —Michael Chen
"We need to be extremely careful testing this new system, especially the security in giving access to parts of the intranet site to the public and clients." —Kevin Dodge and Cindy Dawson

Project Planning

Planning is often the most difficult and unappreciated process in project management. Because planning is not often used to facilitate action, many people view planning negatively. The main purpose of a project plan, however, is *to guide project execution*. To guide execution, plans must be realistic and useful, so a fair amount of time and effort must go into the planning process; people knowledgeable with the work need to plan the work.

Table 3-5 lists the project management knowledge areas, processes, and outputs of project planning according to the PMBOK® Guide 2000. There are many potential outputs from the planning process group, and every knowledge area is included. Just a few planning documents from JWD Consulting's Project Management Intranet Site project are provided in this chapter as examples. Recall that the PMBOK® Guide 2000 is a guide, so many organizations may have different planning outputs based on their particular needs.

Since the Project Management Intranet Site project is relatively small, Erica believes the most important planning documents to create are the following:

- a team contract (not listed in Table 3-5, which is based on the PMBOK® Guide)
- a scope statement
- a work breakdown structure (WBS)
- a project schedule, in the form of a Gantt chart with all dependencies and resources entered
- a list of prioritized risks

Table 3-5: Planning Processes and Outputs

KNOWLEDGE AREA	PROCESS	OUTPUTS
Integration	Project Plan Development	■ Project plan ■ Supporting detail
Scope	Scope Planning	■ Scope statement ■ Supporting detail ■ Scope management plan
	Scope Definition	■ WBS ■ Scope statement updates
Time	Activity Definition	■ Activity list ■ Supporting detail ■ WBS updates
	Activity Sequencing	■ Project network diagram ■ Activity list updates
	Activity Duration Estimating	■ Activity duration estimates ■ Basis of estimates ■ Activity list updates
	Schedule Development	■ Project schedule ■ Supporting detail ■ Schedule management plan ■ Resource requirement updates
Cost	Resource Planning	■ Resource requirements
	Cost Estimating	■ Cost estimates ■ Supporting detail ■ Cost management plan
	Cost Budgeting	■ Cost baseline
Quality	Quality Planning	■ Quality management plan ■ Operational definitions ■ Checklists ■ Inputs to other processes

Table 3-5: Planning Processes and Outputs (continued)

KNOWLEDGE AREA	PROCESS	OUTPUTS
Human Resource	Organizational Planning	■ Role and responsibility assignments ■ Staffing management plan ■ Organizational chart ■ Supporting detail
	Staff Acquisition	■ Project staff assigned ■ Project team directory
Communications	Communications Planning	■ Communications management plan
Risk	Risk Management Planning	■ Risk management plan
	Risk Identification	■ Risks ■ Triggers ■ Inputs to other processes
	Qualitative Risk Analysis	■ Overall risk ranking for the project ■ List of prioritized risks ■ List of risks for additional analysis and management ■ Trends in qualitative risk analysis results
	Quantitative Risk Analysis	■ Prioritized list of quantified risks ■ Probabilistic analysis of the project ■ Probability of achieving the cost and time objectives ■ Trends in quantitative risk analysis results
	Risk Response Planning	■ Risk response plan ■ Residual risks ■ Secondary risks ■ Contractual agreements ■ Contingency reserve amounts needed ■ Inputs to other processes ■ Inputs to a revised project plan
Procurement	Procurement Planning	■ Procurement management plan ■ Statement(s) of work
	Solicitation Planning	■ Procurement documents ■ Evaluation criteria ■ Statement of work updates

All of these documents, as well as other project-related information, will be available to all team members on a project Web site. JWD Consulting has used project Web sites for several years, and has found that they really help facilitate communications and document project information. For larger projects, JWD Consulting also creates project organization charts, a formal communications management plan, a quality management plan, detailed cost estimates, procurement plans, and other planning documents. You will learn more about these documents by knowledge area in the following chapters. Appendix D provides templates for creating many of these documents.

Soon after the project team signed the project charter, Erica organized a kick-off meeting for the Project Management Intranet Site project. An important part of the meeting was helping the project team get to know each other. Erica had met and talked to each member separately, but this was the first time the project team would all meet each other. Jessie Faue worked in the Project Management Office with Erica, so they knew each other well, but Jessie was new to the company and did not know any of the other team members. Michael Chen was a senior consultant and often worked on the highest priority projects for external clients. He attended the kick-off meeting with his assistant, Jill Anderson, who would also support the project when Michael was too busy. Everyone valued Michael's expertise, and he was extremely straightforward in dealing with people. He also knew both of the client representatives from past projects. Kevin Dodge was JWD Consulting's intranet guru who tended to focus on technical details. Cindy Dawson was also from the information technology department and had experience working as a business consultant and negotiating with outside suppliers. Kim Phuong and Page Miller, the two client representatives, were excited about the project, but they were wary of sharing sensitive information about their company.

Erica had everyone introduce him or herself and had an icebreaker activity so everyone would be more relaxed. She then explained the importance of the project, reviewing the signed project charter. She explained that an important tool to help a project team work together was to have members develop a team contract that everyone felt comfortable signing. JWD Consulting believed in using team contracts for all projects to help promote teamwork and clarify team communications. She explained the main topics covered in a team contract and showed them a team contract template. She then had the team members form two smaller groups, with one consultant, one information technology department member, and one client representative in each group. These smaller groups made it easier for everyone to contribute ideas. Each group shared their ideas for what should go into the contract, and then they worked together to form one project team contract. Table 3-6 shows the resulting team contract, which took about ninety minutes to create. Erica could see that there were different personalities on this team, but she felt they all could work together well.

Table 3-6: Team Contract

Code of Conduct: As a project team, we will:

- Work proactively, anticipating potential problems and working to prevent them
- Keep other team members informed of information related to the project
- Focus on what is best for the entire project team

Participation: We will:

- Be honest and open during all project activities
- Encourage diversity in teamwork
- Provide the opportunity for equal participation
- Be open to new approaches and consider new ideas
- Have one discussion at a time
- Let the project manager know well in advance if a team member has to miss a meeting or may have trouble meeting a deadline for a given task

Communication: We will:

- Decide as a team on the best way to communicate. Since a few team members cannot meet often for face-to-face meetings, we will use e-mail, a project Web site, and other technology to assist in communicating
- Have the project manager facilitate all meetings and arrange for phone and videoconferences, as needed
- Work together to create the project schedule and enter actuals into our enterprise-wide project management system by 4 PM every Friday
- Present ideas clearly and concisely
- Keep discussions on track and have one discussion at a time

Problem Solving: We will:

- Encourage everyone to participate in solving problems
- Only use constructive criticism and focus on solving problems, not blaming people
- Strive to build on each other's ideas

Meeting Guidelines: We will:

- Plan to have a face-to-face meeting the first and third Tuesday morning of every month
- Meet more frequently the first month
- Arrange for telephone or videoconferencing for participants as needed
- Hold other meetings as needed
- Record meeting minutes and send them out via e-mail within 24 hours of all project meetings, focusing on decisions made and action items from each meeting
- Develop an agenda before all meetings with our project sponsor and client advisors
- Document major issues and decisions related to the project and send them out via e-mail to all team members and the project sponsor

Erica wanted to keep their meeting to its two-hour time limit. Their next task would be to clarify the scope of the project by developing a scope statement and WBS. She knew it took time to develop these documents, but she wanted to get a feel for what everyone thought were the main deliverables for this project, their roles in producing those deliverables, and what areas of the project scope needed clarification. She reminded everyone what their budget and schedule goals were so they would keep that in mind as they discussed the scope of the project. She also asked each person to provide the number of hours he or she would be available to work on this project each month for the next six months. She then had each person write down his or her answers to the following questions:

1. List one item that is most unclear to you about the scope of this project.
2. What other questions do you have or issues do you foresee about the scope of the project?
3. List what you believe to be the main deliverables for this project.
4. Which deliverables do you think you will help create or review?

Erica collected everyone's inputs. She explained that she would take this information and work with Jessie to develop the first draft of the scope statement that she would e-mail to everyone by the end of the week. She also suggested that they all meet again in one week to develop the scope statement further and to start creating the WBS for the project.

Erica and Jessie reviewed all the information and created the first draft of the scope statement. At their next team meeting, they discussed the scope statement and got a good start on the WBS. Table 3-7 shows a portion of the scope statement that Erica created after a few more e-mails and another team meeting. Note that the scope statement lists the importance of documenting the product characteristics and requirements, summarizes the deliverables, and describes project success criteria.

Table 3-7: Scope Statement (abbreviated)

Project Title: Project Management Intranet Site Project

Date: May 18, 2005   **Prepared by:** Erica Bell, Project Manager, (310)889-5896, erica_bell@jwdconsulting.com

Project Justification: Joe Fleming, CEO of JWD Consulting, requested this project to assist the company in meeting its strategic goals. The new intranet will increase visibility of the company's expertise to current and potential clients through the sections of the intranet to which they will have access. It will also help reduce internal costs and improve profitability by providing standard tools, techniques, templates, and project management knowledge to all internal consultants. The budget for the project is $140,000. An additional $40,000 will be required for operational expenses after the project is completed. Estimated benefits are $200,000 each year. It is important to focus on the system paying for itself within one year.

Product Characteristics and Requirements:

1. Templates and tools: The intranet site will allow authorized users to download files they can use to create project management documents and to help them use project management tools. These files will be in Microsoft Word, Excel, Access, PowerPoint, Project, or in HTML or PDF format, as appropriate.

Table 3-7: Scope Statement (abbreviated) (continued)

2. User submissions: Users will be encouraged to e-mail files with sample templates and tools to the Webmaster. The Webmaster will forward the files to the appropriate person for review and then post the files to the intranet site, if desired.

3. Articles: Articles posted on the intranet site will have appropriate copyright permission. The preferred format for articles will be PDF. The project manager may approve other formats.

4. Requests for articles: The intranet site will include a section for users to request someone from the Project Management Office (PMO) at JWD Consulting to research appropriate articles for them. The PMO manager must first approve the request and negotiate payments, if appropriate.

5. Links: All links to external sites will be tested on a weekly basis. Broken links will be fixed or removed within five working days of discovery.

6. The "Ask the Expert" feature must be user-friendly and capable of soliciting questions and immediately acknowledging that the question has been received in the proper format. The feature must also be capable of forwarding the question to the appropriate expert (as maintained in the system's expert database) and capable of providing the status of questions that are answered. The system must also allow for payment for advice, if appropriate.

7. Security: The intranet site must provide several levels of security. All internal employees will have access to the entire intranet site when they enter their security information to access the main, corporate intranet. Part of the intranet will be available to the public from the corporate Web site. Other portions of the intranet will be available to current clients based on verification with the current client database. Other portions of the intranet will be available after negotiating a fee or entering a fixed payment using preauthorized payment methods.

8. Search feature: The intranet site must include a search feature for users to search by topic, keywords, etc.

9. The intranet site must be accessible using a standard Internet browser. Users must have appropriate application software to open several of the templates and tools.

10. The intranet site must be available 24 hours a day, 7 days a week, with one hour per week for system maintenance and other periodic maintenance, as appropriate.

Summary of Project Deliverables

Project management-related deliverables: Business case, charter, team contract, scope statement, WBS, schedule, cost baseline, status reports, final project presentation, final project report, lessons-learned report, and any other documents required to manage the project.

Product-related deliverables:

1. Survey: Survey current consultants and clients to help determine desired content and features for the intranet site.

2. Files for templates: The intranet site will include templates for at least twenty documents when the system is first implemented, and it will have the capacity to store up to one hundred documents. The project team will decide on the initial twenty templates based on survey results.

3. Examples of completed templates: The intranet site will include examples of projects that have used the templates available on the intranet site. For example, if there is a template for a business case, there will also be an example of a real business case that uses the template.

Table 3-7: Scope Statement (abbreviated) (continued)

4. Files for tools: The intranet site will include information on how to use several project management tools, including the following as a minimum: work breakdown structures, Gantt charts, network diagrams, cost estimates, and earned value management. Where appropriate, sample files will be provided in the application software appropriate for the tool. For example, Microsoft Project files will be available to show sample work breakdown structures, Gantt charts, network diagrams, cost estimates, and applications of earned value management. Excel files will be available for sample cost estimates and earned value management charts.

5. Example applications of tools: The intranet site will include examples of real projects that have applied the tools listed in number 4 above.

6. Articles: The intranet site will include at least ten useful articles about relevant topics in project management. The intranet site will have the capacity to store at least one thousand articles in PDF format with an average length of ten pages each.

7. Links: The intranet site will include links with brief descriptions for at least twenty useful sites. The links will be categorized into meaningful groups.

8. Expert database: In order to deliver an "Ask the Expert" feature, the system must include and access a database of approved experts and their contact information. Users will be able to search for experts by predefined topics.

9. User request: The intranet site will include an application to solicit and process requests from users.

10. Intranet site design: An initial design of the new intranet site will include a site map, suggested formats, appropriate graphics, etc. The final design will incorporate comments from users on the initial design.

11. Intranet site content: The intranet site will include content for the templates and tools section, articles section, article retrieval section, links section, "Ask the Expert" section, User Requests feature, security, and payment features.

12. Test plan: The test plan will document how the intranet site will be tested, who will do the testing, and how bugs will be reported.

13. Promotion: A plan for promoting the intranet site will describe various approaches for soliciting inputs during design. The promotion plan will also announce the availability of the new intranet site.

14. Project benefit measurement plan: A project benefit plan will measure the financial value of the intranet site.

Project Success Criteria: Our goal is to complete this project within six months for no more than $140,000. The project sponsor, Joe Fleming, has emphasized the importance of the project paying for itself within one year after the intranet site is complete. To meet this financial goal, the intranet site must have strong user inputs. We must also develop a method for capturing the benefits while the intranet site is being developed, tested, and after it is rolled out. If the project takes a little longer to complete or costs a little more than planned, the firm will still view it as a success if it has a good payback and helps promote the firm's image as an excellent consulting organization.

As the project team worked on the scope statement, they also developed the Work Breakdown Structure (WBS) for the project. The WBS is a very important tool in project management because it provides the basis for deciding how to do the work. The WBS also provides a basis for creating the project schedule and performing earned value management for measuring and forecasting project performance. Erica and her team decided to use the project management process groups as the main categories for the WBS (Figure 3-3). They included completed work from the initiating process to provide a complete picture of the project's scope. The group also wanted to list several milestones on their schedule, such as the completion of key deliverables, so they prepared a separate list of milestones that they would include on the Gantt chart. You will learn more about creating a WBS in Chapter 5, Project Scope Management.

After preparing the WBS, the project team held another face-to-face meeting to develop the project schedule, following the steps outlined in section 2.5 of the WBS. Several of the project schedule tasks are dependent on one another. For example, the intranet site testing was dependent on the construction and completion of the content tasks. Everyone participated in development of the schedule, especially the tasks on which each would be working. Some of the tasks were broken down further so the team members had a better understanding of what they had to do and when. They also kept their workloads and cost constraints in mind when developing the duration estimates. For example, Erica was scheduled to work twenty hours per week on this project, and the other project team members combined should not spend more than sixty hours per week on average for the project. As team members provided duration estimates, they also estimated how many work hours they would spend on each task.

After the meeting, Erica worked with Jessie to enter all of the information into Microsoft Project. Erica was using the intranet site project to train Jessie in applying several project management tools and templates. They entered all of the tasks, duration estimates, and dependencies to develop the Gantt chart. Erica decided to enter the resource and cost information after reviewing the schedule. Their initial inputs resulted in a completion date a few weeks later than planned. Erica and Jessie reviewed the critical path for the project, and Erica had to shorten the duration estimates for a few critical tasks in order to meet their schedule goal of completing the project within six months. She talked to the team members working on those tasks, and they agreed that they could plan to work more hours each week on those tasks in order to complete them on time. Erica considered adding a project buffer at the end of the project, but she was confident they could meet the schedule goal. You will learn more about project buffers and developing project schedules in Chapter 6, Project Time Management.

1.0 Initiating
    1.1 Determine/assign project manager
    1.2 Identify key stakeholders
    1.3 Prepare business case
    1.4 Prepare project charter
2.0 Planning
    2.1 Hold project kickoff meeting
    2.2 Prepare team contract
    2.3 Prepare scope statement
    2.4 Prepare WBS
    2.5 Prepare schedule and cost baseline
        2.5.1 Determine task resources
        2.5.2 Determine task durations
        2.5.3 Determine task dependencies
        2.5.4 Create draft Gantt chart
        2.5.5 Review and finalize Gantt chart
    2.6 Identify, discuss, and prioritize risks
3.0 Executing
    3.1 Survey
    3.2 User inputs
    3.3 Intranet site content
        3.3.1 Templates and tools
        3.3.2 Articles
        3.3.3 Links
        3.3.4 Ask the Expert
        3.3.5 User requests feature
    3.4 Intranet site design
    3.5 Intranet site construction
    3.6 Intranet site testing
    3.7 Intranet site promotion
    3.8 Intranet site roll-out
    3.9 Project benefits measurement
4.0 Controlling
    4.1 Status reports
5.0 Closing
    5.1 Prepare final project report
    5.2 Prepare final project presentation
    5.3 Lessons learned

Figure 3-3. JWD Consulting Intranet Site Project Work Breakdown Structure (WBS)

Figure 3-4 shows the resulting Gantt chart created in Microsoft Project 2002. Only the executing tasks are expanded to show the subtasks under that category. Figure 3-5 shows part of the network diagram, again emphasizing the executing tasks. The highlighted boxes represent tasks on the critical path. (You can download a series of Project 2002 files for this project from the companion Web site for this text. These files show how the team entered only the WBS information, then the duration estimates and dependencies, and then the cost and resource information. You will learn how to use these features of Project 2002 in Appendix A. Chapter 6, Project Time Management, explains how to use network diagrams to find the critical path and explains Gantt chart symbols.) The baseline schedule projects a completion date of November 1, 2005. The project charter had a planned completion date of November 4, 2005. Erica wanted to complete the project on time, and although three extra days was not much of a buffer, she felt the baseline schedule was very realistic. She would do her best to help everyone meet their deadlines.

Figure 3-4. JWD Consulting Intranet Site Project Baseline Gantt Chart

Figure 3-5. JWD Consulting Intranet Site Project Partial Network Diagram

The majority of the costs for this project were internal labor, and the team kept their labor hour constraints in mind when developing task duration estimates. Erica and Jessie entered each project team member's name and labor rate in the resource sheet for their Microsoft Project file. The client representatives were not being paid for their time, so she left their labor rates at the default value of zero. Erica had also included $10,000 for procurement in the financial analysis she prepared for the business case, and she showed Jessie how to enter that amount as a fixed cost split equally between the "Ask the Expert" and User Requests features, where she thought they would have to purchase some external software and/or services. Erica then helped Jessie assign resources to tasks, entering the projected number of hours everyone planned to work each week on each task. They then ran several cost reports and made a few minor adjustments to resource assignments to make their planned total cost meet their budget constraints. Their cost baseline was very close to their planned budget of $140,000.

The last deliverable her team needed to create within the planning process group was a list of prioritized risks. Erica reviewed the risks she had mentioned in the business case as well as the comments team members made on the project charter and in their team meetings. She held a special meeting for everyone to brainstorm and discuss potential risks. They posted all of the risks they identified on a probability/impact matrix, and then they grouped some of the ideas. There was only one risk in the high probability and high impact category, and several with medium impact in one or both categories. They chose not to list the low probability and low impact risks. After some discussion, the team developed the list of prioritized risks shown in Table 3-8.

Table 3-8: List of Prioritized Risks

RANKING	POTENTIAL RISK
1	Lack of inputs from internal consultants
2	Lack of inputs from client representatives
3	Security of new system
4	Outsourcing for the article retrieval and "Ask the Expert" features
5	Outsourcing for processing online payment transactions
6	Organizing the templates and examples in a useful fashion
7	Providing an efficient search feature
8	Getting good feedback from Michael Chen and other senior consultants
9	Effectively promoting the new system
10	Realizing the benefits of the new system within one year

Project Executing

Executing the project involves taking the actions necessary to ensure that activities in the project plan are completed. The products of the project are produced during project execution, and it usually takes the most resources to accomplish this process. Table 3-9 lists the knowledge areas, processes, and outputs of project execution listed in the PMBOK® Guide 2000. The most important output is work results, or delivery of products and services. Handling change requests is also an important output, as are the various means for distributing project information.

Table 3-9: Executing Processes and Outputs

KNOWLEDGE AREA	PROCESS	OUTPUTS
Integration	Project Plan Execution	Work results Change requests
Quality	Quality Assurance	Quality improvement
Human Resources	Team Development	Performance improvements Inputs to performance Appraisals
Communications	Information Distribution	Project records Project reports Project presentations
Procurement	Solicitation Source Selection Contract Administration	Proposals Contracts Correspondence Contract changes Payment requests

For this relatively small project, Erica could work closely with all the team members to make sure they were producing the desired work results. She also used her networking skills to get input from other people in the firm and from external sources at no additional cost to the project. She knew that providing strong leadership and using good communication skills were crucial to good project execution. The firm did have a formal change request form, but it primarily used it for external projects. The firm also had contract specialists and templates for several procurement documents that the project team would use for the portions of the project they planned to outsource. See Appendix D for sample templates related to change requests and contracts. Chapter 4, Project Integration Management, discusses integrated change control, and Chapter 12, Project Procurement Management, describes various types of contracts and other procurement documents.

As mentioned earlier, Erica knew that the CEO and project sponsor liked to see progress on projects through milestone reports, which are one type of project report. He also wanted Erica to alert him to any potential issues or problems. Table 3-10 shows a sample of a milestone report for the Project Management Intranet Site project that Erica reviewed with Joe in mid-June. Erica met with most of her project team members often, and she talked to Joe about once a week to review progress on completing milestones and to discuss any other project issues. Although Erica could have used Project 2002 to create milestone reports, she instead used word processing software because this project was small, and she could more easily manipulate the report format.

Table 3-10: Milestone Report as of 6/17/05

MILESTONE	DATE	STATUS	RESPONSIBLE	ISSUES/COMMENTS
Initiating				
Determine/assign project manager	5/2/05	Completed	Joe	
Business case completed	5/6/05	Completed	Erica	
Project charter signed	5/10/05	Completed	Erica	
Planning		Completed		
Hold project kickoff meeting	5/13/05	Completed	Erica	
Team contract signed	5/13/05	Completed	Erica	
Scope statement completed	5/27/05	Completed	Erica	
WBS completed	5/31/05	Completed	Erica	
List of prioritized risks completed	6/3/05	Completed	Erica	Review with sponsor
Schedule and cost baseline completed	6/13/05	Completed	Erica	
Executing				
Survey completed	6/28/05		Erica	Poor response so far!
Intranet site design completed	7/26/05		Kevin	
Project benefits measurement completed	8/9/05		Erica	
User inputs collected	8/9/05		Jessie	User inputs collected
Articles completed	8/23/05		Jessie	
Templates and tools completed	9/6/05		Erica	Templates and tools completed
Ask the Expert completed	9/6/05		Michael	
User Requests feature completed	9/6/05		Cindy	
Links completed	9/13/05		Kevin	
Intranet site construction completed	10/4/05		Kevin	
Intranet site testing completed	10/18/05		Cindy	

Table 3-10: Milestone Report as of 6/17/05 (continued)

MILESTONE	DATE	STATUS	RESPONSIBLE	ISSUES/COMMENTS
Intranet site promotion completed	10/25/05		Erica	
Intranet site roll-out completed	10/25/05		Kevin	
Controlling				
Status reports	Every Friday	All		
Closing	10/25/05			
Final project presentation completed	10/27/05		Erica	
Sponsor sign-off on project completion	10/27/05		Joe	
Final project report completed	10/28/05		Erica	
Lessons-learned reports submitted	11/1/05		All	

Human resource issues often occur during project execution, especially conflicts. At several of the team meetings, Erica could see that Michael seemed to be bored and often left the room to make phone calls to clients. She talked to Michael about the situation, and she discovered that Michael was supportive of the project, but he knew he could only spend a minimal amount of time on it. He was much more productive outside of meetings, so Erica agreed to have Michael attend a minimal amount of project team meetings. She could see that Michael was contributing to the team by the feedback he provided and his leadership on the "Ask the Expert" feature for the intranet site. Erica adjusted her communication style to meet his specific needs.

Another problem occurred when Cindy was contacting potential suppliers for software to help with the "Ask the Expert" and User Requests features. Kevin wanted to write all of the software for the project himself, but Cindy knew it made better business sense to purchase these new software capabilities from a reliable source. Cindy had to convince Kevin that it was worth buying some software from other sources.

Cindy also discovered that their estimate of $10,000 was only about half the amount they needed. She discussed the problem with Erica, explaining the need for some custom development no matter which supplier they chose. Erica agreed that they should go with an outside source, and she asked their sponsor to approve the additional funds. Joe agreed, but he stressed the importance of having the system pay for itself within a year.

Erica also had to ask Joe for help when the project team received a low response rate to their survey and requests for user inputs. Joe sent out an e-mail to all of JWD Consulting's consultants describing the importance of the project.

He also offered a free family vacation and five extra vacation days to the person who provided the best examples of how they used tools and templates to manage their projects. Erica then received informative input from the consultants. Having effective communication skills and strong management support are essential to good project execution.

Project Controlling

Controlling is the process of measuring progress toward project objectives, monitoring deviation from the plan, and taking corrective action to match progress with the plan. Controlling affects all other phases of the project life cycle. It also involves seven of the nine project management knowledge areas. Table 3-11 lists the knowledge areas, processes, and outputs of project controlling, according to the PMBOK® Guide 2000.

Table 3-11: Controlling Processes and Outputs

KNOWLEDGE AREA	PROCESS	OUTPUTS
Integration	Integrated Change Control	Project plan corrective actions Lessons learned
Scope	Scope Verification Scope Change Control	Formal acceptance Scope changes Corrective actions Lessons learned Adjusted baseline
Time	Schedule Control	Schedule updates Corrective actions Lessons learned
Cost	Cost Control	Revised cost estimates Budget updates Corrective actions Estimate at completion Project closeout lessons learned
Quality	Quality Control	Quality improvement Acceptance decisions Rework Completed checklists Process adjustments
Communication	Performance Reporting	Performance reports Change requests
Risk	Risk Monitoring and Control	Workaround plans Corrective actions Project change requests Updates to the risk response plan Risk database Updates to risk Identification checklists

On the Project Management Intranet Site project, there were several updates to the project plan to reflect changes made to the project scope, schedule, and budget. Erica and other project team members took corrective action when necessary. For example, when they were not getting many responses to their survey, Erica asked Joe for help. When Cindy had trouble negotiating with a supplier, she got help from another senior consultant who had worked with that supplier in the past. Erica also had to request more funds for that part of the project.

Project team members submitted a brief status report every Friday. They were originally using a company template for status reports, but Erica found that by modifying the old template, she received better information to help her team work effectively. She wanted team members to report what they did but also focus on what was going well or not going well and why. This extra information helped team members reflect on the project's progress and identify areas in need of improvement. Table 3-12 is an example of one of Cindy's status reports.

Table 3-12: Sample Weekly Status Report

Project Name: Project Management Intranet Project

Team Member Name: Cindy Dawson, cindydawson@jwdconsulting.com

Date: August 5, 2005

Work completed this week:

-Worked with Kevin to start the intranet site construction

-Organized all the content files

-Starting developing a file naming scheme for content files

-Continued work on "Ask the Expert" and User Requests features

-Met with preferred supplier

-Verified that their software would meet our needs

-Discovered the need for some customization

Work to complete next week:

-Continue work on intranet site construction

-Prepare draft contract for preferred supplier

-Develop new cost estimate for outsourced work

What's going well and why:

The intranet site construction started well. The design was very clear and easy to follow. Kevin really knows what he's doing.

What's not going well and why:

It is difficult to decide how to organize the templates and examples. Need more input from senior consultants and clients.

Suggestions/Issues:

-Hold a special meeting to decide how to organize the templates and examples on the intranet site.

-Get some sample contracts and help in negotiating with the preferred supplier.

Table 3-12: Sample Weekly Status Report (continued)

Project changes:

I think we can stay on schedule, but it looks like we'll need about $10,000 more for outsourcing. That's doubling our budget in that area.

In addition to status reports, an important tool for controlling the project was using project management software. Each team member submitted his or her actual hours worked on tasks each Friday afternoon by 4 PM via the firm's enterprise-wide project management software. They were using the enterprise version of Microsoft Project 2002, so they could easily update their task information via the Web. Erica worked with Jessie to analyze the information, paying special attention to the critical path and earned value data. (See Chapter 6 on Project Time Management for more information on critical path analysis; Chapter 7 on Project Cost Management for a description of earned value management; and Appendix A for more information on using Project 2002 to help control projects.) Erica wanted to finish the project on time, even if it meant spending more money. Joe agreed with that approach, and approved the additional funding Erica projected they would need based on the earned value projections and the need to make up a little time on critical tasks.

Joe again emphasized the importance of the new system paying for itself within a year. Erica was confident that they could exceed the projected financial benefits, and she decided to begin capturing benefits as soon as the project team began testing the system. When she was not working on this project, Erica was managing JWD Consulting's Project Management Office (PMO), and she could already see how the intranet site would help her staff save time and make their consultants more productive. One of her staff members wanted to move into the consulting group, and she believed the PMO could continue to provide its current services with one less person due to this new system—a benefit she had not considered before. Several of the firm's client contracts were based on performance and not hours billed, so she was excited to start measuring the value of the new intranet site to their consultants as well.

Project Closing

The closing process involves gaining stakeholder and customer acceptance of the final product and bringing the project, or project phase, to an orderly end. It includes verifying that all of the deliverables are complete, and it often includes a final presentation. Even though many information technology projects are canceled before completion, it is still important to formally close any project and reflect on what can be learned to improve future projects. As philosopher George Santayana said, "Those who cannot remember the past are condemned to repeat it."

It is also important to plan for and execute a smooth transition of the project into the normal operations of the company. Most projects produce results that are integrated into the existing organizational structure. For example, JWD Consulting's Project Management Intranet Site project will require staff to support the intranet site after it is operational. Erica did include support costs of $40,000 per year for the projected three-year life of the new system. She also created a transition plan as part of the final report to provide for a smooth transition of the system into the firm's operations. The plan included a list of issues that had to be resolved before the firm could put the new intranet site into production. For example, Michael Chen would not be available to work on the intranet site after the six-month project was complete, so they had to know who would support the "Ask the Expert" feature and plan some time for Michael to work with him or her.

Table 3-13 lists the knowledge areas, processes, and outputs of project closing based on the PMBOK® Guide 2000. During the closing processes of any project, project team members should take the time to communicate project results by documenting the project and sharing lessons learned. If the project team procured items during the project, they must formally complete or close out all contracts.

Table 3-13: Closing Processes and Output

KNOWLEDGE AREA	PROCESS	OUTPUTS
Communications	Administration Closure	Project archives Project closure Lessons learned
Procurement	Contract Close-out	Contract file Formal acceptance and closure

Erica and her team prepared a final report, final presentation, contract files, and lessons-learned report in closing the project. Erica reviewed the confidential, individual lessons-learned report from each team member and wrote one summary lessons-learned report to include in the final documentation, part of which is provided in Table 3-14. Notice the highlighted items, such as the importance of having a good kick-off meeting, working together to develop a team contract, using project management software, and communicating well with the project team and sponsor.

Table 3-14: Sample Project Summary Lessons-Learned Report

Project Name: JWD Consulting Project Management Intranet Site Project

Project Sponsor: Joe Fleming

Project Manager: Erica Bell

Project Dates: May 2, 2005 – November 4, 2005

Final Budget: $150,000

1. Did the project meet scope, time, and cost goals?

    We did meet scope and time goals, but we had to request an additional $10,000, which the sponsor did approve.

2. What was the success criteria listed in the project scope statement?

    Below is what we put in our project scope statement under project success criteria:

    "Our goal is to complete this project within six months for no more than $140,000. The project sponsor, Joe Fleming, has emphasized the importance of the project paying for itself within one year after the intranet site is complete. To meet this financial goal, the intranet site must have strong user input. We must also develop a method for capturing the benefits while the intranet site is in development, tested, and after it is rolled out. If the project takes a little longer to complete or costs a little more than planned, the firm will still view it as a success if it has a good payback and helps promote the firm's image as an excellent consulting organization."

3. Reflect on whether or not you met the project success criteria.

    As stated above, the sponsor was not too concerned about going over budget as long as the system would have a good payback period and help promote our firm's image. We have already documented some financial and image benefits of the new intranet site. For example, we have decided that we can staff the PMO with one less person, resulting in substantial cost savings. We have also received excellent feedback from several of our clients about the new intranet site.

4. In terms of managing the project, what were the main lessons your team learned from this project?

    The main lessons we learned include the following:

    - Having a good project sponsor was instrumental to project success. We ran into a couple of difficult situations, and Joe was very creative in helping us solve problems.

    - Teamwork was essential. It really helped to take time for everyone to get to know each other at the kick-off meeting. It was also helpful to develop and follow a team contract.

    - Good planning paid off in execution. We spent a fair amount of time developing a good project charter, scope statement, WBS, schedules, and so on. Everyone worked together to develop these planning documents, and there was strong buy-in.

    - Project management software was very helpful throughout the project.

5. Describe one example of what went right on this project.

6. Describe one example of what went wrong on this project.

7. What will you do differently on the next project based on your experience working on this project?

Erica also had Joe sign a client acceptance form, one of the sample templates on the new intranet site that the project team suggested all consultants use when closing their projects.

Table 3-15 provides the table of contents for the final project report. The cover page included the project title, date, and team member names. Notice the inclusion of a transition plan and a plan to analyze the benefits of the system each year in the final report. Also, notice that the final report includes attachments for all the project management and product-related documents. Erica knew how important it was to provide good final documentation on projects. The project team produced a hard copy of the final documentation and an electronic copy to store on the new intranet site for other consultants to use as desired.

Table 3-15: Final Project Report Table of Contents

1. Project Objectives

2. Summary of Project Results

3. Original and Actual Start and End Dates

4. Original and Actual Budget

5. Project Assessment: (Why did you do this project? What did you produce? Was the project a success? What went right and wrong on the project?)

6. Transition Plan

7. Annual Project Benefits Measurement Approach

Attachments:

A. Project Management Documentation

- Business case
- Project charter
- Team contract
- Scope statement
- WBS
- Baseline and actual Gantt chart
- List of prioritized risks
- Milestone reports
- Status reports
- Contract files
- Lessons-learned reports
- Final presentation
- Client acceptance form

Table 3-15: Final Project Report Table of Contents (continued)

B. Product-Related Documentation

- Survey and results
- Summary of user inputs
- Intranet site content
- Intranet site design documents
- Test plans and reports
- Intranet site promotion information
- Intranet site roll-out information
- Project benefits measurement information

Erica also planned a project closure luncheon for the project team right after their final project presentation. She used the luncheon to share lessons learned and celebrate a job well done!

The project management process groups—initiating, planning, executing, controlling, and closing—provide a useful framework for understanding project management. They apply to most projects (information technology and non-information technology) and, along with the project management knowledge areas, help project managers see the big picture of managing a project in their particular organization.

CASE WRAP-UP

Erica Bell and her team finished the Project Management Intranet Site project on Friday, November 4, 2005, as planned in their project charter. They did go over budget, however, but Joe had approved Erica's request for additional funds, primarily for purchasing external software and customization. Like any project, they had a few challenges, but they worked together as a team and used good project management to meet their sponsor's and users' needs. They received positive initial feedback from internal consultants and some of their clients on the new intranet site. People were asking for templates, examples, and expert advice even before the system was ready. About a year after the project was completed, Erica worked with a member of the finance department to review the benefits of the new system. The Project Management Office did lose one of its staff members, but it did not request a replacement since the new system helped reduce the PMO's workload. This saved the firm about $70,000 a year for the salary and benefits of that staff position. They also had data to show that the firm saved more than $100,000 on contracts with clients due to the new system. The firm was breaking even with the "Ask the Expert" feature the first year, and Erica estimated that the system provided $30,000 in additional profits the first year by generating new business. Joe was proud of the project team and the system they produced to help make JWD Consulting a world-class organization.

CHAPTER SUMMARY

Project management is often a number of interlinked processes. The five project management process groups are initiating, planning, executing, controlling, and closing. These processes occur at varying levels of intensity throughout each phase of a project, and specific outcomes are produced as a result of each process. Normally the executing processes require the most resources and time, followed by the planning processes.

Mapping the main activities of each project management process group into the nine project management knowledge areas provides a big picture of what activities are involved in project management. Some organizations develop their own information technology project management methodologies, often using the PMBOK® Guide 2000 as a foundation.

The JWD Consulting case study demonstrates how one organization managed an information technology project from its initiation through its closure. The case study provides several samples of outputs produced for initiating, planning, executing, controlling, and closing as follows:

- business case
- project charter
- team contract
- work breakdown structure
- Gantt chart
- network diagram
- list of prioritized risks
- milestone report
- status reports
- lessons-learned report
- final project report

Later chapters in this text provide detailed information on creating these documents and using several of the tools and techniques described in this case study.

DISCUSSION QUESTIONS

1. Briefly describe what happens in each of the five project management process groups (initiating, planning, executing, controlling, and closing).
2. On which process should team members spend the most time? Why?
3. What are some of the project management deliverables of each process group?
4. What are some of the typical challenges project teams face during each of the five process groups?
5. What questions do you have about the sample documents provided?

EXERCISES

1. Study the WBS and Gantt charts provided in Figures 3-3 and 3-4. Enter the WBS into Project 2002, indenting tasks as shown to create the WBS hierarchy. Then enter durations and dependencies to try to reproduce the Gantt chart. Check your work with the files available on the companion Web site for this text.
2. Read the article by William Munro regarding Blue Cross Blue Shield of Michigan's information technology project management methodology (available on the companion Web site for this text). Write a one- to two-page summary of the article, its key conclusions, and your opinion of it. Do you think many other organizations could apply this methodology, or does each organization need to create its own methodology?
3. Read the "ResNet Case Study" described in the second Suggested Reading (available on the companion Web site for this text). Write a two- to three-page paper summarizing the main outputs produced during each project process group in this case. Also, include your opinion of whether or not Peeter was an effective project manager.
4. JWD Consulting wrote a business case during project initiation. Review the contents of this document (Table 3-3) and find another example of a business case for a project. Write a one- to two-page paper comparing the contents of these documents. In addition, describe whether you think all projects should include a business case before the project sponsors officially approve the project.
5. Review one of the case studies that IBM provides on its e-business Web site as described in the third suggested reading. Write a one- to two-page paper summarizing the case study and how IBM justified its investment in an information technology project.

6. Several organizations prepare a business case and do other work before officially starting projects, sometimes describing these activities as pre-project planning. Find an article that discusses the importance of pre-project planning activities and how they influence project selection and success.

MINICASE

Your supervisor liked your presentation on developing a recreational sports application on your company's intranet (see the Chapter 2 Minicase). He also showed you a recent report from the Human Resources department showing the relationship between employee health, weight, and activity level and the company's health insurance costs.

Part 1. Find two articles describing the relationship between employee health and costs to their employers. Try to find specific potential benefits of a corporate-sponsored wellness program. Summarize the findings of these articles in a one- to two-page paper.

Part 2: Your company has 20,000 employees, providing health care coverage to all interested workers. The report from the Human Resources department showed that your company is paying 20 percent more than comparable companies do for health benefits, primarily because many employees are overweight or obese based on the body mass index (BMI) scale. Assume that developing the recreational sports application on your intranet and providing a small monetary award for participation and/or weight reduction will bring your health benefits costs to within 10 percent of industry standards. Write a business case for pursuing the recreational sports application, using your imagination and research to fill in the details.

RUNNING CASES

Review the running cases provided in Appendix C. Many of these running cases ask you to prepare documents similar to those provided in this chapter. Appendix D includes templates for all of these documents plus several more. Appendix A provides a guide to using Project 2002.

SUGGESTED READINGS

1. Munroe, William, "Developing and Implementing an IT Project Management Process," *ISSGI Review* (First Quarter 2001).

   A copy of the article is available on the companion Web site for this text. Blue Cross Blue Shield developed its own information technology project management methodology. They based their methodology on the PMBOK® Guide 2000, *making adjustments as needed to best describe how their organization managed information technology projects.*

2. Schwalbe, Kathy, "ResNet Case Study," *Information Technology Project Management, Second Edition* (Boston: Course Technology, 2001).

   This case is available on the companion Web site for this text. ResNet is the software that Northwest Airlines uses for making reservations. This case study documents the initiating, planning, executing, controlling, and closing process of the ResNet system, providing many examples of real project documents.

3. IBM, E-business Case Studies (*www.ibm.com/e-business/casestudies*) (2003).

   IBM provides many sample case studies of how companies have used business intelligence, customer relationship management, supply chain management, e-business infrastructure, and e-commerce to improve business performance.

4. Hartnett, Ed, "Procedures for Professional Software Engineers," (*www. projectmagazine.com*) (January 15, 2003).

   Every professional follows standard procedures when doing their work. In this article, Harnett describes the most important standard practices to follow when developing software.

KEY TERMS

- **closing processes** — formalizing acceptance of the project or phase and bringing it to an orderly end
- **controlling processes** — actions to ensure that project objectives are met
- **executing processes** — coordinating people and other resources to carry out the project plans and produce the products or deliverables of the project
- **initiating processes** — actions to commit to begin or end projects and project phases
- **planning processes** — devising and maintaining a workable scheme to accomplish the business need that the project was undertaken to address
- **process** — a series of actions directed toward a particular result
- **project management process groups** — the progression of project activities from initiating to planning, executing, controlling, and closing

4

Project Integration Management

Objectives

After reading this chapter, you will be able to:

1. *Describe an overall framework for project integration management as it relates to the other project management knowledge areas and the project life cycle*
2. *Describe project plan development, including project plan content, using guidelines and templates for developing plans, and performing a stakeholder analysis to help manage relationships*
3. *Explain project plan execution, its relationship to project planning, the factors related to successful results, and tools and techniques to assist in project plan execution*
4. *Understand the integrated change control process, planning for and managing changes on information technology projects, and developing and using a change control system*
5. *Describe how software can assist in project integration management*

OPENING CASE

Nick Carson recently became the project manager of a critical biotech project for his company in Silicon Valley. This project involved creating the hardware and software for a DNA-sequencing instrument used in assembling and analyzing the human genome. Each instrument sold for approximately $200,000, and various clients would purchase several instruments. One hundred instruments running 24 hours per day could decipher the entire human genome in less than two years. The biotech project was the company's largest project, and it had tremendous potential for future growth and revenues. Unfortunately, there were problems managing this large project. It had been underway for three years and had already gone through three different project managers. Nick had been the lead software developer on the project before top management made him the project manager. Top management told him to do whatever it took to deliver the first version of the software for the DNA-sequencing instrument in four months and a production version in nine months. Negotiations for a potential corporate buyout with a larger company influenced top management's sense of urgency to complete the project.

Nick was highly energetic and intelligent and had the technical background to make the project a success. He delved into the technical problems and found some critical flaws that were keeping the DNA-sequencing instrument from working. He was having difficulty, however, in his new role as project manager. Although Nick and his team got the product out on time, top management was upset because Nick did not focus on managing all of the project's elements. He never provided top management with accurate schedules or detailed plans of what was happening on the project. Instead of performing the work of project manager, Nick had taken on the role of software integrator and troubleshooter. Nick, however, did not understand top management's problem—he delivered the product, didn't he? Did they not realize how valuable he was?

WHAT IS PROJECT INTEGRATION MANAGEMENT?

Project integration management involves coordinating all of the other project management knowledge areas throughout a project's life cycle. This integration ensures that all the elements of a project come together at the right times to complete a project successfully. There are three main processes involved in project integration management:

- **Project plan development** involves putting the results of other planning processes into a consistent, coherent document—the project plan.
- **Project plan execution** involves carrying out the project plan by performing the activities included in it.
- **Integrated change control** involves coordinating changes across the entire project.

Figure 4-1 provides an overview of the processes, inputs, tools and techniques, and outputs of project integration management based on the PMBOK® Guide 2000. For example, inputs to project plan development include other planning outputs, historical information, organizational policies, constraints, and assumptions. Tools and techniques to assist in project plan development include a project planning methodology, stakeholder skills and knowledge, a project management information system or project management software, and earned value management. Outputs of project plan development are the project plan and supporting detail.

PMBOK® Guide 2000, 42.

Figure 4-1. Project Integration Management Overview

To accomplish project integration management, you must engage in the management of project scope, quality, time, cost, human resources, communications, risk, and procurement. Because it ties together all the project management knowledge areas, project integration management depends on activities from all of the other eight knowledge areas. It also requires commitment from top management in the organization sponsoring the project throughout the life of the project.

Figure 4-2 provides a framework for understanding how project integration management serves as the guiding force in managing a project. Recall from

Chapter 2 that projects pass through the basic phases of concept, development, implementation, and close-out. The x-axis of this figure represents the project life cycle phases, and the y-axis represents the other eight project management knowledge areas. Figure 4-2 represents the project integration management knowledge area as an arrow that becomes more focused as the project progresses through its life cycle. Many projects are somewhat ambiguous at the beginning, often described as having a fuzzy front-end. Project integration management pulls everything together to guide the project toward successful completion. By performing good project planning, execution, and change control, project managers and their teams can meet or exceed stakeholder needs and expectations.

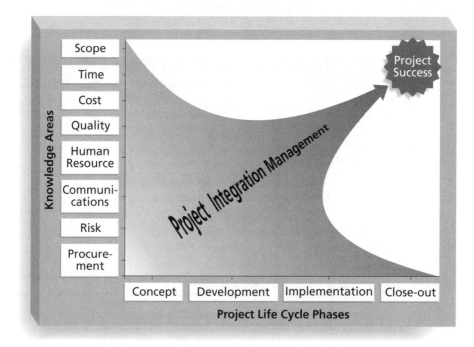

Figure 4-2. Framework for Project Integration Management

Many people consider project integration management the key to overall project success. Someone must take responsibility for coordinating all of the people, plans, and work required to complete a project. Someone must focus on the big picture of the project and steer the project team toward successful completion. Someone must make the final decisions when there are conflicts among project goals or people involved. Someone must communicate key project information to top management. This someone is the project manager, and the project manager's chief tool for accomplishing all these tasks is project integration management.

Good project integration management is critical to providing stakeholder satisfaction. Project integration management includes interface management. **Interface management** involves identifying and managing the points of interaction between various elements of the project. The number of interfaces can increase exponentially as the number of people involved in a project increases. Thus, one of the most important jobs of a project manager is that of establishing and maintaining good communication and relationships across organizational interfaces. The project manager must communicate well with all project stakeholders, including customers, the project team, top management, other project managers, and opponents of the project.

What happens when a project manager does not communicate well with all stakeholders? In the opening case, Nick Carson seemed to ignore a key stakeholder for the DNA-sequencing instrument project—his top management. Nick was comfortable working with other members of the project team, but he was not familiar with his new job as project manager or the needs of the company's top management. Nick continued to do his old job of software developer and took on the added role of software integrator. He mistakenly thought project integration management meant software integration management and focused on the project's technical problems. He totally ignored what project integration management is really about—integrating the work of all of the people involved in the project by focusing on good communication and relationship management. Recall that project management is applying knowledge, skills, tools, and techniques in order to meet project requirements, while also meeting or exceeding stakeholder needs and expectations. Nick did not take the time to find out what top management expected from him as the project manager; he assumed that completing the project on time and within budget was sufficient to make top management happy.

In addition to not understanding project integration management, Nick did not use holistic or systems thinking (see Chapter 2). He burrowed into the technical details of his particular project. He did not stop to think about what it meant to be the project manager, how this project related to other projects in the organization, or what top management's expectations were of him and his team.

Project integration management must occur within the context of the entire organization, not just within a particular project. The project manager must integrate the work of the project with the ongoing operations of the performing organization. In the opening case, Nick's company was negotiating a potential buyout with a larger company. Consequently, top management needed to know when the DNA-sequencing instrument would be ready, how big the market was for the product, and if they had enough in-house staff to continue to manage projects like this one in the future. They wanted to see a project plan and schedule to help them monitor the project's progress and show their potential buyer what was happening. When top managers tried to talk to Nick about these issues, Nick soon returned to discussing the technical details of the project. Even though Nick was very bright, he had no experience or real interest in many of the business aspects of how the company operated. Project managers must always view their projects in the context of the changing needs of their organizations and respond to requests from top management.

Project integration management, therefore, involves integrating the other project management knowledge areas within the project as well as integrating different areas outside the project. In order to integrate across project management knowledge areas and across the organization, there must be a good project plan. Therefore, the first process of project integration management is project plan development.

PROJECT PLAN DEVELOPMENT

A **project plan** is a document used to coordinate all project planning documents and help guide a project's execution and control. Project plans also document project planning assumptions and decisions regarding choices, facilitate communication among stakeholders, define the content, extent, and timing of key management reviews, and provide a baseline for progress measurement and project control. Project plans should be dynamic, flexible, and subject to change when the environment or project changes. Project plans should greatly assist the project manager in leading the project team and assessing project status.

To create and assemble a good project plan, the project manager must practice the art of project integration management, since information is required from all of the project management knowledge areas. Working with the project team and other stakeholders to create a project plan will help the project manager guide the project's execution and understand the overall project.

Project Plan Contents

Just as projects are unique, so are project plans. A small project involving a few people over a couple of months might have a project plan consisting of only a project charter, scope statement, and Gantt chart. A large project involving a hundred people over three years would have a much more detailed project plan. It is important to tailor project plans to fit the needs of specific projects. The project plans should guide the work, so they should be only as detailed as needed for each project.

There are, however, common elements to most project plans. Parts of a project plan include an introduction or overview of the project, a description of how the project is organized, the management and technical processes used on the project, and sections describing the work to be performed, the schedule, and the budget.

The introduction or overview of the project should include, as a minimum, the following information:

- The project name: Every project should have a unique name. Unique names help distinguish each project and avoid confusion among related projects.

- A brief description of the project and the need it addresses: This description should clearly outline the goals of the project and reason for the project. It should be written in layperson's terms, avoid technical jargon, and include a rough time and cost estimate.

- The sponsor's name: Every project needs a sponsor. Include the name, title, and contact information of the sponsor in the introduction.

- The names of the project manager and key team members: The project manager should always be the contact for project information. Depending on the size and nature of the project, names of key team members may also be included.

- Deliverables of the project: This section should briefly list and describe the deliverables that will be produced as part of the project. Software packages, pieces of hardware, technical reports, and training materials are examples of deliverables.

- A list of important reference materials: Many projects have a history preceding them. Listing important documents or meetings related to a project helps project stakeholders understand that history. This section should reference the plans produced for other knowledge areas. (Recall from Chapter 3 that every single knowledge area includes some planning processes.) For example, the overall project plan should reference and summarize important parts of the scope management, schedule management, cost management, quality management, staffing management, communications management, risk management, and procurement management plans, as applicable.

- A list of definitions and acronyms, if appropriate: Many projects, especially information technology projects, involve terminology unique to a particular industry or technology. Providing a list of definitions and acronyms helps to avoid confusion.

The description of how the project is organized should include the following information:

- Organizational charts: In addition to an organizational chart for the company sponsoring the project and for the customer's company (if it is an external customer), there should be a project organizational chart to show the lines of authority, responsibilities, and communication for the project. For example, the Manhattan Project introduced in Chapter 1 had a very detailed organizational chart to show all the people working on the project.

- Project responsibilities: This section of the project plan should describe the major project functions and activities and identify those individuals who are responsible for them. A responsibility assignment matrix (described in Chapter 9) is a tool often used for displaying this information.

- Other organizational or process related information: Depending on the nature of the project, there may be a need to document major processes followed on the project. For example, if the project involves releasing a major software upgrade, it might help everyone involved in the project to see a diagram or timeline of the major steps involved in this process.

The section of the project plan describing management and technical approaches should include the following information:

- Management objectives: It is important to understand top management's view of the project, what the priorities are for the project, and any major assumptions or constraints.

- Project controls: This section describes how to monitor project progress and handle changes. Will there be monthly status reviews and quarterly progress reviews? Will there be specific forms or charts to monitor progress? Will the project use earned value management (described in Chapter 7) to assess and track performance? What is the process for change control? What level of management is required to approve different types of changes? (You will learn more about change control later in this chapter.)

- Risk management: This section briefly addresses how the project team will identify, manage, and control risks. It should refer to the risk management plan, if one is required for the project.

- Project staffing: This section describes the number and types of people required for the project. It should refer to the staffing management plan, if one is required for the project.

- Technical processes: This section describes specific methodologies a project might use and how information is to be documented. For example, many information technology projects follow specific software development methodologies or use particular Computer Aided Software Engineering (CASE) tools. Many companies or customers also have specific formats for technical documentation. It is important to clarify these technical processes in the project plan.

The section of the overall project plan describing the work to be done should reference the scope management plan and summarize the following:

- Major work packages: A project manager usually organizes the project work into several work packages using a work breakdown structure (WBS), and produces a scope statement or statement of work (SOW) to describe the work in more detail (described in Chapter 5). This section should briefly summarize the main work packages for the project and refer to appropriate sections of the scope management plan.

- Key deliverables: This section lists and describes the key deliverables produced as part of the project. It should also describe the quality expectations for the product deliverables.

- Other work-related information: This section highlights key information related to the work performed on the project. For example, it might list specific hardware or software to use on the project or certain specifications to follow. It should document major assumptions made in defining the project work.

The project schedule information section should include the following:

- Summary schedule: It is helpful to see a one-page summary of the overall project schedule. Depending on the size and complexity of the project, the summary schedule might list only key deliverables and their planned completion dates. For smaller projects, it might include all of the work and associated dates for the entire project in a Gantt chart. For example, the Gantt chart and milestone schedule provided in Chapter 3 for JWD Consulting were fairly short and simple. See the last section of this chapter on using project management software for a sample summary schedule from a very large project.

- Detailed schedule: This section provides information on the project schedule that is more detailed. It should reference the schedule management plan and discuss dependencies among project activities that could affect the project schedule. For example, it might explain that a major part of the work cannot be started until funding is provided by an external agency. A project network diagram can show these dependencies (see Chapter 6, Project Time Management).

- Other schedule-related information: Many assumptions are often made in preparing project schedules. This section should document major assumptions and highlight other important information related to the project schedule.

The budget section of the overall project plan should include the following:

- Summary budget: The summary budget includes the total estimate of the overall project's budget. It could also include the budget estimate for each month or year by certain budget categories. It is important to provide some explanation of what these numbers mean. For example, is the total budget estimate a firm number that cannot change, or is it a rough estimate based on projected costs over the next three years?

- Detailed budget: This section summarizes what is in the cost management plan and includes more detailed budget information. For example, what are the fixed and recurring cost estimates for the project each year? What are the projected financial benefits of the project? What types of people are needed to do the work, and how are the labor costs calculated? See Chapter 7, Project Cost Management, for more information on creating cost estimates and budgets.

■ Other budget-related information: This section documents major assumptions and highlights other important information related to financial aspects of the project.

Using Guidelines to Create Project Plans

Many organizations use guidelines to create project plans. Project 2002 and other project management software packages come with several template files to be used as guidelines. (See Appendix A for information on using Project 2002 and Appendix D for sample template files.) Many government agencies also provide guidelines for creating project plans. For example, the Department of Defense (DOD) Standard 2167, Software Development Plan, describes the format for contractors to use in creating a plan for software development for DOD projects. The Institute of Electrical and Electronics Engineers (IEEE) Standard 1058.1 describes the contents of a Software Project Management Plan (SPMP). Table 4-1 provides the format for the IEEE Standard SPMP. Note that this format includes five main sections—introduction, project organization, managerial process, technical process, and work packages, schedule, and budget. Companies working on software development projects for the Department of Defense must follow this standard.

Table 4-1: Sample Outline for a Software Project Management Plan (SPMP)

		PROJECT MANAGEMENT PLAN SECTIONS			
	INTRODUCTION	PROJECT ORGANIZATION	MANAGERIAL PROCESS	TECHNICAL PROCESS	WORK PACKAGES, SCHEDULE, AND BUDGET
Section Topics	Project overview; project deliverables; evolution of the SPMP; reference materials; and definitions and acronyms	Process model; organizational structure; organizational boundaries and interfaces; and project responsibilities	Management objectives and priorities; assumptions, dependencies, and constraints; monitoring and controlling mechanisms; and staffing plan	Methods, tools, and techniques; software documentation; and project support functions	Work packages; dependencies; resource requirements; budget and resource allocation; and schedule

IEEE Standard 1058.1-1987.

In most private organizations, specific documentation standards are not as rigorous; however, there are usually guidelines for developing project plans. It is good practice to follow standards or guidelines for developing project plans in an organization to facilitate the development and execution of those plans. The organization can work more efficiently if all project plans follow a similar format. In fact, research shows that companies that excel in project management develop and deploy standardized project delivery systems.

> "The winners clearly spell out what needs to be done in a project, by whom, when, and how. For this they use an integrated toolbox, including PM tools, methods, and techniques...If a scheduling template is developed and used over and over, it becomes a repeatable action that leads to higher productivity and lower uncertainty. Sure, using scheduling templates is neither a breakthrough nor a feat. But laggards exhibited almost no use of the templates. Rather, in constructing schedules their project managers started with a clean sheet, a clear waste of time."[1]

For example, in the opening case, Nick Carson's top managers were disappointed because he did not provide them with project planning information they needed to make important business decisions. They wanted to see detailed project plans, including schedules and a means for tracking progress. Nick had never created project plans or status reports before, and the organization did not provide templates or examples to follow. If it had, Nick might have been able to deliver the information top management was expecting. See Appendix D for a project plan template and other related templates.

Stakeholder Analysis and Top Management Support

Because the ultimate goal of project management is to meet or exceed stakeholder needs and expectations from a project, it is important to include a stakeholder analysis as part of project planning. A **stakeholder analysis** documents information such as key stakeholders' names and organizations, their roles on the project, unique facts about each stakeholder, their level of interest in the project, their influence on the project, and suggestions for managing relationships with each stakeholder. Since a stakeholder analysis often includes sensitive information, it should not be part of the overall project plan, which would be available for all stakeholders to review. In many cases, only project managers and other key project team members should see the stakeholder analysis. Table 4-2 provides an example of a stakeholder analysis that Nick Carson could have used to help him manage the DNA-sequencing instrument project described in the opening case. It is important for project managers to take the time to perform a stakeholder analysis of some sort to help understand and meet stakeholder needs and expectations.

[1] Milosevic, Dragan and And Ozbay. "Delivering Projects: What the Winners Do." Proceedings of the Project Management Institute Annual Seminars & Symposium (November 2001).

Table 4-2: Sample Stakeholder Analysis

	KEY STAKEHOLDERS				
	AHMED	**SUSAN**	**ERIK**	**MARK**	**DAVID**
Organization	Internal senior management	Project team	Project team	Hardware vendor	Project manager for other internal projects
Role on project	Project sponsor and one of the company's founders	DNA sequencing expert	Lead programmer	Supplies some instrument hardware	Competing for company resources
Unique facts	Quiet, demanding, like details, business focus, Stanford MBA	Ph.D in biology, easy to work with, has toddler	Very smart, Best programmer I know, weird sense of humor	Start-up company, he knows we can make him rich if this works	Nice guy, one of the oldest people at company, has three kids in college
Level of interest	Very high	Very high	High	Very high	Low to medium
Level of influence	Very high; can call the shots	Subject matter expert; critical to success	High; hard to replace	Low; other vendors available	Low to medium
Suggestions on managing relationship	Keep informed, let him lead conversations, do as he says and quickly	Make sure she reviews specifications and leads testing; can do some work from home	Keep him happy so he stays; emphasize stock options; likes Mexican food	Give him enough lead time to deliver hardware	He knows his project takes a back seat to this one, but I can learn from him

A stakeholder analysis will also help the project manager lead the execution of the project plan. For example, Nick Carson should have taken the time to get to know key stakeholders, especially the project sponsor, so he could understand the project sponsor's expectations and working style. Because Nick had a tendency to talk at great lengths about technical details and Ahmed, the project sponsor, was very quiet by nature, Nick should have let him lead conversations about the project to determine his concerns and interests. Once Ahmed told Nick what to do, Nick should have focused on meeting his sponsor's expectations. If Nick did not know how to perform certain project management functions, he should have asked for help. Nick should have realized his own limitations and perhaps hired an assistant with a project management background.

Likewise, Ahmed and other top managers at the company should have taken the time to understand Nick's needs and provided him with more support. Nick heard, "Do whatever it takes to get the job done," so he delved into what he loved to do, technical problem solving. Ahmed did not hire someone to fill Nick's previous technical position, so how could he expect Nick to do both the management work and technical work required for the project? He should have seen that Nick needed help and worked with him to plan the best course of action. Recall from Chapter 2 that executive support contributed the most to successful information technology projects. Top management must work with project managers to help them plan and execute projects successfully. Since Nick was the fourth project manager on this project, top management at his company obviously had some problems choosing and working with project managers in general.

PROJECT PLAN EXECUTION

The second process of project integration management is project plan execution. Project plan execution involves managing and performing the work described in the project plans. The majority of time on a project is usually spent on execution, as is most of the project's budget. The application area of the project directly affects project execution because the products of the project are produced during project execution. For example, the DNA-sequencing instrument from the opening case and all associated software and documentation would be produced during project execution. The project team would need to use their expertise in biology, hardware and software development, and testing to successfully produce the product.

The project manager would also need to focus on leading the project team and managing stakeholder relationships to successfully execute the project plan. Project human resource management and project communications management are crucial to a project's success. See Chapters 9 and 10 respectively for more information on those knowledge areas. If the project involves a significant amount of risk or outside resources, the project manager also needs to be well versed in project risk management and project procurement management. See Chapters 11 and 12 for details on those knowledge areas. Many unique situations occur during project execution, so project managers must be flexible and creative in dealing with them. Review the situation Erica Bell faced during project execution in Chapter 3. Also review the ResNet case study (available on the companion Web site for this text) to understand the execution challenges project manager Peeter Kivestu and his project team faced.

Coordinating Planning and Execution

Project integration management views project planning and execution as intertwined and inseparable activities. The main function of creating project plans is to guide project execution. A good project plan should help produce good products or work results. The project plan should document what good work results consist of. Updates to the project plans should reflect knowledge gained from completing work early in the project. Anyone who has tried to write a computer program from poor specifications appreciates the importance of a good plan. Anyone who has had to document a poorly programmed system appreciates the importance of good execution.

A common-sense approach to improving the coordination between project plan development and execution is to follow this simple rule: Those who will do the work should plan the work. All project personnel need to develop both planning and executing skills and need experience in these areas. In information technology projects, programmers who have had to write detailed specifications and then create the code from their own specifications become better at writing specifications. Likewise, most systems analysts begin their careers as programmers, so they understand what type of analysis and documentation they need to write good code. Although project managers are responsible for developing the overall project plan, they must solicit input from the project team members who are developing plans in each knowledge area.

Providing Strong Leadership and a Supportive Culture

Strong leadership and a supportive organizational culture are crucial during project execution. Project managers must lead by example to demonstrate the importance of creating good project plans and then following them in project execution. Project managers often create plans for things they need to do themselves. If project managers follow through on their own plans, their team members are more likely to do the same.

What Went Wrong?

Many people have a poor view of plans based on past experiences. Top managers often require a project plan, but then no one follows up on whether the project plan was followed. For example, one project manager said he would meet with each project team leader within two months to review their project plans. The project manager created a detailed schedule for these reviews. He cancelled the first meeting due to another business commitment. He rescheduled the next meeting for unexplained personal reasons. Two months later, the project manager had still not met with over half of the project team leaders. Why should project members feel obligated to follow their own plans when the project manager obviously did not follow his?

Although project planning is different from strategic planning, there are many similarities. Henry Mintzberg, former president of the Strategic Management Society, describes the so-called "pitfalls" of planning and shows how the planning process itself can destroy commitment, narrow a company's vision, discourage change, and breed an atmosphere of politics. He also states that planning must improve. "People called planners can sometimes do strange things.... We need to delineate the word carefully if it is not to be eventually dropped from the management literature as hopelessly contaminated."[2]

[2] Mintzberg, Henry. "The Rise and Fall of Strategic Planning." *Free Press* (1993).

Good project plan execution also requires a supportive organizational culture. For example, organizational procedures can help or hinder project plan execution. If an organization has useful guidelines and templates for project management that everyone in the organization follows, it will be easier for project managers and their teams to plan and do their work. If the organization uses the project plans as the basis for performing and monitoring progress during execution, the culture will promote the relationship between good planning and execution. On the other hand, if organizations have confusing or bureaucratic project management guidelines that hinder getting work done or do not measure progress against plans, project managers and their teams will be frustrated.

Even with a supportive organizational culture, project managers may sometimes find it necessary to break the rules to produce project results in a timely manner. When project managers break the rules, politics will play a role in the results. For example, if a particular project requires use of nonstandard software, the project manager must use his or her political skills to convince concerned stakeholders of the need to break the rules on using only standard software. Breaking organizational rules—and getting away with it—requires excellent leadership, communication, and political skills.

Capitalizing on Product, Business, and Application Area Knowledge

In addition to possessing strong leadership, communication, and political skills, project managers also need to posses product, business, and application area knowledge to execute projects successfully. It is often helpful for information technology project managers to have prior technical experience or at least a working knowledge of information technology products. For example, if the project manager were leading a Joint Application Design (JAD) team to help define user requirements, it would be helpful for him or her to understand the language of the business and technical experts on the team. See Chapter 5, Project Scope Management, for more information on JAD.

Many information technology projects are small, so project managers may be required to perform some technical work or mentor team members to complete the project. For example, a three-month project to develop a Web-based application with only three team members would benefit most from a project manager who can complete some of the technical work. On larger projects, however, the project manager's primary responsibility is to lead the team and communicate with key project stakeholders. He or she would not have time to do any of the technical work. In this case, it is usually best that the project manager understand the business and application area of the project more than the technology involved.

What Went Right?

Michael Michalski, information technology manager for BELMARK Inc., provides an excellent example of how a project manager can use product skills and knowledge to help plan and deliver small information technology projects more effectively. BELMARK is a small company that produces all types of labels. Michalski knew that BELMARK could use the Internet to reduce the time it took to create and deliver labels to its customers. He worked with a programmer for about three months to develop an application called "Web-to-Web." This application is an "extranet" that ties together several different companies via the Web. Web-to-Web uses basic Hypertext Markup Language (HTML) codes and other Internet technologies to link information between BELMARK, its customers, and shipping services. For example, customers using Web-to-Web can view and edit their BELMARK label designs using Portable Document Format (PDF) files on the Web, specifying colors, fonts, sizes, and so on, and can then send them back to BELMARK via the Web with an order for a specific quantity. Customers can also select their shipping service (such as Federal Express) and track the status of their shipment of labels. Michalski said this project reduced the time it took for customers to have their final label designs from four days down to as little as fifteen minutes.[3]

However, it is even more important on large projects for the project manager to understand the business and application area of their projects. For example, Northwest Airlines completed a series of projects in the last ten years to develop and upgrade its reservation systems. The company spent millions of dollars and had more than seventy full-time people working on the projects at peak periods. The project manager, Peeter Kivestu, had never worked in an information technology department, but he had extensive knowledge of the airline industry and the reservations process. He carefully picked his team leaders, making sure they had the required technical and product knowledge. ResNet was the first large information technology project at Northwest Airlines led by a business manager instead of a technical expert, and it was a roaring success. Many organizations have found that large information technology projects require experienced general managers who understand the business and application area of the technology, not the technology itself.

ResNet was and continues to be a highly successful information technology project for Northwest Airlines. Peeter Kivestu and his team delivered a new reservation system interface that enabled sales agents to increase direct ticket sales and reduce call handle times. The call centers moved from being a financial drain on the company to being a very profitable part of the company. In 1996, ResNet saved Northwest Airlines more than $15 million, and savings were more than $33 million in 1997. The ResNet projects met stakeholder expectations, and Peeter proved that a large information technology project could be managed well by a non-technical manager with a strong team.[4]

[3] Michalski, Mike. "Using Internet Technologies for Business-to-Business Communications—the Extranet," *Keynote Address at the Information Systems Education Conference* (October, 1998).

[4] Schwalbe, Kathy. *Information Technology Project Management, Second Edition.* Boston: Course Technology (2001).

Project Plan Execution Tools and Techniques

Project plan execution requires specialized tools and techniques, some of which are unique to project management. Tools and techniques that help project managers perform activities that are part of execution processes include:

- **Work Authorization System**. A work authorization system is a method for ensuring proper communications so that qualified people do the work at the right time and in the proper sequence. The work authorization system can be a manual process in which written forms and signatures are used to authorize work to begin on specific project activities or work packages. Automated work authorization systems are also available to streamline this process. For example, you could use project management software to restrict access to information unless you authorize people to use it at a specific time. On smaller projects, verbal authorizations are usually adequate. Using work authorization systems to manage payments is also effective. For example, you could say you will not pay someone to do work unless it is done during a certain period of time.

- **Status Review Meetings**. Status review meetings are regularly scheduled meetings used to exchange project information. Status review meetings are also excellent tools for motivating people to make progress on the project. If project members know they have to give formal reports every month to key project stakeholders, they will make sure to accomplish some work. Sending a written weekly or monthly status report is normally not as effective as having a formal verbal presentation. You will learn more about status review meetings in Chapter 10, Project Communications Management.

- **Project Management Software**. Project management software can greatly assist in creating and executing the project plan. As described in Chapter 1, there are hundreds of project management software products available on the market today. Many large organizations are moving toward powerful enterprise project management systems that are accessible via the Internet. Project managers or other team members can create Gantt charts using software such as Microsoft Project 2002, and these Gantt charts can include hyperlinks to other planning documents. For example, an overall project plan might include a deliverable for creating software test plans. This item on the Gantt chart could include a hyperlink to a Microsoft Word file of that particular software test plan. If a project team member updates the Word file containing the test plan, the hyperlink feature on the Gantt chart automatically links to the updated file. Once a baseline plan is set for the project, project team members can enter information on when activities actually started and ended, how many hours were spent completing them, and so on. The project manager can then use project management software to compare the baseline and actual information by viewing reports on project progress—running a report to show what tasks are past due or what the actual costs are to date, versus what was planned. You will learn how to use Microsoft Project 2002 to assist in various aspects of project management in Appendix A.

Although these tools and techniques can aid in project execution, project managers must remember that positive leadership and strong teamwork are critical to successful project management. Project managers should delegate the detailed work involved in using these tools to other team members and focus on providing leadership for the whole project to ensure project success.

Project managers and their teams are most often remembered for how well they executed a project. As John Madden often says when announcing Super Bowl events, the key to winning is good execution. When one losing coach was asked what he thought about his team's execution, he responded, "I'm all for it!"

INTEGRATED CHANGE CONTROL

Integrated change control involves identifying, evaluating, and managing changes throughout the project life cycle. The three main objectives of integrated change control are:

1. Influencing the factors that create changes to ensure that changes are beneficial: To ensure that changes are beneficial and that a project is successful, project managers and their teams must make trade-offs among key project dimensions such as scope, time, cost, and quality.
2. Determining that a change has occurred: To determine that a change has occurred, the project manager must know the status of key project areas at all times. In addition, the project manager must communicate significant changes to top management and key stakeholders. Top management and other key stakeholders do not like surprises, especially ones that mean the project might produce less, take longer to complete, cost more than planned, or be of lower quality than desired.
3. Managing actual changes as they occur: Managing change is a key role of project managers and their teams. It is important that project managers exercise discipline in managing the project to help minimize the number of changes that occur.

The Integrated Change Control Process

Important inputs to the integrated change control process include project plans, performance reports, and change requests. Important outputs include project plan updates, corrective actions, and documentation of lessons learned. Figure 4-3 provides a schematic of the integrated change control process.

Figure 4-3. Integrated Change Control Process

The project plan provides the baseline for identifying and controlling project changes. For example, the project plan includes a section describing the work to be done on a project. This section of the plan describes the key deliverables for the project, the products of the project, and quality requirements. The schedule section of the project plan lists the planned dates for completing key deliverables, and the budget section of the project plan provides the planned cost for these deliverables. The project team must focus on delivering the work as planned. If the project team makes changes during project execution, they must revise the project plans.

Performance reports provide status information on how project execution is going. The main purpose of these reports is to alert the project manager and project team of issues that might cause problems in the future. The project manager and project team must decide if corrective actions are needed, what the best course of action is, and when to act. For example, suppose one of the key deliverables in a project plan is installation of a new Web server for the project. If one of the project team members reports that he or she is having problems coordinating the purchase and installation of this Web server, the project manager should assess what will happen to the project if this deliverable is a little late. If late installation will cause problems in other areas of the project, the project manager should take necessary actions to help the team member complete this task on time. Perhaps the purchase request is held up because one of the purchasing clerks is on vacation. The project manager could talk to the head of the purchasing department to make sure someone processes the order. If nothing can be done to meet the planned installation date, the project manager should alert the other people who would be affected by this schedule change. The project manager should also look at the big picture of how the project is going. If there is a consistent trend that schedules are slipping, the project manager should alert key stakeholders and negotiate a later completion date for the project. You will learn more about performance reports in Chapter 10, Project Communications Management.

Change requests are common on projects and occur in many different forms. They can be oral or written, formal or informal. For example, the project team member responsible for installing the Web server might ask the project manager at a progress review meeting if it is all right to order a server with a faster processor than planned, from the same manufacturer for the same approximate cost. Since this change is positive and should have no negative effects on the project, the project manager might give a verbal approval at the progress review meeting. Nevertheless, it is still important that the project manager document this change to avoid any potential problems in the future. The appropriate team member should update the section of the scope management plan with the new specifications for the server. Still, keep in mind that many change requests can have a major impact on a project. For example, customers changing their minds about the number of Web servers they want as part of a project will have a definite impact on the scope and cost of the project. Such a change might also affect the project's schedule. The project team must present such significant changes in written form, and there should be a formal review process for analyzing and deciding whether to approve these changes. See Appendix D for a sample change request form.

In addition to updating project plans and taking corrective actions, documenting lessons learned is an important output from the integrated change control process. The project manager and team should share what they have learned with each other and other people in the organization. There should be some documentation of these lessons learned on change control, and discussing them at an open meeting is an effective way to share this information. Discussing lessons learned on change control should help manage the change control process in future projects. See Appendix D for a template for documenting lessons learned.

Change is unavoidable and often expected on many information technology projects. Technologies change, personnel change, organizational priorities change, and so on. Careful change control on information technology projects is a critical success factor. A good change control system is also important for project success.

Change Control on Information Technology Projects

In the 1950s to the 1980s, a widely held view of information technology (then often referred to as data automation) project management was that the project team should strive to do exactly what they planned, on time and within budget. The problem with this view was that project teams could rarely meet original project goals, especially on projects involving new technologies. Stakeholders rarely agreed up front on what the scope of the project really was or what the finished product should really look like. Time and cost estimates created early in a project were rarely accurate.

Beginning in the 1990s, most project managers and top management realized that project management is a process of constant communication and negotiation about project objectives and stakeholder expectations. This view assumes that changes happen throughout the project life cycle and recognizes that changes are often beneficial to some projects. For example, if a project team member discovers a new hardware or software technology that could satisfy the customers' needs for less time and money, the project team and key stakeholders should be open to making major changes in the project.

All projects will have some changes, and managing them is a key issue in project management, especially for information technology projects. Many information technology projects involve the use of hardware and software that is updated frequently. For example, the initial plan for the specifications of the Web server described earlier may have been cutting-edge technology at that time. If the actual ordering of the Web server occurred six months later, it is quite possible that a more powerful server could be ordered at the same cost. This example illustrates a positive change. On the other hand, the manufacturer of the server specified in the project plan could go out of business, which would result in a negative change. Information technology project managers should be accustomed to changes such as these and build some flexibility into their project plans and execution. Customers for information technology projects should also be open to meeting project objectives in different ways.

Even if project managers, teams, and customers are flexible, it is important that projects have a formal change control system. This formal change control system is necessary to plan for managing change.

Change Control System

A **change control system** is a formal, documented process that describes when and how official project documents may be changed. It also describes the people authorized to make changes, the paperwork required for the change, and any automated or manual tracking systems the project will use. A change control system often includes a change control board (CCB), configuration management, and a process for communicating changes.

A **change control board (CCB)** is a formal group of people responsible for approving or rejecting changes to a project. The primary functions of a change control board are to provide guidelines for preparing change requests, evaluating change requests, and managing the implementation of approved changes. An organization could have key stakeholders for the entire organization on this board, and a few members could rotate based on the unique needs of each project. By creating a formal board and process for managing changes, better overall change control should result.

However, CCBs can have some drawbacks. One drawback is the time it takes for decisions to be made on proposed changes. CCBs often meet only once a week or once a month and may not make decisions in one meeting. Some organizations have streamlined processes for making quick decisions on smaller project changes. One company created a "48-hour policy," in which task leaders on a large information technology project would reach agreements on key decisions or changes within their expertise and authority. The person in the area most affected by this decision or change then had 48 hours to go to his or her top management to seek approval. If for some reason the project team's decision could not be implemented, the top manager consulted would have 48 hours to reverse a decision; otherwise, the project team's decision was approved. This type of process is a great way to deal with the many time-sensitive decisions or changes that project teams must make on many information technology projects.

Configuration management is another important technique for integrated change control. **Configuration management** ensures that the descriptions of the project's products are correct and complete, and concentrates on the management of technology by identifying and controlling the functional and physical design characteristics of products and their support documentation. Members of the project team, often called configuration management specialists, are often assigned to perform configuration management for large projects. Their job is to identify and document the functional and physical characteristics of the project's products, control any changes to such characteristics, record and report the changes, and audit the products to verify conformance to requirements.

Another factor in change control is communication. Project managers should use written and oral performance reports to help identify and manage project changes. For example, on software development projects, most programmers must make their edits to one master file in a database that requires the programmers to "check out" the file to edit it. If two programmers check out the same file, they must coordinate their work before they can check the file back in to the database. In addition to written or formal communication methods, oral and informal communications are also important. Some project managers have "stand-up meetings" once a week or even every morning, depending on the nature of the project. The goal of a stand-up meeting is to communicate what is most important on the project quickly. For example, the project manager might have an early morning stand-up meeting every day with all of his or her team leaders. There might be a weekly stand-up meeting every Monday morning with all interested stakeholders. Requiring participants to stand keeps meetings short and forces everyone to focus on the most important project events.

Why is good communication so critical to success? One of the most frustrating aspects of project change is not having everyone coordinated and informed about the latest project information. Again, it is the project manager's responsibility to integrate all project changes so that the project stays on track. The project manager and his or her staff must develop a system for notifying everyone affected by a change in a timely manner. E-mail, real-time databases, and the Web make it easier to disseminate the most current project information. Using special project management software also helps project managers track and communicate project changes. You will learn more about good communication in Chapter 10, Project Communications Management.

Table 4-3 lists suggestions for managing integrated change control. As described earlier, project management is a process of constant communication and negotiation. Project managers should plan for changes and use appropriate tools and techniques such as a change control board, configuration management, and good communication. It is helpful to define procedures for making timely decisions on small changes, use written and oral performance reports to help identify and manage changes, and use software to assist in planning, updating, and controlling projects.

Table 4-3: Suggestions for Managing Integrated Change Control

- View project management as a process of constant communication and negotiation

- Plan for change

- Establish a formal change control system, including a Change Control Board (CCB)

- Use good configuration management

- Define procedures for making timely decisions on smaller changes

- Use written and oral performance reports to help identify and manage change

- Use project management and other software to help manage and communicate changes

Project managers must also provide strong leadership to steer the project to successful completion. They must not get too involved in managing project changes. Project managers should delegate much of the detailed work to project team members and focus on providing overall leadership for the project in general. Remember, project managers must focus on the big picture and perform project integration management well to lead their team and organization to success.

USING SOFTWARE TO ASSIST IN PROJECT INTEGRATION MANAGEMENT

As described throughout this chapter, project teams can use various types of software to assist in project integration management. Project teams can create documents with word processing software, give presentations with presentation software, track information with spreadsheet or database software, and transmit information using various types of communication software such as e-mail and Web authoring tools.

Project management software is also an important tool for developing and integrating project planning documents, executing those project plans, and performing integrated change control. Small project teams can use low-end or mid-range project management software products to coordinate their work. For large projects, however, organizations may benefit most from high-end tools that provide enterprise project management capabilities and integrate all aspects of project management.

Many project teams use Microsoft Project 2002 and Microsoft Office products to meet their software needs. For example, Chapter 3 provides several examples of documents that JWD Consulting used to manage its Project Management Intranet Site project. The companion Web site for this text describes how Northwest Airlines used many software tools to assist the company in managing its large ResNet project. The ResNet project team created planning documents and reports using word processing software, cost estimates using spreadsheets, a detailed project organizational chart using presentation software, and so on. Project team members and other stakeholders could access project files on a shared file server. All of these tools helped the project manager and project team plan, execute, and manage changes.

The ResNet project team also used Microsoft Project to create several Gantt charts and run several reports. The project team's file for the ResNet 1996 project office implementations, for example, included 821 tasks. Obviously, few people would want to see that much information in that level of detail. Microsoft Project and similar products make it easy to filter tasks and run various types of reports to hone in on desired information. Figure 4-4 shows part of the actual Gantt chart used on the ResNet 1996 project. You can see by the numbers next to the task names that only summary information is provided for several sections of the schedules. In Microsoft Project, you can click on the plus symbols to drill down to the next level of detail. See Appendix A for more details on using Microsoft Project 2002 and the companion Web site for a full description of the ResNet project.

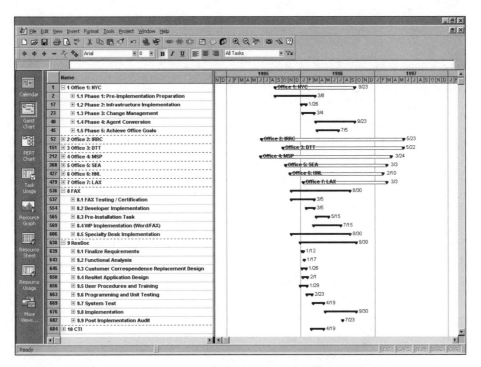

Figure 4-4. Summary Gantt Chart for ResNet 1996 Office Implementation

As you can see, there is a lot of work involved in project integration management. Project managers and their teams must focus on pulling all the elements of a project together to successfully complete projects.

CASE WRAP-UP

Without consulting Nick Carson or his team, Nick's CEO hired a new person, Jim, to act as a middle manager between himself and the people in Nick's department. The CEO and other top managers really liked Jim, the new middle manager. He met with them often, shared ideas, and had a great sense of humor. He started developing standards the company could use to help manage projects in the future. For example, he developed templates for creating project plans and progress reports and put them on the company's intranet. However, Jim and Nick did not get along. Jim accidentally sent an e-mail to Nick that was supposed to go to the CEO. In this e-mail, Jim said that Nick was hard to work with and preoccupied with the birth of his son.

Nick was furious when he read the e-mail and stormed into the CEO's office. The CEO suggested that Nick move to another department, but Nick did not like that option. Without considering the repercussions, the CEO offered Nick a severance package to leave the company. Because of the planned corporate buyout, the CEO knew they might have to let some people go anyway. Nick talked the CEO into giving him a two-month sabbatical he had not yet taken plus a higher percentage on his stock options. After discussing the situation with his wife and realizing that he would get over $70,000 if he resigned, Nick took the severance package. He had such a bad experience as a project manager that he decided to stick with being a technical expert. Jim, however, thrived in his position and helped the company improve their project management practices and ensure success in a highly competitive market.

CHAPTER SUMMARY

Project integration management is often viewed as the most important project management knowledge area, since it ties together all the other areas of project management. A project manager's primary focus should be on project integration management.

Key processes of project integration management include project plan development, project plan execution, and integrated change control.

The main purpose of a project plan is to facilitate action. Project plans should be dynamic, flexible, and subject to change. A project plan includes an introduction or overview of the project, a description of how the project is organized, the management and technical processes used on the project, sections describing the work to be done, schedule information, and budget information. Successful organizations provide and follow standards or guidelines for developing project plans to facilitate the development and execution of those plans.

The project manager should create a stakeholder analysis as part of project planning to understand who all the stakeholders are on a project and how to work with them to satisfy their needs. The project manager should also communicate well with top management and gain their support.

The majority of a project's budget should be spent on project plan execution. Project planning and execution are intertwined. Those who do the work should plan the work. Strong project leadership and an organizational culture that promotes good project management are instrumental to successful execution. It is also important for project managers and team members to have the proper mix of product, business, and application area knowledge.

Integrated change control involves making sure changes are beneficial to the project, determining when changes should occur, and managing those changes. A change control system often includes a change control board (CCB), configuration management, and a process for communicating changes.

Project managers and their teams can use several types of software to assist in project integration management, including word processing software, presentation software, spreadsheets, databases, various types of communications software, and project management software. It is important for project teams to keep project documentation up to date and make it available to everyone who needs it via shared network drives, the Internet, or other means.

DISCUSSION QUESTIONS

1. Describe project integration management in your own words. How does project integration management relate to the project life cycle, stakeholders, and the other project management knowledge areas?

2. What are some crucial elements of a good project plan? Describe what might be in a project plan for a project to develop a Web-based information system for providing transfer credit information for all colleges and universities in the world.

3. What are crucial elements of successful project plan execution? Describe a well-executed project with which you are familiar. Describe a disaster project. What were some of the main differences between these projects?

4. Discuss the importance of following a well-integrated change control process on information technology projects. What do you think of the suggestions made in this chapter? Think of three additional suggestions for integrated change control on information technology projects.

5. Provide an example of how you might use each of the following software tools during project integration management: word processing software, presentation software, spreadsheets, databases, e-mail, the Web, and project management software.

EXERCISES

1. Write a one- to two-page paper based on the opening case. Answer the following questions:

   a. What do you think the real problem was in this case?

   b. Does the case present a realistic scenario? Why or why not?

   c. Was Nick Carson a good project manager? Why or why not?

   d. What could Nick have done to be a better project manager?

   e. What should top management have done to help Nick?

2. Develop an outline (major headings and subheadings only) for a project plan to create a Web site for your class, and then fill in the details for the introduction or overview section. Assume that this Web site would include a home page with links to a syllabus for the class, lecture notes or other instructional information, links to the Web site for this textbook, links to other Web sites with project management information, and links to personal pages for each member of your class and future classes. Also, include a bulletin board and chat room feature where students and the instructor can exchange information. Assume your instructor is the project's sponsor, you are the project manager, your classmates are your project team, and you have up to one year to complete the project.

3. Using the scenario described in Exercise two above, create a stakeholder analysis using Table 4-2 as a guide. Assume that your instructor, your instructor's immediate supervisor, you, at least one other classmate, and your school's Webmaster are key stakeholders. Feel free to make up information, if needed.

4. Review the information in this chapter and read the suggested reading by Milosevich and Ozbay describing what companies that excel in project delivery do differently from other companies. Write a one-page summary of this research and state your opinion of the findings.

5. Interview two to three people to get their views on the differences between staying in a more technical career path and moving into project management. Ask them if they agree with the findings in Johnson's article stating that the best information technology people make the worst project managers. Write a one- to two-page summary of what you find, including your opinions on this subject.

MINICASE

Robin, a senior consultant for a large consulting firm, was just assigned to be the project manager for a $100 million project to upgrade the information systems for the state Department of Corrections. Robin's boss and mentor, Jill, knew that Robin had a lot of potential and wanted to give her a challenging assignment. Robin had a lot of experience with the technology involved in the project, but she had never worked with the state government and she knew little about the state legal system. Her firm assigned its top legal person, Fred, to the project to assist Robin. Fred had worked on several projects with the state and understood the corrections process, but his expertise was in law, not technology. Fred also had never worked for a woman before and he was surprised that Robin was ten years younger than he was.

At the kickoff meeting for the project, Robin could see that understanding and meeting stakeholder needs and expectations for the project would be a huge challenge. The new governor attended the first few minutes of the meeting, an indicator of the high profile of the project. The governor, who was known for not trusting consultants, questioned the value of spending more money on information systems for the Department of Corrections, but the decision to fund this project had been made before he was elected. The federal government would be funding half of the project, since the new system would have to interact with the new federal system. The head of the state Department of Corrections, Donna, had worked in the state legal system for twenty-five years. Donna knew the corrections process and the problems they faced, but she knew very little about information systems and preferred doing work the old-fashioned way. She had heard horror stories about inmates released because of computer errors. Donna's new assistant, Jim, was very computer savvy and seemed most supportive of the project. No federal representatives were invited to the meeting. The meeting did not go well, as it seemed disorganized and highly political.

Part 1: Prepare a stakeholder analysis for this project, using Table 4-2 as a guide. Then write a memo to Robin's boss, Jill, making suggestions on what top management could do to help Robin manage this challenging project.

Part 2: As part of the project planning, Robin and her team decide to emphasize the importance of communications. They decide to document a process for communicating with stakeholders, especially for status review meetings and change control. They plan to discuss these draft documents with their key stakeholders to reach agreement and get off to a good start. Write a draft document describing the process for status review meetings and a process for change control. Include templates of key documents that could be used in both processes, such as an agenda for a monthly status review meeting and a change request form. In addition, review the template files that come with Project 2002 or search for additional Project 2002 templates that you could use for this project. (See the sample templates and related links in Appendix D.)

RUNNING CASES

Several of the running cases in Appendix C provide questions related to project integration management. Appendix A includes information on using Project 2002 to assist in planning and executing projects.

SUGGESTED READINGS

1. IEEE. "Std 1058.1-1987." *IEEE Standard for Software Project Management Plans* (1987).

   This standard describes the format and content of software project management plans. It identifies the minimum set of elements that should appear in all software project management plans.

2. Johnson, Jim. "Turning CHAOS into SUCCESS." *Software Magazine* (*www.softwaremag.com/archive/1999dec/Success.html*) (December 1999).

   This article provides suggestions on how to improve the likelihood of success for information technology projects. Johnson states that many Chief Information Officers say that the best information technology people make the worst project managers. Project managers must have strong business skills and people skills, such as diplomacy and negotiation.

3. Milosevic, Dragan and Ozbay. "Delivering Projects: What the Winners Do." Proceedings of the Project Management Institute Annual Seminars & Symposium (November 2001).

   This report summarizes results of a study to determine what organizations that excel in delivering projects do differently from other organizations. The authors found that using an integrated toolbox of project management tools and templates, growing project leaders within the organization, developing a streamlined project delivery process, and deploying performance metrics to quantify progress helped organizations excel in project management.

4. Stuckenbruck, Linn C. "The Job of the Project Manager: Systems Integration." *The Implementation of Project Management: The Professional's Handbook*. Reading, MA: Addison-Wesley (1996), 141–155.

   This chapter of Stuckenbruck's book provides more details on how project managers perform the integration management function. She stresses the need for maintaining communication links across the organizational interfaces.

5. Ward, James. "Productivity Through Project Management: Controlling the Project Variables." *Information Systems Management* (Winter 1994).

   Ward suggests that effectively planning and controlling three project variables—work, resources, and time—is what ensures a project's success. This article describes the typical problems and opportunities associated with project control and the dynamics and interrelationships of work, resources, and time.

6. Wideman, R. M. "Total Project Management of Complex Projects, Improving Performance with Modern Techniques." New Delhi, India: *Consultancy Development Centre* (January 1990).

> *This article provides more details on the framework for project integration management presented in Figure 4-2. It emphasizes that project integration management, particularly project control, must be applied to every other knowledge area throughout the entire project life cycle.*

KEY TERMS

- **Change Control Board (CCB)** — a formal group of people responsible for approving or rejecting changes on a project
- **change control system** — a formal, documented process that describes when and how official project documents may be changed
- **configuration management** — a process that ensures that the descriptions of the project's products are correct and complete
- **integrated change control** — coordinating changes across the entire project
- **interface management** — identifying and managing the points of interaction between various elements of a project
- **project integration management** — includes the processes involved in coordinating all of the other project management knowledge areas throughout a project's life cycle. The main processes involved in project integration management include project plan development, project plan execution, and integrated change control.
- **project management software** — software specifically designed for project management
- **project plan** — a document used to coordinate all project planning documents and guide project execution and control. Key parts of a project plan include an introduction or overview of the project, a description of how the project is organized, the management and technical processes used on the project, and sections describing the work to be done, schedule information, and budget information.
- **project plan development** — taking the results of other planning processes and putting them into a consistent, coherent document—the project plan
- **project plan execution** — carrying out the project plan by performing the activities it includes
- **stakeholder analysis** — an analysis of information such as key stakeholders' names and organizations, their roles on the project, unique facts about each stakeholder, their level of interest in the project, their influence on the project, and suggestions for managing relationships with each stakeholder
- **status review meetings** — regularly scheduled meetings used to exchange project information
- **work authorization system** — a method for ensuring that qualified people do the work at the right time and in the proper sequence

5

Project Scope Management

┌─ **Objectives** ┐

After reading this chapter, you will be able to:

1. *Understand the elements that make good project scope management important*
2. *Describe the strategic planning process, apply different project selection methods, such as a net present value analysis, a weighted scoring model, and a balanced scorecard, and understand the importance of creating a project charter*
3. *Explain the scope planning process and contents of a scope statement*
4. *Discuss the scope definition process and construct a work breakdown structure using the analogy, top-down, bottom-up, and mind mapping approaches*
5. *Understand the importance of scope verification and scope change control to avoid scope creep on information technology projects*
6. *Describe how software can assist in project scope management*

OPENING CASE

Kim Nguyen was leading a project team meeting to help define the scope of the information technology upgrade project assigned to her last week by top management. This project was necessary for implementing several high-priority, Internet-based applications the company was developing. The upgrade project involved creating and implementing a plan to get all employees' information technology assets to meet new corporate standards within nine months. These standards specified the minimum equipment required for each desktop or laptop computer, namely the type of processor, amount of memory, hard disk size, type of network connection, and software. Kim knew that to perform the upgrades, they would first have to create a detailed inventory of all of the current hardware, networks, and software in the entire company of 2000 employees.

Kim had worked with other stakeholders to develop a project charter explaining the key objectives of the project. She also assisted the Chief Information Officer (CIO) in writing a business case to justify investing in the project. The project charter included rough cost and schedule estimates for the project and signatures of key stakeholders. Kim called a meeting with her project team and other stakeholders to further plan and define the scope of the project. She wanted to get everyone's ideas on what was involved in doing the project, who would do what, and how they could avoid potential scope creep. Kim's boss suggested that the team start by defining the deliverables of the project and then creating a scope statement and work breakdown structure (WBS). Kim understood the importance of defining deliverables, but she was not quite sure where to start in creating a WBS. She had used a WBS before, but this was the first time she was the project manager leading a team to create one. She also knew that her company was getting serious about measuring project performance and that a good WBS was the foundation for scope, time, and cost performance. Where should she begin?

WHAT IS PROJECT SCOPE MANAGEMENT?

Recall from Chapter 2 that several factors are associated with project success. Many of these factors, such as user involvement, clear business objectives, minimized scope, and firm basic requirements, are elements of project scope management. William V. Leban, program manager at Keller Graduate School of Management, cites lack of proper project definition and scope as the main reasons projects fail.[1]

One of the most important and most difficult aspects of project management, therefore, is defining the scope of a project. **Scope** refers to *all* the work involved in creating the products of the project and the processes used to create them. Recall from Chapter 2 that the term **deliverable** describes a product produced as part of a project. Deliverables can be product-related, such as a piece of hardware or a software module, or process-related, such as a planning document or meeting minutes. Project stakeholders must agree on what the products of the project are and, to some extent, how they should produce them in order to define all of the deliverables.

[1] Chalfin, Natalie. "Four Reasons Why Projects Fail," *PM Network*, PMI (June 1998) 7.

Project scope management includes the processes involved in defining and controlling what is or is not included in a project. It ensures that the project team and stakeholders have the same understanding of what products the project will produce and what processes the project team will use to produce them. There are five main processes involved in project scope management:

- **Initiation** involves committing the organization to begin a project or continue to the next phase of a project. An output of initiation processes is a **project charter**, which is a key document for formally recognizing the existence and providing a broad overview of a project.
- **Scope planning** involves developing documents to provide the basis for future project decisions, including the criteria for determining if a project or phase has been completed successfully. The project team creates a scope statement and project scope management plan as a result of the scope planning process.
- **Scope definition** involves subdividing the major project deliverables into smaller, more manageable components. The project team creates a work breakdown structure (WBS) during this process.
- **Scope verification** involves formalizing acceptance of the project scope. Key project stakeholders, such as the customer and sponsor for the project, formally accept the deliverables of the project during this process.
- **Scope change control** involves controlling changes to project scope. Scope changes, corrective action, and lessons learned are outputs of this process.

PROJECT INITIATION: STRATEGIC PLANNING AND PROJECT SELECTION

Successful managers look at the big picture or strategic plan of the organization to determine what types of projects will provide the most value. Therefore, the project initiation process involves identifying potential projects, using realistic methods to select which projects to work on, and then formalizing their initiation by issuing some sort of project charter.

Identifying Potential Projects

The first step in project scope management is deciding what projects to do in the first place. Figure 5-1 shows a four-stage planning process for selecting information technology projects. Note the hierarchical structure of this model and the results produced from each stage. The first step in the information technology planning process, starting at the top of the hierarchy, is to develop an information technology strategic plan based on the organization's overall strategic plan. **Strategic planning** involves determining long-term objectives by analyzing the strengths and weaknesses of an organization, studying opportunities and threats in the business environment, predicting future trends, and projecting the need for new products and services. Many people are familiar with the term "SWOT" analysis—analyzing **S**trengths, **W**eaknesses, **O**pportunities, and **T**hreats—that is used to aid in strategic planning. It is very important to have managers from outside the information technology department assist in the information technology planning process, as they can help information technology personnel understand organizational strategies and identify the business areas that support them.

Figure 5-1. Information Technology Planning Process

After identifying business areas to focus on, the next step in the information technology planning process is to perform a business area analysis. This analysis documents business processes that are central to achieving strategic goals and aids in discovering which ones could most benefit from information technology. Then, the next step is to start defining potential information technology projects, their scope, benefits, and constraints. The last step in the information technology planning process is selecting which projects to do and assigning resources for working on them.

Aligning Information Technology with the Business

It is often easier to identify and select appropriate information technology projects when organizations align their information technology departments more closely with the business. A 2002 study found a direct correlation between closer business and information technology alignment and reporting structure. Hacket Best Practices found that companies in which the Chief Information Officer (CIO) reports directly to the Chief Executive Officer (CEO) lowered their operational support costs per end user by 17 percent. Most CIOs now report to the CEO, increasing from 24 percent in 1999 to 37 percent in 2002. The study also found that companies with centralized or consolidated information technology operations had 24 percent lower operational support costs per end user. The consistent use of information technology standards also lowered companies' application development costs by 41 percent per user.[2]

Information systems can be and often are central to business strategy. Michael Porter, who developed the concept of the strategic value of competitive advantage, has written several books and articles on strategic planning and competition. He and many other experts have emphasized the importance of using information technology to support strategic plans and provide a competitive advantage. Many information systems are classified as "strategic" because they directly support key business strategies. For example, information systems can help an organization support a strategy of being a low-cost producer. Wal-Mart's inventory control system is a classic example of such a system. Information systems can support a strategy of providing specialized products or services that set a company apart from others in the industry. Federal Express's introduction of online package tracking systems created an industry trend that most express carriers now follow. Information systems can also support a strategy of selling to a particular market or occupying a specific project niche. Owens-Corning developed a strategic information system that boosted the sales of its home-insulation products by providing their customers with a system for evaluating the energy efficiency of building designs.

Even though many information technology projects do not produce "strategic information systems" or receive great publicity, it is critical that the information technology planning process start by analyzing the overall strategy of the organization. An organization must develop a strategy for using information technology to define how it will support the organization's objectives. Information technology plans and strategy must align with the organization's strategic plans and strategy. Most organizations face thousands of problems and opportunities for improvement. Therefore, an organization's strategic plan should guide the project selection process. In fact, research shows that supporting explicit business objectives is the number one reason cited for why organizations invest in information technology projects. Other top criteria for investing in information technology projects include supporting implicit business objectives and providing financial incentives, such as a good internal rate of return (IRR) or net present value (NPV). You will learn more about these financial criteria later in this chapter. Table 5-1 summarizes the main reasons why organizations invest in information technology projects.

[2] Cosgrove Ware, Lorraine. "By the Numbers," *CIO Magazine (www.cio.com)* (September 1, 2002).

Table 5-1: Why Organizations Invest in Information Technology Projects

REASON FOR INVESTING IN INFORMATION TECHNOLOGY PROJECTS	RANK BASED ON OVERALL VALUE OF PROJECTS
Supports explicit business objectives	1
Has good internal rate of return (IRR)	2
Supports implicit business objectives	3
Has good net present value (NPV)	4
Has reasonable payback period	5
Used in response to competitive systems	6
Supports management decision making	7
Meets budgetary constraints	8
High probability of achieving benefits	9
Good accounting rate of return	10
High probability of completing project	11
Meets technical/system requirements	12
Supports legal/government requirement	13
Good profitability index	14
Introduces new technology	15

Bacon, James. "The Use of Decision Criteria in Selecting Information Systems/Technology Investments," *MIS Quarterly*, Vol. 16, No. 3 (September 1992).

Methods for Selecting Projects

Organizations identify many potential projects as part of their strategic planning processes, but they need to narrow the list of potential projects to those projects that will be of most benefit. Selecting projects is not an exact science, but it is a critical part of project management. Many methods exist for selecting from among possible projects. Five common techniques are:

- Focusing on broad organizational needs
- Categorizing information technology projects
- Performing net present value or other financial analyses
- Using a weighted scoring model
- Implementing a balanced scorecard

In practice, organizations usually use a combination of these approaches to select projects. Each approach has advantages and disadvantages, and it is up to management to decide the best approach for selecting projects based on their particular organization.

Focusing on Broad Organizational Needs

Top managers must focus on meeting their organization's many different needs when deciding what projects to undertake, when to undertake them, and to what level. Projects that address broad organizational needs are much more likely to be successful because they will be important to the organization. For example, a broad organizational need might be to improve safety, increase morale, provide better communications, or improve customer service. However, it is often difficult to provide a strong justification for many information technology projects related to these broad organizational needs. For example, it is often impossible to estimate the financial value of such projects, but everyone agrees that they do have a high value. As it is said, "It is better to measure gold roughly than to count pennies precisely."

One method for selecting projects based on broad organizational needs is to determine whether they first meet three important criteria: need, funding, and will. Do people in the organization agree that the project needs to be done? Is the organization willing to provide adequate funds to perform the project? Is there a strong will to make the project succeed? For example, many visionary CEOs can describe a broad need to improve certain aspects of their organizations, such as communications. Although they cannot specifically describe how to improve communications, they might allocate funds to projects that address this need. One project's goal might be to develop a strong information technology infrastructure, providing all employees, customers, and suppliers with the hardware and software they need to access information. The information technology upgrade project described in the opening case might fit into this category. As projects progress, the organization must reevaluate the need, funding, and will for each project to determine if the projects should be continued, redefined, or terminated.

Categorizing Information Technology Projects

Another method for selecting projects is based on various categorizations. One type of categorization assesses whether projects provide a response to a problem, an opportunity, or a directive.

- **Problems** are undesirable situations that prevent an organization from achieving its goals. These problems can be current or anticipated. For example, users of an information system may be having trouble logging onto a system or getting information in a timely manner because the system has reached its capacity. In response, the company could initiate a project to enhance the current system by adding more access lines or upgrading the hardware with a faster processor, more memory, or more storage space.

■ **Opportunities** are chances to improve the organization. For example, a company might believe it could enhance sales by selling products directly to customers over the Internet. The company could initiate a project to provide direct sales of products through its Web site.

■ **Directives** are new requirements imposed by management, government, or some external influence. For example, an important customer might require that all of its vendors use a certain form of electronic data interchange (EDI) in order to do business with them. Management would then initiate a project to implement this form of EDI to maintain business with that customer.

Organizations select projects for any of these reasons. It is often easier to get approval and funding for projects that address problems or directives because the organization must respond to these categories of projects to avoid hurting their business. Many problems and directives must be resolved quickly, but managers must also make sure to take a holistic view and seek opportunities for improving the organization through information technology projects.

Another categorization for information technology projects is based on the time it will take to complete a project or the date by which it must be done. For example, some potential projects must be finished within a specific time window. If they cannot be finished by this set date, they are no longer valid projects. Some projects can be completed very quickly—within a few weeks, days, or even minutes. Many organizations have an end user support function to handle very small projects that can be completed quickly. Even though many information technology projects can be completed quickly, it is still important to prioritize them.

A third categorization for project selection is the overall priority of the project. Many organizations prioritize information technology projects as being high, medium, or low priority. The organization should always complete high-priority projects first, even if a low- or medium-priority project could be finished in less time. Usually there are many more potential information technology projects than an organization can undertake at any one time, so it is very important to work on the most important ones first.

Performing Net Present Value Analysis, Return on Investment, and Payback Analysis

Financial considerations are often an important aspect of the project selection process, especially during tough economic times. As authors Dennis Cohen and Robert Graham put it, "Projects are never *ends* in themselves. Financially they are always a means to an end, cash."[3] Many organizations require an approved

---

[3] Cohen, Dennis J. and Robert J. Graham. *The Project Manager's MBA*, San Francisco: Jossey-Bass (2001) 31.

business case before pursuing projects, and financial projections are a critical component of the business case. (See Chapter 3 for a sample business case). Three primary methods for determining the projected financial value of projects include net present value analysis, return on investment, and payback analysis.

Net Present Value Analysis

Net present value (NPV) analysis is a method of calculating the expected net monetary gain or loss from a project by discounting all expected future cash inflows and outflows to the present point in time. An organization should consider only projects with a positive NPV if financial value is a key criterion for project selection. Why? Because a positive NPV means the return from a project exceeds the **cost of capital**—the return available by investing the capital elsewhere. Projects with higher NPVs are preferred to projects with lower NPVs, if all other things are equal.

Figure 5-2 illustrates this concept for two different projects. Note that this example starts discounting right away in Year 1 and uses a 10 percent discount rate. You can use the NPV function in Microsoft Excel to calculate the NPV quickly, as shown in Figure 5-2. Detailed steps on performing this calculation manually are provided later in this section. Note that Figure 5-2 lists the projected benefits first, followed by the costs, and then the calculated cash flow amount. Note that the sum of the **cash flow**—benefits minus costs or income minus expenses—is the same for both projects at $5000. The net present values are different, however, because they account for the time value of money. Money earned today is worth more than money earned in the future. Project 1 had a negative cash flow of $5000 in the first year, while Project 2 had a negative cash flow of only $1000 in the first year. Although both projects had the same total cash flow without discounting, these cash flows are not of comparable financial value. NPV analysis, therefore, is a method for making equal comparisons between cash flow for multi-year projects.

	A	B	C	D	E	F	G
1	Discount rate	10%					
2							
3	PROJECT 1	YEAR 1	YEAR 2	YEAR 3	YEAR 4	YEAR 5	TOTAL
4	Benefits	$0	$2,000	$3,000	$4,000	$5,000	$14,000
5	Costs	$5,000	$1,000	$1,000	$1,000	$1,000	$9,000
6	Cash flow	($5,000)	$1,000	$2,000	$3,000	$4,000	$5,000
7	NPV ⟶	$2,316					
8		Formula =npv(b1,b6:f6)					
9							
10	PROJECT 2	YEAR 1	YEAR 2	YEAR 3	YEAR 4	YEAR 5	TOTAL
11	Benefits	$1,000	$2,000	$4,000	$4,000	$4,000	$15,000
12	Costs	$2,000	$2,000	$2,000	$2,000	$2,000	$10,000
13	Cash flow	($1,000)	$0	$2,000	$2,000	$2,000	$5,000
14	NPV ⟶	$3,201					
15		Formula =npv(b1,b13:f13)					
16							
17	RECOMMEND PROJECT 2 BECAUSE IT HAS THE HIGHER NPV.						

Figure 5-2. Net Present Value Example

There are some items to consider when calculating NPV. Some organizations refer to the investment year(s) for project costs as Year 0 and do not discount costs in Year 0. The discount rate can also vary, often based on the prime rate and other economic considerations. You can enter costs as negative numbers instead of positive numbers, and you can list costs first and then benefits. For example, Figure 5-3 shows the financial calculations JWD Consulting provided in the business case for the Project Management Intranet Site project described in Chapter 3. Note that the discount rate is 8 percent, costs are listed first, and they are entered as positive numbers. The NPV and other calculations are still the same; only the format is different. A project manager must be sure to check with his or her organization to find out its guidelines for when discounting starts, what discount rate to use, and what format is preferred.

Discount rate	8%					
Assume the project is completed in Year 0			Year			
	0	1	2	3	Total	
Costs	140,000	40,000	40,000	40,000		
Discount factor	1	0.93	0.86	0.79		
Discounted costs	**140,000**	**37,200**	**34,400**	**31,600**	**243,200**	
Benefits	0	200,000	200,000	200,000		
Discount factor	1	0.93	0.86	0.79		
Discounted benefits	**0**	**186,000**	**172,000**	**158,000**	**516,000**	
Discounted benefits - costs	(140,000)	148,800	137,600	126,400	272,800	◄—NPV
Cumulative benefits - costs	(140,000)	8,800	146,400	272,800		
		↑				
ROI ─────────────►	112%					
	Payback before Year 1					

Figure 5-3. JWD Consulting Net Present Value Example

To determine NPV, follow these steps:

1. Determine the estimated costs and benefits for the life of the project and the products it produces. For example, JWD Consulting assumed the project would produce a system in about six months that would be used for three years, so costs are included in Year 0, when the system is developed, and ongoing system costs and projected benefits are included for Years 1, 2, and 3.

2. Determine the discount rate. A **discount rate** is the minimum acceptable rate of return on an investment. It is also called the required rate of return, hurdle rate, or opportunity cost of capital. Most organizations use a discount rate based on the return that the organization could expect to receive elsewhere for an investment of comparable risk. In Figure 5-2, the discount rate is 10 percent per year, and in Figure 5-3, the discount rate is 8 percent per year.

3. Calculate the net present value. There are several ways to calculate NPV. Most spreadsheet software has a built-in function to calculate NPV. For example, Figure 5-2 shows the formula that Microsoft Excel uses: =npv(discount rate, range of cash flows), where the discount rate is in cell B3 and the range of cash flows for Project 1 are in cells B6 through F6. (See Chapter 7, Project Cost Management, for more information on cash flow and other cost-related terms). To use the NPV function, there must be a row in the spreadsheet (or column, depending how it is organized) for the cash flow each year, which is the benefit amount for that

year minus the cost amount. The result of the formula yields an NPV of $2316 for Project 1 and $3201 for Project 2. Since both projects have positive NPVs, they are both good candidates for selection. However, since Project 2 has a higher NPV than Project 1, it would be the better choice between the two.

The mathematical formula for calculating NPV is:

$$NPV = \sum t=1...n \; A/(1+r)t$$

where t equals the year of the cash flow, A is the amount of cash flow each year, and r is the discount rate. If you cannot enter the data into spreadsheet software, you can perform the calculations by hand or with a simple calculator. First, determine the annual **discount factor**—a multiplier for each year based on the discount rate and year—and then apply it to the costs and benefits for each year. The formula for the discount factor is $1/(1+r)^t$ where r is the discount rate, such as 8 percent, and t is the year. For example, the discount factors used in Figure 5-3 are calculated as follows:

Year 0: discount factor = $1/(1+0.08)^0 = 1$

Year 1: discount factor = $1/(1+0.08)^1 = .93$

Year 2: discount factor = $1/(1+0.08)^2 = .86$

Year 3: discount factor = $1/(1+0.18)^3 = .79$

After determining the discount factor each year, multiply the costs and benefits each year by the appropriate discount factor. For example, in Figure 5-3, the discounted cost for year 1 is $40,000 * .93 = $37,200. Next, sum all of the discounted costs and benefits each year to get a total. For example, the total discounted costs in Figure 5-3 are $243,200. To calculate the NPV, take the total discounted benefits and subtract the total discounted costs. In this example, the NPV is $516,000 – $243,200 = $272,800.

Return on Investment

Another important financial consideration is return on investment. **Return on investment (ROI)** is the result of subtracting the project costs from the benefits and then dividing by the costs. For example, if you invest $100 today and next year it is worth $110, your ROI is ($110 – 100)/100 or 0.10 (10 percent). Note that the ROI is always a percentage. It can be positive or negative. It is best to consider discounted costs and benefits for multi-year projects when calculating ROI. Figure 5-3 shows an ROI of 112 percent. You calculate this number as follows:

ROI = (total discounted benefits – total discounted costs)/discounted costs

ROI = (516,000 – 243,200) / 243,200 = 112%

The higher the ROI, the better. An ROI of 112 percent is outstanding. Many organizations have a required rate of return for projects. The **required rate of return** is the minimum acceptable rate of return on an investment. The organization bases the required rate of return on what it could expect to receive elsewhere for an investment of comparable risk. For example, an organization might have a required rate of return of at least 10 percent for projects. You can also determine a project's **internal rate of return (IRR)** by finding what discount rate results in an NPV of zero for the project. You can use the Goal Seek function in Excel (use Excel's Help function for more information on Goal Seek) to determine the IRR quickly. Simply set the cell containing the NPV calculation to zero while changing the cell containing the discount rate. For example, in Figure 5-2, you could set cell B7 to zero while changing cell B1 to find that the IRR for Project 1 is 27 percent.

Many organizations use ROI in the project selection process. *Information Week* asked 200 information technology and business professionals how often they measure formal ROI for information technology investments. The poll found that 41 percent of organizations required ROI calculations for all information technology investments, and 59 percent required it for specific initiatives.[4]

Payback Analysis

Payback analysis is another important financial tool to use when selecting projects. **Payback period** is the amount of time it will take to recoup, in the form of net cash inflows, the net dollars invested in a project. In other words, payback analysis determines how much time will lapse before accrued benefits overtake accrued and continuing costs. Payback occurs when the cumulative discounted benefits and costs are greater than zero. Figure 5-3 shows how to find the payback period. The cumulative benefits minus costs for Year 0 are ($140,000). Adding that number to the discounted benefits minus costs for year 1 results in $8800. Since that number is positive, the payback occurs a little before year 1.

Creating a chart helps illustrate more precisely when the payback period occurs. Figure 5-4 charts the cumulative discounted costs and cumulative discounted benefits each year using the numbers from Figure 5-3. Note that the lines cross a little before year 1. The cumulative discounted benefits and costs are greater than zero where the lines cross. A payback in less than a year is normally considered very good.

---

[4] Klein, Paula. "Fickle Yet Focused," *Information Week* (August 3, 2001).

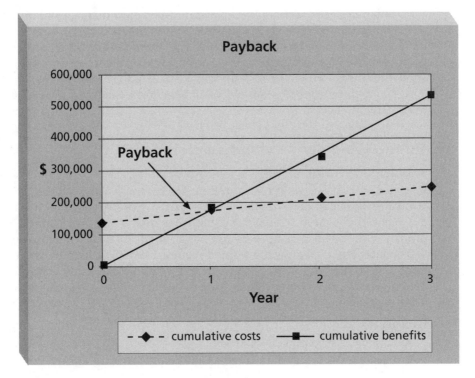

Figure 5-4. Charting the Payback Period

Many organizations have certain recommendations for the length of the payback period of an investment. They might require all information technology
projects to have a payback period of less than two years or even one year,
regardless of the estimated NPV or ROI. Rhonda Hocker, CIO at San Jose–based
BEA Systems Inc., said the middleware vendor has a general rule that its information technology projects should have a payback period of less than one year.
The company also tries to limit project teams to no more than twelve people
doing a job within four months. Given the economic climate and rapid pace of
change in businesses and technology, the company has to focus on delivering
positive financial results quickly.[5] However, organizations must also consider
long-range goals when making technology investments. Many crucial projects
cannot achieve a payback so quickly or be completed in such a short time period.

---

[5] Songini, Marc L. "Tight Budgets Put More Pressure on IT," *Computerworld* (December 2, 2002).

To aid in project selection, it is important for project managers to understand the organization's financial expectations for projects. It is also important for top management to understand the limitations of financial estimates, particularly for information technology projects. For example, it is very difficult to develop good estimates of projected costs and benefits for information technology projects. You will learn more about costs and benefits in Chapter 7, Project Cost Management.

Using a Weighted Scoring Model

A **weighted scoring model** is a tool that provides a systematic process for selecting projects based on many criteria. These criteria can include factors such as meeting broad organizational needs; addressing problems, opportunities, or directives; the amount of time it will take to complete the project; the overall priority of the project; and projected financial performance of the project.

The first step in creating a weighted scoring model is to identify criteria important to the project selection process. It often takes time to develop and reach agreement on these criteria. Holding facilitated brainstorming sessions or using groupware to exchange ideas can aid in developing these criteria. Some possible criteria for information technology projects include:

- Supports key business objectives
- Has strong internal sponsor
- Has strong customer support
- Uses realistic level of technology
- Can be implemented in one year or less
- Provides positive NPV
- Has low risk in meeting scope, time, and cost goals

Next, you assign a weight to each criterion. These weights indicate how much you value each criterion or how important each criterion is. You can assign weights based on percentages, and the sum of all of the criteria's weights must total 100 percent. You then assign numerical scores to each criterion (for example, 0 to 100) for each project. The scores indicate how much each project meets each criterion. At this point, you can use a spreadsheet application to create a matrix of projects, criteria, weights, and scores. Figure 5-5 provides an example of a weighted scoring model to evaluate four different projects. After assigning weights for the criteria and scores for each project, you calculate a weighted score for each project by multiplying the weight for each criterion by its score and adding the resulting values.

For example, you calculate the weighted score for Project 1 in Figure 5-5 as:

25%*90 + 15%*70 + 15%*50 + 10%*25 + 5%*20 + 20%*50 + 10%*20 = 56

	A	B	C	D	E	F
1	Criteria	Weight	Project 1	Project 2	Project 3	Project 4
2	Supports key business objectives	25%	90	90	50	20
3	Has strong internal sponsor	15%	70	90	50	20
4	Has strong customer support	15%	50	90	50	20
5	Uses realistic level of technology	10%	25	90	50	70
6	Can be implemented in one year or less	5%	20	20	50	90
7	Provides positive NPV	20%	50	70	50	50
8	Has low risk in meeting scope, time, and cost goals	10%	20	50	50	90
9	**Weighted Project Scores**	**100%**	**56**	**78.5**	**50**	**41.5**

Weighted Score by Project

Figure 5-5. Sample Weighted Scoring Model for Project Selection

Note that in this example, Project 2 would be the obvious choice for selection because it has the highest weighted score. Creating a bar chart to graph the weighted scores for each project allows you to see the results at a glance. If you create the weighted scoring model in a spreadsheet, you can enter the data, create and copy formulas, and perform a "what-if" analysis. For example, suppose you change the weights for the criteria. By having the weighted scoring model in a spreadsheet, you can easily change the weights to update the weighted scores and charts automatically.

Many teachers use a weighted scoring model to determine grades. Suppose grades for a class are based on two homework assignments and two exams. To calculate final grades, the teacher would assign a weight to each of these items. Suppose Homework One is worth 10 percent of the grade, Homework Two is worth 20 percent of the grade, Test One is worth 20 percent of the grade, and Test Two is worth 50 percent of the grade. Students would want to do well on each of these items, but they should focus on performing well on Test Two since it is 50 percent of the grade.

You can also establish weights by assigning points. For example, a project might receive 10 points if it definitely supports key business objectives, 5 points if it somewhat supports them, and 0 points if it is totally unrelated to key business objectives. With a point model, you can simply add all the points to determine the best projects for selection, without having to multiply weights and scores and sum the results.

You can also determine minimum scores or thresholds for specific criteria in a weighted scoring model. For example, suppose an organization really should not consider a project if it does not score at least 50 out of 100 on every criterion. You can build this type of threshold into the weighted scoring model to reject projects that do not meet these minimum standards. As you can see, weighted scoring models can aid in project selection decisions.

What Went Right?

Virginia Housing Development Authority (VHDA) is a public mortgage finance agency that helps Virginians obtain safe, sound, and decent housing that is otherwise unaffordable to them. VHDA is entirely self-supporting and does not receive any funding from the state. Therefore, it is very careful with its information technology budget.

VHDA developed a planning process for establishing an information technology project list and related budget information. The process began with planning meetings involving all business user groups to identify potential information technology projects. They then developed cost and benefit statements for proposed projects and scored them using a weighted scoring model. The seven criteria in VHDA's scoring model were:

■ The project's net present value

■ Whether the project was mandated and by whom

■ The relationship of the project to VHDA's strategic plan

■ The relationship of the project to the information technology strategic plan

■ Whether the project would replace legacy applications

■ Whether there was an interim solution

■ How the project would affect customer service

The new planning and budgeting process and weighted scoring model for information technology projects is very successful, and people involved in the process agree that it was a worthwhile effort.[6]

[6] Woods, Theodore W. "The Justification, Ranking, and Budgeting Process for IT at the Virginia Housing Development Authority," *Journal of Systems Management* (May/June 1996).

Implementing a Balanced Scorecard

Drs. Robert Kaplan and David Norton developed another approach to help select and manage projects that align with business strategy. A **balanced scorecard** is a methodology that converts an organization's value drivers, such as customer service, innovation, operational efficiency, and financial performance, to a series of defined metrics. Organizations record and analyze these metrics to determine how well projects help them achieve strategic goals. Using a balanced scorecard involves several detailed steps. You can learn more about how balanced scorecards work from the Balanced Scorecard Institute (*www.balancedscorecard.org*) or other sources. Although this concept can work within an information technology department specifically, it is best to implement a balanced scorecard throughout an organization because it helps foster alignment between business and information technology.[7] The Balanced Scorecard Institute, which provides training and guidance to organizations using this methodology, quotes Kaplan and Norton's description of the balanced scorecard as follows:

> "The balanced scorecard retains traditional financial measures. But financial measures tell the story of past events, an adequate story for industrial age companies for which investments in long-term capabilities and customer relationships were not critical for success. These financial measures are inadequate, however, for guiding and evaluating the journey that information age companies must make to create future value through investment in customers, suppliers, employees, processes, technology, and innovation."[8]

As you can see, organizations can use many approaches to select projects. Many project managers have some say in which projects their organization selects for implementation. Even if they do not, they need to understand the motive and overall business strategy for the project.

Project Charters

After top management decides on which projects to pursue, it is important to let the rest of the organization know about these projects. Management needs to create and distribute documentation to authorize project initiation. This documentation can take many different forms, but one common form is a project charter. A **project charter** is a document that formally recognizes the existence of a project and provides direction on the project's objectives and management. Instead of project charters, some organizations initiate projects using a simple letter of agreement, while others use formal contracts. Key project stakeholders should sign a project charter to acknowledge agreement on the need for and intent of the project. A project charter is a key output of the initiation process, as described in Chapter 3.

[7] Berkman, Eric. "How to Use the Balanced Scorecard," *CIO Magazine* (May 15, 2002).
[8] The Balanced Scorecard Institute. "What Is a Balanced Scorecard," *The Balanced Scorecard Institute* (*www.balancedscorecard.org*) (February 2003).

Table 5-2 provides a sample project charter for the information technology upgrade project described in the opening case. Notice that the key parts of this project charter are:

- The project's title and date of authorization
- The project manager's name and contact information
- A brief description of the project objectives
- A summary of the planned approach for managing the project
- A roles and responsibilities matrix
- A sign-off section for signatures of key project stakeholders
- A comments section in which stakeholders can provide important comments related to the project

Table 5-2: Sample Project Charter

Project Title: Information Technology (IT) Upgrade Project

Project Start Date: March 4, 2005   **Projected Finish Date:** December 4, 2005

Project Manager: Kim Nguyen, 691-2784, *knguyen@abc.com*

Project Objectives: Upgrade hardware and software for all employees (approximately 2,000) within 9 months based on new corporate standards. See attached sheet describing the new standards. Upgrades may affect servers and midrange computers, as well as network hardware and software. Budgeted $1,000,000 for hardware and software costs and $500,000 for labor costs.

Approach:

- Update the information technology inventory database to determine upgrade needs

- Develop detailed cost estimate for project and report to CIO

- Issue a request for quotes to obtain hardware and software

- Use internal staff as much as possible to do the planning, analysis, and installation

ROLES AND RESPONSIBILITIES:

NAME	ROLE	RESPONSIBILITY
Walter Schmidt	CEO, Project Sponsor	Provide direction and funding
Mike Zwack	CIO	Monitor project, provide staff
Kim Nguyen	Project Manager	Plan and execute project
Jeff Johnson	Director of Information, Technology Operations	Mentor Kim
Nancy Reynolds	VP, Human Resources	Provide staff, issue memo to all employees about project
Steve McCann	Director of Purchasing	Assist in purchasing hardware and software

Table 5-2: Sample Project Charter (continued)

Sign-off: (Signatures of all the above stakeholders)

Walter Schmidt	*Jeff Johnson*
Mike Zwack	*Nancy Reynolds*
Kim Nguyen	*Steve McCann*

Comments: (Typed or handwritten comments from above stakeholders, if applicable)

"This project must be done within ten months at the absolute latest." Mike Zwack, CIO

"We are assuming that adequate staff will be available and committed to supporting this project. Some work must be done after hours to avoid work disruptions, and overtime will be provided." Jeff Johnson and Kim Nguyen, Information Technology Department

Notice that this sample project charter fits on one page. A project charter can be as simple as a one-page form or a memo from a senior manager briefly describing the project and listing the responsibilities and authority of the new project manager and stakeholders. Project charters may be much longer, however, depending on the nature of the project. In some cases, a contract serves as a project charter. You will learn more about contracts in Chapter 12, Project Procurement Management.

Project charters are usually not difficult to write. What is difficult is getting people with the proper knowledge and authority to write and sign the project charters. Even though senior managers verbally approve a project, it is still important to have a formal, written project charter to clarify roles and expectations. Many people assume, therefore, that senior managers must write the project charter, but that does not need to be the case. As Paula Martin and Karen Tate, writing in *PM Network*, note: "What's that you said? It would be a cold day in where before your sponsor provides you with a charter? Not a problem. You can create a charter with your project team and then review it with your sponsor. That way you'll clarify the sponsor's expectations for the project."[9]

Since many projects fail because of unclear requirements and expectations, starting with a simple project charter makes a lot of sense. If project managers are having difficulty obtaining support from project stakeholders, for example, they can refer to what everyone agreed to in the project charter.

After formally recognizing the existence of a project, the next step in project scope management is detailed scope planning.

[9] Martin, Paula K. and Tate, Karen. "Ready, Set, . . . Whoa," *PM Network*, PMI (June 1998): 22.

SCOPE PLANNING AND THE SCOPE STATEMENT

Project scope planning involves developing documents to provide the basis for future project decisions, including the criteria for determining if a project or phase has been completed successfully. The project charter, descriptions of the products involved in the project, project constraints, and project assumptions are inputs to the scope planning process. The main output of this process is the creation of a written scope statement, including supporting detail, and a project scope management plan.

A **scope statement** is a document used to develop and confirm a common understanding of the project scope. It describes in detail the work to be accomplished on the project and is an important tool for preventing scope creep, as described later in this chapter. The scope statement should include a project justification, a brief description of the project's products, a summary of all project deliverables, a statement of what determines project success, and references to supporting documents.

- The project justification describes the business need that sparked creation of the project. For example, in the opening case, Kim Nguyen's company initiated the information technology upgrade project in order to support several high-priority, Internet-based applications it was developing.

- The brief description of the project's products summarizes the characteristics of the products or services that the project will produce. For example, this section of the scope statement of the information technology upgrade project would describe the minimum hardware and software requirements for all of the company's computers.

- The summary of project deliverables would list the deliverables of the project. Deliverables in the information technology upgrade project might include documentation such as a project plan, a Work Breakdown Structure, a detailed cost estimate, a communications management plan, performance reports, and the like. Other deliverables would include an updated inventory of all hardware and software, upgraded hardware and software, status presentations, and so on.

- The section of the project scope management plan describing what determines project success lists the quantifiable criteria to meet for project success, such as cost, schedule, and quality measures. For example, the information technology upgrade project might be considered a success if 90 percent of all employees have computers capable of using the new Internet systems under development within nine months and for no more than $1.5 million. Even though the goal is to have all employees' hardware and software upgraded, stakeholders might be happy with 90 percent. This section would also stress the need to finish within ten months. It might also mention that the budget could go up to $1.7 million, if the project exceeded nine months.

- The references to supporting documents provide the names of related documents, such as product specifications, organizational charts, methodology descriptions, and any other information that will help clarify the scope of the project.

Scope statements also vary by type of project. Very large, complex projects have very long scope statements whereas smaller projects have shorter scope statements. See Chapter 3 for a sample scope statement for JWD Consulting's Project Management Intranet Site project. Government projects often include a scope statement known as a Statement of Work (SOW). Some SOWs are hundreds of pages long, particularly if they include detailed product specifications. As with many other project management documents, the scope statement should be tailored to meet the needs of the particular project.

SCOPE DEFINITION AND THE WORK BREAKDOWN STRUCTURE

After completing scope planning, the next step in project scope management is to define further the work required for the project and to break it into manageable pieces. The process of breaking work into manageable pieces is called scope definition. Good scope definition is very important to project success because it helps improve the accuracy of time, cost, and resource estimates, it defines a baseline for performance measurement and project control, and it aids in communicating clear work responsibilities. The output of the scope definition process is the Work Breakdown Structure for the project.

The Work Breakdown Structure

A **work breakdown structure (WBS)** is a deliverable-oriented grouping of the work involved in a project that defines the total scope of the project. It is a useful tool for breaking down a lot of work into manageable pieces. Because most projects involve many people and many different deliverables, it is important to organize and divide the work into logical parts based on how the work will be performed. The WBS is a foundation document in project management because it provides the basis for planning and managing project schedules, costs, resources, and changes. Since the WBS defines the total scope of the project, some project management experts believe that work should not be done on a project if it is not included in the WBS. Therefore, it is crucial to develop a good WBS.

A WBS is often depicted as a task-oriented family tree of activities, similar to an organizational chart. A project manager often organizes the WBS around project products, project phases, or using the project management process groups. Many people like to create a WBS in chart form first to help them visualize the whole project and all of its main parts. Figure 5-6 shows a WBS for an intranet project. Notice that product areas provide the basis for its organization. In this case, there are main boxes or groupings on the WBS for developing the Web site design, the home page for the intranet, the marketing department's pages, and the sales department's pages.

Figure 5-6. Sample Intranet WBS Organized by Product

In contrast, a WBS for the same intranet project can be organized around project phases, as shown in Figure 5-7.[10] Notice that project phases of concept, Web site design, Web site development, roll out, and support provide the basis for its organization.

A WBS can also be shown in tabular form as an indented list of tasks that shows the same groupings of the work. For example, Table 5-3 shows the WBS from Figure 5-7 in tabular form. The items on the WBS are the same, but the numbering scheme and indentation of tasks show the structure. Many documents, such as contracts, use this tabular form. Project management software also uses this form. The WBS is actually the contents of the Task Name column in Project 2002.

---

[10] This particular structure is based on a sample Project 98 file.

Figure 5-7. Sample Intranet WBS Organized by Phase

Table 5-3: Intranet WBS in Tabular Form

1.0 Concept

    1.1 Evaluate current systems

    1.2 Define Requirements

        1.2.1 Define user requirements

        1.2.2 Define content requirements

        1.2.3 Define system requirements

        1.2.4 Define server owner requirements

    1.3 Define specific functionality

    1.4 Define risks and risk management approach

    1.5 Develop project plan

    1.6 Brief Web development team

2.0 Web Site Design

3.0 Web Site Development

4.0 Roll Out

5.0 Support

Figure 5-8 shows this phase-oriented intranet WBS, using the same numbering scheme from Table 5-3, in the form of a Gantt chart created in Microsoft Project 2002. You can see from this figure that the WBS is the basis for project schedules. Notice that the WBS is in the left part of the figure under the Task Name column. The resulting schedule is in the right part of the figure. You will learn more about Gantt charts in Chapter 7, Project Time Management.

Figure 5-8. Intranet Gantt Chart in Microsoft Project

The work breakdown structures in Figures 5-6, 5-7, and 5-8 and in Table 5-3 present information in hierarchical form. The 0 level of a WBS represents the entire project and is the top level (Note the labels on the left side of Figure 5-7. Also, note that some sources call the entire project level 1 instead of level 0.) The next level down is level 1, which represents the major products or phases of the project. Level 2 includes the major subsets of level 1. For example, in Figure 5-7 the level 2 items under the level 1 item called "Concept" include: evaluate current system, define requirements, define specific functionality, define risks and risk management approach, develop project plan, and brief Web development team. Under the level 2 item called "Define Requirements" are four level 3 items on the WBS: define user requirements, define content requirements, define server requirements, and define server owner requirements.

In Figure 5-7, the lowest level is level 3. The lowest level of the WBS represents work packages. A **work package** is a task at the lowest level of the WBS. Generally, each work package in a WBS should represent roughly eighty hours of effort. You can also think of work packages in terms of accountability and reporting. If a project has a relatively short time frame and requires weekly progress reports, a work package might represent 40 hours of work or less. On the other hand, if a project has a very long time frame and requires quarterly progress reports, a work package might represent more than a hundred hours of work. Another way to think of work packages relates to entering data into project management software. You can only enter duration estimates for work packages. The rest of the WBS items are just groupings or summary tasks for the work packages. The software automatically calculates duration estimates for various WBS levels based on data entered for each work package and the WBS hierarchy.

The sample WBSs shown here seem somewhat easy to construct and understand. *Nevertheless, it is very difficult to create a good WBS.* To create a good WBS, you must understand both the project and its scope and incorporate the needs and knowledge of the stakeholders. The project manager and the project team must decide as a group how to organize the work and how many levels to include in the WBS. Many project managers have found that it is better to focus on getting the top levels done well before getting too bogged down in more detail.

Another concern when creating a WBS is how to organize it so that it provides the basis for the project schedule. Some project managers suggest creating a WBS using the project management process groups of initiating, planning, executing, controlling, and closing as level 1 in the WBS. By doing this, not only does the project team follow good project management practice, but the WBS tasks can also be mapped more easily against time. For example, Figure 5-9 shows a WBS and Gantt chart for the intranet project, organized by the product phases described earlier. Tasks under initiating include selecting a project manager, forming the project team, and developing the project charter. Tasks under planning include developing a scope statement, creating a WBS, and developing and refining other plans, which would be broken down in more detail for a real project. The tasks of concept, Web site design, Web site development, and roll out, which were WBS level 1 items in Figure 5-7, now become WBS level 2 items under executing. The executing tasks vary the most from project to project, but many of the tasks under the other project management process groups would be similar for all projects.

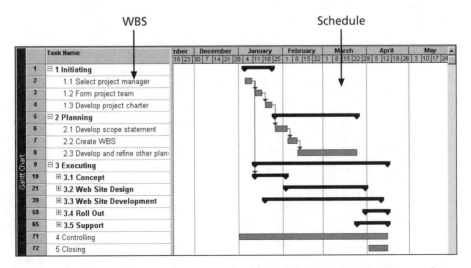

Figure 5-9. Intranet Gantt Chart Organized by Project Management Process Groups

JWD Consulting used the project management process groups for the level 1 items in its WBS for the Project Management Intranet Site project (see Chapter 3). The project team focused on the product deliverables they had to produce for the project in breaking down the executing task. Table 5-4 shows the categories they used for that part of the WBS. Some project teams like to list every deliverable they need to produce and then use those as the basis for creating all or part of their WBS. Recall that the scope statement should list and describe all of the deliverables required for the project. It is important to ensure consistency between the project charter, scope statement, WBS, and Gantt chart to define the scope of the project accurately.

Table 5-4: Executing Tasks for JWD Consulting's WBS

3.0 Executing

    3.1 Survey

    3.2 User inputs

    3.3 Intranet site content

        3.3.1 Templates and Tools

        3.3.2 Articles

        3.3.3 Links

        3.3.4 Ask the Expert

        3.3.5 User requests feature

    3.4 Intranet site design

    3.5 Intranet site construction

    3.6 Site testing

    3.7 Site promotion

    3.8 Site roll out

    3.9 Project benefits measurement

It is important to involve the entire project team and customer in creating and reviewing the WBS. People who will do the work should help to plan the work by creating the WBS. Having group meetings to develop a WBS helps everyone understand what work must be done for the entire project and how it should be done, given the people involved. It also helps to identify where coordination between different work packages will be required.

Approaches to Developing Work Breakdown Structures

There are several approaches you can use to develop work breakdown structures. These approaches include:

- Using guidelines
- The analogy approach
- The top-down approach
- The bottom-up approach
- The mind-mapping approach

Using Guidelines

If guidelines for developing a WBS exist, it is very important to follow them. Some organizations, for example, the U.S. Department of Defense (DOD), prescribe the form and content for WBSs for particular projects. Many DOD projects require contractors to prepare their proposals based on the DOD-provided WBS. These proposals must include cost estimates for each task in the WBS at a detailed and summary level. The cost for the entire project must be calculated by summing the costs of all of the lower-level WBS tasks. When DOD personnel evaluate cost proposals, they must compare the contractors' costs with the DOD's estimates. A large variation in costs for a certain WBS task often indicates confusion as to what work must be done.

Consider a large automation project for the U.S. Air Force. In the mid-1980s, the Air Force developed a request for proposals for the Local On-Line Network System (LONS) to automate fifteen Air Force Systems Command bases. This $250 million project involved providing the hardware and developing software for sharing documents such as contracts, specifications, requests for proposals, and so on. The Air Force proposal guidelines included a WBS that contractors were required to follow in preparing their cost proposals. Level 1 WBS items included hardware, software development, training, project management, and the like. The hardware item was composed of several level 2 items, such as servers, workstations, printers, network hardware, and so on. Air Force personnel reviewed the contractors' cost proposals against their internal cost estimate, which was also based on this WBS. Having a prescribed WBS helped contractors to prepare their cost proposals and the Air Force to evaluate them.

Many organizations provide guidelines and templates for developing WBSs as well as examples of WBSs from past projects. At the request of many of its members, the Project Management Institute recently developed a WBS Practice Standard to provide guidance for developing and applying the WBS to project management (see Suggested Readings). This practice standard is available free to all members from PMI's Web site (*www.pmi.org*). This document includes sample WBSs for a wide variety of projects in various industries, including projects for Web design, telecom, service industry outsourcing, and software implementation.

Project managers and their teams should review appropriate information to develop their unique project WBSs more efficiently. For example, Kim Nguyen and key team members from the opening case should review their company's WBS guidelines, templates, and other related information before the team meetings to create their WBS.

The Analogy Approach

Another approach for constructing a WBS is the analogy approach. In the **analogy approach**, you use a similar project's WBS as a starting point. For example, Kim Nguyen from the opening case might learn that one of her organization's suppliers did a similar information technology upgrade project last year. She could ask them to share their WBS for that project to provide a starting point for her own project.

McDonnell Aircraft Company, now part of Boeing, provides an example of using an analogy approach when creating WBSs. McDonnell Aircraft Company designed and manufactured several different fighter aircraft. When creating a WBS for a new aircraft design, it started by using 74 predefined subsystems for building a fighter aircraft based on past experience. There was a level 1 WBS item for the airframe that was composed of level 2 items such as a forward fuselage, center fuselage, aft fuselage, and wings. This generic product-oriented WBS provided a starting point for defining the scope of new aircraft projects and developing cost estimates for new aircraft designs.

Some organizations keep a repository of WBSs and other project documentation on file to assist people working on projects. Project 2002 and many other software tools include sample files to assist users in creating a WBS and Gantt chart. Viewing examples of other similar projects' WBSs allows you to understand different ways to create a WBS. See the companion Web site for this text for sample WBSs.

The Top-Down and Bottom-Up Approaches

Two other approaches for creating WBSs are the top-down and bottom-up approaches. Most project managers consider the top-down approach of WBS construction to be conventional. To use the **top-down approach**, start with the largest items of the project and break them into their subordinate items. This process involves refining the work into greater and greater levels of detail. For example, Figure 5-7 shows how work was broken down to level 3 for part of the intranet project. After finishing the process, all resources should be assigned at the work package level. The top-down approach is best suited to project managers who have vast technical insight and a big-picture perspective.

In the **bottom-up approach**, team members first identify as many specific tasks related to the project as possible. They then aggregate the specific tasks and organize them into summary activities, or higher levels in the WBS. For example, a group of people might be responsible for designing a WBS to create an e-commerce application. Instead of looking for guidelines on how to create a WBS or viewing similar projects' WBSs, they could begin by listing detailed tasks they think they would need to do in order to create the application. After listing detailed tasks, they would group the tasks into categories. Then, they would group these categories into higher-level categories. Some people have found that writing down all possible tasks on notes and then placing them on a wall helps them to see all the work required for the project and to develop logical groupings for performing the work. For example, a business analyst on the project team might know that they had to define user requirements and content requirements for the e-commerce application. These tasks might be part of the requirements documents they would have to create as one of the project deliverables. A hardware specialist might know they had to define system requirements and server requirements, which would also be part of a requirements document. As a group, they might decide to put all four of these tasks under a higher-level item called "define requirements" that would result in the delivery of a requirements document. Later, they might realize that defining requirements should fall under a broader category of concept design for the e-commerce application, along with other groups of tasks related to the concept design. The bottom-up approach can be very time-consuming, but it can also be a very effective way to create a WBS. Project managers often use the bottom-up approach for projects that represent entirely new systems or approaches to doing a job, or to help create buy-in and synergy with a project team.

Mind Mapping

Some project managers like to use mind mapping to help develop WBSs. **Mind mapping** is a technique that uses branches radiating out from a core idea to structure thoughts and ideas. Instead of writing down tasks in a list or immediately trying to create a structure for tasks, mind mapping allows people to write and even draw pictures of ideas in a non-linear format. This more visual, less structured approach to defining and then grouping tasks can unlock creativity among individuals and increase participation and morale among teams. [11]

Figure 5-10 shows a diagram with an example of using mind mapping to create a WBS. The circle in the center represents the entire project. Each of the four main branches radiating from the center represents the main tasks or groupings for the WBS. Branching off from Main task 1 are two subtasks. Branching off from Subtask 1 are two deliverables, labeled Deliverable 1 and Deliverable 2. Of course, you would write appropriate text to describe the

---

[11] Mindjet Visual Thinking. "About Mind Maps," (*www.mindjet.com/about/about_mindmapping.shtml*) (2002).

tasks, subtasks, and deliverables or draw pictures, if desired. After creating the diagram using the mind-mapping technique, you could then translate the information into chart or tabular form, as described earlier.

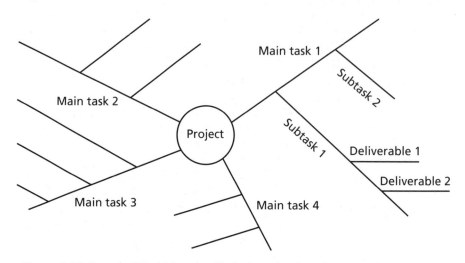

Figure 5-10. Sample Mind-Mapping Technique for Creating a WBS

Mind mapping can be used for developing WBSs using the top-down or bottom-up approach. For example, you could conduct mind mapping for an entire project by listing the whole project in the center of a document, adding the main categories on branches radiating from the center, and then adding branches for appropriate subcategories. You could also develop a separate mind-mapping document for each deliverable of a project by putting the deliverable in the center of a document and then surrounding it with other mind-mapping documents to create one for the entire project. You can also add items anywhere on a mind-mapping document without following a strict top-down or bottom-up approach. After the mind-mapping documents are complete, you can convert them into a chart or tabular WBS form.

Advice for Creating a WBS

As stated previously, creating a good WBS is no easy task and usually requires several iterations. Often, it is best to use a combination of approaches to create a project WBS. There are some basic principles, however, that apply to creating any good WBS.

- A unit of work should appear at only one place in the WBS.
- The work content of a WBS item is the sum of the WBS items below it.
- A WBS item is the responsibility of only one individual, even though many people may be working on it.

- The WBS must be consistent with the way in which work is actually going to be performed; it should serve the project team first, and other purposes only if practical.
- Project team members should be involved in developing the WBS to ensure consistency and buy-in.
- Each WBS item must be documented to ensure accurate understanding of the scope of work included and not included in that item.
- The WBS must be a flexible tool to accommodate inevitable changes while properly maintaining control of the work content in the project according to the scope statement.[12]

SCOPE VERIFICATION AND SCOPE CHANGE CONTROL

It is difficult to create a good scope statement and WBS for a project. It is even more difficult, especially on information technology projects, to verify the project scope and minimize scope changes. Many information technology projects suffer from **scope creep**—the tendency for project scope to keep getting bigger and bigger. There are many horror stories about information technology projects failing due to scope problems such as scope creep. For this reason, it is very important to verify the project scope and develop a process for controlling scope changes.

What Went Wrong?

A project scope that is too broad and grandiose can cause severe problems. Scope creep and an overemphasis on technology for technology's sake resulted in the bankruptcy of a large pharmaceutical firm, Texas-based FoxMeyer Drug. In 1994, the CIO was pushing for a $65 million system to manage the company's critical operations. He did not believe in keeping things simple, however. The company spent nearly $10 million on state-of-the-art hardware and software and contracted the management of the project to a prestigious (and expensive) consulting firm. The project included building an $18 million robotic warehouse, which looked like something out of a science fiction movie, according to insiders. The scope of the project kept getting bigger and more impractical. The elaborate warehouse was not ready on time, and the new system generated erroneous orders that cost FoxMeyer Drug more than $15 million in unrecovered excess shipments. In July of 1996, the company took a $34 million charge for its fourth fiscal quarter, and by August of that year, FoxMeyer Drug filed for bankruptcy.[13]

---

[12] Cleland, David I. *Project Management: Strategic Design and Implementation, 2nd ed.* New York: McGraw-Hill, 1994.

[13] James, Geoffrey. "Information Technology fiascoes . . . and how to avoid them," *Datamation* *www.datamation.com/PlugIn/issues/1997/november/11disas.html* (November 1997).

Another major source of information technology project scope problems is a lack of user involvement. At Grumman in the late 1980s, for example, an information technology project team became convinced that it could, and should, automate the review and approval process of government proposals. They implemented a powerful workflow system to manage the whole process. Unfortunately, the end users for the system were aerospace engineers who preferred to work in a more casual, ad hoc fashion. They dubbed the system "Naziware" and refused to use it. This example illustrates an information technology project that wasted millions of dollars developing a system that was not in touch with the way end users did their work.[14]

Failing to follow good project management processes and use off-the-shelf software often results in scope problems. 21st Century Insurance Group in Woodland Hills, California, paid Computer Sciences Corporation $100 million on a project to develop a system for managing business applications, including managing insurance policies, billing, claims, and customer service. After five years, the system was still in development and used to support less than 2 percent of the company's business in 2002. Joshua Greenbaum, an analyst at Enterprise Applications Consulting, called the project a "huge disaster" and questioned the insurance company's ability "to manage a process that is pretty well known these days...I'm surprised that there wasn't some way to build what they needed using off-the-shelf components and lower their risk."[15]

Scope verification involves formal acceptance of the project scope by the stakeholders. This acceptance is often achieved by a customer sign-off on key deliverables and the entire project. To receive formal acceptance of the project scope, the project team must develop clear documentation of the project's products and procedures to evaluate if they were completed correctly and satisfactorily. In contrast, scope change control involves controlling changes to the project scope. In order to minimize scope change control, it is crucial to do a good job of verifying project scope.

The 1995 Standish Group CHAOS study found that key factors associated with information technology project success include user involvement and a clear statement of requirements. Likewise, the lack of either of these factors is strongly associated with information technology project failure. Table 5-5 lists factors reported to cause the most problems on information technology projects, in order of importance. Notice that the top three factors directly relate to scope verification and change control.

[14] Ibid.
[15] Songini, Marc L. "21st Century Insurance apps in limbo despite $100M investment," *Computerworld* (December 6, 2002).

Table 5-5: Factors Causing Information Technology Project Problems

FACTOR	RANK
Lack of user input	1
Incomplete requirements and specifications	2
Changing requirements and specifications	3
Lack of executive support	4
Technology incompetence	5
Lack of resources	6
Unrealistic expectations	7
Unclear objectives	8
Unrealistic time frames	9
New Technology	10

Johnson, Jim. "CHAOS: The Dollar Drain of Information Technology Project Failures," *Application Development Trends* (January 1995).

Research and practice, therefore, indicate that in order to verify project scope and control scope change, you need to improve user input and reduce incomplete and changing requirements and specifications. Suggestions for accomplishing these two tasks are in the next section.

Suggestions for Improving User Input

Lack of user input is among the more important factors contributing to project failure and can lead to problems with managing scope creep and controlling change. How can you manage this important issue? Following are suggestions for improving user input.

- Develop a good project selection process for information technology projects. Insist that all projects have a sponsor from the user organization. The sponsor should not be someone in the information technology department, nor should the sponsor be the project manager. Make information technology project information, including the project charter, scope statement, and WBS, easily available in the organization. Making basic project information available will help avoid duplication of effort and ensure that the most important projects are the ones on which people are working.

- Have users on the project team. Some organizations require project managers to come from the business area of the project instead of the information technology group. Some organizations assign co-project managers to information technology projects, one from information technology and one from the main business group. Users should be assigned full-time to large information technology projects and part-time to smaller projects. A key success factor in Northwest Airline's ResNet project (see the companion Web site for this text for an entire case study on this project) was training reservation agents—the users—in how to write programming code for their new reservation system. Because the sales agents had intimate knowledge of the business, they provided excellent input and actually created most of the software.

- Have regular meetings. Meeting regularly sounds obvious, but many information technology projects fail because the information technology team members do not have regular interaction with users. They assume they understand what users need without getting direct user feedback. To encourage this interaction, users should sign off on key deliverables presented at meetings.

- Deliver something to project users and sponsors on a regular basis. If it is some sort of hardware or software, make sure it works first.

- Co-locate users with the developers. People often get to know each other better by being in close proximity. If the users cannot be physically moved to be near developers during the entire project, they could set aside certain days for co-location.

Suggestions for Reducing Incomplete and Changing Requirements

Some requirement changes are expected on information technology projects, but many projects have too many changes to their requirements, especially during later stages of the project life cycle when it is more difficult to implement them. The following are suggestions for improving the requirements process.

- Develop and follow a requirements management process that includes procedures for initial requirements determination. (See the suggested reading by Robertson for detailed information on managing requirements.)

- Employ techniques such as prototyping, use case modeling, and Joint Application Design to understand user requirements thoroughly. **Prototyping** involves developing a working replica of the system or some aspect of the system. These working replicas may be throwaways or an incremental component of the deliverable system. Prototyping is an effective tool for gaining an understanding of requirements, determining the feasibility of requirements, and resolving user interface uncertainties. **Use case modeling** is a process for identifying and modeling business events, who initiated them, and how the system should respond to them. It is an effective tool for understanding requirements for object-oriented systems. **Joint Application Design (JAD)** uses highly organized and intensive workshops to bring together project stakeholders—the sponsor, users, business analysts, programmers, and so on—to jointly define and design information systems. The techniques also help users become more active in defining system requirements.

■ Put all requirements in writing and keep them current and readily available. Several tools are available to automate this function. For example, a type of software, called a requirements management tool, aids in capturing and maintaining requirements information, provides immediate access to the information, and assists in establishing necessary relationships between requirements and information created by other tools.

■ Create a requirements management database for documenting and controlling requirements. Computer Aided Software Engineering (CASE) tools or other technologies can assist in maintaining a repository for project data.

■ Provide adequate testing to verify that the project's products perform as expected. Conduct testing throughout the project life cycle.

■ Use a process for reviewing requested requirements changes from a systems perspective. For example, ensure that project scope changes include associated cost and schedule changes. Require approval by appropriate stakeholders. For example, at PanEnergy Corporation in Houston, Bruce Woodland completed a $10 million electronic commerce project on time and under budget. He says, "Whenever someone wanted to add or change something, we'd tell them how long it would take and then ask them how they wanted to deal with it . . . . The first few times, the users almost took us out behind the dumpster and hanged us. But when we could match specific functionalities they wanted with the time and dollars they required, we didn't have a problem."[16]

■ Emphasize completion dates. For example, a project manager at Farmland Industries Inc., in Kansas City, Missouri, kept her 15-month, $7 million integrated supply-chain project on track by setting the project deadline. She says, "May 1 was the drop-dead date, and everything else was backed into it. Users would come to us and say they wanted something, and we'd ask them what they wanted to give up to get it. Sticking to the date is how we managed scope creep."[17]

■ Allocate resources specifically for handling change requests. For example, Peeter Kivestu and his ResNet team at Northwest Airlines knew that users would request enhancements to the reservations system they were developing. They provided a special function key on the ResNet screen for users to submit their requests, and the project included three full-time programmers to handle these requests. Users made over 11,000 enhancement

[16] King, Julia. "IS reins in runaway projects," *Computerworld* (September 24, 1997).
[17] Ibid.

requests. The managers who sponsored the four main software applications had to prioritize the software enhancement requests and decide as a group what changes to approve. The three programmers then implemented as many items as they could, in priority order, given the time they had. Although they only implemented 38 percent of the requested enhancements, they were the most important ones, and the users were very satisfied with the system and process.

USING SOFTWARE TO ASSIST IN PROJECT SCOPE MANAGEMENT

Project managers and their teams can use several types of software to assist in project scope management. As shown in several of the figures and tables in this chapter, you can use word-processing software to create scope-related documents, and most people use spreadsheet software to perform financial calculations, create weighting scoring models, and develop various charts and graphs related to scope management. Project stakeholders also transmit project scope management information using various types of communication applications such as e-mail and Web-based applications.

Project management software helps develop a WBS, which serves as a basis for creating Gantt charts, assigning resources, allocating costs, and so on. Templates that come with various project management software products can also be used to help you create a WBS for your project. See the section on project scope management in Appendix A for detailed information on using Project 2002 and Appendix D for templates related to project scope management.

You can also use many types of specialized software to assist in project scope management. For example, some companies sell software for applying the balanced scorecard method for project selection. Other companies provide software for mind mapping that you can use to help create a WBS. Many information technology projects use special software for requirements management, prototyping, modeling, and other scope-related work. Because scope is such a crucial part of project management, there are many software products available to assist in managing project scope.

Project scope management is very important, especially on information technology projects. Organizations must first select important projects, plan what is involved in performing the work of the project, break down the work into manageable pieces, verify the scope with project stakeholders, and manage changes to project scope. Using the basic project management concepts, tools, and techniques discussed in this chapter can help you successfully perform project scope management.

CASE WRAP-UP

Kim Nguyen reviewed guidelines for developing scope statements and WBSs provided by her company and other sources. She had a meeting with the three team leaders for her project to get their input on how to proceed. They reviewed several sample documents and decided to have major groupings for their project based on updating the inventory database, acquiring the necessary hardware and software, and installing the hardware and software. After they decided on a basic approach, Kim led a meeting with the entire project team of twelve people. She reviewed the project charter, described the basic approach to defining the project scope, and reviewed sample scope statements and WBSs. Kim opened the floor for questions, which she answered confidently. Kim then let each team leader work with his or her people to start writing the scope statement and WBS. Everyone participated in the meeting, sharing their individual expertise and openly asking questions. Kim could see that the project was off to a good start.

CHAPTER SUMMARY

Project scope management includes the processes required to ensure that the project addresses all the work required, and only the work required, to complete the project successfully. The main processes involve initiation, scope planning, scope definition, scope verification, and scope change control.

The first step in project scope management is deciding on which projects to work. To decide on which projects to work, organizations should review the organization's strategic plan and then develop an information technology strategic plan that supports the organization's overall strategy.

Organizations initiate information technology projects for many different reasons. Common techniques for selecting projects include focusing on broad organization needs, categorizing projects, performing financial analyses, developing weighted scoring models, and using balanced scorecards.

Net present value (NPV) analysis is the preferred financial measure for selecting projects. A positive NPV is good, and the higher the NPV, the better.

Other important financial considerations are the projected return on investment (ROI) of a project and the payback period.

A project charter is a document that formally recognizes the existence of a project. Key project stakeholders must sign it to acknowledge agreement on the need for and intent of the project.

A scope statement is created in the scope planning process. This document includes a project justification, a brief description of the project's products, a summary of all project deliverables, and a statement of what determines project success.

A work breakdown structure (WBS) is the main output of the scope definition process. It is a deliverable-oriented grouping of the work involved in a project and defines the total scope of the project. The WBS forms the basis for planning and managing project schedules, costs, teams, and changes. Project management software cannot be used without first creating a good WBS. Good WBSs are difficult to create. Approaches for developing a WBS include using guidelines, the analogy approach, the top-down approach, and the bottom-up approach. Mind mapping can facilitate the process of creating a WBS.

Scope verification involves formal acceptance of the project scope by the stakeholders. Scope change control involves controlling changes to the project scope.

Poor project scope management is one of the key reasons projects fail. For information technology projects, it is important for good project scope management to have strong user involvement, a clear statement of requirements, and a process for managing scope changes.

There are many software products available to assist in project scope management. The WBS is a key concept in properly using project management software since it provides the basis for entering tasks.

DISCUSSION QUESTIONS

1. What is involved in project scope management, and why is good project scope management so important on information technology projects?
2. Provide examples of using various project selection methods, such as focusing on broad organizational needs, categorizing projects, and using financial analysis techniques.
3. What are the main parts of a scope statement? Why do you need to have a scope statement if you have a good project charter and WBS?
4. Describe different ways to develop a WBS and why it is often so difficult to do.
5. Describe a project that suffered from scope creep. Could it have been avoided? How? Can scope creep be a good thing? When?
6. Why do you need a good WBS to use project management software? What other types of software can you use to assist in project scope management?

EXERCISES

1. Use spreadsheet software to create Figures 5-2 through 5-5 in this text. Make sure your formulas work correctly.

2. Use PowerPoint, Visio, or similar software to create a WBS in chart form (similar to an Organizational chart—see the intranet sample in Figure 5-6) for the information technology upgrade project described in the opening case of this chapter. Break down the work to at least the third level for one of the WBS items. Make notes of questions you had while completing this exercise.

3. Create a WBS for one of the following projects:

   ■ Building your dream house
   ■ Planning a traditional wedding
   ■ Creating a new information system for your school or company

   Break down the work to at least the third level for one of the items on the WBS. Enter the WBS into Project 2002 and print out the Gantt chart. Do not enter any durations or dependencies. Make notes of questions you had while completing this exercise.

4. Review the files in the Microsoft Project 2002 templates folder. What do you think about the WBSs for them? Try drawing a WBS in chart form for one of the sample templates. Draw the WBS by hand or try using the Visio WBS Chart Wizard add-in tool described in Appendix A.

5. Perform a financial analysis for a project using the format provided in Figure 5-3. Assume the projected costs and benefits for this project are spread over four years as follows: Estimated costs are $1,500,000 in year 1 and $300,000 each year in years 2, 3, and 4. Estimated benefits are $0 in year 1 and $1,000,000 each year in years 2, 3, and 4. Use a 10 percent discount rate. Create a spreadsheet to calculate and clearly display the NPV, ROI, and year in which payback occurs. In addition, write a paragraph explaining whether you would recommend investing in this project, based on your financial analysis.

6. Read one of the suggested readings or find an article related to project scope management. Write a one-page summary of the article, its key conclusions, and your opinion.

MINICASE

A financial services company has a long list of potential projects to consider this year. Managers at this company must decide which projects to pursue and how to define the scope of the projects selected for approval. The company has decided to use a weighted scoring model to help in project selection, using criteria that map to corporate objectives. All projects selected must develop a WBS using corporate guidelines.

Part 1: You are part of a team that will analyze proposals and recommend which projects to pursue. Your team has decided to create a weighted scoring model using the following criteria and weights:

Criteria	Weight
1. Enhances new product development	20%
2. Streamlines operations	20%
3. Increases cross-selling	25%
4. Has good NPV	35%

To determine the score for the last criterion, your team has developed the following scoring system:

- NPV is less than 0, the score is 0
- NPV is between 0 and $100,000, the score is 25
- NPV is between $100,000 and $200,000, the score is 50
- NPV is between $200,000 and $400,000, the score is 75
- NPV is above $400,000, the score is 100

The following is information for three potential projects:

- Project 1: Scores for criteria 1, 2, and 3 are 10, 20, and 80, respectively. Estimated costs the first year are $500,000, and costs for years 2 and 3 are $100,000 each. Estimated benefits for years 1, 2, and 3 are $200,000, $400,000, and $600,000, respectively.
- Project 2: Scores for criteria 1, 2, and 3 are all 50. Estimated costs the first year are $700,000, and costs for the second year are $200,000. Estimated benefits for years 1 and 2 are $300,000 and $700,000, respectively.
- Project 3: Scores for criteria 1, 2, and 3 are 0, 50, and 80, respectively. Estimated costs the first year are $300,000, and costs for years 2, 3, and 4 are $100,000 each. Estimated benefits for years 1, 2, 3, and 4 are $0, $600,000, $500,000, and $400,000, respectively.

Develop a spreadsheet to calculate the NPVs and weighted scores for the three projects. Use a 10 percent discount rate for the NPV calculations.

Part 2: One project selected for initiation is development of an expert system that current customers can use to give them advice on other financial products and services that would meet their needs. Corporate guidelines for creating project WBSs are to use the five project management process groups as level 1 in the WBS. Each process group must have at least one major deliverable, such as a signed project charter for initiating, an approved project plan for planning, monthly status reports for controlling, and a final report, presentation, and lessons learned for closing. The deliverables for execution vary by project.

Use the corporate guidelines to develop a WBS in tabular form. The categories of initiating, planning, executing, controlling, and closing are the level 1 categories. For the executing section, include level 2 categories of analysis, design, prototyping, testing, implementation, and support. Assume the support category includes level 3 categories of training, documentation, user support, and enhancements. Create the WBS in Project 2002, indenting categories appropriately. Use the outline numbering feature to display the outline numbers (under Tools, Options, Show outline number.) For example, your WBS should start with 1.0 Initiating. See Appendix A or Project 2002's Help for assistance. Print out the resulting Gantt chart, being sure to display the entire Task Name column.

RUNNING CASES

See the running cases in Appendix C for additional activities related to project scope management. Appendix A also provides information and exercises on using Project 2002 for project scope management. Appendix D provides several templates related to project scope management.

SUGGESTED READINGS

1. Abramovici, Adrian. "Controlling Scope Creep." *PM Network* (January 2000).

   This article states that scope creep is one of the most common problems that project managers face. The author illustrates the problems with scope creep in a short case study and provides practical advice on how to control it.

2. Cohen, Dennis J. and Robert J. Graham. *The Project Manager's MBA: How to Translate Project Decisions into Business Success*. Jossey-Bass (2001).

   This book emphasizes the importance of selecting and managing projects to meet organizations' financial goals. It summarizes important financial concepts that project managers need to understand in today's competitive market.

3. Kaplan, Robert S. and David P. Norton. *The Balanced Scorecard: Translating Strategy into Action*. Harvard Business School Press (1996).

   In this book, the authors describe how senior executives in various industries are using the Balanced Scorecard both to guide current performance and to target future performance. They show how to use measures in four categories—financial performance, customer knowledge, internal business processes, and learning and growth—to align individual, organizational, and cross-departmental initiatives and to identify entirely new processes for meeting customer and shareholder objectives.

4. Project Management Institute. "Project Management Institute Practice Standard for Work Breakdown Structures" (2001).

The WBS Practice Standard is intended to provide guidance for developing and applying the WBS to project management. PMI members can download a free copy of this document from PMI's Web site (www.pmi.org). Non-members can purchase the document from PMI. As well as providing guidelines, this document includes several sample WBSs for a variety of projects from various industries.

5. Robertson, Susan and James Robertson. *Mastering the Requirements Process*. Addison-Wesley (1999).

This book does an outstanding job of explaining the requirements management process. The heart of this book is the Volere Requirements Process Model, an industry-tested and adaptable template for gathering and verifying requirements for software products.

6. The Standish Group. "Unfinished Voyages" (1996) (*www.standishgroup.com*).

The Standish Group used the project success criteria from the 1995 CHAOS research to create a weighted scoring model for estimating information technology project success potential. The most important success criteria, user involvement, was given 19 success points, while the least important, hard-working, focused staff, was given 3 success points. You can use this checklist to decide whether a project is worth starting or continuing.

KEY TERMS

- **analogy approach** — creating a WBS by using a similar project's WBS as a starting point
- **balanced scorecard** — a methodology that converts an organization's value drivers, such as customer service, innovation, operational efficiency, and financial performance, to a series of defined metrics
- **bottom-up approach** — creating a WBS by having team members identify as many specific tasks related to the project as possible and then grouping them into higher-level categories
- **cash flow**—benefits minus costs or income minus expenses
- **cost of capital** —the return available by investing the capital elsewhere
- **deliverable** — a product, such as a report or segment of software code, produced as part of a project
- **directives** — new requirements imposed by management, government, or some external influence
- **discount factor** — a multiplier for each year based on the discount rate and year

- **discount rate** — the minimum acceptable rate of return on an investment; also called the required rate of return, hurdle rate, or opportunity cost of capital, shown in a percentage format, such as 8 percent

- **initiation** — committing the organization to begin a project or continue to the next phase of a project

- **internal rate of return (IRR)** — the discount rate that results in an NPV of zero for a project

- **Joint Application Design (JAD)** — using highly organized and intensive workshops to bring together project stakeholders—the sponsor, users, business analysts, programmers, and so on—to jointly define and design information systems

- **mind mapping** — a technique that can be used to develop WBSs by using branches radiating from a central core idea to structure thoughts and ideas

- **net present value (NPV) analysis** — a method of calculating the expected net monetary gain or loss from a project by discounting all expected future cash inflows and outflows to the present point in time

- **opportunities** — chances to improve the organization

- **payback period** — the amount of time it will take to recoup, in the form of net cash inflows, the net dollars invested in a project

- **problems** — undesirable situations that prevent the organization from achieving its goals

- **project charter** — a document that formally recognizes the existence of a project and provides direction on the project's objectives and management

- **project scope management** — the processes involved in defining and controlling what is or is not included in a project

- **prototyping** — developing a working replica of the system or some aspect of the system to help define user requirements

- **required rate of return** — the minimum acceptable rate of return on an investment

- **return on investment (ROI)** — (benefits minus costs) divided by costs

- **scope** — all the work involved in creating the products of the project and the processes used to create them

- **scope change control** — controlling changes to project scope

- **scope creep** — the tendency for project scope to keep getting bigger and longer

- **scope definition** — subdividing the major project deliverables into smaller, more manageable components

- **scope planning** — developing documents to provide the basis for future project decisions, including the criteria for determining if a project or phase has been completed successfully

- **scope statement** — a document used to develop and confirm a common understanding of the project scope
- **scope verification** — formalizing acceptance of the project scope, sometimes by customer sign-off
- **strategic planning** — determining long-term objectives by analyzing the strengths and weaknesses of an organization, studying opportunities and threats in the business environment, predicting future trends, and projecting the need for new products and services
- **top-down approach**—creating a WBS by starting with the largest items of the project and breaking them into their subordinate items
- **use case modeling** — a process for identifying and modeling business events, who initiated them, and how the system should respond to them
- **weighted scoring model** — a technique that provides a systematic process for basing project selection on numerous criteria
- **work breakdown structure (WBS)** — a deliverable-oriented grouping of the work involved in a project that defines the total scope of the project
- **work package** — a task at the lowest level of the WBS

6

Project Time Management

Objectives

After reading this chapter, you will be able to:

1. *Understand the importance of project schedules and good project time management*
2. *Define activities as the basis for developing project schedules*
3. *Describe how project managers use network diagrams and dependencies to assist in activity sequencing*
4. *Explain how various tools and techniques help project managers perform activity duration estimating and schedule development*
5. *Use a Gantt chart for schedule planning and tracking schedule information*
6. *Understand and use critical path analysis*
7. *Describe how to use several techniques for shortening project schedules*
8. *Explain the basic concepts behind critical chain scheduling and Program and Review Technique (PERT)*
9. *Discuss how reality checks and people issues are involved in controlling and managing changes to the project schedule*
10. *Describe how software can assist in project time management*

OPENING CASE

Sue Johnson was the project manager for a consulting company contracted to provide a new online registration system at a local college. This system absolutely had to be operational by May 1 so students could use it to register for the fall semester. Her company's contract had a stiff penalty clause if the system was not ready by then, and Sue and her team would get nice bonuses for doing a good job on this project. Sue knew that it was her responsibility to meet the schedule and manage scope, cost, and quality expectations. She and her team developed a detailed schedule and network diagram to help organize the project.

Developing the schedule turned out to be the easy part while keeping the project on track was more difficult. Managing people issues and resolving schedule conflicts were two of the bigger challenges. Many of the customers' employees took unplanned vacations and missed or rescheduled project review meetings. These changes made it difficult for the project team to follow their planned schedule for the system because they had to have customer sign-off at various stages of the systems development life cycle. One senior programmer on her project team quit, and she knew it would take extra time for a new person to get up to speed. It was still early in the project, but Sue knew they were falling behind. What could she do to meet the operational date of May 1?

IMPORTANCE OF PROJECT SCHEDULES

Many information technology projects are failures in terms of meeting scope, time, and cost projections. Managers often cite delivering projects on time as one of their biggest challenges. According to the 1995 CHAOS report, the average time overrun on unsuccessful information technology projects was 222 percent of the original estimate.[1] A 222 percent time overrun means that a project that was planned to take one year ended up taking 2.2 years to complete. The 2001 CHAOS report showed that time overruns decreased between 1995 and 2000—down to 163 percent—but most completed projects still finished significantly behind schedule.

Managers also cite schedule issues as the main reason for conflicts on projects throughout the project life cycle. Figure 6-1 shows the results of research on causes of conflict on projects. This figure shows that, overall, schedule issues cause the most conflict over the life of a project. (Note that this study calls the project phases project formation, early phases, middle phases, and end phases. You can interpret these phase names as concept, development, implementation, and close-out, as described in Chapter 2). During the project formation or concept phase, priorities and procedures cause more conflict than schedules. During the early or development phase, only priorities cause more conflict than schedules. During the middle or implementation phase and end or close-out phase, schedule issues are the predominant source of conflict.

[1] Johnson, Jim, "CHAOS — The Dollar Drain of IT Project Failures," *Application Development Trends* (January 1995).

Perhaps part of the reason schedule problems are so common is that time is easily and simply measured. You can debate scope and cost overruns and make actual numbers appear closer to estimates, but once a project schedule is set, anyone can quickly estimate schedule performance by subtracting the original time estimate from how long it really took to complete the project. People often compare planned and actual project completion times without taking into account approved changes in the project. Time is also the one variable that has the least amount of flexibility. Time passes no matter what happens on a project.

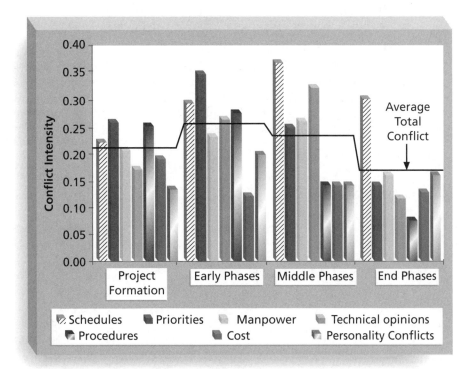

Figure 6-1. Conflict Intensity over the Life of a Project [2]

In that case, what *is* involved in project time management, and how can project managers help improve performance in this area? **Project time management**, simply defined, involves the processes required to ensure timely completion of a project. Achieving timely completion of a project, however, is by no means simple. There are five main processes involved in project time management:

■ **Activity definition** involves identifying the specific activities that the project team members and stakeholders must perform to produce the project deliverables. An **activity** or **task** is an element of work normally

[2] Thamhain, H. J., and Wilemon, D. L., "Conflict Management in Project Life Cycles," *Sloan Management Review* (Summer 1975).

found on the work breakdown structure (WBS) that has an expected duration, a cost, and resource requirements.

■ **Activity sequencing** involves identifying and documenting the relationships between project activities.

■ **Activity duration estimating** involves estimating the number of work periods that are needed to complete individual activities.

■ **Schedule development** involves analyzing activity sequences, activity duration estimates, and resource requirements to create the project schedule.

■ **Schedule control** involves controlling and managing changes to the project schedule.

You can improve project time management by performing these processes and by using some basic project management tools and techniques. Every manager is familiar with some form of scheduling, but most managers have not used several of the tools and techniques unique to project time management, such as Gantt charts, network diagrams, and critical path analysis.

ACTIVITY DEFINITION

Project schedules grow out of the basic documents that initiate a project. The project charter often mentions planned project start and end dates, which serve as the starting points for a more detailed schedule. The project manager starts with the project charter and develops a detailed project scope statement and WBS, as discussed in Chapter 5, Project Scope Management. The project charter should also include some estimate of how much money will be allocated to the project. Using this information with the scope statement, WBS, and budget information, the project manager and project team begin developing a detailed project schedule and estimated completion date. If the estimated completion date varies significantly from top management's or the customer's original plans, the project manager must negotiate changes in scope or cost to meet schedule expectations.

Recall the triple constraint of project management—balancing scope, time, and cost goals—and note the order of these items. Ideally, the project team and key stakeholders first define the project scope, then the time or schedule for the project, and then the project's cost. The order of these three items reflects the basic order of the first three processes in project time management: activity definition (further defining the scope), activity sequencing (further defining the time), and activity duration estimating (further defining the cost). These three project time management processes are the basis for creating a project schedule.

Activity definition usually results in the project team developing a more detailed WBS and supporting explanations. The goal of this process is to ensure that the project team has complete understanding of all the work they must do as part of the project scope. For example, a WBS item might be "Produce report." The project team would have to understand what that means. How long should the report be? Does it require a survey or extensive research to produce it? What skill level does the report writer need to have? Further defining that task will help the project team determine how long it will take to do and who should do it.

The WBS is often dissected further during the activity definition process as the project team members further define the activities required for performing the work. For example, the task "Produce report" might be broken down into several subtasks describing the steps involved in producing the report, such as developing a survey, administering the survey, analyzing the survey results, performing research, writing a draft report, editing the report, and finally producing the report.

As stated earlier, activities or tasks are elements of work performed during the course of a project; they have expected durations, costs, and resource requirements. Activity definition also results in supporting detail to document important product information as well as assumptions and constraints related to specific activities. The project team should review the revised WBS and supporting detail with project stakeholders before moving on to the next step in project time management. If they do not review the WBS, they could produce an unrealistic schedule and deliver unacceptable results. For example, if a project manager simply estimated that it would take one day for the "Produce report" task and had an intern write a ten-page report to complete that task, the result could be a furious customer who expected extensive research, surveys, and a one-hundred-page report. Clearly defining the work is crucial to all projects.

In the opening case, Sue Johnson and her project team had a contract and detailed specifications for the college's new online registration system. They also had to focus on meeting the May 1 date for an operational system so the college could start using the new system for fall registration. To develop a project schedule, Sue and her team had to review the contract, detailed specifications, and desired operational date, create a more detailed WBS, and highlight the most important product information, assumptions, and constraints. After developing more detailed definitions of project activities, Sue and her team would review them with their customers to ensure that they were on the right track.

ACTIVITY SEQUENCING

After defining project activities, the next step in project time management is activity sequencing. Activity sequencing involves reviewing the activities in the detailed WBS, detailed product descriptions, assumptions, and constraints to determine the relationships between activities. It also involves evaluating the reasons for dependencies and the different types of dependencies.

Dependencies

A **dependency** or **relationship** shows the sequencing of project activities or tasks. For example, does a certain activity have to be finished before another one can start? Can the project team do several activities in parallel? Can some overlap? Determining these relationships or dependencies between activities has a significant impact on developing and managing a project schedule.

There are three basic reasons for creating dependencies among project activities:

- **Mandatory dependencies** are inherent in the nature of the work being done on a project. They are sometimes referred to as hard logic. For example, you cannot test code until after the code is written.
- **Discretionary dependencies** are defined by the project team. For example, a project team might follow good practice and not start the detailed design of a new information system until the users sign off on all of the analysis work. Discretionary dependencies are sometimes referred to as soft logic and should be used with care since they may limit later scheduling options.
- **External dependencies** involve relationships between project and non-project activities. The installation of a new operating system and other software may depend on delivery of new hardware from an external supplier. Even though the delivery of the new hardware may not be in the scope of the project, you should add an external dependency to it because late delivery will affect the project schedule.

As with activity definition, it is important that project stakeholders work together to define the activity dependencies that exist on their project. Some organizations have guidelines based on the activity dependencies of similar projects. Some organizations rely on the expertise of the people working on the project and their contacts with other employees and colleagues in the profession. Some people like to write each activity letter or name on a sticky note or some other moveable paper to determine dependencies or sequencing. Still others work directly in project management software to establish relationships. Just as it is easier to write a research paper by first having some thoughts on paper before typing into a word processor, it is usually easier to do some manual form of activity sequencing before entering the information into project management software.

Many organizations do not understand the importance of defining activity dependencies and do not use them *at all* in project time management. If you do not define the sequence of activities, you cannot use some of the most powerful schedule tools available to project managers: network diagrams and critical path analysis.

Network Diagrams

Network diagrams are the preferred technique for showing activity sequencing. A project **network diagram** is a schematic display of the logical relationships among, or sequencing of, project activities. Figure 6-2 shows a sample network diagram for Project X, which uses the arrow diagramming method (ADM) or activity-on-arrow (AOA) approach.

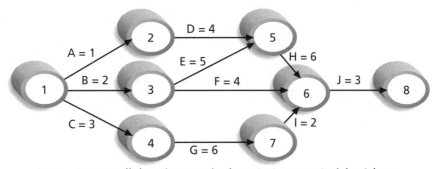

Note: Assume all durations are in days; A=1 means Activity A has a duration of 1 day.

Figure 6-2. Sample Activity-on-Arrow (AOA) Network Diagram for Project X

Note the main elements on this network diagram. The letters A, B, C, D, E, F, G, H, I, and J represent activities with dependencies that are required to complete the project. These activities come from the WBS and activity definition process described earlier. The arrows represent the activity sequencing or relationships between tasks. For example, Activity A must be done before Activity D; Activity D must be done before Activity H, and so on.

The format of this network diagram uses the **activity-on-arrow (AOA)** or the **arrow diagramming method (ADM)**—a network diagramming technique in which activities are represented by arrows and connected at points called nodes to illustrate the sequence of activities. A **node** is simply the starting and ending point of an activity. The first node signifies the start of a project, and the last node represents the end of a project.

Keep in mind that the network diagram represents activities that must be done to complete the project. It is not a race to get from the first node to the last node. Every activity on the project network diagram must be completed in order for the project to finish.

It is also important to note that not every single item on the WBS needs to be on the network diagram, especially on large projects. Sometimes it is enough to put summary tasks on a network diagram or to break the project down into several smaller network diagrams. Some items that the team knows must be done and will be done regardless of other activities do not need to be included in the network diagram.

Assuming you have a list of the project activities and their start and finish nodes, follow these steps to create an AOA network diagram:

1. Find all of the activities that start at Node 1. Draw their finish nodes, and draw arrows between Node 1 and each of those finish nodes. Put the activity letter or name on the associated arrow. If you have a duration estimate, write that next to the activity letter or name, as shown in Figure 6-2. For example, A = 1 means that the duration of Activity A is one day, week, or other standard unit of time. Also be sure to put arrowheads on all arrows to signify the direction of the relationships.

2. Continue drawing the network diagram, working from left to right. Look for bursts and merges. **Bursts** occur when two or more activities follow a single node. A **merge** occurs when two or more nodes precede a single node. For example, in Figure 6-2, Node 1 is a burst since it goes into Nodes 2, 3, and 4. Node 5 is a merge preceded by Nodes 2 and 3.

3. Continue drawing the project network diagram until all activities are included on the diagram.

4. As a rule of thumb, all arrowheads should face toward the right, and no arrows should cross on an AOA network diagram. You may need to redraw the diagram to make it look presentable.

Even though AOA or ADM network diagrams are generally easy to understand and create, a different method is more commonly used: the precedence diagramming method.

The **precedence diagramming method (PDM)** is a network diagramming technique in which boxes represent activities. It is particularly useful for visualizing certain types of time relationships.

Figure 6-3 illustrates the types of dependencies that can occur among project activities, based on a Microsoft Project Help screen. After you determine the reason for a dependency between activities (mandatory, discretionary, or external), you must determine the type of dependency. Note that the terms activity and task are used interchangeably, as are relationship and dependency. The four types of dependencies or relationships between activities include:

■ **Finish-to-start:** a relationship where the "from" activity must finish before the "to" activity can start. For example, you cannot provide user training until after software, or a new system, has been installed. Finish-to-start is the most common type of relationship, or dependency, and the AOA method uses only finish-to-start dependencies.

- **Start-to-start:** a relationship in which the "from" activity cannot start until the "to" activity is started. For example, on several information technology projects, a group of activities all start simultaneously, such as the many tasks that occur when a new system goes live.

- **Finish-to-finish:** a relationship where the "from" activity must be finished before the "to" activity can be finished. One task cannot finish before another finishes. For example, quality control efforts cannot finish before production finishes, although the two activities can be performed at the same time.

- **Start-to-finish:** a relationship where the "from" activity must start before the "to" activity can be finished. This type of relationship is rarely used.

Task dependencies

The nature of the relationship between two linked tasks. You link tasks by defining a dependency between their finish and start dates. For example, the "Contact caterers" task must finish before the start of the "Determine menus" task. There are four kinds of task dependencies in Microsoft Project:

Task dependency	Example	Description
Finish-to-start (FS)		Task (B) cannot start until task (A) finishes.
Start-to-start (SS)		Task (B) cannot start until task (A) starts.
Finish-to-finish (FF)		Task (B) cannot finish until task (A) finishes.
Start-to-finish (SF)		Task (B) cannot finish until task (A) starts.

Figure 6-3. Task Dependency Types

Figure 6-4 illustrates Project X using the precedence diagramming method. Notice that the activities are placed inside boxes, which represent the nodes on this diagram. Arrows and arrowheads show the relationships between activities. This figure was created using Microsoft Project 2002, and the application automatically places additional information inside each node. Each task box includes the start and finish date, labeled Start and Finish, the task ID number, labeled ID, the task's duration, labeled Dur, and the names of resources, if any, assigned to the task, labeled Res. The border of the boxes for tasks on the critical path appears automatically in red in the Project 2002 network diagram view. In Figure 6-4, the boxes for critical tasks have a thicker border and are not shaded.

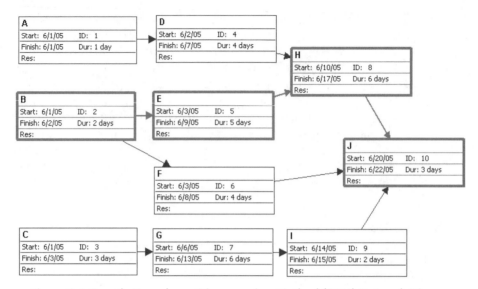

Figure 6-4. Sample Precedence Diagramming Method (PDM) Network Diagram for Project X

The precedence diagramming method is used more often than AOA network diagrams and offers a number of advantages over the AOA technique. First, most project management software uses the precedence diagramming method. Second, the precedence diagramming method avoids the need to use dummy activities. **Dummy activities** have no duration and no resources but are occasionally needed on AOA network diagrams to show logical relationships between activities. They are represented with dashed arrow lines, and you put in zero for the duration estimate. For example, looking back at Figure 6-2, activity A precedes activity D and activity B precedes activities E and F. If there was also a dependency between activity B and D, you would draw a dashed arrow line between node 3 and node 2. Third, the precedence diagramming method shows different dependencies among tasks, whereas AOA network diagrams use only finish-to-start dependencies. You will learn more about activity sequencing using Project 2002 in Appendix A.

ACTIVITY DURATION ESTIMATING

After working with key stakeholders to define activities and determine their sequence, the next process in project time management is to estimate the duration of activities. It is important to note that **duration** includes the actual amount of time worked on an activity *plus* elapsed time. For example, even though it might take one workweek or five workdays to do the actual work, the duration estimate might be two weeks to allow extra time needed to obtain outside information. The resources assigned to a task will also affect the task duration estimate. Do not confuse duration with **effort**, which is the number of workdays or work hours required to complete a task. A duration estimate of one day could be based on eight hours of work or eighty hours of work. Duration relates to the time estimate, not the effort estimate. Of course, the two are related, so project team members must document their assumptions when creating duration estimates and update the estimates as the project progresses. As in defining activities and their sequences, it is important for the project stakeholders to discuss activity duration estimates. The people who will actually do the work, in particular, should have a lot of say in these duration estimates, since they are the ones whose performance will be based on meeting them. If changes are made to the scope of the project, the duration estimates should be updated to reflect those changes. It is also helpful to review similar projects and seek the advice of experts in estimating activity durations.

There are several inputs to activity duration estimating. The detailed activity list and sequencing provide a basis for estimates. It is also important to review constraints and assumptions related to the estimates. Historical information is also useful. In addition to reviewing past project information, the team should also review the accuracy of the duration estimates thus far on the project. For example, if they find that all of their estimates have been much too long or short, the team should update the estimates to reflect what they have learned. One of the most important considerations in making activity duration estimates is the availability of resources, especially human resources. What specific skills do people need to do the work? What are the skill levels of the people assigned to the project? How many people are expected to be available to work on the project at any one time? (You will learn more about resources in Chapter 9, Project Human Resource Management.)

The outputs of activity duration estimating include duration estimates for each activity, a document describing the basis of the estimates, and updates to the WBS. Updates to the WBS can occur when stakeholders change requirements or when project team members decide to dissect certain activities further to produce more detailed duration estimates.

SCHEDULE DEVELOPMENT

Schedule development uses the results of all the preceding project time management processes to determine the start and end date of the project. There are often several iterations of all the project time management processes before a project schedule is finalized. The ultimate goal of schedule development is to create a realistic project schedule that provides a basis for monitoring project progress for the time dimension of the project.

Several tools and techniques assist in the schedule development process:

- A Gantt chart is a common tool for displaying project schedule information.
- Critical path analysis is a very important tool for developing and controlling project schedules.
- Critical chain scheduling is a technique that accounts for resource constraints and uses project and feeding buffers.
- PERT analysis is a means for evaluating schedule risk on projects.

The following sections provide samples of each of these tools and techniques and a discussion of their advantages and disadvantages.

Gantt Charts

Gantt charts provide a standard format for displaying project schedule information by listing project activities and their corresponding start and finish dates in a calendar format. As mentioned in Chapter 1, Henry Gantt developed the first Gantt chart during World War I for scheduling work in factories. Early versions simply listed project activities or tasks in one column to the left, calendar time units such as months to the right, and horizontal bars under the calendar units to illustrate when activities should start and end. However, Gantt charts normally do not show relationships among project activities, as network diagrams do. Today, most people use project management software to create versions of Gantt charts that are more sophisticated and allow for easy updates of information. They can also show dependencies in some format using project management software.

Figure 6-5 shows a simple Gantt chart for Project X. Figure 6-6 shows a Gantt chart that is more sophisticated based on a software launch project. Recall that the activities on the Gantt chart should coincide with the activities on the WBS. Notice that the software launch project's Gantt chart contains milestones, summary tasks, individual task durations, and arrows showing task dependencies.

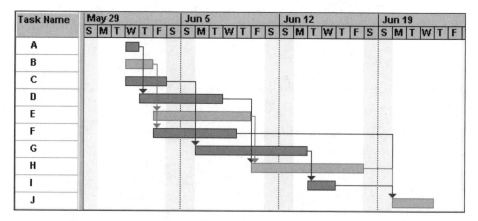

Figure 6-5. Gantt Chart for Project X

Notice the different symbols on the software launch project's Gantt chart (Figure 6-6):

■ The black diamond symbol represents a **milestone**—a significant event on a project with zero duration. In Figure 6-6, Task 1, "Marketing Plan distributed," is a milestone that occurs on March 17. Tasks 3, 4, 8, 9, 14, 25, 27, 43, and 45 are also milestones. For very large projects, top managers might want to see only milestones on a Gantt chart. Project 2002 allows you to filter information displayed on a Gantt chart so you can easily show specific tasks, such as milestones.

■ The thick black bars with arrows at the beginning and end represent summary tasks. For example, Activities 12 through 15—"Develop creative briefs," "Develop concepts," "Creative concepts," and "Ad development"—are all subtasks of the summary task called Advertising, Task 11. WBS activities are referred to as tasks and subtasks in most project management software.

■ The light gray horizontal bars such as those found in Figure 6-6 for tasks 5, 6, 7, 10, 12, 13, 15, 22, 24, 26, and 44, represent the duration of each individual task. For example, the light gray bar for Subtask 5, "Packaging," starts in mid-February and extends until early May.

■ Arrows connecting these symbols show relationships or dependencies between tasks. Gantt charts often do not show dependencies, which is their major disadvantage. If dependencies have been established in Project 2002, they are automatically displayed on the Gantt chart.

Figure 6-6. Gantt Chart for Software Launch Project

Adding Milestones to Gantt Charts

Milestones can be a particularly important part of schedules, especially for large projects. Many people like to focus on meeting milestones, so you can create milestones to emphasize important events or accomplishments on projects. Normally you create milestones by entering tasks with zero duration. In Project 2002, you can also mark any task as a milestone by checking the appropriate box in the Advanced tab of the Task Information dialog box. The duration of the task will not change to zero, but the Gantt chart will show the milestone symbol to represent that task. See Appendix A for more information.

To make milestones meaningful, some people use the SMART criteria to help define them. The **SMART criteria** are guidelines suggesting that milestones should be:

- **S**pecific
- **M**easurable
- **A**ssignable
- **R**ealistic
- **T**ime-framed

For example, distributing a marketing plan is specific, measurable, and assignable if everyone knows what should be in the marketing plan, how it should be distributed, how many copies should be distributed and to whom, and who is responsible for the actual delivery. Distributing the marketing plan is realistic and able to be time-framed if it is an achievable event and scheduled at an appropriate time.

Using Tracking Gantt Charts to Compare Planned and Actual Dates

You can use a special form of a Gantt chart to evaluate progress on a project by showing actual schedule information. Figure 6-7 shows a **tracking Gantt chart**—a Gantt chart that compares planned and actual project schedule information. The planned schedule dates for activities are called the **baseline dates**. The tracking Gantt chart includes columns (hidden in Figure 6-7) labeled "Start" and "Finish" to represent actual start and finish dates for each task, as well as columns labeled "Baseline Start" and "Baseline Finish" to represent planned start and finish dates for each task. In this example, the project is completed, but several tasks missed their planned start and finish dates.

Planned dates

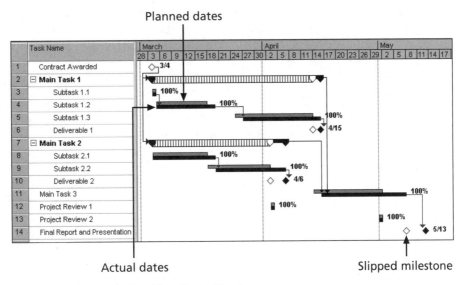

Actual dates   Slipped milestone

Figure 6-7. Sample Tracking Gantt Chart

To serve as a progress evaluation tool, a tracking Gantt chart uses a few additional symbols.

- Notice that the Gantt chart in Figure 6-7 often shows two horizontal bars for tasks. The top horizontal bar represents the planned or baseline duration for each task. The bar below it represents the actual duration. Subtasks 1.2 and 1.3 illustrate this type of display. If these two bars are the same length and start and end on the same date, then the actual schedule was the same as the planned schedule for that task. This scheduling occurred for Subtask 1.1, where the task started and ended as planned on March 4. If the bars do not start and end on the same date, then the actual schedule differed from the planned or baseline schedule. If the top horizontal bar is shorter than the bottom one, the task took longer than planned, as you can see for Subtask 1.2. If the top horizontal bar is longer than the bottom one, the task took less time than planned. A striped horizontal bar, illustrated by Main Task 1, represents the planned duration for summary tasks. The black bar adjoining it shows progress for summary tasks. For example, Main Task 1 shows the actual duration took three days longer than what was planned.

- A white diamond on the tracking Gantt chart represents a **slipped milestone**. A slipped milestone means the milestone activity was actually completed later than originally planned. For example, the last task provides an example of a slipped milestone since the final report and presentation were completed later than planned.

- Percentages to the right of the horizontal bars display the percentage of work completed for each task. For example, 100 percent means the task is finished. Fifty percent means the task is still in progress and is 50 percent completed.

- In the columns to the left of the tracking Gantt chart, you can display baseline and actual start and finish dates.

A tracking Gantt chart is based on the percentage of work completed for project tasks or the actual start and finish dates. It allows the project manager to monitor schedule progress on individual tasks and the whole project. For example, Figure 6-7 shows that this project is completed. It started on time, but it finished a little late, on May 13 (5/13) versus May 8.

The main advantage of using Gantt charts is that they provide a standard format for displaying planned and actual project schedule information. In addition, they are easy to create and understand. The main disadvantage of Gantt charts is that they do not usually show relationships or dependencies between tasks. If Gantt charts are created using project management software and tasks are linked, then the dependencies will be displayed, but not as clearly as they would be displayed on a network diagram.

Critical Path Method

Many projects fail to meet schedule expectations. **Critical path method (CPM)**—also called **critical path analysis**—is a project network analysis technique used to predict total project duration. This important tool will help you combat project schedule overruns. A **critical path** for a project is the series of activities that determine the earliest time by which the project can be completed. It is the longest path through the network diagram and has the least amount of slack or float. **Slack** or **float** is the amount of time an activity may be delayed without delaying a succeeding activity or the project finish date.

Calculating the Critical Path

To find the critical path for a project, you must first develop a good network diagram, which, in turn, requires a good activity list based on the WBS. Once you create a network diagram, you must also estimate the duration of each activity to determine the critical path. Calculating the critical path involves adding the durations for all activities on each path through the network diagram. The longest path is the critical path.

Figure 6-8 shows the AOA network diagram for Project X again. Note that you can use either the AOA or precedence diagramming method to determine the critical path on projects. Figure 6-8 shows all of the paths—a total of four— through the network diagram. Note that each path starts at the first node (1) and ends at the last node (8) on the AOA network diagram. This figure also shows the length or total duration of each path through the network diagram. These lengths are computed by adding the durations of each activity on the path. Since path B-E-H-J at 16 days has the longest duration, it is the critical path for the project.

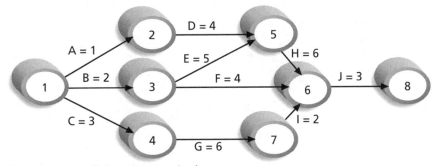

Note: Assume all durations are in days.

Path 1:	A-D-H-J	Length = 1+4+6+3 = 14 days
Path 2:	B-E-H-J	Length = 2+5+6+3 = 16 days
Path 3:	B-F-J	Length = 2+4+3 = 9 days
Path 4:	C-G-I-J	Length = 3+6+2+3 = 14 days

Since the critical path is the longest path through the network diagram, Path 2, B-E-H-J, is the critical path for Project X.

Figure 6-8. Determining the Critical Path for Project X

What does the critical path really mean? The critical path shows the shortest time in which a project can be completed. Even though the critical path is the *longest* path, it represents the *shortest* time it takes to complete a project. If one or more of the activities on the critical path takes longer than planned, the whole project schedule will slip *unless* the project manager takes corrective action.

Growing Grass Can Be on the Critical Path

People are often confused about what the critical path is for a project or what it really means. Some people think the critical path includes the most critical activities. However, the critical path is concerned only with the time dimension of a project. The fact that its name includes the word *critical* does not mean that it includes all critical activities. For example, Frank Addeman, Executive Project Director at Walt Disney Imagineering, explained in a keynote address at the May 2000 PMI-ISSIG Professional Development Seminar that growing grass was on the critical path for building Disney's Animal Kingdom theme park! This 500-acre park required special grass for its animal inhabitants, and some of the grass took years to grow. Another misconception is that the critical path is the shortest path through the network diagram. In some areas, such as transportation modeling, similar network diagrams are drawn in which identifying the shortest path is the goal. For a project, however, each activity must be done in order to complete the project. It is not a matter of choosing the shortest path.

Other aspects of critical path analysis may cause confusion. Can there be more than one critical path on a project? Does the critical path ever change? In the Project X example, suppose that Activity A has a duration estimate of three days instead of one day. This new duration estimate would make the length of Path 1 equal to sixteen days. Now the project has two longest paths of equal duration, so there are two critical paths. Therefore, there *can* be more than one critical path on a project. Project managers should closely monitor performance of activities on the critical path to avoid late project completion. If there is more than one critical path, project managers must keep their eyes on all of them.

The critical path on a project can change as the project progresses. For example, suppose everything is going as planned at the beginning of the project. In this example, suppose Activities A, B, C, D, E, F, and G all start and finish as planned. Then suppose Activity I runs into problems. If Activity I takes more than four days, it will cause path C-G-I-J to be longer than the other paths, assuming they progress as planned. This change would cause path C-G-I-J to become the new critical path. Therefore, the critical path can change on a project.

Using Critical Path Analysis to Make Schedule Trade-Offs

It is important to know what the critical path is throughout the life of a project so the product manager *can* make trade-offs. If the project manager knows that one of the tasks on the critical path is behind schedule, he or she needs to decide what to do about it. Should the schedule be renegotiated with stakeholders? Should more resources be allocated to other items on the critical path to make up for that time? Is it okay if the project finishes behind schedule? By keeping track of the critical path, the project manager and his or her team take a proactive role in managing the project schedule.

A technique that can help project managers make schedule trade-offs is determining the free slack and total slack for each project activity. **Free slack** or **free float** is the amount of time an activity can be delayed without delaying the early start date of any immediately following activities. The **early start date** for an activity is the earliest possible time an activity can start based on the project network logic. **Total slack** or **total float** is the amount of time an activity can be delayed from its early start without delaying the planned project finish date.

Product managers calculate free slack and total slack by doing a forward and backward pass through a network diagram. A **forward pass** determines the early start and early finish dates for each activity. The **early finish date** for an activity is the earliest possible time an activity can finish based on the project network logic. The project start date is equal to the early start date for the first network diagram activity. Early start plus the duration of the first activity is equal to the early finish date of the first activity. It is also equal to the early start date of each subsequent activity. When an activity has multiple predecessors, its early start date is the latest of the early finish dates of those predecessors. For example, Tasks D and E immediately precede Task H in Figure 6-8. The early start date for Task H, therefore, is the early finish date of Task E, since it occurs later than the early finish date of Task D. A **backward pass** through the network diagram determines the late start and late finish dates for each activity in a similar fashion. The **late start date** for an activity is the latest possible time an activity might begin without delaying the project finish date. The **late finish date** for an activity is the latest possible time an activity can be completed without delaying the project finish date.

Project managers can determine the early and late start and finish dates of each activity by hand. For example, Figure 6-9 shows a simple network diagram with three tasks, A, B, and C. Tasks A and B both precede Task C. Assume all duration estimates are in days. Task A has an estimated duration of 5 days, Task B has an estimated duration of 10 days, and Task C has an estimated duration of 7 days. There are only two paths through this small network diagram: path A-C has a duration of 12 days (5+7), and path B-C has a duration of 17 days (10+7). Since path B-C is longest, it is the critical path. There is no float or slack on this path, so the early and late start and finish dates are the same. However, Task A has 5 days of float or slack. Its early start date is day 0, and its late start date is day 5. Its early finish date is day 5, and its late finish date is day 10. Both the free and total float amounts for Task A are the same at 5 days.

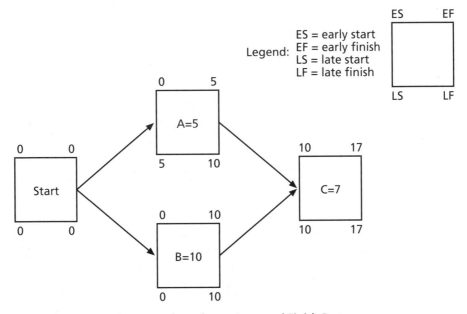

Figure 6-9. Calculating Early and Late Start and Finish Dates

A much faster and easier way to determine early and late start and finish dates and free and total slack amounts for activities is by using project management software. Table 6-1 shows the free and total slack for all activities on the network diagram for Project X. The data in this table was created by selecting the Schedule Table view in Project 2002. Knowing the amount of float or slack allows project managers to know whether the schedule is flexible and how flexible it might be. For example, at 7 days (7d), Task F has the most free and total slack. The most slack on any other activity is only 2 days (2d). Understanding how to create and use slack information provides a basis for negotiating (or not negotiating) project schedules.

Table 6-1: Free and Total Float or Slack for Project X

TASK	START	FINISH	LATE START	LATE FINISH	FREE SLACK	TOTAL SLACK
A	6/2/05	6/2/05	6/4/05	6/4/05	0d	2d
B	6/2/05	6/3/05	6/2/05	6/3/05	0d	0d
C	6/2/05	6/4/05	6/4/05	6/6/05	0d	2d
D	6/3/05	6/6/05	6/5/05	6/10/05	2d	2d
E	6/4/05	6/10/05	6/4/05	6/10/05	0d	0d
F	6/4/05	6/9/05	6/13/05	6/18/05	7d	7d
G	6/5/05	6/12/05	6/9/05	6/16/05	0d	2d
H	6/11/05	6/18/05	6/11/05	6/18/05	0d	0d
I	6/13/05	6/16/05	6/17/05	6/18/05	2d	2d
J	6/19/05	6/23/05	6/19/05	6/23/05	0d	0d

Using the Critical Path to Shorten a Project Schedule

It is common for stakeholders to want to shorten a project schedule estimate. By knowing the critical path, the project manager and his or her team can use several duration compression techniques to shorten the project schedule. One technique is to reduce the duration of activities on the critical path. The project manager can shorten the duration of critical path activities by allocating more resources to those activities or by changing their scope.

Recall that Sue Johnson in the opening case was having schedule problems with the online registration project because several users missed important project review meetings and one of the senior programmers quit. If Sue and her team created a realistic project schedule, produced accurate duration estimates, and established dependencies between tasks, they could analyze their status in terms of meeting the May 1 deadline. If some activities on the critical path had already slipped and they did not build in extra time at the end of the project, then they would have to take corrective actions to finish the project on time. Sue could request that her company or the college provide more people to work on the project in an effort to make up time. She could also request that the scope of activities be reduced to complete the project on time. Sue could also use project time management techniques, such as crashing or fast tracking, to shorten the project schedule.

Crashing is a technique for making cost and schedule trade-offs to obtain the greatest amount of schedule compression for the least incremental cost. For example, suppose one of the items on the critical path for the online registration project was entering course data for the fall term into the new system. If this task is yet to be done and was originally estimated to take two weeks based on the college providing one part-time data entry clerk, Sue could suggest that the college have the clerk work full-time to finish the task in one week instead of two. This change would not cost Sue's company more money, and it could shorten the project end date by one week. If the college could not meet this request, Sue could consider hiring a temporary data entry person for one week to help get the task done faster. By focusing on tasks on the critical path that could be done more quickly for no extra cost or a small cost, the project schedule can be shortened.

The main advantage of crashing is shortening the time it takes to finish a project. The main disadvantage of crashing is that it often increases total project costs. You will learn more about project costs in Chapter 7, Project Cost Management.

Another technique for shortening a project schedule is fast tracking. **Fast tracking** involves doing activities in parallel that you would normally do in sequence. For example, Sue Johnson's project team may have planned not to start any of the coding for the online registration system until *all* of the analysis was done. Instead, they could consider starting some coding activity before the analysis is completed.

The main advantage of fast tracking, like crashing, is that it can shorten the time it takes to finish a project. The main disadvantage of fast tracking is that it can end up lengthening the project schedule since starting some tasks too soon often increases project risk and results in rework.

What Went Wrong?

Even though many information technology projects seem to be rushed, some project managers have realized that a slower approach is required for some projects to be successful. For example, project managers continue to face considerable problems with Customer Relationship Management (CRM) software. Scott Nelson, a Gartner Inc. analyst, said that based on Gartner's survey results last year, about 55 percent of the respondents characterized their CRM rollouts as failures. In fact, 42 percent of the CRM end-user licenses bought last year are not even being used.

Why are CRM rollouts so troublesome? Most people say that adoption of the tools by end users can be very slow. End users are interested in the new capabilities, but the technology "causes fear", said Shawn Kaplan, director of marketing and business development at New York-based financial data provider Reuters America, Inc. Reuters rolled out Siebel Systems, Inc.'s CRM applications internally and plans to provide the functionality to financial advisers who use Reuters data by using Web services integration technology. That way, financial advisers will not even know they are using the system, according to Kaplan. Fairchild Semiconductor International, Inc. plans to take up to three years to complete its CRM project. American Trans Air, Inc. (ATA) also started small with its CRM software. "We have been taking baby steps," noted Robert Ellison, director of e-business and network services. "In fact, it was probably more like a crawl."[3]

Importance of Updating Critical Path Data

In addition to finding the critical path at the beginning of a project, it is important to update the schedule with actual data. After the project team completes activities, the project manager should document the actual durations of those activities. He or she should also document revised estimates for activities in progress or yet to be started. These revisions often cause a project's critical path to change, resulting in a new estimated completion date for the project. Again, proactive project managers and their teams stay on top of changes so they can make informed decisions and keep stakeholders informed of, and involved in, major project decisions.

Critical Chain Scheduling

Another technique that addresses the challenge of meeting or beating project finish dates is an application of the theory of constraints called critical chain scheduling. The Theory of Constraints is a management philosophy developed by Eliyahu M. Goldratt and introduced in his book *The Goal*. The **Theory of Constraints (TOC)** is based on the fact that, like a chain with its weakest link, any complex system at any point in time often has only one aspect or constraint that limits its ability to achieve more of its goal. For the system to attain any significant improvements, that constraint must be identified, and the whole system must be managed with it in mind. **Critical chain scheduling** is a method of scheduling that considers limited resources when creating a project schedule and includes buffers to protect the project completion date.

You can find the critical path for a project without considering resource allocation. For example, task duration estimates and dependencies can be made without considering the availability of resources. In contrast, an important concept in critical chain scheduling is the availability of resources. If a particular resource is needed full time to complete two tasks that were originally planned to occur simultaneously, critical chain scheduling acknowledges that you must either delay one of those tasks until the resource is available or find another

[3] Songini, Marc L. "CRM Projects Continue to Inspire Caution, Users Say," *Computerworld* (3/10/03).

resource. In this case, accounting for limited resources often extends the project finish date, which is not most people's intent. Other important concepts related to critical chain scheduling include multitasking and time buffers.

Multitasking occurs when a resource works on more than one task at a time. This situation occurs frequently on projects. People are assigned to multiple tasks within the same project or different tasks on multiple projects. For example, suppose someone is working on three different tasks, Task 1, Task 2, and Task 3, for three different projects, and each task takes ten days to complete. If the person did not multitask, and instead completed each task sequentially, starting with Task 1, then Task 1 would be completed after day ten, Task 2 would be completed after day twenty, and Task 3 would be completed after day thirty, as shown in Figure 6-10a. However, because many people in this situation try to please all three people who need their tasks completed, they often work on the first task for some time, then the second, then the third, then go back to the first task, and so on, as shown in Figure 6-10b. In this example, the tasks were all half-done one at a time, then completed one at a time. Task 1 is now completed at the end of day twenty instead of day ten, Task 2 is completed at the end of day twenty-five instead of day twenty, and Task 3 is still completed on day thirty. This example illustrates how multitasking can delay task completions. Multitasking also often involves wasted setup time, which increases total duration.

Figure 6-10a. Three Tasks Without Multitasking

Figure 6-10b. Three Tasks With Multitasking

Critical chain scheduling assumes that resources do not multitask. Someone cannot be assigned to two tasks simultaneously on the same project, when critical chain scheduling is in effect. Likewise, critical chain theory suggests that projects be prioritized so people working on more than one project at a time know which tasks take priority. Preventing multitasking avoids resource conflicts and wasted setup time caused by shifting between multiple tasks over time.

An essential concept to improving project finish dates with critical chain scheduling is to change the way people make task estimates. Many people add a safety or **buffer**, which is additional time to complete a task, to an estimate to account for various factors. These factors include the negative effects of multitasking, distractions and interruptions, fear that estimates will be reduced, Murphy's Law, and so on. **Murphy's Law** states that if something can go wrong, it will. Critical chain scheduling removes buffers from individual tasks and instead creates a **project buffer**, which is additional time added before the project's due date. Critical chain scheduling also protects tasks on the critical chain from being delayed by using **feeding buffers**, which are additional time added before tasks on the critical chain that are preceded by non-critical-path tasks.

Figure 6-11 provides an example of a network diagram constructed using critical chain scheduling. Note that the critical chain accounts for a limited resource, X, and the schedule includes use of feeding buffers and a project buffer in the network diagram. The tasks marked with an X are part of the critical chain, which can be interpreted as being the critical path using this technique. The task estimates in critical chain scheduling should be shorter than traditional estimates because they do not include their own buffers. Not having task buffers should mean less occurrence of **Parkinson's Law**, which states that work expands to fill the time allowed. The feeding and project buffers protect the date that really needs to be met—the project completion date.

X = Tasks done by limited resource
FB = Feeding buffer

Figure 6-11. Example of Critical Chain Scheduling [4]

[4] Goldratt, Eliyahu, *Critical Chain*, Great Barrington, MA: The North River Press, 1997, 218.

Several companies have reported successes with critical chain scheduling:

■ Lucent Technology's Outside Plant Fiber Optic Cable Business Unit used critical chain scheduling to reduce its product introduction interval by 50 percent, improve on-time delivery, and increase the organization's capacity to develop products.

■ Synergis Technologies Group successfully implemented critical chain scheduling to manage more than 200 concurrent projects in nine locations, making on-time delivery their top priority.

■ The Antarctic Support Associates project team switched to critical chain scheduling to enable the Laurence M. Gould research vessel to pass sea-trial tests and be ready to embark on its voyage to Antarctica on schedule in January 1998, rather than four months late as had been anticipated.[5]

■ The U.S. health care industry is learning to think differently by using the Theory of Constraints. "Consider a relatively simple system of a physician's office or clinic. The steps in the process could be patients checking in, filling out forms, having vital signs taken by a nurse, seeing the physician, seeing the nurse for a prescribed procedure such as vaccination, and so forth. These steps could take place in a simple linear sequence or chain...this process or chain only can produce an average of eight per hour. The chain is only as strong as its weakest link and the rate of the slowest resource in this example, the weakest link, is eight. This is true regardless of how fast each of the other resources can process individually or how much work is stuffed into the pipeline, how complex the process or set of interconnected processes is to complete. Moreover, improving the performance of any link besides the constraint does nothing to improve the system as a whole."[6]

As you can see, critical chain scheduling is a fairly complicated yet powerful tool that involves critical path analysis, resource constraints, and changes in how task estimates are made in terms of buffers. Some consider critical chain scheduling one of the most important new concepts in the field of project management.

Program Evaluation and Review Technique (PERT)

Another project time management technique is the **Program Evaluation and Review Technique (PERT)**—a network analysis technique used to estimate project duration when there is a high degree of uncertainty about the individual activity duration estimates. PERT applies the critical path method to a weighted average duration estimate.

[5] Avraham Y. Goldratt Institute Web site, *www.goldratt.com* (January 2001).

[6] Breen, Anne M., Tracey Burton-Houle, and David C. Aron, "Applying the Theory of Constraints in Health Care: Part 1 - The Philosophy," *Quality Management in Health Care*, Volume 10, Number 3 (Spring 2002) (*www.goldratt.com/for-cause/applyingtocinhcpt1fco.htm*).

PERT uses **probabilistic time estimates**—duration estimates based on using optimistic, most likely, and pessimistic estimates of activity durations—instead of one specific or discrete duration estimate. Like the critical path method, PERT is based on a network diagram, normally the precedence diagramming method. To use PERT, you calculate a weighted average for the duration estimate of each project activity using the following formula:

PERT weighted average = $\dfrac{\text{optimistic time} + 4 \times \text{most likely time} + \text{pessimistic time}}{6}$

Suppose Sue Johnson's project team in the opening case used PERT to determine the schedule for the online registration system project. They would have to collect numbers for the optimistic, most likely, and pessimistic duration estimates for each project activity. Suppose one of the activities was to design an input screen for the system. Someone might estimate that it would take about two weeks or 10 workdays to do this activity. Without using PERT, the duration estimate for that activity would be 10 workdays. Using PERT, the project team would also need to estimate the pessimistic and optimistic times for completing this activity. Suppose an optimistic estimate is that the input screen can be designed in 8 workdays, and a pessimistic time estimate is 24 workdays. Applying the PERT formula, you get the following:

PERT weighted average = $\dfrac{\text{8 workdays} + 4 \times \text{10 workdays} + \text{24 workdays}}{6}$ = 12 workdays

Instead of using the most likely duration estimate of 10 workdays, the project team would use 12 workdays when doing critical path analysis.

The main advantage of PERT is that it attempts to address the risk associated with duration estimates. PERT has three disadvantages: it involves more work since it requires several duration estimates, there are better probabilistic methods for assessing risk (See Chapter 11, Project Risk Management), and it is rarely used in practice. In fact, many people confuse PERT with network diagrams since the latter are often referred to as PERT charts.

CONTROLLING CHANGES TO THE PROJECT SCHEDULE

There are many issues involved in controlling changes to project schedules. It is important first to ensure that the project schedule is realistic. Many projects, especially in information technology, have very unrealistic schedule expectations. It is also important to use discipline and leadership to emphasize the importance of following and meeting project schedules. Although the various tools and techniques assist in developing and managing project schedules, project managers must manage several people-related issues to keep projects on track. "Most projects fail because of people issues, not from failure to draw a good PERT chart."[7] Project managers can perform a number of reality checks that will help them manage changes to project schedules. Several skills help project managers control schedule changes related to people issues.

Reality Checks on Scheduling

One of the first reality checks a project manager should make is to review the draft schedule usually included in the project charter. Although this draft schedule might include only a project start and end date, the project charter sets some initial schedule expectations for the project. Next, the project manager and his or her team should prepare a more detailed schedule and get stakeholders' approval. To establish the schedule, it is critical to get involvement and commitment from all project team members, top management, the customer, and other key stakeholders.

Several experts have written about the lack of realistic schedule estimates for information technology projects. Ed Yourdon, a well-known and respected software development expert, describes "death march" projects as projects doomed for failure from the start, due to unrealistic expectations, especially for meeting time constraints.[8] Many schedules for embedded software systems are "propped up by prayer and hopeful wishes."[9] Yielding to pressure from top management or other groups, such as marketing, to determine project schedules is nothing short of lying.[10] Experts in chaos theory suggest that project managers must account for the complex behavior of the organization by arranging for additional resources to be committed to the project. Using an analogy of how to avoid traffic jams on highways, they suggest that project managers schedule resources on the project so that no single resource is utilized more than 75 percent.[11] It is very important, therefore, to set realistic project schedules and allow for contingencies throughout the life of a project.

Another type of reality check comes from progress meetings with stakeholders. The project manager is responsible for keeping the project on track, and key stakeholders like to stay informed, often through high-level periodic reviews. Managers like to see progress made on projects approximately each month. Project managers often illustrate progress with a tracking Gantt chart showing key deliverables and activities. The project manager needs to understand the schedule and why activities are or are not on track and take a proactive approach to meeting stakeholder expectations. Top management hates

[7] Bolton, Bart, "IS Leadership," *Computerworld* (May 19, 1997).

[8] Yourdon, Ed, "Surviving a Death March Project," *Software Development* (July 1997).

[9] Ganssle, Jack G., "Lies, Damn Lies, and Schedule," *Embedded Systems Programming* (December 1997).

[10] Ibid.

[11] Monson, Robert, *The Role of Complexity and Chaos in Project Management.* Working paper (February 1999).

surprises, so it is very important for the project manager to be clear and honest in communicating project status. By no means should project managers create the illusion that the project is going fine when, in fact, it is having serious problems. When serious conflicts arise that could affect the project schedule, the project manager must alert top management and work with them to resolve the conflicts.

Working with People Issues

As in any endeavor involving many people, it is often not the technical aspects that are most difficult but rather the people. Creating a good network diagram and detailed project schedule are significant skills for project managers, but good project managers realize that their main job is to lead the people involved in the project. Depending on the size of the project, many project managers have one or more staff members responsible for coordinating input from many other people to create and update the project schedule. Delegating the details of a project schedule allows the project manager to focus on the big picture and lead the entire project, ensuring it remains on schedule. Several leadership skills that help project managers control schedule changes include:

■ Empowerment
■ Incentives
■ Discipline
■ Negotiation

It is important for the project manager to empower project team members to take responsibility for their activities. Having team members help create a detailed schedule and provide timely status information empowers them to take responsibility for their actions. As a result, they should feel more committed to the project.

The project manager can also use financial or other incentives to encourage people to meet schedule expectations. Alternatively, it is sometimes effective to use coercive power or negative incentives to stop people from missing deadlines. For example, one project manager started charging her team members twenty-five cents every time they were late in submitting their weekly time sheets for a project. She was amazed at how few people were late after implementing that simple policy.

Project managers must also use discipline to control project schedules. Several information technology project managers have discovered that setting firm dates for key project milestones helps minimize schedule changes. It is very easy for scope creep to raise its ugly head on information technology projects. Insisting that important schedule dates be met and that proper planning and analysis be completed up front helps everyone focus on doing what is most important for the project. This discipline results in meeting project schedules.

What Went Right?

Chris Higgins used the discipline he learned in the Army to transform project management into a cultural force at Bank of America. Higgins learned that taking time on the front end of a project could save significant time and money on the back end. As a quartermaster in the Army, when Higgins's people had to pack tents, he devised a contest to find the best way to fold a tent and determine the precise spots to place the pegs and equipment for the quickest possible assembly. Higgins used the same approach when he led an interstate banking initiative to integrate incompatible check processing, checking account, and savings account platforms in various states.

Law mandated that the banks solve the problem in one year or less. Higgins's project team was pushing to get to the coding phase of the project quickly, but Higgins held them back. He made the team members analyze, plan, and document requirements for the system in such detail that it took six months just to complete that phase. However, the discipline up front enabled the software developers on the team to do all of the coding in only three months, and the project was completed on time.[12]

Project managers and their team members also need to practice good negotiation skills. Customers and management often pressure people to shorten project schedules. Some people blame information technology schedule overruns on poor estimating, but others state that the real problem is that software developers and other information technology professionals are terrible at defending their estimates. Unlike the general population, most information technology professionals are introverts and have a difficult time holding their own when extroverts in marketing, sales, or top management tell them to shorten their schedule estimates. Many information technology professionals are also younger than the stakeholders pushing for shorter schedules, making them less apt to question senior members' opinions. It is important, therefore, for the project manager and team members to defend their estimates and learn to negotiate with demanding stakeholders.

[12] Melymuke, Kathleen, "Spit and Polish," *Computerworld* (February 16, 1998).

USING SOFTWARE TO ASSIST IN PROJECT TIME MANAGEMENT

Several types of software are available to assist in project time management. Software for facilitating communications helps project managers exchange schedule-related information with project stakeholders. Decision support models can help project managers analyze various trade-offs that can be made related to schedule issues. However, project management software, such as Microsoft Project 2002, was designed specifically for performing project management tasks. You can use project management software to draw network diagrams, determine the critical path for a project, create Gantt charts, and report, view, and filter specific project time management information.

Many projects involve hundreds of tasks with complicated dependencies. After you enter the necessary information, project management software automatically generates a network diagram and calculates the critical path(s) for the project. It also highlights the critical path in red on the network diagram. Project management software also calculates the free and total float or slack for all activities. For example, the data in Table 6-1 was created using Project 2002 by selecting a particular view from the menu bar. Using project management software eliminates the need to manually perform cumbersome calculations and allows for "what if" analysis as activity duration estimates or dependencies change. Recall that knowing which activities have the most slack gives the project manager an opportunity to reallocate resources or make other changes to compress the schedule or help keep it on track.

Project 2002 easily creates Gantt charts and tracking Gantt charts, which make tracking actual schedule performance versus the planned or baseline schedule much easier to do. It is important, however, to enter actual schedule information in a timely manner in order to benefit from using the Tracking Gantt chart feature. Some organizations use e-mail or other communications software to send up-to-date task and schedule information to the person responsible for updating the schedule. He or she can then quickly authorize these updates and enter them directly into the project management software. This process provides an accurate and up-to-date project schedule in Gantt chart form.

Project 2002 also includes many built-in reports, views, and filters to assist in project time management. Table 6-2 lists some of these features. For example, a project manager can quickly run a report to list all tasks that are to start soon. He or she could then send out a reminder to the people responsible for these tasks. If the project manager were presenting project schedule information to top management, he or she could create a Gantt chart showing only summary tasks or milestones. You can also create custom reports, views, tables, and filters. You will learn more about the project time management features of Project 2002 in Appendix A.

Table 6-2: Project 2002 Features Related to Project Time Management

REPORTS	VIEWS AND TABLE VIEWS	FILTERS
Overview reports: critical tasks and milestones	Gantt chart, network diagram, Tracking Gantt, schedule, tracking, variance, constraint dates, and delay	All tasks, completed critical tasks, incomplete tasks, and milestone tasks
Current activities reports: unstarted tasks, tasks starting soon, tasks in progress, completed tasks, should have started tasks, and slipping tasks		
Assignment reports: who does what and when		

Words of Caution on Using Project Management Software

Many people misuse project management software because they do not understand the concepts behind creating a network diagram, determining the critical path, or setting a schedule baseline. They might also rely too heavily on sample files or templates in developing their own project schedules. Understanding the underlying concepts (even being able to work with the tools manually) is critical to successful use of project management software, as is understanding the specific needs of your project.

Many top managers, including software professionals, have made blatant errors using various versions of Microsoft Project and similar tools. For example, one top manager did not know about establishing dependencies among project activities and entered every single start and end date for hundreds of activities. When asked what would happen if the project started a week or two late, she responded that she would have to reenter all of the dates.

This manager did not understand the importance of establishing relationships among the tasks. Establishing relationships among tasks allows the software to update formulas automatically when the inputs change. If the project start date slips by one week, the project management software will update all the other dates automatically, as long as they are not hard-coded into the software. (Hard-coding involves entering all activity dates manually instead of letting the software calculate them based on durations and relationships.) If one activity cannot start before another ends, and the first activity's actual start date is two days late, the start date for the succeeding activity will automatically be moved back two days. To achieve this type of functionality, tasks that have relationships must be linked in project management software.

Another top manager on a large information technology project did not know about setting a baseline in Microsoft Project. He spent almost one day every week copying and pasting information from Microsoft Project into a spreadsheet and using complicated "IF" statements to figure out what activities were behind schedule. He had never received any training on Microsoft Project and did not know about many of its capabilities. To use any software effectively, users must have adequate training in the software and an understanding of its underlying concepts.

Many project management software programs also come with templates or sample files. It is very easy to use these files without considering unique project needs. For example, a project manager for a software development project can use the Project 2002 Software Development template file, the files containing information from similar projects done in the past, or purchase sample files from other companies. All of these files include suggested tasks, durations, and relationships. There are benefits to using templates or sample files, such as less setup time and a reality check if the project manager has never managed this type of project before. However, there are also drawbacks to this approach. There are many assumptions in these template files that may not apply to the project, such as a design phase that takes three months to complete or the performance of certain types of testing. Project managers and their teams may overrely on templates or sample files and ignore unique concerns for their particular projects.

As you can see, there are many terms, concepts, tools, and techniques related to project time management. It is very important for project managers and their teams to develop realistic schedules and apply good project time management to complete projects within schedule constraints.

CASE WRAP-UP

It was now March 15, just a month and a half before the new online registration system was supposed to go live. The project was in total chaos. Sue Johnson thought she could handle all of the conflicts that kept appearing on the project, and she was too proud to admit to her top management or the college president that things were not going well. She spent a lot of time preparing a detailed schedule for the project, and she thought she was using their project management software well enough to keep up with project status. However, the five main programmers on the project all figured out a way to generate automatic updates for their tasks every week, saying that everything was completed as planned. They paid very little attention to the actual plan and hated filling out status information. Sue did not verify most of their work to check that it was actually completed. In addition, the head of the Registrar's Office was uninterested in the project and delegated sign-off responsibility to one of his clerks who really did not understand the entire registration process. When Sue and her team started testing the new system, she learned they were using last year's course data. Using last year's course data caused additional problems because the college was moving from quarters to semesters in the fall. How could they have missed that requirement? Sue hung her head in shame as she walked into a meeting with her manager to ask for help. She learned the hard way how difficult it was to keep a project on track. She wished she had spent more time talking face-to-face with key project stakeholders, especially her programmers and the Registrar's Office representatives, to verify that the project was on schedule and that the schedule was updated accurately.

CHAPTER SUMMARY

Project time management is often cited as the main source of conflict on projects. Most information technology projects exceed time estimates. The main processes involved in project time management include activity definition, activity sequencing, activity duration estimating, schedule development, and schedule control.

Activity definition involves identifying the specific activities that must be done to produce the project deliverables. It usually results in a more detailed WBS and supporting detail.

Activity sequencing determines the relationships or dependencies between activities. Three reasons for creating relationships are that they are mandatory based on the nature of the work, discretionary based on the project team's experience, or external based on non-project activities. Activity sequencing must be done in order to use critical path analysis.

Network diagrams are the preferred technique for showing activity sequencing. The two methods used for creating these diagrams are the arrow diagramming method and the precedence diagramming method. There are four types of relationships between tasks: finish-to-start, finish-to-finish, start-to-start, and start-to-finish.

Activity duration estimating creates estimates for the amount of time it will take to complete each activity. These time estimates include the actual amount of time worked plus elapsed time.

Schedule development uses results from all of the other project time management processes to determine the start and end dates for the project. Project managers often use Gantt charts to display the project schedule. Tracking Gantt charts show planned and actual schedule information.

The critical path method predicts total project duration. The critical path for a project is the series of activities that determines the earliest completion date for the project. It is the longest path through a network diagram. If any activity on the critical path slips, the whole project will slip unless the project manager takes corrective action.

Crashing and fast tracking are two techniques for shortening project schedules. Project managers and their team members must be careful about accepting unreasonable schedules, especially for information technology projects.

Critical chain scheduling is an application of the theory of constraints that uses critical path analysis, resource constraints, and buffers to help meet project completion dates.

Program Evaluation and Review Technique (PERT) is a network analysis technique used to estimate project duration when there is a high degree of uncertainty about the individual activity duration estimates. It uses optimistic, most likely, and pessimistic estimates of activity durations. PERT is seldom used today.

Even though scheduling techniques are very important, most projects fail because of people issues, not from a poor network diagram. Project managers must involve all stakeholders in the schedule development process. It is critical to set realistic project schedules and allow for contingencies throughout the life of a project. Several leadership skills that help project managers control schedule changes include empowerment, discipline, using incentives, and negotiation.

Project management software can assist in project scheduling if used properly. With project management software, you can avoid the need to perform cumbersome calculations manually and perform "what if" analysis as activity duration estimates or dependencies change. Many people misuse project management software because they do not understand the concepts behind creating a network diagram, determining the critical path, or setting a schedule baseline. People must also avoid overrelying on sample files or templates when creating their unique project schedules.

DISCUSSION QUESTIONS

1. Why do you think schedule issues often cause the most conflicts on projects?
2. Why is activity definition the first process involved in project time management?
3. Why is it important to determine activity sequencing on projects? Discuss diagrams you have seen that are similar to network diagrams. How are they similar, and how are they different?
4. Explain the difference between activity duration estimating and estimating the effort required to perform an activity.
5. Explain the following schedule development tools and concepts: Gantt charts, critical path method, PERT, and critical chain scheduling.

6. How can you minimize or control changes to project schedules?

7. List some of the reports you can generate with Project 2002 to assist in project time management.

8. Why is it difficult to use project management software well?

EXERCISES

1. Using Figure 6-2, enter the activities, their durations, and their relationships in Project 2002. Use a project start date of 6/01/05. View the network diagram. Does it look like Figure 6-2? It should be very similar, but appear in PDM form, as shown in Figure 6-4. Print the network diagram on one page. Return to Gantt chart view. Select View, Table, Schedule from the menu bar to recreate Table 6-1. You may need to move the split bar to the right to reveal all of the table columns. Write a few paragraphs explaining what the network diagram and schedule table show concerning Project X's schedule.

2. Consider Table 6-3, Network Diagram Data for a Small Project. All times are in days; the network proceeds from Node 1 to Node 9.

Table 6-3: Network Diagram Data for a Small Project

ACTIVITY	INITIAL NODE	FINAL NODE	ESTIMATED TIME
A	1	2	2
B	2	3	2
C	2	4	3
D	2	5	4
E	3	6	2
F	4	6	3
G	5	7	6
H	6	8	2
I	6	7	5
J	7	8	1
K	8	9	2

a. Draw an AOA network diagram representing the project. Put the node numbers in circles and draw arrows from node to node, labeling each arrow with the activity letter and estimated time.

b. Identify all of the paths on the network diagram and note how long they are.

c. What is the critical path for this project and how long is it?

d. What is the shortest possible time it will take to complete this project?

3. Enter the information from Exercise 2 into Project 2002. View the network diagram and task schedule table to see the critical path and float or slack for each activity. Print out the Gantt chart and network diagram views and the task schedule table. Write a short paper that interprets this information for someone unfamiliar with project time management.

4. Perform the following steps:

   a. Using the Information Technology Upgrade Project WBS you created in Chapter 5, Exercise 2, manually create a rough Gantt chart for the project. Make sure the project does not take more than nine months to complete. Document any assumptions you made in creating the Gantt chart.

   b. Determine the relationships between activities listed on your WBS and the Gantt chart. Manually build a network diagram for this project. Estimate the duration of each activity, and then enter the data into Project 2002. Make sure you can complete the project in nine months or less. Print the Gantt chart and network diagram and write a one-page paper interpreting the results.

5. Interview someone who uses some of the techniques discussed in this chapter. How does he or she feel about network diagrams, critical path analysis, Gantt charts, critical chain scheduling, using project management software, and managing the people issues involved in project time management? Write a paper describing the responses.

6. Review two different articles about critical chain scheduling. Write a two- to three-page paper describing how this technique can help improve project schedule management.

MINICASE

You have been asked to determine a rough schedule for a nine-month Billing System Conversion project, as part of your job as a consultant to a Fortune 500 firm. The firm's old system was written in COBOL on a mainframe computer, and the maintenance costs are prohibitive. The new system will run on an off-the-shelf application. You have identified several high-level activities that must be done in order to initiate, plan, execute, control, and close the project. The following table shows your analysis of the project's tasks and schedule so far.

Table 6-4: Billing System Conversion Project Schedule

TASKS	MAR	APR	MAY	JUN	JUL	AUG	SEP	OCT	NOV
Initiating									
Develop project charter									
Meet with stakeholders									
Planning									
Create detailed WBS and schedule									
Estimate project costs									
Create project team									
Create communication plan									
Organize a comprehensive project plan									
Executing									
Award and manage contract for software conversion									
Install new software on servers									
Install new hardware and software on client's machines									
Test new billing system									
Train users on new system									
Controlling									
Closing									

Part 1: Using the information in the table, draw horizontal bars to illustrate when you think each task would logically start and end. Then use Microsoft Project 2002 to create a Gantt chart and network diagram based on this information.

Part 2: Identify at least two milestones that could be included under each of the process groups in the table. Then write a detailed description of each of these milestones that meets the SMART criteria.

RUNNING CASES

See Appendixes A and C for several running case studies that apply various concepts from this chapter.

SUGGESTED READINGS

1. Abel-Hamid, Tarek K. "Investigating the Cost/Schedule Trade-off in Software Development." *IEEE Software* (1990).

   Abel-Hamid analyzes current models for estimating the effect of schedule compression on total project cost. He then presents his own system-dynamics model to illustrate the integrated nature of projects by including subsystems related to software production, planning, control, and human resource management.

2. Breen, Anne M., Tracey Burton-Houle, and David C. Aron. "Applying the Theory of Constraints in Health Care: Part 1 - The Philosophy," *Quality Management in Health Care* (Volume 10, Number 3) (Spring 2002) (*www.goldratt.com/for-cause/applyingtocinhcpt1fco.htm*).

   This article summarizes some of the basic concepts behind critical chain scheduling and shows how it can be used to improve various health care processes. It explains how the existence of constraints represents an excellent opportunity for improvement by allowing organizations to focus their efforts in the most productive areas.

3. Ganssle, Jack G. "Lies, Damn Lies, and Schedule." *Embedded Systems Programming* (December 1997).

   Ganssle writes an amusing article about the way in which many schedules for embedded software systems are "propped up by prayer and hopeful wishes." He suggests that yielding to pressure from management or other groups, such as marketing, in determining project schedules is nothing short of lying.

4. Goldratt, Eliyahu. *Critical Chain*. Great Barrington, MA: The North River Press (1997).

   This text is written as a business novel about a college professor and his project management class who together develop the concept of critical chain scheduling. Many other books and articles have since been written about critical chain. Goldratt now runs the Avraham Y. Goldratt Institute (www.goldratt.com), the world's largest consulting firm specializing in the application of theory of constraints (TOC) to achieve significant bottom-line results in organizations.

5. Leach, Larry P. "Critical Chain Project Management Improves Project Performance," Project Management Institute, *Project Management Journal* (June 1999).

   This article summarizes some of the key concepts involved in critical chain scheduling and how it can improve project planning and performance. The author suggests that critical chain scheduling can reduce total project schedules by 50 percent.

6. McConnell, Steve. "How to Defend an Unpopular Schedule." *IEEE Software* (May 1996).

> *This article emphasizes the fact that it is just as important, if not more so, to defend project schedule estimates as it is to create them. Most software developers must learn to defend their schedule estimates by practicing good negotiation techniques.*

KEY TERMS

- **activity** — an element of work, normally found on the WBS, that has an expected duration, cost, and resource requirements; also called **task**
- **activity definition** — identifying the specific activities that the project team members and stakeholders must perform to produce the project deliverables
- **activity duration estimating** — estimating the number of work periods that are needed to complete individual activities
- **activity-on-arrow (AOA)** or **arrow diagramming method (ADM)** — a network diagramming technique in which activities are represented by arrows and connected at points called nodes to illustrate the sequence of activities
- **activity sequencing** — identifying and documenting the relationships between project activities
- **backward pass** — a project network diagramming technique that determines the late start and late finish dates for each activity in a similar fashion
- **baseline dates** — the planned schedule dates for activities in a Gantt chart
- **buffer** — additional time to complete a task, added to an estimate to account for various factors
- **burst** — when a single node is followed by two or more activities on a network diagram
- **crashing** — a technique for making cost and schedule tradeoffs to obtain the greatest amount of schedule compression for the least incremental cost
- **critical chain scheduling** — a method of scheduling that takes limited resources into account when creating a project schedule and includes buffers to protect the project completion date
- **critical path** — the series of activities in a network diagram that determines the earliest completion of the project. It is the longest path through the network diagram and has the least amount of slack or float
- **critical path method (CPM)** or **critical path analysis** — a project network analysis technique used to predict total project duration
- **dependency** — the sequencing of project activities or tasks; also called a **relationship**

- **discretionary dependencies** — sequencing of project activities or tasks defined by the project team and used with care since they may limit later scheduling
- **dummy activities** — activities with no duration and no resources used to show a logical relationship between two activities in the arrow diagramming method of project network diagrams
- **duration** — the actual amount of time worked on an activity *plus* elapsed time
- **early finish date** — the earliest possible time an activity can finish based on the project network logic
- **early start date** — the earliest possible time an activity can start based on the project network logic
- **effort** — the number of workdays or work hours required to complete a task
- **external dependencies** — sequencing of project activities or tasks that involve relationships between project and non-project activities
- **fast tracking** — a schedule compression technique in which you do activities in parallel that you would normally do in sequence
- **feeding buffers** — additional time added before tasks on the critical path that are preceded by non-critical-path tasks
- **finish-to-finish dependency** — a relationship on a network diagram where the "from" activity must be finished before the "to" activity can be finished
- **finish-to-start dependency** — a relationship on a project network diagram where the "from" activity must be finished before the "to" activity can be started
- **forward pass** — a network diagramming technique that determines the early start and early finish dates for each activity
- **free slack (free float)** — the amount of time an activity can be delayed without delaying the early start of any immediately following activities
- **Gantt chart** — a standard format for displaying project schedule information by listing project activities and their corresponding start and finish dates in a calendar format
- **late finish date** — the latest possible time an activity can be completed without delaying the project finish date
- **late start date** — the latest possible time an activity may begin without delaying the project finish date
- **mandatory dependencies** — sequencing of project activities or tasks that are inherent in the nature of the work being done on the project
- **merge** — when two or more nodes precede a single node on a network diagram
- **milestone** — a significant event on a project with zero duration
- **multitasking** — when a resource works on more than one task at a time
- **Murphy's Law** — if something can go wrong, it will
- **network diagram** — a schematic display of the logical relationships or sequencing of project activities

- **node** — the starting and ending point of an activity on an activity-on-arrow diagram
- **Parkinson's Law** — work expands to fill the time allowed
- **PERT weighted average** = $\dfrac{\text{optimistic time} + 4 \times \text{most likely time} + \text{pessimistic time}}{6}$
- **Precedence Diagramming Method (PDM)** — a network diagramming technique in which boxes represent activities
- **probabilistic time estimates** — duration estimates based on using optimistic, most likely, and pessimistic estimates of activity durations instead of using one specific or discrete estimate
- **Program Evaluation and Review Technique (PERT)** — a project network analysis technique used to estimate project duration when there is a high degree of uncertainty with the individual activity duration estimates
- **project buffer** — additional time added before the project's due date
- **project time management** — the processes required to ensure timely completion of a project
- **schedule control** — controlling and managing changes to the project schedule
- **schedule development** — analyzing activity sequences, activity duration estimates, and resource requirements to create the project schedule
- **slack** — the amount of time a project activity may be delayed without delaying a succeeding activity or the project finish date; also called **float**
- **slipped milestone** — a milestone activity that is completed later than planned
- **SMART criteria** — guidelines to help define milestones that are specific, measurable, assignable, realistic, and time-framed
- **start-to-finish dependency** — a relationship on a project network diagram where the "from" activity cannot start before the "to" activity is finished
- **start-to-start dependency** — a relationship in which the "from" activity cannot start until the "to" activity starts
- **Theory of constraints** (TOC) — a management philosophy that states that any complex system at any point in time often has only one aspect or constraint that is limiting its ability to achieve more of its goal
- **total slack (total float)** — the amount of time an activity may be delayed from its early start without delaying the planned project finish date
- **tracking Gantt chart** — a Gantt chart that compares planned and actual project schedule information

7

Project Cost Management

Objectives

After reading this chapter, you will be able to:

1. *Understand the importance of good project cost management*
2. *Explain basic project cost management principles, concepts, and terms*
3. *Describe how resource planning relates directly to project cost management*
4. *Explain cost estimating using definitive, budgetary, and rough order of magnitude (ROM) estimates*
5. *Understand the processes involved in cost budgeting and preparing a cost estimate for an information technology project*
6. *Understand the benefits of earned value management and project portfolio management to assist in cost control*
7. *Describe how software can assist in project cost management*

OPENING CASE

Juan Gonzales was a systems analyst and network specialist for a major city's waterworks department. He enjoyed helping his country, Mexico, develop its infrastructure. His next career objective was to become a project manager so he could have even more influence. One of his colleagues invited him to attend an important project review meeting for large government projects, including the surveyor's assistant project, in which Juan was most interested. The surveyor's assistant project was a concept for developing a sophisticated information system that included expert systems, object-oriented databases, and wireless communications. The system would provide instant, graphical information to government surveyors to help them do their jobs. For example, after a surveyor touched a map on the screen of a hand-held device, the system would prompt him or her for the type of information needed for that area. This system would help in planning and implementing many projects, from laying fiber-optic cable to laying water lines.

Juan was very surprised, however, when the majority of the meeting was spent discussing cost-related issues. The government officials were reviewing many existing projects to evaluate their performance to date and the potential impact on their budgets before discussing the funding for any new projects. Juan did not understand many of the terms and charts the presenters were showing. What was this earned value term they kept referring to? How were they estimating what it would cost to complete projects or how long it would take? Juan thought he would learn more about the new technologies the surveyor's assistant project would use, but he discovered that the cost estimate and projected benefits were of most interest to the government officials at the meeting. It also seemed as if a lot of effort would be spent on detailed financial studies before any technical work could even start. Juan wished he had taken some accounting and finance courses so he could understand the acronyms and concepts people were discussing. Although Juan had a degree in electrical engineering, he had no formal education, and little experience, in finance. If Juan could understand information systems and networks, he was confident that he could understand financial issues on projects, too. He jotted down questions to discuss with his colleagues after the meeting.

THE IMPORTANCE OF PROJECT COST MANAGEMENT

Just as information technology projects have a poor track record in meeting schedule goals, they also have a poor track record in meeting budget goals. The 1995 CHAOS report showed that the average cost **overrun**—the additional percentage or dollar amount by which actual costs exceed estimates—for information technology projects was 189 percent of their original estimates. This means a project that was estimated to cost $100,000 ended up costing $189,000. In 1995, more than 31 percent of information technology projects were canceled before completion, costing U.S. companies and government agencies over $81 billion.[1]

[1] Johnson, Jim, "CHAOS: The Dollar Drain of IT Project Failures," *Application Development Trends*(January 1995).

The 2001 CHAOS report showed that companies have made great improvements in reducing cost overruns since 1995—from 189 percent to 145 percent—but that most completed projects still exceeded their approved budgets. This chapter describes important concepts in project cost management, particularly the use of earned value management (EVM) to assist in cost control.

What Went Wrong?

According to the *San Francisco Chronicle* front-page story, "Computer Bumbling Costs the State $1 Billion," the state of California had a series of expensive information technology project failures in the late 1990s, costing taxpayers nearly $1 billion. Some of the poorly managed projects included the Department of Motor Vehicles registration and driver's license databases, a statewide child support database, the State Automated Welfare System, and a Department of Corrections system for tracking inmates. Senator John Vasconcellos thought it was ironic that the state that was leading in the creation of computers was also the state most behind in using computer technology to improve its services.[2]

Consider also the Internal Revenue Service's expensive reengineering and information technology project failures. The Internal Revenue Service (IRS) managed a series of project failures in the 1990s that cost taxpayers over $50 billion a year—roughly as much money as the annual net profit of the entire computer industry that year.[3]

Project cost overruns often occur when companies try to purchase new systems from an outside source. Connecticut General Life Insurance Co. sued PeopleSoft, Inc. over an aborted installation of a finance system. PeopleSoft increased the original project estimate from $5.16 million to $11.7 million and indicated that the final cost could go even higher. Connecticut General wanted PeopleSoft to refund the $5.16 million since they had to develop a homegrown application.[4]

What Is Cost?

A popular cost accounting textbook states, "Accountants usually define cost as a resource sacrificed or foregone to achieve a specific objective."[5] Webster's dictionary defines cost as "something given up in exchange." Costs are often measured in monetary amounts, such as dollars, that must be paid to acquire goods and services. Because projects cost money and consume resources that could be used elsewhere, it is very important for project managers to understand project cost management.

Many information technology professionals, however, often react to cost overrun information with a smirk. They know that many of the original cost

[2] Lucas, Greg, "Computer Bumbling Costs the State $1 Billion," *San Francisco Chronicle* (2/21/99) 1.

[3] James, Geoffrey, "IT Fiascoes . . . and How to Avoid Them," *Datamation* (November 1997).

[4] Songini, Marc L., "PeopleSoft project ends in court,". *Computerworld* (September 2001).

[5] Horngren, Charles T., George Foster, and Srikanti M. Datar, *Cost Accounting*, 8th ed. Englewood Cliffs, NJ: Prentice-Hall, 1994.

estimates for information technology projects are low to begin with or based on very unclear project requirements, so naturally there will be cost overruns. Not emphasizing the importance of realistic project cost estimates from the outset is only one part of the problem. In addition, many information technology professionals think preparing cost estimates is beneath them—a job for accountants. On the contrary, preparing good cost estimates is a very demanding, important skill that all information technology project professionals need to acquire.

Another perceived reason for cost overruns is that many information technology projects involve new technology or business processes. Any new technology or business process is untested and has inherent risks. Thus, costs grow and failures are to be expected. Right? Wrong. Using good project cost management can change this false perception. In fact, the 1995 Standish Group study mentions that many of the information technology projects in their study were as mundane as a new accounting package or an order-entry system. The state of California and the IRS examples also demonstrate that many information technology projects cannot blame untested technology for their cost problems. What they needed was better project cost management.

What Is Project Cost Management?

Recall from Chapter 1 that the triple constraint of project management involves balancing scope, time, and cost goals. Chapters 5 and 6 discuss project scope and time management, and this chapter describes **project cost management**. Project cost management includes the processes required to ensure that a project team completes a project within an approved budget. Notice two crucial phrases in this definition: "a project" and "approved budget." Project managers must make sure *their* projects are well defined, have accurate time and cost estimates, and have a realistic budget that *they* were involved in approving. It is the project manager's job to satisfy project stakeholders while continuously striving to reduce and control costs. There are four project cost management processes:

- **Resource planning** involves determining what resources (people, equipment, and materials) a project team should use to perform project activities and the quantities of each resource. The output of the resource planning process is a list of resource requirements.
- **Cost estimating** involves developing an approximation or estimate of the costs of the resources needed to complete a project. The main outputs of the cost estimating process are cost estimates, supporting detail, and a cost management plan.
- **Cost budgeting** involves allocating the overall cost estimate to individual work items to establish a baseline for measuring performance. The main output of the cost budgeting process is a cost baseline.
- **Cost control** involves controlling changes to the project budget. The main outputs of the cost control process are revised cost estimates, budget updates, corrective action, estimate at completion, and lessons learned.

To understand each of the project cost management processes, you must first understand the basic principles of cost management. Many of these principles are not unique to project management; however, project managers need to understand how these principles relate to their specific projects.

BASIC PRINCIPLES OF COST MANAGEMENT

Many information technology projects are never initiated because information technology professionals do not understand the importance of knowing basic accounting and finance principles. Important concepts such as net present value analysis, return on investment, and payback analysis were discussed in Chapter 5, Project Scope Management. Likewise, many projects that are started never finish because of cost management problems. Most members of an executive board understand and are more interested in financial terms than information technology terms. Therefore, information technology project managers need to be able to present and discuss project information in financial terms as well as in technical terms. In addition to net present value analysis, return on investment, and payback analysis, project managers must understand several other cost management principles, concepts, and terms. This section describes general topics such as profits, life cycle costing, cash flow analysis, internal rate of return, tangible and intangible costs and benefits, direct costs, sunk costs, learning curve theory, and reserves. Another important topic and one of the key tools and techniques for controlling project costs—earned value management—is described in detail in the section on cost control.

Profits are revenues minus expenses. To increase profits, a company can increase revenues, decrease expenses, or try to do both. Most executives are more concerned with profits than with other issues. When justifying investments in new information systems and technology, it is important to focus on the impact on profits, not just revenues or expenses. Consider an e-commerce application that you estimate will increase revenues for a $100 million company by 10 percent. You cannot measure the potential benefits of the e-commerce application without knowing the profit margin. **Profit margin** is the ratio between revenues and profits. If revenues of $100 generate $2 in profits, there is a 2 percent profit margin. If the company loses $2 for every $100 in revenue, there is a –2 percent profit margin.

Life cycle costing allows you to see a big-picture view of the cost of a project throughout its life cycle. This helps you develop an accurate projection of a project's financial benefits. Life cycle costing considers the total cost of ownership, or development plus support costs, for a project. For example, a company might complete a project to develop and implement a new customer service system in one or two years, but the new system could be in place for ten years. Project managers should create estimates of the costs and benefits of the project for its entire life cycle, or ten years in the preceding example. Recall that the net present value analysis for the project would include the entire ten-year period of costs and benefits (see Chapter 5). Top management and project managers need to consider the life cycle costs of projects when they make financial decisions.

Organizations have a history of not spending enough money in the early phases of information technology projects, which impacts total cost of ownership. For example, it is much more cost-effective to spend money on defining user requirements and doing early testing on information technology projects than to wait for problems to appear after implementation. Table 7-1 summarizes the typical costs of correcting software defects during different stages of the system development life cycle. If you identified and corrected a software defect during the user requirements phase of a project, you would need to add $100 to $1,000 to the project's total cost of ownership. In contrast, if you waited until after implementation to correct the same software defect, you would need to add possibly millions of dollars to the project's total cost of ownership.

Table 7-1: Costs of Software Defects[6]

When Defect Is Detected	Typical Costs of Correction
User Requirements	$100 – $1,000
Coding/Unit Testing	$1,000 or more
System Testing	$7,000 – $8,000
Acceptance Testing	$1,000 – $100,000
After Implementation	Up to millions of dollars

Cash flow analysis is a method for determining the estimated annual costs and benefits for a project and the resulting *annual* cash flow. Project managers must conduct cash flow analysis to determine net present value (see Chapter 5, Project Scope Management). Most consumers understand the basic concept of cash flow. If they do not have enough money in their wallets or checking accounts, they cannot purchase something. Top management must consider cash flow concerns when selecting projects in which to invest. If top management selects too many projects that have high cash flow needs in the same year, the company will not be able to support all of its projects and maintain its profitability. It is important to define clearly the year on which the company bases the dollar amounts. For example, if a company bases all costs on 2004 estimates, it would need to account for inflation and other factors when projecting costs and benefits in future-year dollars.

Internal rate of return (IRR) is the discount rate that makes net present value equal to zero. It is also called the **time-adjusted rate of return**. Some

[6] Collard, Ross, *Software Testing and Quality Assurance*, working paper (1997).

companies prefer to estimate the internal rate of return instead of, or in addition to, net present value and set minimum values that must be achieved for projects to be selected or continued. For example, assume a three-year project had projected costs in year one of $100 and projected benefits in years two and three of $100 each year. Assume there were no benefits in year one and no costs in years two and three. Using a 10-percent discount rate, you could compute the net present value to be about $67. (See Chapter 5 for details on performing net present value calculations.) The internal rate of return for this project is the discount rate that makes the net present value equal to zero. In this example, the internal rate of return is 62 percent.

Tangible and intangible costs and benefits are categories for determining how definable the estimated costs and benefits are for a project. **Tangible costs or benefits** are those costs or benefits that an organization can easily measure in dollars. For example, suppose the surveyor's assistant project described in the opening case included a preliminary feasibility study. If a firm completed this study for $100,000, the tangible cost of the study is $100,000. If Juan's government estimated that it would have cost them $150,000 to do the study themselves, the tangible benefits of the study would be $50,000 if they assigned the people who would have done the study to other projects. Conversely, **intangible costs or benefits** are costs or benefits that are difficult to measure in monetary terms. Suppose Juan and a few other people, out of personal interest, spent some time using government-owned computers, books, and other resources to research areas related to the study. Although their hours and the government-owned materials were not billed to the project, they could be considered intangible costs. Intangible benefits for projects often include items like goodwill, prestige, and general statements of improved productivity that an organization cannot easily translate into dollar amounts. Because intangible costs and benefits are difficult to quantify, they are often harder to justify.

Direct costs are costs related to a project that an organization can trace back in a cost-effective way. You can attribute direct costs directly to a certain project. For example, the salaries of people working full-time on the project and the cost of hardware and software purchased specifically for the project are direct costs. Project managers should focus on direct costs, since they can control them.

Indirect costs are costs related to a project that an organization cannot trace back in a cost-effective way. For example, the cost of electricity, paper towels, and so on in a large building housing a thousand employees who work on many projects would be indirect costs. Indirect costs are allocated to projects, and project managers have very little control over them.

Sunk cost is money that has been spent in the past. Consider it gone, like a sunken ship that can never be returned. When deciding what projects to invest in or continue, you should not include sunk costs. For example, in the opening case, suppose Juan's office had spent $1 million on a project over the past three years to create a geographic information system, but they never produced anything valuable. If his government were evaluating what projects to fund next year and someone suggested that they keep funding the geographic information system project because they had already spent $1 million on it, he or she would be incorrectly making sunk cost a key factor in the project selection decision. Many people fall into the trap of considering how much money has been spent on a failing project and, therefore, hate to stop spending money on it. This trap is similar to gamblers not wanting to stop gambling because they have already lost money. Sunk costs should be forgotten.

Learning curve theory states that when many items are produced repetitively, the unit cost of those items decreases in a regular pattern as more units are produced. For example, suppose the surveyor's assistant project would potentially produce 1000 hand-held devices that could run the new software and access information via satellite. The cost of the first hand-held device or unit would be much higher than the cost of the 1000th unit. Learning curve theory should help estimate costs on projects involving the production of large quantities of items. Learning curve theory also applies to the amount of time it takes to complete some tasks. For example, the first time a new employee performs a specific task, it will probably take longer than the tenth time that employee performs a very similar task.

Reserves are dollars included in a cost estimate to mitigate cost risk by allowing for future situations that are difficult to predict. **Contingency reserves** allow for future situations that may be partially planned for (sometimes called **known unknowns**) and are included in the project cost baseline. For example, if an organization knows it has a 20 percent rate of turnover for information technology personnel, it should include contingency reserves to pay for recruiting and training costs for information technology personnel. **Management reserves** allow for future situations that are unpredictable (sometimes called **unknown unknowns**). For example, if a project manager gets sick for two weeks or an important supplier goes out of business, management reserve could be set aside to cover the resulting costs.

RESOURCE PLANNING

Resource planning is a crucial part of project cost management. To estimate, budget, and control costs, project managers and their teams must determine what physical resources (people, equipment, and materials) and what quantities of those resources are required to complete the project. The nature of the project and the organization will affect resource planning. Expert judgment and the availability of alternatives are the only real tools available to assist in resource planning. It is important that the people who help determine what resources are necessary include people who have experience and expertise in similar projects and with the organization performing the project.

Important questions to answer in resource planning include:

- How difficult will it be to do specific tasks on this project?
- Is there anything unique in this project's scope statement that will affect resources?
- What is the organization's history in doing similar tasks? Have similar tasks been done before? What level of personnel did the work?
- Does the organization have people, equipment, and materials that are capable and available for performing the work?
- Does the organization need to acquire more resources to accomplish the work? Would it make sense to outsource some of the work?
- Are there any organizational policies that might affect the availability of resources?

A project's work breakdown structure, scope statement, historical information, resource information, and policies are all important inputs to answering these questions.

A large percentage of total project costs are often labor costs. Many organizations estimate the number of people they need by department or skill over the course of a project. For example, when Northwest Airlines developed cost estimates for its reservation system project, ResNet, it determined the maximum number of people it could assign to the project each year by department. Table 7-2 shows this information. (Figure 9-6 in Chapter 9, Project Human Resource Management, provides similar resource information in graphical form, where the number of resources are provided by job category, such as business analyst, programmer, and so on.) Note the small number of contractors Northwest Airlines planned to use. Labor costs are often much higher for contractors, so it is important to distinguish between internal and external resources. (See this text's companion Web site for a detailed case study on ResNet, including cost estimates.)

Table 7-2: Maximum Departmental Headcounts by Year

DEPARTMENT	1994	1995	1996	1997	1998	TOTALS
Information Systems	24	31	35	13	13	116
Marketing Systems	3	3	3	3	3	15
Reservations	12	29	33	9	7	90
Contractors	2	3	1	0	0	6
Totals	41	66	72	25	23	227

It is important to thoroughly brainstorm and evaluate alternatives related to resources, especially on projects that involve people from multiple disciplines and companies. Since most projects involve many human resources, it is often effective to solicit ideas from different people to help develop alternatives and address resource and cost-related issues early in a project.

What Went Right?

Organizations realize that teamwork helps reduce costs. Ford Motor Company had great success with its Taurus/Sable project. Ford excels in team product development and seeks to remove barriers among design, engineering, production, marketing, sales, and purchasing. However, the Taurus team went far beyond removing barriers by creating a car that excelled in design and quality at half the typical development cost. They brought together all the relevant disciplines as a team and took the various steps in designing, producing, marketing, and selling the cars simultaneously as well as sequentially. Ford even involved people outside the company for advice—car dealerships, insurance companies, and suppliers. One supplier even offered the services of its own drafting department to prepare initial designs for Ford's approval. Bringing together all the key players provided extremely valuable ideas at the conceptual stage, the time when the company can make changes without much extra cost. "Not only were there substantial savings in cost and design time, but major production contracts were being negotiated and set up some three years ahead of production, with the duration of the contract some five years. This also led to cost economies."[7]

Organizations often purchase information technology products and services to help reduce costs and improve operations. For example, using J.D. Edwards software, Drug Trading Company, Limited, a Canadian retail pharmacy services company, reduced information technology costs and achieved tighter customer and partner relationships. The company can now respond faster to customer requests, with a lower total cost of ownership (TCO). "We're targeting a very quick ROI and a significant reduction in TCO—by as much as 40 percent—once we're in full production with J.D. Edwards. Our strategic objectives are to improve the supply pipeline and provide convenient 24×7 communications channels for customers and vendors," says Grant Schwartz, Corporate Director of IT for Drug Trading Company.[8]

The main output of the resource planning process is a list of resource requirements, including people, equipment, and materials. In addition to providing the basis for cost estimates, budgets, and cost controls, the resource requirements list provides vital information for project human resource management (Chapter 9) and project procurement management (Chapter 12).

[7] Pinto, Jeffrey K. and O.P. Kharbanda, *Successful Project Managers*. New York: Van Nostrand Reinhold, 1995.

[8] Customer Testimonial, "Drug Trading Company Targets Reduced IT Costs and Tighter Customer and Partner Relationships," Bitpipe, Inc. (2002).

COST ESTIMATING

Project managers must take cost estimates seriously if they want to complete projects within budget constraints. After developing a good resource requirements list, project managers and their project teams must develop several estimates of the costs for these resources. This section describes various types of cost estimates, tools and techniques for cost estimation, typical problems associated with information technology cost estimates, and a detailed example of a cost estimate for an information technology project.

Types of Cost Estimates

One of the main outputs of project cost management is a cost estimate. Project managers normally prepare several types of cost estimates for most projects. Rough order of magnitude, budgeting, and definitive estimates vary in date of completion, how they will be used, and accuracy levels.

- A **rough order of magnitude (ROM) estimate** provides a rough idea of what a project will cost. ROM estimates can also be referred to as a ballpark estimate, a guesstimate, a swag, or a broad gauge. This type of estimate is done very early in a project or even before a project is officially started. Project managers and top management use this estimate to help make project selection decisions. The timeframe for this type of estimate is often three or more years prior to project completion. A ROM estimate's accuracy is typically –25 percent to +75 percent, meaning the project's actual costs could be 25 percent below the ROM estimate or 75 percent above. For information technology project estimates, this accuracy range is often much wider. Many information technology professionals automatically double estimates for software development because of the history of cost overruns on information technology projects.

- A **budgetary estimate** is used to allocate money into an organization's budget. Many organizations develop budgets at least two years into the future. Budgetary estimates are made one to two years prior to project completion. The accuracy of budgetary estimates is typically –10 percent to +25 percent, meaning the actual costs could be 10 percent less or 25 percent more than the budgetary estimate.

- A **definitive estimate** provides an accurate estimate of project costs. Definitive estimates are used for making many purchasing decisions for which accurate estimates are required and for estimating final project costs. For example, if a project involves purchasing 1000 personal computers from an outside supplier in the next three months, a definitive estimate would be required to aid in evaluating supplier proposals and allocating the funds to pay the chosen supplier. Definitive estimates are made one year or less prior to project completion. A definitive estimate should be the most accurate of the three types of estimates. The accuracy of this type of estimate is normally –5 percent to +10 percent, meaning the actual costs could be 5 percent less or 10 percent more than the definitive estimate. Table 7-3 summarizes the three basic types of cost estimates.

Table 7-3: Types of Cost Estimates

TYPE OF ESTIMATE	WHEN DONE	WHY DONE	HOW ACCURATE
Rough Order of Magnitude (ROM)	Very early in the project life cycle, often 3–5 years before project completion	Provides estimate of cost for selection decisions	–25%, +75%
Budgetary	Early, 1–2 years out	Puts dollars in the budget plans	–10%, +25%
Definitive	Later in the project, less than 1 year out	Provides details for purchases, estimates actual costs	–5%, +10%

Two additional outputs of the cost estimating process are supporting details and a cost management plan. It is very important to include supporting details with all cost estimates. The supporting details include the ground rules and assumptions used in creating the estimate, a description of the project (scope statement, WBS, and so on) used as a basis for the estimate, and details on the cost estimation tools and techniques used to create the estimate. These supporting details should make it easier to prepare an updated estimate or similar estimate as needed.

A **cost management plan** is a document that describes how the organization will manage cost variances on the project. For example, if a definitive cost estimate provides the basis for evaluating supplier cost proposals for all or part of a project, the cost management plan describes how to respond to proposals that are higher or lower than the estimates. Some organizations assume that a cost proposal within 10 percent of the estimate is fine and only negotiate items that are more than 10 percent higher or 20 percent lower than the estimated costs. The cost management plan is part of the overall project plan described in Chapter 4, Project Integration Management.

Cost Estimation Tools and Techniques

There are three basic techniques for cost estimating: analogous estimates, bottom-up estimating, and parametric modeling. Many computerized tools are also available to help implement these techniques.

Analogous estimates, also called **top-down estimates**, use the actual cost of a previous, similar project as the basis for estimating the cost of the current project. This technique requires a good deal of expert judgment and is generally less costly than others are, but it is also less accurate. Analogous estimates are most reliable when the previous projects are similar in fact, not just in appearance. In addition, the groups preparing cost estimates must have the needed expertise to determine whether certain parts of the project will be more or less expensive than analogous projects. However, if the project to be estimated involves a new programming language or working with a new type of hardware or network, the analogous estimate technique could easily result in too low an estimate.

Bottom-up estimating involves estimating individual work items and summing them to get a project total. The size of the individual work items and the experience of the estimators drive the accuracy of the estimates. If a detailed WBS is available for a project, the project manager could have each person responsible for a work package develop his or her own cost estimate for that work package. The project manager would then add all of the cost estimates to create cost estimates for each higher-level WBS item and finally for the entire project. See Appendix A's section on project cost management for detailed information on entering resource costs and assigning resources to tasks to create a bottom-up estimate. Using smaller work items increases the accuracy of the cost estimate because the people assigned to do the work develop the cost estimate instead of someone unfamiliar with the work. The drawback with bottom-up estimates is that they are usually time-intensive and therefore expensive to develop.

Parametric modeling uses project characteristics (parameters) in a mathematical model to estimate project costs. A parametric model might provide an estimate of $50 per line of code for a software development project based on the programming language the project is using, the level of expertise of the programmers, the size and complexity of the data involved, and so on. Parametric models are most reliable when the historical information that was used to create the model is accurate, the parameters are readily quantifiable, and the model is flexible in terms of the size of the project. For example, in the 1980s, engineers at McDonnell Douglas Corporation (now part of Boeing) developed a parametric model for estimating aircraft costs based on a large historical database. The model included the following parameters: the type of aircraft (fighter aircraft, cargo aircraft, or passenger aircraft), how fast the plane would fly, the thrust-to-weight ratio of the engine, the estimated weights of various parts of the aircraft, the number of aircraft produced, the amount of time available to produce them, and so on. In contrast to this sophisticated model, some parametric models involve very simple heuristics or rules of thumb. For example, a large office automation project might use a ballpark figure of $10,000 per workstation based on a history of similar office automation projects developed during the same time period. Parametric models that are more complicated are usually computerized.

One popular parametric model is the **Constructive Cost Model (COCOMO)**, which is used for estimating software development costs based on parameters such as the source lines of code or function points. Barry Boehm, a well-known expert in the field of software development and cost estimating, developed COCOMO. **Function points** are technology-independent assessments of the functions involved in developing a system. For example, the number of inputs and outputs, the number of files maintained, and the number of updates are examples of function points. **COCOMO II** is a newer, computerized version of Boehm's model that allows you to estimate the cost, effort, and schedule when planning a new software development activity. Boehm suggests that only algorithmic or parametric models do not suffer from the limits of human decision-making capability. Therefore, Boehm and many other software experts favor using algorithmic models when estimating software project costs.

Computerized tools, such as spreadsheets and project management software, can make working with different cost estimates and cost estimation tools easier. Computerized tools, when used properly, can also help improve the accuracy of estimates. In addition to spreadsheets and project management software, more sophisticated tools are available for estimating software project costs.

Typical Problems with Information Technology Cost Estimates

Although there are many tools and techniques to assist in creating project cost estimates, many information technology project cost estimates are still very inaccurate, especially those involving new technologies or software development. Tom DeMarco, a well-known author on software development, suggests four reasons for these inaccuracies and some ways to overcome them.[9]

1. Developing an estimate for a large software project is a complex task requiring a significant amount of effort. Many estimates must be done quickly and before clear system requirements have been produced. For example, the surveyor's assistant project described in the opening case involves a lot of complex software development. Before fully understanding what information surveyors really need in the system, someone would have to create a ROM estimate and budgetary estimates for this project. Rarely are the more precise, later estimates less than the earlier estimates for information technology projects. It is important to remember that estimates are done at various stages of the project, and project managers need to explain the rationale for each estimate.

2. The people who develop software cost estimates often do not have much experience with cost estimation, especially for large projects. There is also not enough accurate, reliable project data available on which to base estimates. If an organization uses good project management and develops a history of keeping reliable project information, including estimates,

[9] DeMarco, Tom, *Controlling Software Projects*. New York: Yourdon Press, 1982.

it should help improve the organization's estimates. Enabling information technology people to receive training and mentoring on cost estimating will also improve cost estimates.

3. Human beings are biased toward underestimation. For example, senior information technology professionals or project managers might make estimates based on their own abilities and forget that many junior people will be working on a project. Estimators might also forget to allow for extra costs needed for integration and testing on large information technology projects. It is important for project managers and top management to review estimates and ask important questions to make sure the estimates are not biased.

4. Management might ask for an estimate, but really desire a more accurate number to help them create a bid to win a major contract or get internal funding. This problem is similar to the situation discussed in Chapter 6, Project Time Management, in which top managers or other stakeholders want project schedules to be shorter than the estimates. It is important for project managers to help develop good cost and schedule estimates and to use their leadership and negotiation skills to stand by those estimates.

Sample Cost Estimates and Supporting Detail for an Information Technology Project

One of the best ways to learn how the cost estimating process works is by studying a sample cost estimate. Every cost estimate is unique, just as every project is unique. This section provides information from an actual information technology project cost estimate. The entire estimate was much longer, but this chapter provides parts of it here to illustrate what a good cost estimate entails. You can also see a short sample cost estimate in Chapter 3 for JWD Consulting's Project Management Intranet Site project, and view the ResNet project cost estimate on the companion Web site for this text.

Before beginning the creation of any cost estimate, you must first know how the organization plans to use the cost estimate. If the cost estimate will be the basis for contract awards and performance reporting, it should be a definitive estimate and as accurate as possible, as described earlier. In the example presented here, a budgetary estimate was done for replacing **legacy systems**, or older information systems that run on old mainframe computers. Legacy systems support basic business processes such as general ledger, accounts payable and receivable, and project accounting. The name of this project was the Business Systems Replacement (BSR) project. The total cost estimate for this three-year project was about $7.5 million. The life of the new system was estimated to be eight years.

The main objective of the BSR project was to replace several legacy systems with a suite of off-the-shelf financial applications software (provided by Oracle).

The suite of financial applications would provide more timely information for management decision making, easier access to data by end users, and reduced costs through productivity improvements throughout the company. The legacy systems were all written in-house specifically for that company. By purchasing off-the-shelf application software, the company would no longer need its old mainframe computer and would benefit from having additional features and regular upgrades to the software without needing to write its own code.

Table 7-4 shows the overview for the BSR cost estimate. Notice the main categories found in the cost estimate overview: objective, scope, assumptions, and cost/benefit analysis and internal rate of return (IRR). The overview highlights information important to top management and others interested in the cost estimate for the project.

Table 7-4: Business Systems Replacement Project Cost Estimate Overview

CATEGORY	DESCRIPTION
Objective	Install a suite of off-the-shelf financial applications software that will enable more timely information for management decision making, provide easier access to data by the ultimate end user, and allow for cost savings through productivity improvements throughout the company.
Scope	Oracle financial applications will replace the core financial systems and include: ■ General Ledger ■ Fixed Assets ■ Operations Report ■ Accounts Payable ■ Accounts Receivable ■ Project Accounting ■ Project Management
Assumptions	Oracle's financial applications provide: ■ Minimal customization ■ No change in procurement systems during accounts payable implementation
Cost/Benefit Analysis & Internal Rate of Return (IRR)	BSR was broken down into a three-year cash outlay without depreciation. Costs are represented in thousands. Capital and expenses are combined in this example.

Another important part of the BSR project cost estimate was the cash flow analysis. Table 7-5 shows the estimated costs, benefits or savings, and net cost or savings from fiscal year (FY) 1995 through 1997. Numbers in parentheses represent negative numbers or costs. The table also shows the total project costs and savings and projected annual savings in future years beyond 1997. Notice that this company chose to focus on the first three years of the project in this summary cash flow analysis and estimated all future annual savings in one column. Estimates are less reliable the farther out in time you go, so some estimates use this approach.

Table 7-5: Business Systems Replacement Project Cash Flow Analysis

	FY95 ($000)	FY96 ($000)	FY97 ($000)	3 YEAR TOTAL ($000)	FUTURE ANNUAL COSTS/SAVINGS ($000)
COSTS					
Oracle/PM Software (List Price)	992	500	0	1492	0
60% Discount	(595)			(595)	
Oracle Credits	(397)	0		(397)	
Net Cash for Software	0	500		500	
Software Maintenance	0	90	250	340	250
Hardware & Maintenance	0	270	270	540	270
Consulting & Training	205	320	0	525	0
Tax & Acquisition	0	150	80	230	50
Total Purchased Costs	205	1330	600	2135	570
Information Services & Technology (IS&T)	500	1850	1200	3550	0
Finance/Other Staff	200	990	580	1770	
Total Costs	905	4170	2380	7455	570
SAVINGS					
Mainframe		(101)	(483)	(584)	(597)
Finance/Asset/PM		(160)	(1160)	(1320)	(2320)
IS&T Support/Data Entry		(88)	(384)	(472)	(800)
Interest		0	(25)	(25)	(103)
Total Savings		(394)	(2052)	(2401)	(3820)
Net Cost (Savings)	905	3821	328	5054	(3250)
8-Year Internal Rate of Return	35%				

In Table 7-5, you can see that the main cost categories for this estimate include the purchase cost for Oracle's software, hardware and software maintenance related to the project, consulting and training, tax and acquisition, Information Services and Technology (IS&T) effort, and Finance and other departments' effort. These cost categories are the same as summary tasks from the project's WBS. In this example, the main supplier, Oracle, provided detailed information for determining the cost of purchasing its software. The company received discounts and credits from previous purchases that made the net cash cost for the purchased software equal to zero. Oracle also provided estimates for software maintenance, consulting, and training costs. Note that the largest item in this cost estimate is the Information Services and Technology (IS&T) effort, which involves the people in the company's IS&T department who worked with Oracle and other internal departments to plan, implement, and support the new system. Major savings or benefit items were the savings from replacing the mainframe system, savings in various departments, a reduction in IS&T staff needed to support the legacy system, and interest savings from loans related to the old system.

The cash flow analysis provided the basis for determining the internal rate of return (IRR) for the life of the project. In this example, the IRR was 35 percent over eight years, which is a very good return.

Estimates provide the basis for cost budgeting and cost control, as described in the following sections. To use cost estimates as the basis for budgeting, the estimator must understand how his or her organization prepares budgets. Likewise, to use cost estimates as the basis for cost control, the estimator must understand how the organization performs cost control.

COST BUDGETING

Project cost budgeting involves allocating the project cost estimate to individual work items. These work items are based on the work breakdown structure for the project. The WBS, therefore, is a required input to the cost budgeting process. Likewise, the project schedule is required to allocate costs over time. The main goal of the cost budgeting process is to prepare budgetary estimates and to produce a cost baseline for measuring project performance.

Most organizations have a well-established process for preparing budgets. Continuing with the BSR project example, the company involved required budget estimates to include the number of full-time equivalent (FTE) staff, often referred to as headcount, for each year of the project. This number provided the basis for estimating total compensation costs each year, as shown in Table 7-6. The budget also required input in the categories of consultant/purchased services, travel, depreciation, rents/leases, and other supplies and expenses. Notice that the total cost for the IS&T budget for FY97, $1,800,000, is based on the costs highlighted in the cash flow analysis in Table 7-5 ($600,000 for total purchased costs plus $1,200,000 for IS&T effort in FY97).

Table 7-6: Business Systems Replacement Project Budget Estimates for FY97 and Explanations

BUDGET CATEGORY	ESTIMATED COSTS	EXPLANATION
Headcount (FTE)	13	Included are 9 programmer/analysts, 2 database analysts, and 2 infrastructure technicians
Compensation	$1,008,500	Calculated by employee change notices (ECNs) and assumed a 4% pay increase in June; Overload support was planned at $10,000
Consultant/Purchased Services	$424,500	Expected consulting needs in support of the Project Accounting and Cascade implementation efforts; maintenance expenses associated with the Hewlett-Packard (HP) computing platforms; maintenance expenses associated with the software purchased in support of the BSR project
Travel	$25,000	Incidental travel expenses incurred in support of the BSR project
Depreciation	$91,000	Included is the per-head share of workstation depreciation, the Cascade HP platform depreciation, and the depreciation expense associated with capitalized software purchases
Rents/Leases	$98,000	Expenses associated with the Mach1 computing platforms
Other Supplies and Expenses	$153,000	Incidental expenses associated with things such as training, reward and recognition, long-distance phone charges, and miscellaneous office supplies
Total Costs	$1,800,000	

Table 7-6 also includes explanations to support the budget estimates. Staff from the IS&T department included nine programmer/analysts, two database analysts, and two infrastructure technicians. Their compensation is based on employee change notices (ECNs), which provide actual salary and benefit information for employees when they are hired or transferred within the company. Notice that the compensation costs are the main part of the budget estimate. Having the largest part of cost estimates come from compensation costs is typical of estimates on many information technology projects. Also notice that the budget amount for compensation includes an allowance for raises and overload support or overtime pay. Other explanations include brief descriptions of the travel, depreciation, rents/leases, and other supplies and expenses for the project. It is very important to document assumptions and explanations when preparing cost estimates and cost budgets.

In addition to providing input for budgetary estimates, cost budgeting provides a cost baseline. A **cost baseline** is a time-phased budget that project managers use to measure and monitor cost performance. Estimating costs for each major project activity over time provides project managers and top management with a foundation for project cost control. You will learn more about entering cost data and establishing a cost baseline in Appendix A.

COST CONTROL

Project cost control includes monitoring cost performance, ensuring that only appropriate project changes are included in a revised cost baseline, and informing project stakeholders of authorized changes to the project that will affect costs. The cost baseline, performance reports, change requests, and the cost management plan are inputs to the cost control process. Outputs of this process are revised cost estimates, budget updates, corrective action, revised estimates for project completion, and lessons learned.

Several tools and techniques assist in project cost control. As shown in Appendix A, Project 2002 has many cost management features to help you enter budgeted costs, set a baseline, enter actuals, calculate variances, and run various cost reports. In addition to using software, however, there must be some change control system to define procedures for changing the cost baseline. This cost change control system is part of the integrated change control system described in Chapter 4, Project Integration Management. Since many projects do not progress exactly as planned, new or revised cost estimates are often required, as are estimates to evaluate alternate courses of action. Another very important tool for cost control is performance measurement. Although many general accounting approaches are available for measuring cost performance, earned value management (EVM) is a very powerful cost control technique that is unique to the field of project management.

Earned Value Management

Earned value management (EVM) is a project performance measurement technique that integrates scope, time, and cost data. Given a cost performance baseline, project managers and their teams can determine how well the project is meeting scope, time, and cost goals by entering actual information and then comparing it to the baseline. A **baseline** is the original project plan plus approved changes. Actual information includes whether or not a WBS item was completed or approximately how much of the work was completed, when the work actually started and ended, and how much it actually cost to do the completed work.

Figure 7-1 shows the BSR project's input form for collecting information for cost control and earned value management. This internal company form includes the following information:

- Descriptive information: The top line of this form lists the WBS number for an activity, a description of the activity, a revision number, and a revision date. This example of an input form reports information for WBS Item 6.8.1.2, a Level 4 WBS activity (based on the numbering scheme) called "Design Interface Process-Customer Information."

- Assignment information: This activity is assigned to someone with the initials SMC. Other assignment information includes the responsible person's role and number of hours available per day for the activity.

- Forecast information: The BSR project used a PERT weighted average to estimate the number of hours it would take to complete activities. Recall from Chapter 5 that the PERT weighted average uses most likely, optimistic, and pessimistic estimates. In this case, the PERT calculation results in a planned effort of thirty hours. The planned duration is five days.

- Description: This section provides a detailed description of the activity.

- Assumptions: This section documents major assumptions related to this particular WBS item.

- Results/Deliverables: This section briefly lists the main outcomes from the WBS item.

- Dependencies: This section lists the WBS number for any predecessor or successor activities.

WBS#: 6.8.1.2	Description:	Design Interface Process-Customer Information	Revision:	Revision Date:

Assignments	Forecast

| | Hours per day | Effort (in hours) | Calculated |

Responsible: SMC   Role: PA   Availability: 6

Involved: _____   Role: _____   Availability: _____
Involved: _____   Role: _____   Availability: _____

Involved: _____   Role: _____

Optimistic: 20
Most Likely: 30

Pessimistic: 40

Delay (Days): _____

Plan Effort: 30 Hrs

Plan Duration: 5 Days

Description	Assumptions
Develop an operational process design for the Customer Information interface from the Invoicing System to Oracle Receivables. This task will accept as input the business/functional requirements developed during the tactical analysis phase and produce as output a physical operational design, which provides the specifications, required for code development.	- All business rules and issues will be resolved prior to this task. - The Entity Relationship Diagram (ERD) and data model for Oracle Receivables and any Oracle extension required will be completed and available prior to this task. - The ERD for the Invoicing System will be completed and available prior to this task. - Few iterations of the review/modify cycle will be required.

Results / Deliverables	Dependencies

Process Design Document - Technical
 - Operation/Physical Data Flow Diagram
 - Process Specifications
 - Interface Data Map

Predecessors (WBS#): 4.7   Successors (WBS#):

Figure 7-1. Cost Control Input Form for Business Systems Replacement Project

This particular activity has not yet started, so it does not include the actual number of hours or the actual duration for the activity. Actual cost, actual duration, and percentage complete information is required in order to perform earned value management.

Earned value management involves calculating three values for each activity or summary activity from a project's WBS.

1. The **planned value (PV)**, formerly called the budgeted cost of work scheduled (BCWS), also called the budget, is that portion of the approved total cost estimate planned to be spent on an activity during a given period. Table 7-7 shows an example of earned value calculations. Suppose a project included a summary activity of purchasing and installing a new Web server. Suppose further that according to the plan, it would take one week and cost a total of $10,000 for the labor hours, hardware, and software involved. The planned value (PV) for that activity that week is, therefore, $10,000.

2. The **actual cost (AC)**, formerly called the actual cost of work performed (ACWP), is the total direct and indirect costs incurred in accomplishing work on an activity during a given period. For example, suppose it actually took two weeks and cost $20,000 to purchase and install the new Web server. Assume that $15,000 of these actual costs were incurred during Week 1 and $5,000 was incurred during Week 2. These amounts are the actual cost (AC) for the activity each week.

3. The **earned value (EV)**, formerly called the budgeted cost of work performed (BCWP), is an estimate of the value of the physical work actually completed. In Table 7-7, the earned value estimate after one week is $7500.

Table 7-7: Earned Value Calculations for One Activity After Week One

ACTIVITY	WEEK 1
Earned Value (EV)	7,500
Planned Value (PV)	10,000
Actual Cost (AC)	15,000
Cost Variance (CV)	−7,500
Schedule Variance (SV)	−2,500
Cost Performance Index (CPI)	50%
Schedule Performance Index (SPI)	75%

The earned value calculations in Table 7-7 are carried out as follows:

CV = 7,500 − 15,000	= −7,500
SV = 7,500 − 10,000	= −2,500
CPI = 7,500/15,000	= 50%
SPI = 7,500/10,000	= 75%

Table 7-8 summarizes the formulas used in earned value management. Note that the formulas for variances and indexes start with EV, the earned value. Variances are calculated by subtracting the actual cost or planned value from EV, and indexes are calculated by dividing EV by the actual cost or planned value. After you total the EV, AC, and PV data for all activities on a project, you can use the CPI and SPI to project how much it will cost. Given the Budget at Completion and original time estimate, you can divide by the appropriate index to calculate the Estimate at Completion (EAC) and estimated time to complete, assuming performance remains the same. There are no standard acronyms for the term estimated time to complete or original time estimate.

Table 7-8: Earned Value Formulas

TERM	FORMULA
Earned value	EV = PV to date X percent complete
Cost variance	CV = EV – AC
Schedule variance	SV = EV – PV
Cost performance index	CPI = EV/AC
Schedule performance index	SPI = EV/PV
Estimate at completion (EAC)	EAC = BAC/CPI
Estimated time to complete	Original time estimate/SPI

Cost variance (CV) is the earned value minus the actual cost. In other words, cost variance shows the difference between the estimated cost of an activity and the actual cost of that activity. If cost variance is a negative number, it means that performing the work cost more than planned. If cost variance is a positive number, it means that performing the work cost less than planned.

Schedule variance (SV) is the earned value minus the planned value. Schedule variance shows the difference between the scheduled completion of an activity and the actual completion of that activity. A negative schedule variance means that it took longer than planned to perform the work, and a positive schedule variance means that it took less time than planned to perform the work.

The **cost performance index (CPI)** is the ratio of earned value to actual cost and can be used to estimate the projected cost of completing the project. If the cost performance index is equal to one or 100 percent, then the planned and actual costs are equal, or the costs are exactly as budgeted. If the cost performance index is less than one or less than 100 percent, the project is over budget. If the cost performance index is greater than one or more than 100 percent, the project is under budget.

The **schedule performance index (SPI)** is the ratio of earned value to planned value and can be used to estimate the projected time to complete the project. Similar to the cost performance index, a schedule performance index of one or 100 percent means the project is on schedule. If the schedule performance index is greater than one or 100 percent, then the project is ahead of schedule. If the schedule performance index is less than one or 100 percent, the project is behind schedule.

Note that in general, *negative numbers for cost and schedule variance indicate problems in those areas.* Negative numbers mean the project is costing more than planned or taking longer than planned. Likewise, *CPI and SPI less than one or less than 100 percent also indicate problems.*

Earned value calculations for all project activities (or summary level activities) are required to estimate the earned value for the entire project. Some activities may be over budget or behind schedule, but others may be under budget and ahead of schedule. By adding all of the earned values for all project activities, you can determine how the project as a whole is performing.

Figure 7-2 provides sample earned value information for a one-year project. This project had a planned total cost of $100,000. The spreadsheet shows actual cost and percentage complete information for the first five months, or through the end of May. Notice the "% Complete" column at the upper-right side, or in column O, of the spreadsheet. The earned value (EV) for each activity in this example is calculated by multiplying the percent complete value by the planned value. The "Monthly Actual Cost" row (row 15 of the spreadsheet) shows the actual cost each month for the project's activities through May. By calculating the total planned costs, the total actual costs, and the earned value costs, you can determine the cost variance, schedule variance, cost performance index, and schedule performance index for the entire project (see cells A18 through B25 in Figure 7-2).

	A	B	C	D	E	F	G	H	I	J	K	L	M	N	O	P
		Jan	Feb	Mar	Apr	May	Jun	Jul	Aug	Sep	Oct	Nov	Dec	PV	% Complete	EV
1	Activity															
2	Plan and staff project	4,000	4,000											8,000	100	8,000
3	Analyze requirements		6,000	6,000										12,000	100	12,000
4	Develop ERDs			4,000	4,000									8,000	100	8,000
5	Design database tables				6,000	4,000								10,000	100	10,000
6	Design forms, reports, and queries					8,000	4,000							12,000	50	6,000
7	Construct working prototype						10,000							10,000	-	-
8	Test/evaluate prototype						2,000	6,000						8,000	-	-
9	Incorporate user feedback							4,000	6,000	4,000				14,000	-	-
10	Test system								4,000	4,000	2,000			10,000	-	-
11	Document system										3,000	1,000		4,000	-	-
12	Train users											4,000	4,000	4,000	-	-
13	Monthly Planned Value (PV)	4,000	10,000	10,000	10,000	12,000	16,000	10,000	6,000	8,000	4,000	5,000	5,000	100,000		44,000
14	Cumulative Planned Value (PV)	4,000	14,000	24,000	34,000	46,000	62,000	72,000	78,000	86,000	90,000	95,000	100,000			
15	Monthly Actual Cost (AC)	4,000	11,000	11,000	12,000	15,000										
16	Cumulative Actual Cost (AC)	4,000	15,000	26,000	38,000	53,000										
17	Monthly Earned Value (EV)	4,000	10,000	10,000	10,000	10,000										
18	Cumulative Earned Value (EV)	4,000	14,000	24,000	34,000	44,000										
19	Project EV as of May 31	44,000														
20	Project PV as of May 31	46,000														
21	Project AC as of May 31	$ 53,000														
22	CV=EV-AC	$ (9,000)														
23	SV=EV-PV	$ (2,000)														
24	CPI=EV/AC	83%														
25	SPI=EV/PV	96%														
26	Estimate at Completion (EAC)	$120,455	(original plan of $100,000 divided by CPI of 83%)													
27	Estimated time to complete	12.55	(original plan of 12 months divided by SPI of 96%)													

Figure 7-2. Earned Value Calculations for a One-Year Project After Five Months

In this example, the cost variance is –$9,000, and the schedule variance is –$2,000. These values mean that the project is both over budget and behind schedule after five months. The cost performance and schedule performance indexes are 83 percent and 96 percent, respectively. The cost performance index can be used to calculate the **estimate at completion (EAC)**—an estimate of what it will cost to complete the project based on performance to date. Similarly, the schedule performance index can be used to calculate an estimated time to complete the project. For example, in Figure 7-2 the estimate at completion is $120,455, or $100,000/83 percent. The estimated time to complete the project is 12.55 months, or twelve months/96 percent (see cells A26 through B27).

You can graph earned value information to track project performance. Figure 7-3 shows an earned value chart for the one-year sample project from Figure 7-2. The chart includes three lines and two points, as follows:

■ Planned value (PV), the cumulative planned amounts for all activities by month. Note that the planned value line extends for the estimated length of the entire project.

■ Actual cost (AC), the cumulative actual amounts for all activities by month

■ Earned value (EV), the cumulative earned value amounts for all activities by month

■ **Budget at Completion (BAC)**, the original total budget for the project, or $100,000 in this example. The BAC point is plotted on the chart at the original time estimate of twelve months.

■ Estimate at Completion (EAC), estimated to be $120,455, as described earlier. This EAC point is plotted on the chart at the estimated time to complete of 12.55 months.

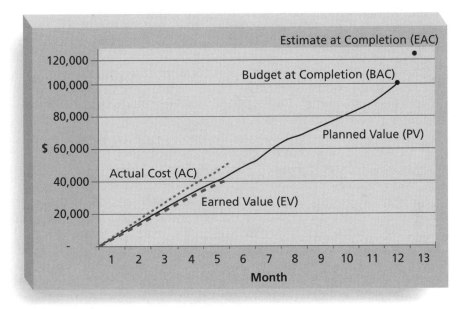

Figure 7-3. Earned Value Chart for Project After Five Months

Viewing earned value information in chart form helps you visualize how the project is performing. For example, you can see the planned performance by looking at the planned value line. If the project goes as planned, it will finish in twelve months and cost $100,000, represented by budget at completion (BAC). Notice in the example that the actual cost line is always right on or above the earned value line. When the actual cost line is right on or above the earned value line, costs are equal to or more than planned. The planned value line is pretty close to the earned value line, just slightly higher in the last month. This relationship means that the project has been right on schedule until the last month, when the project fell behind schedule.

Top managers overseeing multiple projects often like to see performance information in a graphical form such as the earned value chart in Figure 7-3. For example, in the opening case, the government officials were reviewing earned value charts and estimates at completion for several different projects. Earned value charts allow you to quickly see how projects are performing. If there are serious cost and schedule performance problems, top management may decide to terminate projects or take other corrective action. The estimates at completion are important inputs to budget decisions, especially if total funds are limited. Earned value management is an important technique because, when used effectively, it helps top management and project managers evaluate progress and make sound management decisions.

However, apart from government agencies and their contractors, many organizations do not use earned value management on many projects. Two reasons why organizations do not widely use earned value management are its focus on tracking actual performance versus planned performance and the importance of percentage completion data in making calculations. Many projects, particularly information technology projects, do not have good planning information, so tracking performance against a plan might produce misleading information. Several cost estimates are usually made on information technology projects, and keeping track of the most recent cost estimate and the actual costs associated with it could be cumbersome. In addition, estimating percentage completion of tasks might produce misleading information. What does it really mean to say that a task is 75 percent complete after three months? Such a statement is often not synonymous with saying the task will be finished in one more month or after spending an additional 25 percent of the planned budget.

To make earned value management simpler to use, organizations can modify the level of detail and still reap the benefits of the technique. For example, you can use percentage completion data such as 0 percent for items not yet started, 50 percent for items in progress, and 100 percent for completed tasks. As long as the project is defined in enough detail, this simplified percentage completion data should provide enough summary information to allow managers to see how well a project is doing overall. You can get very accurate total project performance information using these simple percentage complete amounts. For example, using simplified percentage complete amounts for a one-year project with weekly reporting and an average task or work packet size of one week, you can expect about a 1 percent error rate.[10]

You can also only enter and collect earned value data at summary levels of your WBS. Quentin Fleming, co-author of the book *Earned Value Project*

[10] Brandon, Daniel M., Jr, "Implementing Earned Value Easily and Effectively," *Project Management Journal* (June 1998) 29(2), 11–18.

Management, Second Edition, often gives presentations about earned value management. Many people express their frustration in trying to collect such detailed information. Quentin explains that you do not have to collect information at the work package level in order to use earned value management. It is most important to have a deliverable-oriented WBS, and many WBS items can summarize several sub-deliverables. For example, you might have a WBS for a house that includes items for each room in the house. Just collecting earned value data for each room would provide meaningful information instead of trying to collect the detailed information on each component of the room, such as flooring, furniture, lighting, and so on.

Earned value management is the primary method available for integrating performance, cost, and schedule data. It can be a powerful tool for project managers and top management to use in evaluating project performance. Project management software, such as Project 2002, includes tables for collecting earned value data and reports that calculate variance information. See the cost management section of Appendix A for an example of using earned value management. Another approach to evaluating the performance of multiple projects is project portfolio management.

Project Portfolio Management

As mentioned in Chapter 1, many organizations now collect and control an entire suite of projects or investments as one set of interrelated activities in one place—a portfolio. There can be a portfolio for information technology projects, for example, and portfolios for other types of projects. An organization can view project portfolio management as having five levels, from simplest to most complex, as follows:

1. Put all your projects in one database.
2. Prioritize the projects in your database.
3. Divide your projects into two or three budgets based on type of investment, such as utilities or required systems to keep things running, incremental upgrades, and strategic investments.
4. Automate the repository.
5. Apply modern portfolio theory, including risk-return tools that map project risk on a curve.

For example, Jane Walton, the project portfolio manager for information technology projects at Schlumberger, saved the company $3 million in one year by organizing the organization's 120 information technology projects into a portfolio. Manufacturing companies used project portfolio management in the 1960s, and Walton anticipated the need to justify investments in information technology projects just as managers have to justify capital investment projects. She found that 80 percent of the organization's projects overlapped, and fourteen separate projects were trying to accomplish the same thing.

Other managers, such as Douglas Hubbard, president of a consulting firm, see the need to use project portfolio management, especially for information technology projects. Hubbard suggests, "IT investments are huge, risky investments. It's time we do this."[11]

Project portfolio managers can use spreadsheet software to develop and manage project portfolios, or use sophisticated software designed to help manage project portfolios. Several software tools available today help project portfolio managers summarize earned value and project portfolio information, as described in the following section.

USING SOFTWARE TO ASSIST IN PROJECT COST MANAGEMENT

Most organizations use software to assist in various activities related to project cost management. Spreadsheets are a common tool for performing resource planning, cost estimating, cost budgeting, and cost control. Many companies also use more sophisticated and centralized financial applications software to provide important cost-related information to accounting and finance personnel. This section will focus specifically on how you can use project management software in cost management. Appendix A includes a section on using the cost management features in Microsoft Project 2002.

Project management software can be a very helpful tool during each project cost management process. It can help you study overall project information or focus on tasks that are over a specified cost limit. You can use the software to assign costs to resources and tasks, prepare cost estimates, develop cost budgets, and monitor cost performance. Microsoft Project 2002 has several standard cost reports: cash flow, budget, over budget tasks, over budget resources, and earned value reports. For several of these reports, you must enter percentage completion information and actual costs, just as you need this information when manually calculating earned value or other analyses.

Although Microsoft Project 2002 has a fair amount of cost management features, many information technology project managers use other tools to manage cost information; they do not know that they can use Microsoft Project 2002 for cost management. As with most software packages, users need training to use the software effectively and understand the features available. Instead of using dedicated project management software for cost management,

[11] Berinato, Scott. "Do the Math," *CIO Magazine* (October 1, 2001).

some information technology project managers use company accounting systems; others use spreadsheet software to achieve more flexibility. Project managers who use other software often do so because these other systems are more generally accepted in their organizations and more people know how to use them. On the other hand, some companies have developed methods to link data between their project management software and their main accounting software systems.

Many organizations are starting to use enterprise project management software to organize and analyze all types of project data into project portfolios and across the entire enterprise. Enterprise project management software, such as Microsoft Solution for Enterprise Project Management (a different product than Project 2002 Standard), Pacific Edge's The Edge and The Edge for IT, and PlanView's Web Software, integrate information from multiple projects to show the project's status, or health. For example, Figure 7-4 provides a sample screen from PlanView. Note the charts and text in the upper half of the screen showing the number and percentage of projects in this project portfolio that are on target and in trouble in terms of schedule and cost variance. The bottom half of the screen lists the names of individual projects, percent complete, schedule variance, cost variance, budget variance, and risk percentage. Providing integrated data in a graphical format helps top managers make decisions concerning projects.

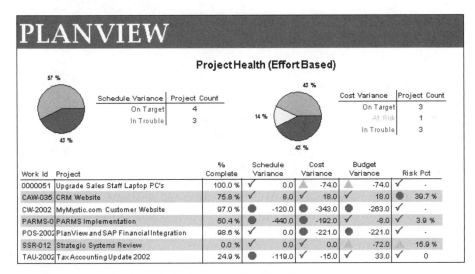

Figure 7-4. ProjectHealth Screen from PlanView Software

Project cost management is an important part of a project manager's job. Because top management often focus on the bottom line, it is crucial for project managers to speak their language and understand how their projects contribute to the organization's financial success.

CASE WRAP-UP

After talking to his colleagues about the meeting, Juan had a better idea of the importance of project cost management. He understood the value of doing detailed studies before making major expenditures on new projects, especially after learning about the high cost of correcting defects late in a project. He also learned how important it is to develop good cost estimates and keep costs on track. Government officials cancelled several projects when the project managers showed how poorly the projects were performing and admitted that they did not do much planning and analysis early in the projects. Juan knew that he could not focus on just the technical aspects of projects if he wanted to move ahead in his career. He began to wonder whether several projects the city was considering were really worth the taxpayers' money. Issues of cost management added a new dimension to Juan's job.

CHAPTER SUMMARY

Project cost management is a traditionally weak area of information technology projects. Information technology project managers must acknowledge the importance of cost management and take responsibility for improving resource planning, cost estimating, budgeting, and cost control.

Project managers must understand several basic principles of cost management in order to be effective in managing project costs. Important concepts include profits and profit margins, life cycle costing, cash flow analysis, sunk costs, and learning curve theory.

Resource planning involves determining what people, equipment, and materials to use to perform project activities. It is important to evaluate alternatives and use expert judgment in determining resource requirements.

Cost estimating is a very important part of project cost management. There are several types of cost estimates, including rough order of magnitude (ROM), budgetary, and definitive. Each type of estimate is done during different stages of the project life cycle, and each has a different level of accuracy. There are four basic tools and techniques for developing cost estimates: analogous estimating, bottom-up estimating, parametric modeling, and computerized tools.

The main parts of a cost estimate include an objective statement, scope, assumptions, cost/benefit analysis, cash flow analysis, budget breakdown, and explanations or supporting detail.

Cost budgeting involves allocating costs to individual work items over time. It is important to understand how particular organizations prepare budgets so estimates are made accordingly.

Project cost control includes monitoring cost performance, reviewing changes, and notifying project stakeholders of changes related to costs. Many basic accounting and finance principles relate to project cost management. However, earned value management is the main method used for measuring project performance. Earned value management integrates scope, cost, and schedule information. Project portfolio management allows organizations to collect and control an entire suite of projects or investments as one set of inter-related activities.

Several software products assist with project cost management. Microsoft Project 2002 has many cost management features, including earned value management. Enterprise project management helps managers evaluate inte-grated project data in a graphical format.

DISCUSSION QUESTIONS

1. Discuss why many information technology professionals have a poor atti-tude toward project cost management and how this attitude might affect the poor track record for completing projects within budget.

2. Explain some of the basic principles of cost management, such as profits, life cycle costs, tangible and intangible costs and benefits, direct and indi-rect costs, reserves, and so on.

3. Give examples of when you would prepare rough order of magnitude, budgetary, and definitive cost estimates for an information technology project. Give an example of using each of the following techniques for creating a cost estimate: analogous, parametric, bottom-up, and comput-erized tools.

4. Explain what happens during the cost budgeting process.

5. Explain why earned value management is the preferred method for mea-suring project performance and speculate as to why it is not used more often. What are some general rules of thumb for deciding if cost variance, schedule variance, cost performance index, and schedule performance index numbers are good or bad?

6. What is project portfolio management? Can project managers use it with earned value management?

7. Describe several types of software that project managers can use to sup-port project cost management.

EXERCISES

1. Given the following information for a one-year project, answer the following questions. Recall that PV is the planned value, EV is the earned value, AC is the actual cost, and BAC is the budget at completion.

   PV= $23,000

   EV= $20,000

   AC= $25,000

   BAC= $120,000

   a. What is the cost variance, schedule variance, cost performance index (CPI), and schedule performance index (SPI) for the project?
   b. How is the project doing? Is it ahead of schedule or behind schedule? Is it under budget or over budget?
   c. Use the CPI to calculate the estimate at completion (EAC) for this project. Is the project performing better or worse than planned?
   d. Use the schedule performance index (SPI) to estimate how long it will take to finish this project.

2. Using the data in Figure 7-2 for cumulative PV, AC, and EV, create an earned value chart similar to the one shown in Figure 7-3.

3. Create a cost estimate for building a new, state-of-the-art multimedia classroom for your organization within the next six months. The classroom should include twenty high-end personal computers with appropriate software for your organization, a network server, Internet access for all machines, a teacher station, and a projection system. Be sure to include personnel costs associated with the project management for this project. Document the assumptions you made in preparing the estimate and provide explanations for key numbers.

4. Research online information, textbooks, and local classes for using the cost management features of Microsoft Project 2002. Ask three different people in different information technology organizations that use Microsoft Project 2002 if they use the cost management features and in what ways they use them. Write a brief report of what you learned from your research.

5. Research three different products to assist organizations in project portfolio management. Microsoft, Pacific Edge, PlanView, Artemis, and Welcom are just a few companies that provide products for project portfolio management. Write a one- to two-page paper summarizing your results.

6. Perform the project cost management part of one of the exercises or case studies provided in Appendix A or Appendix C. Print the reports or information required and write a one-page paper interpreting the results.

MINICASE

You have decided that you are now an expert in project cost estimating, and you want to hold some public classes or workshops on this topic. You completed some training in the past, which has given you confidence in your abilities. Your first project is to develop materials, then market and hold your first seminar. You plan to do all of the work within the next six months.

You want to create a good cost estimate for this project. Below are some of your assumptions:

- You will charge $600 per person for a two-day class.
- You estimate that 30 people will attend.
- Your fixed costs include $500 to rent a room for both days, setup fees of $400 for registration, and $300 for designing a postcard for advertising.
- You will not include any of your labor costs for this estimate, but you estimate that you will spend at least 150 hours developing materials, managing the project, and giving the actual class. You would like to know what your time is worth given different scenarios.
- You will order 5,000 postcards, mail 4,000, and distribute the rest to friends and colleagues.

Your variable costs include the following:

a. $5 per person for registration plus 4 percent of the class fee per person to handle credit card processing. Assume everyone pays by credit card.

b. $.40 per postcard if you order 5,000 or more

c. $.25 per postcard for mailing and postage

d. $25 per person for beverages and lunch

e. $30 per person for class handouts

Assume 30 people register for and attend the class.

Part 1: Using Excel, create a spreadsheet to calculate your projected total costs, total revenues, and total profits given the preceding information. Be sure to have input cells for any variables that might change, such as the cost of postage, handouts, and so on. Calculate your profits based on the following number of people who attend: 10, 20, 30, 40, 50, and 60. In addition, calculate what your time would be worth per hour based on the number of students. Try to use the Excel Scenario feature (under the Tools menu) to create a scenario pivot table showing total profits based on the number of people who attend. You could also create a data table showing the profits based on the number of students. If you are unfamiliar with the Scenario feature or data tables, just repeat the calculations for each possibility of 10, 20, 30, 40, 50, and 60 students.

Part 2: Research the local market (or the location of your choice) for short classes on topics similar to project cost estimating. Find at least three organizations that offer estimating courses. You can search the Registered Education

Provider's database for organizations from PMI's Web site (*www.pmi.org*), using "estimating" in the title. Given your research into similar classes and the work you think is required for this project, what assumptions, variables, or costs would you change from the information provided above? Justify your changes to the estimate, and create a spreadsheet with your best estimate for this project. Then write a paragraph describing whether you think this particular project is worth pursuing.

Part 3: Review Appendix A, especially the section on project human resource and cost management. Do you think you could create a cost estimate for this workshop using Project 2002? Why or why not? Write a one- to two-page paper describing why this type of project (giving a workshop) is or is not a good candidate for using Project 2002 cost management features.

Suggested Readings

1. Berinato, Scott. "Do the Math," *CIO Magazine*, (*www.cio.com/archive/100101/math_content.html*) (October 1, 2001).

   This article explains the benefits of applying project portfolio management to information technology projects. It summarizes the five levels of portfolio management and provides several examples of how companies use it to help align information technology projects with the business.

2. Brandon, Daniel M., Jr. "Implementing Earned Value Easily and Effectively." *Project Management Journal* (June 1998): 29(2), 11–18.

   Brandon describes the benefits of using earned value management and the reasons that more organizations do not use it. This paper provides suggestions for overcoming some of the problems associated with this technique.

3. Fleming, Quentin W. and Joel M. Koppelman. *Earned Value Project Management, Second Edition*. Project Management Institute (2000).

   This book provides a detailed description of the basic concepts involved in earned value management and a brief history of its use. The authors show how earned value management can be used to assess the status of any kind of project as well as forecast project outcomes.

4. Hamaker, Joseph W. "But What Will It Cost? The History of NASA Cost Estimating." NASA Web site (*http://www.jsc.nasa.gov/bu2/hamaker.html*) (2000).

   This article provides a wealth of information on how NASA develops cost estimates. Sections of the Web site include cost estimating models, handbooks to assist in cost estimating, salaries, and many links to great references and software.

5. Kemerer, Chris F. "Software Cost Estimation Models." *Software Project Management Readings and Cases*. Chicago: McGraw-Hill (1997).

> *This article summarizes the three basic types of software cost estimating models. The earliest models are sometimes called the economic stream and are based on work done at TRW and IBM. Boehm's COCOMO is the most popular of these economic stream models. The second stream, called the Rayleigh stream, is based on the work of Putnam. The third stream, called the function point stream, first appeared in 1979 and is based on work done by Albrecht.*

6. Roetzheim, William. "Estimating Software Costs," Cost Xpert Group, Inc. (*www.costxpert.com/resource_center/articles.html*) (2003).

> *This article is the first in a series of four articles written by the CEO of the Cost Xpert Group, Inc. It summarizes the traditional measures of estimating the size of a program, such as Lines of Code and Function Points, and introduces other approaches.*

KEY TERMS

- **actual cost (AC)** — the total of direct and indirect costs incurred in accomplishing work on an activity during a given period, formerly called the actual cost of work performed (ACWP)
- **analogous estimates** — a cost estimating technique that uses the actual cost of a previous, similar project as the basis for estimating the cost of the current project, also called **top-down estimates**
- **baseline** — the original project plan plus approved changes
- **bottom-up estimates** — a cost estimating technique based on estimating individual work items and summing them to get a project total
- **budget at completion (BAC)** — the original total budget for a project
- **budgetary estimate** — a cost estimate used to allocate money into an organization's budget
- **cash flow analysis** — a method for determining the estimated *annual* costs and benefits for a project
- **COCOMO II** — a newer, computerized cost-estimating tool based on Boehm's original model that allows one to estimate the cost, effort, and schedule when planning a new software development activity
- **computerized tools** — cost-estimating tools that use computer software, such as spreadsheets and project management software
- **Constructive Cost Model (COCOMO)** — a parametric model developed by Barry Boehm for estimating software development costs
- **contingency reserves** — dollars included in a cost estimate to allow for future situations that may be partially planned for (sometimes called **known unknowns**) and are included in the project cost baseline

- **cost baseline** — a time-phased budget that project managers use to measure and monitor cost performance
- **cost budgeting** — allocating the overall cost estimate to individual work items to establish a baseline for measuring performance
- **cost control** — controlling changes to the project budget
- **cost estimating** — developing an approximation or estimate of the costs of the resources needed to complete the project
- **cost management plan** — a document that describes how cost variances will be managed on the project
- **cost performance index (CPI)** — the ratio of earned value to actual cost; can be used to estimate the projected cost to complete the project
- **cost variance (CV)** — the earned value minus the actual cost
- **definitive estimate** — a cost estimate that provides an accurate estimate of project costs
- **direct costs** — costs that are related to a project and can be traced back in a cost-effective way
- **earned value (EV)** — the percentage of work actually completed multiplied by the planned cost, formerly called the **budgeted cost of work performed (BCWP)**
- **earned value management (EVM)** — a project performance measurement technique that integrates scope, time, and cost data
- **estimate at completion (EAC)** — an estimate of what it will cost to complete the project based on performance to date
- **function points** — technology independent assessments of the functions involved in developing a system
- **indirect costs** — costs that are related to the project but cannot be traced back in a cost-effective way
- **intangible costs or benefits** — costs or benefits that are difficult to measure in monetary terms
- **internal rate of return (IRR)** — the discount rate that makes the net present value equal to zero; also called **time-adjusted rate of return**
- **learning curve theory** — a theory that states that when many items are produced repetitively, the unit cost of those items normally decreases in a regular pattern as more units are produced
- **legacy systems** — older information systems that usually ran on an old mainframe computer
- **life cycle costing** — considers the total cost of ownership, or development plus support costs, for a project
- **management reserves** — dollars included in a cost estimate to allow for future situations that are unpredictable (sometimes called **unknown unknowns**)
- **overrun** — the additional percentage or dollar amount by which actual costs exceed estimates

- **parametric modeling** — a cost-estimating technique that uses project characteristics (parameters) in a mathematical model to estimate project costs
- **planned value (PV)** — that portion of the approved total cost estimate planned to be spent on an activity during a given period, formerly called the budgeted cost of work scheduled (BCWS)
- **profit margin** — the ratio between revenues and profits
- **profits** — revenues minus expenses
- **project cost management** — the processes required to ensure that the project is completed within the approved budget
- **reserves** — dollars included in a cost estimate to mitigate cost risk by allowing for future situations that are difficult to predict
- **resource planning** — determining what resources (people, equipment, and materials) and what quantities of each resource should be used to perform project activities
- **rough order of magnitude (ROM) estimate** — a cost estimate prepared very early in the life of a project to provide a rough idea of what a project will cost
- **schedule performance index (SPI)** — the ratio of earned value to planned value; can be used to estimate the projected time to complete a project
- **schedule variance (SV)** — the earned value minus the planned value
- **sunk cost** — money that has been spent in the past
- **tangible costs** or **benefits** — costs or benefits that can be easily measured in dollars

8

Project Quality Management

Objectives

After reading this chapter, you will be able to:

1. *Understand the importance of project quality management for information technology products and services*
2. *Define project quality management and understand how quality relates to various aspects of information technology projects*
3. *Describe quality planning and its relationship to project scope management*
4. *Discuss the importance of quality assurance*
5. *List the three outputs of the quality control process*
6. *Understand the tools and techniques for quality control, such as Pareto analysis, statistical sampling, Six Sigma, quality control charts, and testing*
7. *Describe important concepts related to Six Sigma and how it helps organizations improve quality and reduce costs*
8. *Summarize the contributions of noteworthy quality experts to modern quality management*
9. *Understand how the Malcolm Baldrige Award and ISO 9000 standard promote quality in project management*
10. *Describe how leadership, cost, organizational influences, and maturity models relate to improving quality in information technology projects*
11. *Discuss how software can assist in project quality management*

OPENING CASE

A large medical instruments company just hired Scott Daniels, a senior consultant from a large consulting firm, to lead a project to resolve the quality problems with the company's new Executive Information System (EIS). A team of internal programmers and analysts worked with several company executives to develop this new system. Many executives who had never used computers before were hooked on the new, user-friendly EIS. They loved the way the system allowed them to track sales of various medical instruments quickly and easily by product, country, hospital, and sales representative. After successfully testing the new EIS with several executives, the company decided to make the system available to all levels of management.

Unfortunately, several quality problems developed with the new EIS after a few months of operation. People were complaining that they could not get into the system. The system started going down a couple of times a month, and the response time was reportedly getting slower. Users complained when they could not access information within a few seconds. Several people kept forgetting how to log in to the system, thus increasing the number of calls to the company's Help Desk. There were complaints that some of the reports in the system gave inconsistent information. How could a summary report show totals that were not consistent with a detailed report on the same information? The executive sponsor of the EIS wanted the problems fixed quickly and accurately, so he decided to hire an expert in quality from outside the company whom he knew from past projects. Scott Daniels' job was to lead a team of people from both the medical instruments company and his own firm to identify and resolve quality-related issues with the EIS and to develop a plan to help prevent quality problems from happening on future information technology projects.

QUALITY OF INFORMATION TECHNOLOGY PROJECTS

Most people have heard jokes about what cars would be like if they followed a development history similar to that of computers. A well-known joke that has been traveling around the Internet goes as follows:

"At a recent computer exposition (COMDEX), Bill Gates, the founder and CEO of Microsoft Corporation, stated: "If General Motors had kept up with technology like the computer industry has, we would all be driving twenty-five dollar cars that got 1000 miles to the gallon." In response to Gates' comments, General Motors issued a press release stating: "If GM had developed technology like Microsoft, we would all be driving cars with the following characteristics:

1. For no reason whatsoever your car would crash twice a day.
2. Every time they repainted the lines on the road, you would have to buy a new car.
3. Occasionally, your car would die on the freeway for no reason, and you would just accept this, restart, and drive on.
4. Occasionally, executing a maneuver such as a left turn would cause your car to shut down and refuse to restart, in which case you would have to reinstall the engine.
5. Only one person at a time could use the car, unless you bought 'Car95' or 'CarNT.' But then you would have to buy more seats.
6. Macintosh would make a car that was powered by the sun, reliable, five times as fast, and twice as easy to drive, but would run on only five percent of the roads.
7. The oil, water temperature, and alternator warning lights would be replaced by a single 'general car default' warning light.
8. New seats would force everyone to have the same size hips.
9. The airbag system would say 'Are you sure?' before going off.
10. Occasionally, for no reason whatsoever, your car would lock you out and refuse to let you in until you simultaneously lifted the door handle, turned the key, and grabbed hold of the radio antenna.
11. GM would require all car buyers to also purchase a deluxe set of Rand McNally road maps (now a GM subsidiary), even though they neither need them nor want them. Attempting to delete this option would immediately cause the car's performance to diminish by 50 percent or more. Moreover, GM would become a target for investigation by the Justice Department.
12. Every time GM introduced a new model car, buyers would have to learn how to drive all over again because none of the controls would operate in the same manner as the old car.
13. You would press the Start button to shut off the engine."[1]

Most people simply accept poor quality from many information technology products. So what if your computer crashes a couple of times a month? Just make sure you back up your data. So what if you cannot log in to the corporate intranet or the Internet right now? Just try a little later when it is less busy. So what if the latest version of your word-processing software was shipped with several known bugs? You like the software's new features, and all new software has bugs. Is quality a real problem with information technology projects?

[1] This joke was found on hundreds of Web sites and printed in the *Consultants in Minnesota Newsletter*, Independent Computer Consultants Association, December 1998.

Yes, it is! Information technology is not just a luxury available in some homes, schools, or offices. Many companies throughout the world provide all employees with access to computers, and about 75 percent of U.S. homes have computers. The majority of people in the United States now use the Internet, and usage in other countries continues to grow rapidly. It took only five years for fifty million people to use the Internet compared to twenty-five years for fifty million people to use telephones. Many aspects of our daily lives depend on high-quality information technology products. Food is produced and distributed with the aid of computers; cars have computer chips to track performance; children use computers to help them learn in school; corporations depend on computers for many business functions; and millions of people rely on computers for entertainment and personal communications. Many information technology projects develop mission-critical systems that are used in life-and-death situations, such as navigation systems on board airplanes and computer components built into medical equipment. When one of these systems does not function correctly, it is much more than a slight inconvenience.

What Went Wrong?

- In 1981, a small timing difference caused by a computer program change created a 1/67 (or 1 in 67) chance that the space shuttle's five on-board computers would not synchronize. The error caused a launch abort.[2]

- In 1986, two hospital patients died after receiving fatal doses of radiation from a Therac 25 machine. A software problem caused the machine to ignore calibration data.[3]

- In one of the biggest software errors in banking history, Chemical Bank mistakenly deducted about $15 million from more than 100,000 customer accounts one evening. The problem resulted from a single line of code in an updated computer program that caused the bank to process every withdrawal and transfer at its automated teller machines (ATMs) twice. For example, a person who withdrew $100 from an ATM had $200 deducted from his or her account, though the receipt only indicated a withdrawal of $100. The mistake affected 150,000 transactions from Tuesday night through Wednesday afternoon.[4]

- In 1996, Apple Computer's PowerBook 5300 model had problems with lithium-ion battery packs catching fire, causing Apple to halt shipments and replace all the packs with nickel-metal-hydride batteries. Other quality problems also surfaced, such as cracks in the PowerBook's plastic casing and a faulty electric power adapter.[5]

[2] *Design News* (February 1988).
[3] *Datamation* (May 1987).
[4] *The New York Times* (February 18, 1994).
[5] "Apple to Try, Try Again," *Information Week* (July 29,1996) 15–16.

■ Hundreds of newspapers and Web sites ran stories about the "Melissa" virus in March of 1999. The rapidly spreading computer virus forced several large corporations to shut down their e-mail servers as the virus rode the Internet on a global rampage, according to several leading network security companies.[6]

Before you can improve the quality of information technology projects, it is important to understand the basic concepts of project quality management and the development of modern quality management.

WHAT IS PROJECT QUALITY MANAGEMENT?

Project quality management is a difficult knowledge area to define. The International Organization for Standardization (ISO) defines **quality** as "the totality of characteristics of an entity that bear on its ability to satisfy stated or implied needs." Many people spent many hours coming up with this definition, yet it is still very vague. Other experts define quality based on conformance to requirements and fitness for use. **Conformance to requirements** means the project's processes and products meet written specifications. For example, if the project scope statement requires delivery of one hundred Pentium 4 computers, you could easily check whether the correct computers had been delivered. **Fitness for use** means a product can be used as it was intended. If these Pentium 4 computers were delivered without monitors or keyboards and were left in boxes on the customer's shipping dock, the customer might not be satisfied because the computers would not be fit for use. The customer may have assumed that the delivery included monitors and keyboards, unpacking the computers, and installation so they would be ready to use.

The purpose of **project quality management** is to ensure that the project will satisfy the needs for which it was undertaken. Recall that project management involves meeting or exceeding stakeholder needs and expectations. The project team must develop good relationships with key stakeholders, especially the main customer for the project, to understand what quality means to them. *After all, the customer ultimately decides if quality is acceptable.* Many technical projects fail because the project team focuses only on meeting the written requirements for the main products being produced and ignores other stakeholder needs and expectations for the project. For example, the project team should know what successfully delivering one hundred Pentium 4 computers means to the customer.

Quality, therefore, must be on an equal level with project scope, time, and cost. If a project's stakeholders are not satisfied with the quality of the project

___

[6] Richtel, Matt, "New Fast-Spreading Virus Takes Internet by Storm," *The New York Times* on the Web (March 28, 1999) *http://www.nytimes.com.*

management or the resulting products of the project, the project team will need to adjust scope, time, and cost to satisfy stakeholder needs and expectations. Meeting only written requirements for scope, time, and cost is not sufficient. To achieve stakeholder satisfaction, the project team must develop a good working relationship with all stakeholders and understand their stated or implied needs.

Project quality management involves three main processes:

- **Quality planning** includes identifying which quality standards are relevant to the project and how to satisfy those standards. Incorporating quality standards into project design is a key part of quality planning. For an information technology project, quality standards might include allowing for system growth, planning a reasonable response time for a system, or ensuring that the system produces consistent and accurate information. Quality standards can also apply to information technology services. For example, you can set standards for how long it should take to get a reply from a Help Desk or how long it should take to ship a replacement part for a hardware item under warranty.

- **Quality assurance** involves periodically evaluating overall project performance to ensure the project will satisfy the relevant quality standards. The quality assurance process involves taking responsibility for quality during the project as well as at the end of the project. Top management must take the lead in emphasizing the roles all employees play in quality assurance, especially senior managers' roles.

- **Quality control** involves monitoring specific project results to ensure that they comply with the relevant quality standards while identifying ways to improve overall quality. This process is often associated with the technical tools and techniques of quality management such as Pareto charts, quality control charts, and statistical sampling. You will learn more about these tools and techniques later in this chapter.

QUALITY PLANNING

Project managers today have a vast knowledge base of information related to quality, and the first step to ensuring project quality management is planning. Quality planning implies the ability to anticipate situations and prepare actions that bring about the desired outcome. The current thrust in modern quality management is the prevention of defects through a program of selecting the proper materials, training and indoctrinating people in quality, and planning a process that ensures the appropriate outcome. In project quality planning, it is important to identify relevant quality standards for each unique project and to design quality into the products of the project and the processes involved in managing the project.

Design of experiments is a quality planning technique that helps identify which variables have the most influence on the overall outcome of a process. Understanding which variables affect outcome is a very important part of quality planning. For example, computer chip designers might want to determine which combination of materials and equipment will produce the most reliable chips at a reasonable cost. You can also apply design of experiments to project management issues such as cost and schedule trade-offs. For example, junior programmers or consultants cost less than senior programmers or consultants, but you cannot expect them to complete the same level of work in the same amount of time. An appropriately designed experiment to compute project costs and durations for various combinations of junior and senior programmers or consultants can allow you to determine an optimal mix of personnel, given limited resources.

Quality planning also involves communicating the correct actions for ensuring quality in a format that is understandable and complete. In quality planning for projects, it is important to describe important factors that directly contribute to meeting the customer's requirements. Organizational policies related to quality, the particular project's scope statement and product descriptions, and related standards and regulations are all important inputs to the quality planning process. The main outputs of quality planning are a quality management plan and checklists for ensuring quality throughout the project life cycle.

As mentioned in the discussion of project scope management (see Chapter 5), it is often difficult to completely understand the performance dimension of information technology projects. Even if hardware, software, and networking technology would stand still for a while, it is often difficult for customers to explain exactly what they want in an information technology project. Important scope aspects of information technology projects that affect quality include functionality and features, system outputs, performance, and reliability and maintainability.

- **Functionality** is the degree to which a system performs its intended function. **Features** are the system's special characteristics that appeal to users. It is important to clarify what functions and features the system *must* perform, and what functions and features are *optional*. In the EIS example in the opening case, the mandatory functionality of the system might be that it allows users to track sales of specific medical instruments by predetermined categories such as the product group, country, hospital, and sales representative. Mandatory features might be a graphical user interface with icons, drop-down menus, online help, and so on. One of the problems users experienced was forgetting passwords. Perhaps a required feature was missed that could alleviate this problem.

- **System outputs** are the screens and reports the system generates. It is important to define clearly what the screens and reports look like for a system. Can the users easily interpret these outputs? Can users get all of the reports they need in a suitable format?

- **Performance** addresses how well a product or service performs the customer's intended use. In order to design a system with high-quality performance, project stakeholders must address many issues. What volumes of data and transactions should the system be capable of handling? How many simultaneous users should the system be designed to handle? What is the projected growth rate in the number of users? What type of equipment must the system run on? How fast must the response time be for different aspects of the system under different circumstances? For the EIS, several of the quality problems appear to relate to performance issues. The system is going down a couple of times a month, and users are unsatisfied with the response time. The project team may not have had specific performance requirements or tested the system under the right conditions to deliver the expected performance. Buying faster hardware might address these performance issues. Another performance problem that might be more difficult to fix is the fact that some of the reports are generating inconsistent results. This could be a software quality problem that may be difficult and costly to correct since the system is already in operation.

- **Reliability** is the ability of a product or service to perform as expected under normal conditions without unacceptable failures. **Maintainability** addresses the ease of performing maintenance on a product. Most information technology products cannot reach 100 percent reliability, but stakeholders must define what their expectations are. For the EIS, what are the normal conditions for operating the system? Should reliability tests be based on one hundred people accessing the system at once and running simple queries? Maintenance for the EIS might include uploading new data into the system or performing maintenance procedures on the system hardware and software. Are the users willing to have the system be unavailable several hours a week for system maintenance? Providing Help Desk support could also be a maintenance function. How fast a response do users expect for Help Desk support? How often can users tolerate system failure? Are the stakeholders willing to pay more for higher reliability and less failures?

These aspects of project scope are just a few of the requirement issues related to quality planning. Project managers and their teams need to consider all of these project scope issues in determining quality goals for the project. The main customers for the project must also realize their role in defining the most critical quality needs for the project and constantly communicate these needs and expectations to the project team. Since most information technology projects involve requirements that are not set in stone, it is important for all project stakeholders to work together to balance the quality, scope, time, and cost dimensions of the project. *Project managers, however, are ultimately responsible for quality management on their projects.*

QUALITY ASSURANCE

It is one thing to develop a plan for ensuring quality on a project; it is another to ensure delivery of quality products and services. **Quality assurance** includes all of the activities related to satisfying the relevant quality standards for a project. Another goal of quality assurance is continual quality improvement.

Top management and project managers can have the greatest impact on the quality of projects by doing a good job of quality assurance. The importance of leadership in improving information technology project quality is discussed in more detail later in this chapter.

Several tools used in quality planning can also be used in quality assurance. Design of experiments, as described under quality planning, can also help ensure and improve product quality. **Benchmarking** generates ideas for quality improvements by comparing specific project practices or product characteristics to those of other projects or products within or outside the performing organization. For example, if a competitor has an EIS with an average down time of only one hour a week, that might be a benchmark for which to strive. Fishbone or Ishikawa diagrams, as described later in this chapter, can assist in ensuring and improving quality by finding the root causes of quality problems.

Several organizations use templates for developing various plans related to quality assurance. Table 8-1 provides the table of contents for a Quality Assurance Plan for the Desktop, Telecommunications, and LAN environment developed by the U.S. Department of Energy (DOE). Note the main categories with a description of the plan itself, the management structure, required documentation, quality assurance procedures, problem-reporting procedures, quality assurance metrics, and quality assurance check list forms. You can visit the referenced Web site for more information on creating a quality assurance plan.

Table 8-1: Table of Contents for a Quality Assurance Plan

1.0 Draft Quality Assurance Plan
    1.1 Introduction
    1.2 Purpose
    1.3 Policy Statement
    1.4 Scope
2.0 Management
    2.1 Organizational Structure
    2.2 Roles and Responsibilities
        2.2.1 Technical Monitor/Senior Management
        2.2.2 Task Leader
        2.2.3 Quality Assurance Team
        2.2.3 Technical Staff
3.0 Required Documentation
4.0 Quality Assurance Procedures
    4.1 Walkthrough Procedure
    4.2 Review Process
        4.2.1 Review Procedures

Table 8-1: Table of Contents for a Quality Assurance Plan (continued)

U.S. Department of Energy Office of the Chief Information Officer, Software Quality & Systems Engineering, Templates, "XXX YYYY Quality Assurance Plan For the Desktop, Telecommunications, and LAN Environment" (*http://cio.doe.gov/ITReform/sqse/template.htm*) (March 15, 2002).

An important tool for quality assurance is a quality audit. A **quality audit** is a structured review of specific quality management activities that help identify lessons learned that could improve performance on current or future projects. In-house auditors or third parties with expertise in specific areas can perform quality audits, and quality audits can be scheduled or random. Industrial engineers often perform quality audits by helping to design specific quality metrics for a project and then applying and analyzing the metrics throughout the project. For example, the Northwest Airlines ResNet project (available on the companion Web site for this text) provides an excellent example of using quality audits to emphasize the main goals of a project and then track progress in reaching those goals. The main objective of the ResNet project was to develop a new reservation system to increase direct airline ticket sales and reduce the time it took for sales agents to handle customer calls. The measurement techniques for monitoring these goals helped ResNet's project manager and project team supervise various aspects of the project by focusing on meeting those goals. Measuring progress toward increasing direct sales and reducing call times also helped the project manager justify continued investments in ResNet.

QUALITY CONTROL

Many people only think of quality control when they think of quality management. Perhaps it is because there are many popular tools and techniques in this area. Before describing some of these tools and techniques, it is important to distinguish quality control from quality planning and quality assurance.

Although one of the main goals of quality control is to improve quality, the main outputs of this process are acceptance decisions, rework, and process adjustments.

- **Acceptance decisions** determine if the products or services produced as part of the project will be accepted or rejected. If project stakeholders reject some of the products or services produced as part of the project, there must be rework. For example, the executive who sponsored development of the EIS in the opening case was obviously not satisfied with the system and hired an outside consultant, Scott Daniels, to lead a team to address and correct the quality problems.

- **Rework** is action taken to bring rejected items into compliance with product requirements or specifications or other stakeholder expectations. Rework can be very expensive, so the project manager must strive to do a good job of quality planning and quality assurance to avoid this need. Since the EIS did not meet all of the stakeholders' expectations for quality, the medical instruments company was spending a lot of money for rework.

- **Process adjustments** correct or prevent further quality problems based on quality control measurements. For example, Scott Daniels, the consultant in the opening case, might recommend that the medical instruments company purchase a faster server for the EIS to correct the response-time problems. The company also hired Scott to develop a plan to help prevent future information technology project quality problems.

TOOLS AND TECHNIQUES FOR QUALITY CONTROL

Quality control includes many general tools and techniques. This section focuses on just a few of them—Pareto analysis, statistical sampling, Six Sigma, and quality control charts—and discusses how they can be applied to information technology projects. The section concludes with a discussion on testing, since information technology projects use testing extensively to ensure quality.

Pareto Analysis

Pareto analysis involves identifying the vital few contributors that account for most quality problems in a system. It is sometimes referred to as the 80-20 rule, meaning that 80 percent of problems are often due to 20 percent of the causes. **Pareto diagrams** are histograms, or column charts representing a frequency distribution that help identify and prioritize problem areas. The variables described by the histogram are ordered by frequency of occurrence.

For example, suppose there was a history of user complaints about the EIS described in the opening case. Many organizations log user complaints, but many do not know how to analyze them. A Pareto diagram is an effective tool for analyzing user complaints and helping project managers decide which ones to address first. Figure 8-1 shows a Pareto diagram of user complaints by category that could apply to the EIS project. The bars represent the number of complaints for each category, and the line shows the cumulative percent of the complaints. For example, Figure 8-1 shows that the most frequent user complaint was log-in problems, followed by the system locking up, the system being too slow, the system being hard to use, and the reports being inaccurate. The first complaint accounts for 55 percent of the total complaints. The first and second complaints have a cumulative percentage of almost 80 percent, meaning these two areas account for 80 percent of the complaints. Therefore, the company should focus on making it easier to log in to the system to improve quality, since the majority of complaints fall under that category. The company should also address why the system locks up. Because Figure 8-1 shows that the problem of inaccurate reports is rarely mentioned, the project manager should investigate who made this complaint before spending a lot of effort on addressing that potentially critical problem with the system. The project manager should also find out if complaints about the system being too slow were actually due to the user not being able to log in or the system locking up.

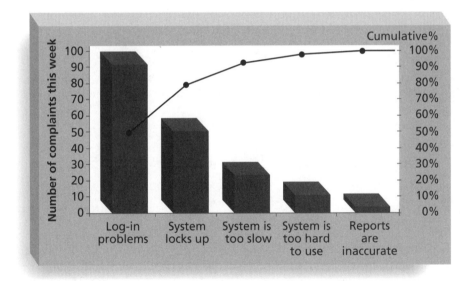

Figure 8-1. Sample Pareto Diagram

Statistical Sampling

Statistical sampling is a key concept in project quality management. Members of a project team who focus on quality control must have a strong understanding of statistics, but other project team members need to understand only the basic concepts. These concepts include statistical sampling, certainty factor, standard deviation, and variability. Standard deviation and variability are fundamental concepts for understanding quality control charts. This section briefly describes these concepts and describes how a project manager might apply them to information technology projects.

Statistical sampling involves choosing part of a population of interest for inspection. For example, suppose a company wants to develop an electronic data interchange (EDI) system for handling data on invoices from all of its suppliers. Assume also that in the past year, the total number of invoices was fifty thousand from two hundred different suppliers. It would be very time-consuming and expensive to review every single invoice to determine data requirements for the new system. Even if the system developers did review all two hundred invoice forms, the data might be entered differently on every form. It is impractical to study every member of a population, such as all fifty thousand invoices, so statisticians have developed techniques to help determine an appropriate sample size. If the system developers used statistical techniques, they might find that by studying only one hundred invoices, they would have a good sample of the type of data they would need in designing the system.

The size of the sample depends on how representative you want the sample to be. A simple formula for determining sample size is:

$$\text{Sample size} = .25 \times (\text{certainty factor/acceptable error})^2$$

The certainty factor denotes how certain you want to be that the data sampled will not include variations that do not naturally exist in the population. You calculate the certainty factor from tables available in statistics books. Table 8-2 shows some commonly used certainty factors.

Table 8-2: Commonly Used Certainty Factors

DESIRED CERTAINTY	CERTAINTY FACTOR
95%	1.960
90%	1.645
80%	1.281

For example, suppose the developers of the EDI system described earlier would accept a 95 percent certainty that a sample of invoices would contain no variation unless it was present in the population of total invoices. They would then calculate the sample size as:

$$\text{Sample size} = 0.25 \times (1.960/.05)^2 = 384$$

If the developers wanted 90 percent certainty, they would calculate the sample size as:

$$\text{Sample size} = 0.25 \times (1.645/.10)^2 = 68$$

If the developers wanted 80 percent certainty, they would calculate the sample size as:

$$\text{Sample size} = 0.25 \times (1.281/.20)^2 = 10$$

Assume the developers decide on 90 percent for the certainty factor. Then they would need to examine 68 invoices to determine the type of data the EDI system would need to capture.

Six Sigma

The work of many project quality experts contributed to the development of today's Six Sigma principles. There has been some confusion in the past few years about the term Six Sigma.[7] This section summarizes recent information about this important concept and explains how organizations worldwide use Six Sigma principles to improve quality, decrease costs, and better meet customer needs.

In their book, *The Six Sigma Way*, authors Peter Pande, Robert Neuman, and Roland Cavanagh define **Six Sigma** as "a comprehensive and flexible *system* for achieving, sustaining and maximizing business success. Six Sigma is uniquely driven by close understanding of customer needs, disciplined use of facts, data, and statistical analysis, and diligent attention to managing, improving, and reinventing business processes."[8]

Six Sigma's target for perfection is the achievement of no more than 3.4 defects, errors, or mistakes per million opportunities. This target number is explained in more detail later in this section. An organization can apply the Six Sigma principles to the design and production of a product, or a Help Desk or other customer-service process.

Projects that use Six Sigma principles for quality control normally follow a five-phase improvement process called **DMAIC** (pronounced de-MAY-ick), which stands for Define, Measure, Analyze, Improve, and Control. DMAIC is a systematic, closed-loop process for continued improvement that is scientific and fact based. The following are brief descriptions of each phase of the DMAIC improvement process:

1. **Define:** Define the problem/opportunity, process, and customer requirements. Important tools used in this phase include a project charter (very similar to the project charters described in Chapters 3 and 4 of this text), a description of customer requirements, process maps, and Voice of the Customer (VOC) data. Examples of VOC data include complaints, surveys,

[7] The previous version of this text included incorrect data on what Six Sigma meant, thinking it related directly to the statistical term.

[8] Pande, Peter S., Robert P. Neuman, and Roland R. Cavanagh, *The Six Sigma Way*. New York: McGraw-Hill, 2000, p. xi.

comments, and market research that represent the views and needs of the organization's customers.

2. **Measure:** Define measures, collect, compile, and display data. Measures are defined in terms of defects per opportunity, as defined later in this section.

3. **Analyze:** Scrutinize process details to find improvement opportunities. A project team working on a Six Sigma project, normally referred to as a Six Sigma team, investigates and verifies data to prove the suspected root causes of quality problems and substantiates the problem statement. An important tool in this phase is the Fishbone or Ishikawa diagram, as described later in this chapter.

4. **Improve:** Generate solutions and ideas for improving the problem. A final solution is verified with the project sponsor, and the Six Sigma team develops a plan to pilot test the solution. The Six Sigma team reviews the results of the pilot test to refine the solution, if needed, and then implements the solution where appropriate.

5. **Control:** Track and verify the stability of the improvements and the predictability of the solution. Control charts are one tool used in the control phase, as described later in this chapter.

How is Six Sigma Quality Control Unique?

How does using Six Sigma principles differ from using previous quality control initiatives? Many people remember other quality initiatives from the past few decades such as Total Quality Management (TQM) and Business Process Reengineering (BPR), to name a few. The origins of many Six Sigma principles and tools are found in these previous initiatives; however, there are several new ideas included in Six Sigma principles that help organizations improve their competitiveness and bottom-line results. Below are a few of these principles:

■ Using Six Sigma principles is an organization-wide commitment. CEOs, top managers, and all levels of employees in an organization that embraces Six Sigma principles (often referred to as a Six Sigma organization) have seen remarkable improvements due to its use. There are often huge training investments, but these investments pay off as employees practice Six Sigma principles and produce higher-quality goods and services at lower costs. Individuals can earn Green, Black, and Master Black belts based on the depth of their training and experience on Six Sigma projects.

■ Organizations that successfully implement Six Sigma principles have the ability and willingness to adopt two seemingly contrary objectives at the same time. Authors James Collins and Jerry Porras describe this as the "We can do it all" or "Genius of the And" approach in their book, *Built to Last*. For example, Six Sigma organizations believe that they can be creative *and* rational, focus on the big picture *and* minute details, reduce errors *and* get things done faster, and make customers happy *and* make a lot of money.

■ Six Sigma is not just a program or a discipline to organizations that have benefited from it. Six Sigma is an operating philosophy that is customer-focused and strives to drive out waste, raise levels of quality, and improve financial performance at *breakthrough* levels. A Six Sigma organization sets high goals and uses the DMAIC improvement process to achieve extraordinary quality improvements.

Many organizations do some of what now fits under the definition of Six Sigma, and many Six Sigma principles are not brand-new. What *is* new is its ability to bring together many different themes, concepts, and tools into a coherent management process that can be used on an organization-wide basis.

What Went Right?

Many organizations have reported dramatic improvements as a result of implementing Six Sigma principles.

■ Motorola, Inc. pioneered the adoption of Six Sigma in the 1980s. Its reason for developing and implementing Six Sigma was clear—to stay in business. Japanese competitors were putting several U.S. and European companies out of business. Using Six Sigma provided Motorola with a simple, consistent way to track and compare performance to customer requirements and meet ambitious quality goals in terms of defect reduction. Chairman Bob Galvin set a goal of ten times improvement in defect reduction every two years, or 100 times improvement in four years. Motorola did stay in business, achieving excellent growth and profitability in the 1980s and 1990s. Motorola estimates the cumulative savings based on Six Sigma efforts to be about $14 billion.[9]

■ Allied Signal/Honeywell, a technology and manufacturing company, began several quality improvement activities in the early 1990s. By 1999, the company reported saving more than $600 million a year, thanks to training in and application of Six Sigma principles. Six Sigma teams reduced the costs of reworking defects and applied the same principles to designing new products like aircraft engines. They reduced the time it took from design to certification of engines from 42 to 33 months. Allied Signal/Honeywell credits Six Sigma with a 6 percent productivity increase in 1998 and record profit margins of 13 percent. As one of the company's directors said, Six Sigma "changed the way we think and the way we communicate. We never used to talk about the process or the customer; now they're part of our everyday conversation."[10]

■ General Electric Company (GE) uses Six Sigma to focus on achieving customer satisfaction. Chairman Jack Welch urged his top managers to become "passionate lunatics" about Six Sigma. Welch said, "Six Sigma has forever changed GE. Everyone from the Six Sigma zealots emerging from their Black Belt teams, to the engineers, the auditors, and the scientists, to the senor leadership that will take this Company into its new millennium—is a true believer in Six Sigma, the way this Company now works."[11]

---

[9] Ibid, p. 7.
[10] Ibid, p. 9.
[11] Address to General Electric Company Annual Meeting, Cleveland, Ohio, April 21, 1999.

Six Sigma and Project Management

Joseph M. Juran stated, "All improvement takes place project by project, and in no other way."[12] Organizations implement Six Sigma by selecting and managing projects. An important part of project management is good project selection. This statement is especially true for Six Sigma projects. Pande, Neuman, and Cavanagh conducted an informal poll to find out what the most critical and most commonly mishandled activity was in launching Six Sigma, and the unanimous answer was project selection. "It's a pretty simple equation, really: Well-selected and -defined improvement projects equal better, faster results. The converse equation is also simple: Poorly selected and defined projects equal delayed results and frustration."[13]

Organizations must also be careful to apply higher quality where it makes sense. A recent article in *Fortune* states that companies that have implemented Six Sigma have not necessarily boosted their stock values. Although some companies such as GE boasted savings of more than $2 billion in 1999 due to its use of Six Sigma, other companies, such as Whirlpool, could not clearly demonstrate the value of its investments. Why can't all companies benefit from Six Sigma? Because minimizing defects does not matter if an organization is making a product that no one wants to buy. As one of Six Sigma's biggest supporters, Mikel Harry, puts it, "I could genetically engineer a Six Sigma goat, but if a rodeo is the marketplace, people are still going to buy a Four Sigma horse."[14]

As described in Chapter 5, Project Scope Management, there are several methods for selecting projects. However, what makes a project a potential Six Sigma project? First, there must be a quality problem or gap between the current and desired performance. Many projects do not meet this first criterion, such as building a house, merging two corporations, or providing an information technology infrastructure for a new organization. Six Sigma projects are also best suited for projects that do not have a clearly understood problem. A third criterion for Six Sigma projects is that the solution is not predetermined, and an optimal solution is not apparent.

Once a project is selected as a good candidate for Six Sigma, many project management concepts, tools, and techniques described in this text come into play. For example, Six Sigma projects usually have a business case, a project charter, requirements documents, a schedule, a budget, and so on. Six Sigma projects are done in teams and sponsored by sponsors or champions. There are also, of course, project managers, often called team leaders in Six Sigma organizations. In other words, Six Sigma projects are simply types of projects that focus on supporting the Six Sigma philosophy by being customer-focused and striving to drive out waste, raise levels of quality, and improve financial performance at breakthrough levels.

[12] "What You Need to Know About Six Sigma," Productivity Digest (December 2001), 38.
[13] Pande, Peter S., Robert P. Neuman, and Roland R. Cavanagh, *The Six Sigma Way*. New York: McGraw-Hill, 2000, 137.
[14] Clifford, Lee, "Why You Can Safely Ignore Six Sigma," *Fortune* (January 22, 2001), 140.

Six Sigma and Statistics

An important concept in Six Sigma is improving quality by reducing variation. The term sigma means standard deviation. **Standard deviation** measures how much variation exists in a distribution of data. A small standard deviation means that data clusters closely around the middle of a distribution and there is little variability among the data. A large standard deviation means that data is spread out around the middle of the distribution and there is relatively greater variability. Statisticians use the Greek symbol σ (sigma) to represent the standard deviation.

Figure 8-2 provides an example of a **normal distribution**—a bell-shaped curve that is symmetrical regarding the **mean** or average value of the population (the data being analyzed). In any normal distribution, 68.3 percent of the population is within one standard deviation (1σ) of the mean. Ninety-five point five percent of the population is within two standard deviations (2σ), and 99.7 percent of the population is within three standard deviations (3σ) of the mean.

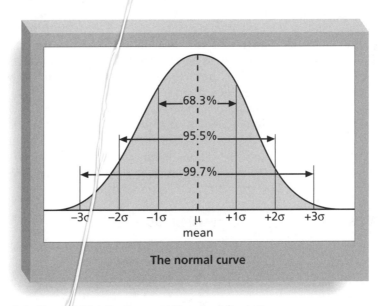

The normal curve

Figure 8-2. Normal Distribution and Standard Deviation

Standard deviation is a key factor in determining the acceptable number of defective units found in a population. Table 8-3 illustrates the relationship between sigma, the percentage of the population within that sigma range, and the number of defective units per billion. Note that this table shows that being plus or minus six sigma in pure statistical terms means only *two* defective units per *billion*. Why, then, is the target for Six Sigma programs 3.4 defects per million opportunities, as stated earlier in this chapter?

Table 8-3: Sigma and Defective Units

SPECIFICATION RANGE (IN +/− SIGMAS)	PERCENT OF POPULATION WITHIN RANGE	DEFECTIVE UNITS PER BILLION
1	68.27	317,300,000
2	95.45	45,400,000
3	99.73	2,700,000
4	99.0037	63,000
5	99.999943	57
6	99.9999998	2

Based on Motorola's original work on Six Sigma in the 1980s, the convention used for Six Sigma is a scoring system that accounts for more variation in a process than you would typically find in a few weeks or months of data gathering. This 1.5 sigma shift is a key cause of disagreement among statistical experts regarding how Six Sigma measures are defined. Table 8-4 shows the Six Sigma conversion table applied to Six Sigma projects. The **yield** represents the number of units handled correctly through the process steps. A **defect** is any instance where the product or service fails to meet customer requirements. Because most products or services have multiple customer requirements, there can be several opportunities to have a defect. For example, suppose a company is trying to reduce the number of errors on customer billing statements. There could be several errors on a billing statement due to a misspelled name, incorrect address, wrong date of service, calculation error, and so on. There might be 100 opportunities for a defect to occur on one billing statement. Instead of measuring the number of defects per unit or billing statement, Six Sigma measures the number of defects based on the number of opportunities.

Table 8-4: Six Sigma Conversion Table

SIGMA	YIELD	DEFECTS PER MILLION OPPORTUNITIES (DPMO)
1	31.0%	690,000
2	69.2%	308,000
3	93.3%	66,800
4	99.4%	6,210
5	99.97%	230
6	99.99966%	3.4

Another term used in the telecommunications industry is **six 9s of quality**. Six 9s of quality is a measure of quality control equal to 1 fault in 1 million opportunities. In the telecommunications industry, it means 99.9999 percent service availability or *30 seconds of downtime a year*. This level of quality has also been stated as the target goal for the number of errors in a communications circuit, system failures, or errors in lines of code. To achieve six 9s of quality requires continual testing to find and eliminate errors or enough redundancy and back-up equipment in systems to reduce the overall system failure rate to that low a level.

Quality Control Charts and the Seven Run Rule

A **control chart** is a graphic display of data that illustrates the results of a process over time. The main use of control charts is to prevent defects, rather than to detect or reject them. Quality control charts allow you to determine whether a process is in control or out of control. When a process is in control, any variations in the results of the process are created by random events. Processes that are in control do not need to be adjusted. When a process is out of control, variations in the results of the process are caused by nonrandom events. When a process is out of control, you need to identify the causes of those nonrandom events and adjust the process to correct or eliminate them. Control charts are often used to monitor manufactured lots, but they can also be used to monitor the volume and frequency of change requests, errors in documents, cost and schedule variances, and other items related to project quality management.

Figure 8-3 illustrates an example of a control chart for a process that manufactures twelve-inch rulers. Assume that these are wooden rulers created by machines on an assembly line. Each point on the chart represents a length measurement for a ruler that comes off the assembly line. The scale on the vertical axis goes from 11.90 to 12.10. These numbers represent the lower and upper specification limits for the ruler. In this case, this would mean that the customer for the rulers has specified that all rulers purchased must be between 11.90 and 12.10 inches long, or twelve inches plus or minus 0.10 inches. The lower and upper control limits on the quality control chart are 11.91 and 12.09 inches, respectively. This means the manufacturing process is designed to produce rulers between 11.91 and 12.09 inches long.

Figure 8-3. Sample Quality Control Chart

Note the dotted lines on the chart at 12.03 inches, 12.06 inches, and 12.09 inches. These dotted lines represent the points at one, two, and three standard deviations above the mean. The dotted lines at 11.97 inches, 11.94 inches, and 11.91 inches represent the points at one, two, and three standard deviations below the mean. Based on the statistical definition of $\pm 3\sigma$ described earlier, 99.73 percent of the manufactured rulers should come off the assembly line with a measurement between 11.91 and 12.09 inches, if the manufacturing process is operating in control.

Looking for and analyzing patterns in process data is an important part of quality control. You can use quality control charts and the seven run rule to look for patterns in data. The **seven run rule** states that if seven data points in a row are all below the mean, above the mean, or are all increasing or decreasing, then the process needs to be examined for nonrandom problems. In Figure 8-3, data points that violate the seven run rule are starred. Note that you include the first point in a series of points that are all increasing or decreasing. In the ruler manufacturing process, these data points may indicate that a calibration device may need adjustment. For example, the machine that cuts the wood for the rulers might need to be adjusted or the blade on the machine might need to be replaced.

Testing

Many information technology professionals think of testing as a stage that comes near the end of information technology product development. Instead of putting serious effort into proper planning, analysis, and design of information technology projects, some organizations rely on testing just before a product ships to ensure some degree of quality. In fact, testing needs to be done during almost every phase of the systems development life cycle, not just before the organization ships or hands over a product to the customer.

Figure 8-4 shows one way of portraying the systems development life cycle. This example includes seventeen main phases involved in a software development project and shows their relationship to each other. For example, every project should start by initiating the project, then conducting a feasibility study, and then performing project planning. The figure then shows that the work involved in preparing detailed requirements and the detailed architecture for the system can be performed simultaneously. The oval-shaped phases represent areas that should be tested to help ensure quality on software development projects.[15]

Several of the phases in Figure 8-4 include specific work related to testing. A **unit test** is done to test each individual component (often a program) to ensure it is as defect-free as possible. Unit tests are performed before moving on to the integration test. **Integration testing** occurs between unit and system testing to test functionally grouped components. It ensures a subset(s) of the entire system works together. **System testing** tests the entire system as one entity. It focuses on the big picture to ensure the entire system is working properly. **User acceptance testing** is an independent test performed by end users prior to accepting the delivered system. It focuses on the business fit of the system to the organization, rather than technical issues.

[15] Hollstadt & Associates, Inc., *Software Development Project Life Cycle Testing Methodology User's Manual.* Burnsville: MN, August 1998, 13.

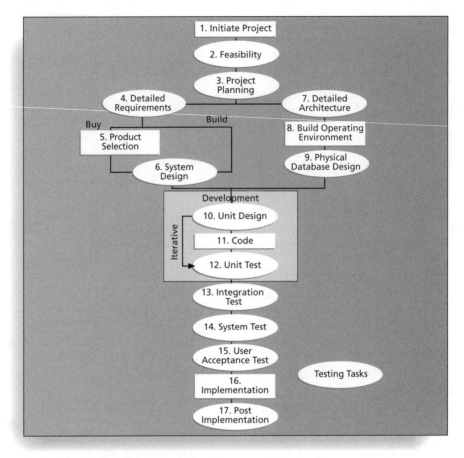

Figure 8-4. Testing Tasks in the Systems Development Life Cycle

Figure 8-5 presents a Gantt chart that shows testing tasks that are appropriate for different phases of the systems development life cycle.[16] This is a simplified version of a detailed testing management plan used by a consulting firm. To help improve the quality of software development projects, it is important for organizations to follow a thorough and disciplined testing methodology. System developers and testers must also establish a partnership with all project stakeholders to make sure the system meets their needs and expectations and the tests are done properly. As described in the next section, there are tremendous costs involved in failure to perform proper testing.

16 Ibid, 54–57.

Figure 8-5. Gantt Chart for Building Testing into a Systems Development Project Plan

MODERN QUALITY MANAGEMENT

Modern quality management requires customer satisfaction, prefers prevention to inspection, and recognizes management responsibility for quality. Several noteworthy people helped develop the following theories, tools, and techniques that define modern quality management.[17] The suggestions from these quality experts led to many projects to improve quality and provided the foundation for today's Six Sigma projects. This section summarizes major contributions made by Deming, Juran, Crosby, Ishikawa, Taguchi, and Feigenbaum.

Deming and his Fourteen Points for Management

Dr. W. Edwards Deming is primarily known for his work on quality control in Japan. Dr. Deming went to Japan after World War II, at the request of the Japanese government, to assist them in improving productivity and quality.

---

[17] Kerzner, Harold, *Project Management*, 6th ed. New York: Van Nostrand Reinhold, 1998, 1048.

Deming, a statistician and former professor at New York University, taught Japanese manufacturers that higher quality meant greater productivity and lower cost. Deming's theories were not recognized by American industry until Japanese manufacturers started producing products that seriously challenged American products, particularly in the auto industry. Ford Motor Company then adopted Deming's quality methods and experienced dramatic improvement in quality and sales thereafter. By the 1980s, after seeing the excellent work coming out of Japan, several U.S. corporations vied for Deming's expertise to help them establish quality improvement programs in their own factories. Many people are familiar with the Deming Prize, an award given to recognize high-quality organizations, and Deming's Cycle for Improvement: plan, do, check, and act. Most Six Sigma principles described earlier are based on the plan-do-check-act model created by Deming. Many are also familiar with Deming's fourteen Points for Management, summarized below from Deming's text *Out of the Crisis*:

1. Create constancy of purpose for improvement of product and service.
2. Adopt the new philosophy.
3. Cease dependence on inspection to achieve quality.
4. End the practice of awarding business based on price tag alone. Instead, minimize total cost by working with a single supplier.
5. Improve constantly and forever every process for planning, production, and service.
6. Institute training on the job.
7. Adopt and institute leadership.
8. Drive out fear.
9. Break down barriers between staff areas.
10. Eliminate slogans, exhortations, and targets for the workforce.
11. Eliminate numerical quotas for the workforce and numerical goals for management.
12. Remove barriers that rob people of workmanship. Eliminate the annual rating or merit system.
13. Institute a vigorous program of education and self-improvement for everyone.
14. Put everyone in the company to work to accomplish the transformation.

Juran and the Importance of Top Management Commitment to Quality

Joseph M. Juran, like Deming, helped Japanese manufacturers learn how to improve their productivity. U.S. companies later discovered him as well. He wrote the first edition of the *Quality Control Handbook* in 1974, stressing the importance of top management commitment to continuous product quality improvement. In 1999, at the age of 94, Juran published the fifth edition of this famous handbook. He also developed the Juran Trilogy: quality improvement, quality planning, and quality control. Juran stressed the difference between the manufacturer's view of quality and the customer's view. Manufacturers often focus on conformance to requirements, but customers focus on fitness for use. Most definitions of quality now use fitness for use to stress the importance of satisfying stated or implied needs and not just meeting stated requirements or specifications. Juran developed ten steps to quality improvement:

1. Build awareness of the need and opportunity for improvement.
2. Set goals for improvement.
3. Organize to reach the goals (establish a quality council, identify problems, select projects, appoint teams, designate facilitators).
4. Provide training.
5. Carry out projects to solve problems.
6. Report progress.
7. Give recognition.
8. Communicate results.
9. Keep score.
10. Maintain momentum by making annual improvement part of the regular systems and processes of the company.

Crosby and Striving for Zero Defects

Philip B. Crosby wrote *Quality Is Free* in 1979 and is best known for suggesting that organizations strive for zero defects. He stressed that the costs of poor quality should include all the costs of not doing the job right the first time, such as scrap, rework, lost labor hours and machine hours, customer ill will and lost sales, and warranty costs. Crosby suggested that the cost of poor quality is so understated that companies can profitably spend unlimited amounts of money on improving quality. Crosby developed the following fourteen steps for quality improvement:

1. Make it clear that management is committed to quality.
2. Form quality improvement teams with representatives from each department.
3. Determine where current and potential quality problems lie.
4. Evaluate the cost of quality and explain its use as a management tool.
5. Raise the quality awareness and personal concern of all employees.
6. Take actions to correct problems identified through previous steps.

7. Establish a committee for the zero-defects program.

8. Train supervisors to actively carry out their part of the quality improvement program.

9. Hold a "zero-defects day" to let all employees realize that there has been a change.

10. Encourage individuals to establish improvement goals for themselves and their groups.

11. Encourage employees to communicate to management the obstacles they face in attaining their improvement goals.

12. Recognize and appreciate those who participate.

13. Establish quality councils to communicate on a regular basis.

14. Do it all over again to emphasize that the quality improvement program never ends.

Ishikawa and the Fishbone Diagram

Kaoru Ishikawa is best known for his 1972 book *Guide to Quality Control*. He developed the concept of quality circles and pioneered the use of Fishbone diagrams. **Quality circles** are groups of nonsupervisors and work leaders in a single company department who volunteer to conduct group studies on how to improve the effectiveness of work in their department. Ishikawa suggested that Japanese managers and workers were totally committed to quality, but that most U.S. companies delegated the responsibility for quality to a few staff members. **Fishbone diagrams**, sometimes called **Ishikawa diagrams**, trace complaints about quality problems back to the responsible production operations. In other words, they help find the root cause of quality problems.

Figure 8-6 provides an example of a fishbone or Ishikawa diagram that Scott Daniels, the consultant in the opening case, might create to trace the real cause of the problem of users not being able to log in to the EIS. Notice that it resembles the skeleton of a fish, hence its name. This fishbone diagram lists the main areas that could be the cause of the problem: the EIS system's hardware, the individual user's hardware, software, or training. This figure describes two of these areas, the individual user's hardware and training, in more detail. The root cause of the problem would have a significant impact on the actions taken to solve the problem. If many users could not get into the system because their computers did not have enough memory, the solution might be to upgrade memory for those computers. If many users could not get into the system because they forgot their passwords, there might be a much quicker, less expensive solution.

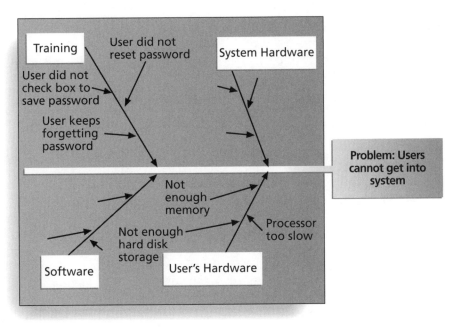

Figure 8-6. Sample Fishbone or Ishikawa Diagram

Taguchi and Robust Design Methods

Genichi Taguchi is best known for developing the Taguchi methods for optimizing the process of engineering experimentation. Key concepts in the Taguchi methods are that quality should be designed into the product and not inspected into it and that quality is best achieved by minimizing deviation from the target value. For example, if the target response time for accessing the EIS described in the opening case is half a second, there should be little deviation from this time. A 1998 article in *Fortune* states, "Japan's Taguchi is America's new quality hero."[18] Many companies, including Xerox, Ford, Hewlett-Packard, and Goodyear, have recently used Taguchi's Robust Design methods to design high-quality products. **Robust Design methods** focus on eliminating defects by substituting scientific inquiry for trial-and-error methods.

[18] Bylinsky, Gene, "How to Bring Out Better Products Faster," *Fortune* (November 23, 1998), 238[B].

Feigenbaum and Workers Responsibility for Quality

Armand V. Feigenbaum developed the concept of total quality control (TQC) in his 1983 book *Total Quality Control: Engineering and Management*. He proposed that the responsibility for quality should rest with the people who do the work. In TQC, product quality is more important than production rates, and workers are allowed to stop production whenever a quality problem occurs.

Malcolm Baldrige National Quality Award

The **Malcolm Baldrige National Quality Award** originated in 1987 to recognize companies that have achieved a level of world-class competition through quality management. The award was started in honor of Malcolm Baldrige, who was the U.S. Secretary of Commerce from 1981 until his death in a rodeo accident in July 1987. Baldrige was a proponent of quality management as a key element in improving the prosperity and long-term strength of U.S. organizations. The Baldrige Award is given by the President of the United States to businesses—manufacturing and service, small and large—and to education and health-care organizations. Organizations must apply for the award, and they must be judged outstanding in seven areas: leadership, strategic planning, customer and market focus, information and analysis, human resource focus, process management, and business results. The awards recognize achievements in quality and performance and raise awareness about the importance of quality as a competitive edge. The award is not given for specific products or services. Three awards may be given annually in each of these categories: manufacturing, service, small business and, starting in 1999, education and health care.

ISO 9000

The International Organization for Standardization (ISO) is a network of national standards institutes from 145 countries that work in partnership with international organizations, governments, industries, businesses, and consumer representatives. **ISO 9000**, a quality system standard developed by the ISO, is a three-part, continuous cycle of planning, controlling, and documenting quality in an organization. ISO 9000 provides minimum requirements needed for an organization to meet its quality certification standards. The ISO 9000 family of international quality management standards and guidelines has earned a global reputation as the basis for establishing quality management systems. The ISO organization's Web site includes testimonials from many business leaders throughout the world explaining the benefits of following these standards.

■ A government official in Malaysia believes that the universally accepted ISO 9000 standard can contribute significantly to improving quality and enhancing development of an excellent work culture. ISO 9000 leads to a more systematic management of quality and provides a means of consolidating quality management systems in public service.

■ Managers in the United Kingdom say that ISO 9000 has become a firmly established cornerstone of their organizations' drive toward quality improvement. It leads to substantial cost savings and contributes strongly to customer satisfaction.

■ A Brazilian newspaper reported that following ISO 9000 helped their home delivery service improve so much that the complaints dropped to a mere 0.06 percent of the total copies distributed.[19]

The contributions of several quality experts, quality awards, and quality standards are important parts of project quality management. The Project Management Institute was proud to announce in 1999 that their certification department had become the first certification department in the world to earn ISO 9000 certification, and that the PMBOK® Guide 1996 had been recognized as a national standard. Emphasizing quality in project management helps ensure that projects produce products or services that meet customer needs and expectations.

IMPROVING INFORMATION TECHNOLOGY PROJECT QUALITY

In addition to some of the suggestions provided for using good quality planning, quality assurance, and quality control, there are several other important issues involved in improving the quality of information technology projects. Strong leadership, understanding the cost of quality, providing a good workplace to enhance quality, and working toward improving the organization's overall maturity level in software development and project management can all assist in improving quality.

Leadership

As Joseph M. Juran said in 1945, "It is most important that top management be quality-minded. In the absence of sincere manifestation of interest at the top, little will happen below."[20] Juran and many other quality experts argue that the main cause of quality problems is a lack of leadership.

As globalization continues to increase and customers become more and more demanding, creating quality products quickly at a reasonable price is essential for staying in business. Having good quality programs in place helps organizations

---

[19] International Organization for Standardization (Market feedback), (*www.iso.ch/iso/en/iso9000-14000/feedback.html*) (March 2003).

[20] American Society for Quality (ASQ), Web site, *http://www.asqc.org/about/history/juran.html*.

remain competitive. To establish and implement effective quality programs, top management must lead the way. A large percentage of quality problems are associated with management, not technical issues. Therefore, top management must take responsibility for creating, supporting, and promoting quality programs.

Motorola provides an excellent example of a high-technology company that truly emphasizes quality. Leadership is one of the factors that helped Motorola achieve its great success in quality management and Six Sigma. Top management emphasized the need to improve quality and helped all employees take responsibility for customer satisfaction. Strategic objectives in Motorola's long-range plans included managing quality improvement in the same way that new products or technologies were managed. Top management stressed the need to develop and use quality standards and provided resources such as staff, training, and customer inputs to help improve quality.

Leadership provides an environment conducive to producing quality. Management must publicly declare the company's philosophy and commitment to quality, implement company-wide training programs in quality concepts and principles, implement measurement programs to establish and track quality levels, and actively demonstrate the importance of quality. When every employee understands and insists on producing high-quality products, then top management has done a good job in promoting the importance of quality.

The Cost of Quality

The **cost of quality** is the cost of conformance plus the cost of nonconformance. **Conformance** means delivering products that meet requirements and fitness for use. Examples of these costs include the costs associated with developing a quality plan, costs for analyzing and managing product requirements, and costs for testing. The **cost of nonconformance** means taking responsibility for failures or not meeting quality expectations.

A 2002 study by RTI International reported that software bugs cost the U.S. economy $59.6 billion each year, about .6 percent of gross domestic product. Most of the costs are borne by software users, and the rest by developers and vendors. RTI International also suggested that more than one-third of these costs could be eliminated by an improved testing infrastructure to enable earlier and more effective identification and removal of software defects.[21]

Other studies estimate the costs per hour of down time for systems. Table 8-5 summarizes the net bottom-line costs per hour of down time caused by software defects for different businesses.[22] Both American Airlines and United Airlines have estimated that reservation system down time costs them more than $20,000 per minute in lost income. Table 8-5 illustrates more examples of the cost of nonconformance.

21 RTI International, "Software Bugs Cost U.S. Economy $59.6 Billion Annually, RTI Study Finds," July 1, 2002.
22 Contingency Planning Research, White Plains, NY, 1995.

Table 8-5: Costs Per Hour of Down Time Caused by Software Defects

BUSINESS	COST PER HOUR OF DOWN TIME
Automated teller machines (medium-sized bank)	$14,500
Package shipping service	$28,250
Telephone ticket sales	$69,000
Catalog sales center	$90,000
Airline reservation center (small airline)	$89,500

The five major cost categories related to quality include:

1. **Prevention cost**—the cost of planning and executing a project so that it is error-free or within an acceptable error range. Preventive actions such as training, detailed studies related to quality, and quality surveys of suppliers and subcontractors fall under this category. Recall from the discussion of cost management (see Chapter 7) that detecting defects in information systems during the early phases of the system development life cycle is much less expensive than during the later phases. One hundred dollars spent refining user requirements could save millions by finding a defect before implementing a large system. The Year 2000 (Y2K) issue provides a good example of these costs. If organizations had decided during the 1960s, 1970s, and 1980s that all dates would need four characters to represent the year instead of two characters, they would have saved billions of dollars.

2. **Appraisal cost**—the cost of evaluating processes and their outputs to ensure that a project is error-free or within an acceptable error range. Activities such as inspection and testing of products, maintenance of inspection and test equipment, and processing and reporting inspection data all contribute to appraisal costs of quality.

3. **Internal failure cost**—a cost incurred to correct an identified defect before the customer receives the product. Items such as scrap and rework, charges related to late payment of bills, inventory costs that are a direct result of defects, costs of engineering changes related to correcting a design error, premature failure of products, and correcting documentation all contribute to internal failure cost.

4. **External failure cost**—a cost that relates to all errors not detected and not corrected before delivery to the customer. Items such as warranty cost, field service personnel training cost, product liability suits, complaint handling, and future business losses are examples of external failure costs.

5. **Measurement and test equipment costs**—the capital cost of equipment used to perform prevention and appraisal activities.

Many industries tolerate a very low cost of nonconformance, but not the information technology industry. Tom DeMarco is famous for several studies he conducted on the cost of nonconformance in the information technology industry. DeMarco found that the average large company devoted more than 60 percent of its software development efforts to maintenance. Around 50 percent of development costs are typically spent on testing and debugging software.[23]

Top management is primarily responsible for the high cost of nonconformance in information technology. Top managers often rush their organizations to develop new systems and do not give project teams enough time or resources to do a project right the first time. To correct these quality problems, top management must create a culture that embraces quality.

Organizational Influences, Workplace Factors, and Quality

A study done by Tom DeMarco and Timothy Lister produced interesting results related to organizations and relative productivity. Starting in 1984, DeMarco and Lister conducted "Coding War Games" over several years. Over the years, more than six hundred software developers from ninety-two organizations have participated in these games. The games are designed to examine programming quality and productivity over a wide range of organizations, technical environments, and programming languages. The study demonstrated that organizational issues had a much greater influence on productivity than the technical environment or programming languages.

For example, DeMarco and Lister found that productivity varied by a factor of about one to ten across all participants. That is, one team may have finished a coding project in one day while another team took ten days to finish the same project. In contrast, productivity varied by an average of only 21 percent between pairs of software developers from the same organization. If one team from a specific organization finished the coding project in one day, the longest time it took for another team from the same organization to finish the project was 1.21 days.

DeMarco and Lister also found *no correlation* between productivity and programming language, years of experience, or salary. Furthermore, the study showed that providing a dedicated workspace and a quiet work environment were key factors in improving productivity. The results of the study suggest that top managers must focus on workplace factors to improve productivity and quality.[24]

DeMarco and Lister wrote a booked called *Peopleware* in 1987, and the second edition was published in 1999. The underlying thesis of their book is that major problems with work performance and project failures are not technological in

[23] DeMarco, Tom, *Controlling Software Projects: Management, Measurement and Estimation.* New York: Yourdon Press, 1982.

[24] DeMarco, Tom and Timothy Lister, *Peopleware: Productive Projects and Teams.* New York: Dorset House, 1987.

nature; they are sociological in nature. They suggest minimizing politics and giving smart people physical space, intellectual responsibility, and strategic direction. And then just let them work. The manager's function is not to make people work but to make it possible for people to work by removing political roadblocks.

Maturity Models

Another approach to improving quality in software development projects and project management in general is the use of **maturity models**, which are frameworks for helping organizations improve their processes and systems. Three popular maturity models include the Software Quality Function Deployment (SQFD) model, the Capability Maturity Model (CMM), and the Project Management Maturity Model.

Software Quality Function Deployment Model

The Software Quality Function Deployment (SQFD) model is an adaptation of the quality function deployment model suggested in 1986 as an implementation vehicle for total quality management. SQFD focuses on defining user requirements and planning software projects. The end result of SQFD is a set of measurable technical product specifications and their priorities. Having clearer requirements can lead to fewer design changes, increased productivity, and, ultimately, software products that are more likely to satisfy stakeholder requirements. The idea of introducing quality early in the design stage was based on Taguchi's emphasis on robust quality.[25]

Capability Maturity Model

Another popular maturity model is in continuous development at the Software Engineering Institute at Carnegie Mellon University. The Software Engineering Institute (SEI) is a federally funded research and development center established in 1984 by the U.S. Department of Defense with a broad mandate to address the transition of software engineering technology. The **Capability Maturity Model (CMM)** is a five-level model laying out a generic path to process improvement for software development in organizations. Watts Humphrey's book *Managing the Software Process* (1989) describes an earlier version of the CMM.[26]

---

[25] Yilmaz, R. R. and Sangit Chatterjee, "Deming and the Quality of Software Development," *Business Horizons*, Foundation for the School of Business at Indiana University, 40(6) (November-December 1997), 51(8).

[26] Humphrey, W. S, *Managing the Software Process*. Reading, MA: Addison-Wesley, 1989.

The five levels of the CMM are:

1. *Initial*: The software development processes for organizations at this maturity level are ad hoc and occasionally even chaotic. Few processes are defined, and success often depends on individual effort.

2. *Repeatable*: Organizations at this maturity level have established basic project management processes to track cost, schedule, and functionality for software projects. Process discipline is in place to repeat earlier successes on similar projects.

3. *Defined*: At this level, the software processes for both management and software engineering activities are documented, standardized, and integrated into a standard software process for the organization. All projects use an approved, tailored version of the standard process in the organization.

4. *Managed*: At this maturity level, organizations collect detailed measures of the software process and product quality. Both the software processes and products are quantitatively understood and controlled.

5. *Optimizing*: Operating at the highest level of the maturity model, organizations can enable continuous process improvement by using quantitative feedback from the processes and from piloting innovative ideas and technologies.[27]

Project Management Maturity Models

In the late 1990s, several organizations began developing project management maturity models based on the Capability Maturity Model. Just as organizations realized the need to improve their software development processes and systems, they also realized the need to enhance their project management processes and systems.

The PMI Standards Development Program has made substantial progress on an Organizational Project Management Maturity Model (OPM3). More than 200 volunteers from around the world are part of the OPM3 team. John Schlichter, the program director, said the target date for delivering the OPM3 to PMI Headquarters for publishing is June 2003. The model is based on market research surveys sent to more than 30,000 project management professionals and incorporates 180 best practices and more than 2,400 capabilities, outcomes, and key performance indicators.[28] The OPM3 Web site (see Suggested Readings) provides information to assist organizations in implementing organization strategy through the successful, consistent, and predictable delivery of projects. The OPM3 will include a method for assessing organizations' project

---

[27] Paulk, Mark C., Bill Curtis, Mary Beth Chrissis, and Charles V. Weber, *Capability Maturity Model for Software*, Version 1.1, Technical Report, CMU/SEI-93-TR-024, ESC-TR-93-177 (February 1993).

[28] Schlichter, John, The Project Management Institute's Organizational Project Management Maturity Model: An Update on the PMI's OPM3 Program (September 2002), (*www.pmforum.org/library/opm3update1.htm*).

management maturity levels as well as a step-by-step method for increasing and maintaining an organization's ability to deliver projects as promised.

One sample project management maturity model developed by Micro-Frame Technologies, Inc. and Project Management Technologies, Inc. in 1997 has the following basic levels:

1. *Ad-Hoc*: The project management process is described as disorganized and occasionally even chaotic. The organization has not defined systems and processes, and project success depends on individual effort. There are chronic cost and schedule problems.

2. *Abbreviated*: There are some project management processes and systems in place to track cost, schedule, and scope. Project success is largely unpredictable and cost and schedule problems are common.

3. *Organized*: There are standardized, documented project management processes and systems that are integrated into the rest of the organization. Project success is more predictable, and cost and schedule performance is improved.

4. *Managed*: Management collects and uses detailed measures of the effectiveness of project management. Project success is more uniform, and cost and schedule performance conforms to plan.

5. *Adaptive*: Feedback from the project management process and from piloting innovative ideas and technologies enables continuous improvement. Project success is the norm, and cost and schedule performance is continuously improving.[29]

Several other companies provide similar project management maturity models. The International Institute for Learning, Inc. calls the five levels in its model common language, common processes, singular methodology, benchmarking, and continuous improvement. The ESI International Inc. model's five levels are called ad hoc, consistent, integrated, comprehensive, and optimizing. Regardless of the names of each level, the goal is clear: organizations want to improve their ability to manage projects.

William Ibbs and Young H. Kwak have been researching project management maturity in a series of studies, partially funded by the PMI Educational Foundation. They developed an assessment methodology to measure project management maturity, consisting of 148 multiple-choice questions based on the project management knowledge areas and process groups. Developing this systematic assessment methodology was a major contribution to the study of project management maturity. Organizations can benchmark their project management maturity using factual, impartial techniques such as those contained in this questionnaire, giving them a good reference point from which to begin making process improvements. Preliminary findings of Ibbs's and Kwak's research show that certain industries, such as engineering and construction,

[29] Enterprise Planning Associates, Project Management Maturity Model, Interactive Quick Look (1998).

have a higher maturity level than other industries, such as information systems/software development.[30]

Many organizations are assessing where they stand in terms of project management maturity, just as they did for software development maturity. Organizations are recognizing that they must make a commitment to the discipline of project management in order to improve project quality.

USING SOFTWARE TO ASSIST IN PROJECT QUALITY MANAGEMENT

This chapter provides examples of several tools and techniques used in project quality management. Software can be used to assist with several of these tools and techniques. For example, you can create Pareto diagrams using spreadsheet and charting software. You can use statistical software packages to help you determine standard deviations and perform many types of statistical analyses. You can create Gantt charts using project management software to help you plan and track work related to project quality management. There are also several specialized software products to assist people with managing Six Sigma projects or creating quality control charts. Project teams need to decide what types of software will help them manage their particular projects.

As you can see, quality itself is a very broad topic, and it is only one of the nine project management knowledge areas. Project managers must focus on defining how quality relates to their specific projects and ensure that those projects will satisfy the needs for which they were undertaken.

CASE WRAP-UP

Scott Daniels assembled a team to identify and resolve quality-related issues with the EIS and to develop a plan to help the medical instruments company prevent future quality problems. The first thing Scott's team did was to research the problems with the EIS. They created a fishbone diagram similar to the one in Figure 8-6. They also created a Pareto diagram (Figure 8-1) to help analyze the many complaints the Help Desk received and documented about the EIS. After further investigation, Scott and his team found out that many of the managers using the system were very inexperienced in using computer systems.

[30] Ibbs, C. William and Young Hoon Kwak, "Assessing Project Management Maturity," *Project Management Journal* (March 2000).

The system was designed to automatically generate a user password that included letters and numbers. Many managers did not know how to change their passwords to something they could remember or to have the system save their password. Scott also found out that several people who had just received passwords had misinterpreted the number one in their password for the lowercase letter L. There did not appear to be any major problems with the hardware for the EIS or the users' individual computers. The complaints about reports not giving consistent information all came from one manager, who had actually misread the reports, so there were no problems with the way the software was designed. Scott was very impressed with the quality of the entire project, except for the training. Scott reported his team's findings to the project sponsor of the EIS, who was relieved to find out that the quality problems were not as serious as many people feared they were.

CHAPTER SUMMARY

Many news headlines regarding the poor quality of information technology projects demonstrate that quality is a serious issue in information technology projects. Several mission-critical information technology systems have caused deaths, and quality problems in many business systems have resulted in major financial losses.

Customers are ultimately responsible for defining quality. Important quality concepts include satisfying stated or implied stakeholder needs, conforming to requirements, and delivering items that are fit for use.

Project quality management includes quality planning, quality assurance, and quality control. Quality planning identifies which quality standards are relevant to the project and how to satisfy them. Quality assurance involves evaluating overall project performance to ensure the project will satisfy the relevant quality standards. Quality control includes monitoring specific project results to ensure that they comply with quality standards and identifying ways to improve overall quality.

There are many tools and techniques related to project quality management. Fishbone diagrams help find the root cause of problems. Pareto charts help identify the vital few contributors that account for the most quality problems. Statistical sampling helps define a realistic number of items to include in analyzing a population. Six Sigma is helping companies improve quality by reducing defects. Standard deviation measures the variation in data. Control charts display data to help keep processes in control. Testing is very important in developing and delivering quality information technology products.

Many people contributed to the development of modern quality management. Deming, Juran, Crosby, Ishikawa, Taguchi, and Feigenbaum all made significant contributions to the field. Their works are being used today by many organizations and influenced Six Sigma principles. The Malcolm Baldrige National Quality Award and ISO 9000 have also helped organizations emphasize the importance of improving quality.

There is much room for improvement in information technology project quality. Strong leadership helps emphasize the importance of quality. Understanding the cost of quality provides an incentive for its improvement. Providing a good workplace can improve quality and productivity. Developing and following maturity models can help organizations systematically improve their project management processes to increase the quality and success rate of projects.

There are several types of software available to assist in project quality management. It is important for project teams to decide which software will be most helpful for their particular projects.

DISCUSSION QUESTIONS

1. Discuss some of the examples of poor quality in information technology projects presented in the "What Went Wrong?" section. Could most of these problems have been avoided? Why do you think there are so many examples of poor quality in information technology projects?

2. What are the main processes included in project quality management?

3. How do functionality, system outputs, performance, reliability, and maintainability requirements affect quality planning?

4. What type of information goes in a quality assurance plan?

5. What are the three main outputs of quality control?

6. Provide examples of when you would use a Pareto diagram, statistical sampling, quality control charts, and testing on an information technology project.

7. Discuss the history of modern quality management. How have experts such as Deming, Juran, Crosby, and Taguchi affected the quality movement and today's use of Six Sigma?

8. Discuss three suggestions for improving information technology project quality that were not made in this chapter.

9. Describe three different types of software that can assist in project quality management.

EXERCISES

1. Research progress that has been made in improving the quality of information technology projects. Write a one- to two-page paper discussing your findings.

2. Assume your organization wants to hire new instructors for your project management course. Develop a list of quality standards that you could use in making this hiring decision.

3. To illustrate a normal distribution, shake and roll a pair of dice thirty times and graph the results. It is more likely for someone to roll a six, seven, or eight than a two or twelve, so these numbers should come up more often. To create the graph, use graph paper or draw a grid. Label the x-axis with the numbers two through twelve. Label the y-axis with the numbers one through eight. Fill in the appropriate grid for each roll of the dice. Do your results resemble a normal distribution? Why or why not?

4. Research the criteria for the Malcolm Baldrige National Quality Award. Investigate a company other than Motorola that has received this award. What steps did the company take to earn this prestigious quality award? What are the benefits of earning this award?

5. Review the information in this chapter about Six Sigma principles and Six Sigma organizations. Brainstorm ideas for a potential Six Sigma project that could improve quality on your campus, at your workplace, or in your community. Write a one- to two-page paper describing one project idea and why it would be a Six Sigma project. Review and discuss how you could use the DMAIC process on this project.

6. Review the concepts in this chapter related to improving the quality of software. Write a one- to two-page paper describing how you could apply these concepts to software development projects.

MINICASE

You are part of a team analyzing quality problems that you have been having with your call center/customer service area. After studying customer complaints, you have categorized and logged them for a week as follows:

Part 1: Create a Pareto diagram based on the information in the table below. First, create a spreadsheet in Excel, using the data in the table below. List the most frequent customer problems first. Add a column called "% of Total" and another one called "Cumulative %." Then enter formulas to calculate those items. Next, use the Excel Chart Wizard to create a Pareto diagram based on this data. Use the Line—Column on 2 Axis custom type chart so your resulting chart looks similar to the one in Figure 8-1.

CUSTOMER COMPLAINTS	FREQUENCY/WEEK
Customer is on hold too long	90
Customer gets transferred to wrong area or cut off	20
Service rep cannot answer customer's questions	120
Service rep does not follow through as promised	40

Part 2: Your team has decided to add some customer service features to your Web site. You believe that having "Frequently Asked Questions" and answers will help reduce the complaints about service reps not being able to answer customers' questions and will lower the volume of calls. Customers and service reps could access information through this new Web page. However, you know that just as people complain about call centers, they also complain about Web sites. Develop a list of at least four potential complaints that your customers and service reps might have about this new Web page, and then describe your plans for preventing those problems. Describe also how this new Web page could continuously help to improve the quality of service provided by your call center/customer service area.

RUNNING CASES

Several questions in the running cases include activities related to project quality management. One of the running cases involves a Six Sigma project. See Appendix C for details.

SUGGESTED READINGS

1. Bylinksky, Gene. "How to Bring Out Better Products Faster." *Fortune* (November 23, 1998), 238[B].

    This article describes how companies such as Xerox, Hewlett-Packard, Ford Motor Company, and Goodyear have developed outstanding new products by using Taguchi's Robust Design method. This method goes directly to the basic physics and thermodynamics of product design— torque, electrical charge, heat flux, and so on—that are causing trouble, and solves problems during the early design phase.

2. Deming, W. Edwards. *Out of the Crisis.* Cambridge, MA: Massachusetts Institute of Technology, 1988.

    Despite Deming's final book, The New Economics for Industry, Government, Education *(1995),* Out of the Crisis *is Deming's best-known classic and is the source of constant reference. Deming uses real-life case studies to share with the reader the "simplicity" of his profound knowledge, his fourteen points, and the Deadly Diseases of quality. Deming has authored and coauthored several other books on quality.*

3. Ireland, Lewis R. *Quality Management for Projects and Programs.* Upper Darby, PA: Project Management Institute, 1991.

    This PMI publication includes seven chapters and several appendices with information on project quality management. Lew Ireland is well known for his expertise in quality and project management and his contributions to PMI. In 1999, he served as the president of PMI.

4. Juran, Joseph. *Juran's Quality Handbook*, 5th ed. New York: McGraw Hill, 1999.

> *Juran published the fifth edition of this famous handbook at the age of 94. This major revision of the classic reference on quality provides a comprehensive body of knowledge needed to engineer and manage for quality into the twenty-first century. There is also an excellent Web site with information about quality and Juran at www.juran.com.*

5. Pande, Peter S., Neuman, Robert P., and Cavanagh, Roland R. *The Six Sigma Way*. New York: McGraw-Hill (2000).

> *Many books and articles have been written in the past few years about Six Sigma.* The Six Sigma Way, *published in 2000, continues to be a top-seller. This comprehensive book is a self-help guide to adapting and using Six Sigma under various conditions. Part one of the book provides an executive summary of Six Sigma. Part two describes how to gear up and adapt Six Sigma to various organizations. Part three provides detailed information on implementing Six Sigma, including a detailed roadmap and description of tools. The appendix includes several checklists and a detailed Six Sigma conversion table.*

6. Project Management Institute, Organizational Project Management Maturity Model (OPM3), (*www.pmi.org/opm3*) (2003).

> *This Web site provides information about PMI's Organizational Project Management Maturity Model (OPM3) program, created to assist organizations in implementing organizational strategy through the successful, consistent, and predictable delivery of projects. The OPM3 will include a method for assessing organizations' project management maturity levels as well as a step-by-step method for increasing and maintaining an organization's ability to deliver projects as promised.*

KEY TERMS

- **acceptance decisions** — decisions that determine if the products or services produced as part of the project will be accepted or rejected
- **appraisal cost** — the cost of evaluating processes and their outputs to ensure that a project is error-free or within an acceptable error range
- **benchmarking** — a technique used to generate ideas for quality improvements by comparing specific project practices or product characteristics to those of other projects or products within or outside the performing organization
- **Capability Maturity Model (CMM)** — a five-level model laying out a generic path to process improvement for software development in organizations
- **conformance** — delivering products that meet requirements and fitness for use
- **conformance to requirements** — the project processes and products meet written specifications

- **control chart** — a graphic display of data that illustrates the results of a process over time
- **cost of nonconformance** — taking responsibility for failures or not meeting quality expectations
- **cost of quality** — the cost of conformance plus the cost of nonconformance
- **defect** — any instance where the product or service fails to meet customer requirements
- **Define, Measure, Analyze, Improve, Control (DMAIC)** —a systematic, closed-loop process for continued improvement that is scientific and fact based
- **design of experiments** — a quality technique that helps identify which variables have the most influence on the overall outcome of a process
- **external failure cost** — a cost related to all errors not detected and corrected before delivery to the customer
- **features** — the special characteristics that appeal to users
- **Fishbone diagrams** — diagrams that trace complaints about quality problems back to the responsible production operations; sometimes called Ishikawa diagrams
- **fitness for use** — a product can be used as it was intended
- **functionality** — the degree to which a system performs its intended function
- **integration testing** — testing that occurs between unit and system testing to test functionally grouped components to ensure a subset(s) of the entire system works together
- **internal failure cost** — a cost incurred to correct an identified defect before the customer receives the product
- **ISO 9000** — a quality system standard developed by the International Organization for Standardization (ISO) that includes a three-part, continuous cycle of planning, controlling, and documenting quality in an organization
- **maintainability** — the ease of performing maintenance on a product
- **Malcolm Baldrige National Quality Award** — an award started in 1987 to recognize companies that have achieved a level of world-class competition through quality management
- **maturity model** — a framework for helping organizations improve their processes and systems
- **mean** — the average value of a population
- **measurement and test equipment cost** — the capital cost of equipment used to perform prevention and appraisal activities
- **normal distribution** — a bell-shaped curve that is symmetrical about the mean of the population
- **Pareto analysis** — identifying the vital few contributors that account for most quality problems in a system

- **Pareto diagrams** — histograms that help identify and prioritize problem areas
- **performance** — how well a product or service performs the customer's intended use
- **prevention cost** — the cost of planning and executing a project so that it is error-free or within an acceptable error range
- **process adjustments** — adjustments made to correct or prevent further quality problems based on quality control measurements
- **project quality management** — ensuring that a project will satisfy the needs for which it was undertaken
- **quality** — the totality of characteristics of an entity that bear on its ability to satisfy stated or implied needs
- **quality assurance** — periodically evaluating overall project performance to ensure the project will satisfy the relevant quality standards
- **quality audits** — structured reviews of specific quality management activities that help identify lessons learned and can improve performance on current or future projects
- **quality circles** — groups of nonsupervisors and work leaders in a single company department who volunteer to conduct group studies on how to improve the effectiveness of work in their department
- **quality control** — monitoring specific project results to ensure that they comply with the relevant quality standards and identifying ways to improve overall quality
- **quality planning** — identifying which quality standards are relevant to the project and how to satisfy them
- **reliability** — the ability of a product or service to perform as expected under normal conditions without unacceptable failures
- **rework** — action taken to bring rejected items into compliance with product requirements or specifications or other stakeholder expectations
- **Robust Design methods** — methods that focus on eliminating defects by substituting scientific inquiry for trial-and-error methods
- **seven run rule** — if seven data points in a row on a quality control chart are all below the mean, above the mean, or are all increasing or decreasing, then the process needs to be examined for nonrandom problems
- **Six Sigma** — a comprehensive and flexible system for achieving, sustaining, and maximizing business success that is uniquely driven by close understanding of customer needs, disciplined use of facts, data, statistical analysis, and diligent attention to managing, improving, and reinventing business processes
- **six 9s of quality** — a measure of quality control equal to 1 fault in 1 million opportunities
- **standard deviation** — a measure of how much variation exists in a distribution of data

- **statistical sampling** — choosing part of a population of interest for inspection
- **system outputs** — the screens and reports the system generates
- **system testing** — testing the entire system as one entity to ensure it is working properly
- **unit test** — a test of each individual component (often a program) to ensure it is as defect-free as possible
- **user acceptance testing** — an independent test performed by end users prior to accepting the delivered system
- **yield** — represents the number of units handled correctly through the development process

9

Project Human Resource Management

Objectives

After reading this chapter, you will be able to:

1. *Explain the importance of good human resource management on projects, especially on information technology projects*
2. *Define project human resource management and understand its processes*
3. *Summarize key concepts for managing people by understanding the theories of Abraham Maslow, Frederick Herzberg, David McClelland, and Douglas McGregor on motivation, H. J. Thamhain and D. L. Wilemon on influencing workers, and Stephen Covey on how people and teams can become more effective*
4. *Discuss organizational planning and be able to create a project organizational chart, responsibility assignment matrix, and resource histogram*
5. *Understand important issues involved in project staff acquisition and explain the concepts of resource assignments, resource loading, and resource leveling*
6. *Assist in team development with training, team-building activities, and reward systems*
7. *Describe how project management software can assist in project human resource management*

OPENING CASE

This was the third time someone from the Information Technology Department tried to work with Ben, the head of the F-44 aircraft program. Ben, who had been with the company for almost thirty years, was known for being rough around the edges and very demanding. The company was losing money on the F-44 upgrade project because the upgrade kits were not being delivered on time. The Canadian government had written severe late penalty fees into its contract, and other customers were threatening to take their business elsewhere. Ben blamed it all on the Information Technology Department for not letting his staff access the F-44 upgrade project's information system directly so they could work with their customers and suppliers more effectively. The information system was based on very old technology that only a couple of people in the company knew how to use. It often took days or even weeks for Ben's group to get the information they needed.

Ed Davidson, a senior programmer, attended a meeting with Sarah Ellis, an internal information technology business consultant. Sarah was in her early thirties and had advanced quickly in her company, primarily due to her keen ability to work well with all types of people. Sarah's job was to uncover the real problems with the F-44 aircraft program's information technology support, and then develop a solution with Ben and his team. If she found it necessary to invest in more information technology hardware, software, or staff, Sarah would write a business case to justify these investments and then work with Ed, Ben, and his group to implement the suggested solution as quickly as possible. Ben and three of his staff entered the conference room. Ben threw his books on the table and started yelling at Ed and Sarah. Ed could not believe his eyes or ears when Sarah stood nose to nose with Ben and started yelling right back at him.

THE IMPORTANCE OF HUMAN RESOURCE MANAGEMENT

Many corporate executives have said, "People are our most important asset." People determine the success and failure of organizations and projects. Most project managers agree that managing human resources effectively is one of the toughest challenges they face. Project human resource management is a vital component of project management, especially in the information technology field—in which qualified people are often hard to find and keep. It is important to understand the current state of human resource management in the information technology industry and its implications for the future.

Current State of Human Resource Management

In the 1990s, there was a growing shortage of personnel in information technology. The Information Technology Association of America (ITAA) calculated that in 1997 there were more than 345,000 job openings for programmers, systems analysts, computer scientists, and computer engineers in information technology and non-information technology companies with one hundred or more employees. These estimated job openings represented about ten percent of current employment in these four occupations.[1] In April 2000, ITAA published its "Bridging the Gap" workforce study, which clearly documented the continued staffing crisis facing the information technology industry. This study revealed that one in fourteen American workers is involved in information technology, and that 844,000 information technology jobs were unfilled in 2000, with even more vacancies anticipated in future years.

Unfortunately, ITAA and other groups projecting employment in the first decade of the new millennium were wrong about these estimated vacancies. No one could foresee how changes in the global economy, the effects of terrorism, or the declining stock market would affect job openings. A more recent ITAA report showed that the U.S. information technology workforce was at least stabilizing. Below are some findings from this December 2002 report:

- The total number of people working in information technology in the United States was more than 10.1 million, compared to 9.9 million in January of 2002.

- U.S. companies hired 359,000 information technology workers in the last quarter of 2002 but dismissed 211,000 workers, which is a net gain of 147,000 workers.

- Annual dismissals dropped 68% in 2002, from 2.6 million in 2001 to 844,000 in 2002.

- Information technology managers predict they will need to hire an additional 1.2 million workers in the near future.

- Technical support workers represent 42% of all information technology hires in the first quarter of 2002. The overall number of network administrators declined by 8.5% since the beginning of 2002, while the number of Web developers and database developers increased more than 5% each since January 2002.

- Hiring by non-information technology companies outpaces hiring by information technology companies by a ratio of twelve to one.[2]

[1] Information Technology Association of America (ITAA). "Help Wanted: The IT Workforce Gap at the Dawn of a New Century," 1997.

[2] Information Technology Association of America, "New ITAA Report Shows U.S. IT Workforce Continues to Stabilize in Third Quarter," December 18, 2002, *www.itaa.org*.

In 2000, the World Information Technology and Services Alliance (WITSA) released its study, "Digital Planet 2000: The Global Information Economy," which revealed that the global high-tech industry generated more than $2.1 trillion in 1999, and is expected to surpass $3 trillion in 2003.[3] The 2002 study reported that the global high-tech industry did grow, but only from $2.3 trillion in 2000 to $2.4 trillion in 2001. Digital Planet 2002 reports that the Internet and e-commerce remains a notable bright spot in the global economy, despite flattened spending in most of the developed world. Global e-commerce (business-to-business and business-to-consumer) reached $633 billion in 2001, an increase of 79 percent from the previous year. "Even an economic downturn cannot stifle the value and convenience of electronic commerce to businesses and consumers," said WITSA President Harris N. Miller at a press conference on February 28, 2002. The United States continues to lead in overall information and communication technology (ICT) spending, but its overall percentage of global spending continues to decline. "Another positive story Digital Plan 2002 tells is that as the benefits and efficiency of ICT spread across the globe, mature markets such as the U.S. will become less dominant. China, Poland, and other developing countries are already playing an increasing role in the global ICT marketplace, showing the upside opportunities for global ICT spending are still enormous," Miller said.[4]

These studies highlight the continued need for skilled information technology workers and project managers to lead their projects. As the job market changes, however, people should upgrade their skills to remain marketable and flexible. Many highly qualified information technology workers lost their jobs in the past few years. Workers need to develop a support system and financial reserves to make it through difficult economic times. They also need to consider different types of jobs and industries where they can use their skills and earn a living. Negotiation and presentation skills also become crucial in finding and keeping a good job. Workers need to know how they personally contribute to an organization's bottom line.

However, even as organizations demand increased profits and available jobs become harder to find, it is still crucial that organizations do not take advantage of their workers' time. Jeannette Cabanis-Brewin, Editor-in-Chief of the Center for Business Practices publications and former editor for *PM Network* and the *Project Management Journal*, specializes in the organizational and human side of project management. Cabanis-Brewin, also a proud grandmother, wrote an article complaining that her grandchildren did not visit for Christmas in 2000 because their parents could not take a few days off work.[5] Cabanis-Brewin's

---

[3] Information Technology Association of America, "Global IT Spending to Rocket from Current $2 Trillion to $3 Trillion, New Study Finds," *Update for IT Executives* (2001), 6 (15), *www.itaa.org*.

[4] World Information Technology and Services Alliance (WITSA), *WITSA Global Research Shows World's Largest Consumer of Technology Increased Spending Less Than 1% Last Year, Internet Spending Up, Despite Worldwide Economic Condition*, February 28, 2002, *www.witsa.org*.

[5] Cabanis-Brewin, Jeannette, "An Open Letter to Scrooge, Inc.," *Best Practices e-Advisor* (January 7, 2001), 16.

editorial stressed that too many people in top management have not learned their lessons from watching the movie *Scrooge*.

Excessive working hours are not the only problem in the information technology industry. In addition, the field of information technology has become unappealing to certain groups, such as women. The number of women entering the information technology field peaked in 1984. Today, that number continues to decline. More than half of all students attending colleges today are women, yet less than 20 percent of the graduates awarded Bachelor's degrees in computer science and related fields are female. Anita Borg, a computer engineer at Xerox Palo Alto Research Center and founder of the Institute for Women and Technology, states that if women had been going into computer fields at the same rate as men since 1984, the shortage of information technology workers in the late 1990s would not have existed.[6]

White House findings in 2000 showed that employment in the U.S. information technology industry grew 81 percent. Women, who represent 47 percent of the work force at large, made up only 29 percent of several information technology categories. Despite the demand for computing professionals outstripping the supply, more men than women are earning undergraduate, Master's, and doctoral degrees in computer science. The Association for Computing Machinery Committee on Women in Computing (ACM-W), which surveyed the shrinking pipeline of women in computer science in 1998, released a study in 2000 that finds the pipeline is continuing to shrink. Females in high school, college, graduate school, and beyond are not choosing computer science or related degrees. Why not? Denise Gurer, 3Com Corporation technologist and ACM-W co-chair says, "Girls and women are not turned off by technology, but by how it is used in our society….We can attract more girls and women to technology areas if we make classrooms more inclusive of technology uses that interest them."[7] Girls and women are interested in using technology to save time, to be more productive and creative, and to develop relationships. Girls are often turned off by technology when they see computers used to play unproductive games, which are often violent or sexist, to track sports statistics, to download unappealing music, or to avoid having real relationships with other people.

The ACM-W studies found that factors contributing to the gender gap include the following:

- The stereotype that girls are not good at math
- Computer games that target boys

---

[6] Olsen, Florence, "Institute for Women and Technology Works to Bridge Computing's Gender Gap," *The Chronicle of Higher Education* (February 25, 2000).

[7] DeBlasi, Patrick J., "The Shrinking Pipeline Unlikely to Reverse: Nearly 30 Percent Decline of Women Choosing Undergraduate Degrees in Computer Science," *Association of Computing Machinery (ACM) Pressroom* (June 5, 2000).

- Online discussion groups that are largely male-oriented and sexist
- The fact that few women hold positions of power in the computer industry[8]

Implications for the Future of Human Resource Management

It is crucial for organizations to practice what they preach about human resources. If people truly are their greatest asset, organizations must work to fulfill their human resource needs *and* the needs of individual people in their organizations, regardless of the job market. If organizations want to be successful at implementing information technology projects, they need to understand the importance of project human resource management and take actions to make effective use of people. Proactive organizations are addressing current and future human resource needs, by, for example, improving benefits, redefining work hours and incentives, and finding future workers.

Many organizations have changed their benefits policies to meet worker needs. Most workers assume that their companies provide some perks, such as casual dress codes, flexible work hours, and tuition assistance. Other companies provide on-site day care, dry-cleaning services, and staff to run errands such as walking your dog. The January 20, 2003 issue of *Fortune* magazine included its annual list of the best companies to work for, stressing the importance of treating workers well. SAS, a software development firm headquartered in Cary, North Carolina, continues to make the list. It has a very low turnover rate because it treats employees very well, with on-site child care for only $300 per month, a pianist in the cafeteria, a gym, and well-groomed walking paths, plus challenging work and good pay. Some organizations go even further and use special perks as a competitive advantage. For example, an organization may promote policies such as allowing workers to bring their pets to work, providing free vacation rentals, paying for children's college tuition, or providing workers with nice cars. Jeff Finney, a computer programmer in suburban Atlanta, accepted delivery of a BMW from his employer, Revenue Systems, Inc. All 45 employees at this firm get to lease BMWs at the bosses' expense. Finney, who normally drives a pickup, says, "It is probably one of the best perks I've seen any company give."[9] Although some of these perks disappear in hard economic times, successful organizations continue to reward their workers with bonuses or profit sharing when they can prove their direct contributions to the bottom line.

Other implications for the future of human resource management relate to the hours organizations expect many information technology professionals to

[8] Ohlson, Kathleen, "ACM launches gender gap study," *Online News* (November 13, 1998).
[9] Irvine, Martha, "Executive-style perks spread to rank and file," *Minneapolis Star Tribune* (September 29, 1998), D-2.

work and how they reward performance. Today people brag about the fact that they can work less than 40 hours a week or work from home several days a week—not that they haven't had a vacation in years. If companies plan their projects well, they can avoid the need for overtime, or they can make it clear that overtime is optional. Companies can also provide incentives that use performance, not hours worked, as the basis of rewards. If performance can be objectively measured, as it can in many aspects of information technology jobs, then it should not matter where employees do their work or how long it takes them to do it. For example, if a technical writer can produce a high-quality publication at home in one week, the company and the writer are better off than if the company insisted that the writer come to the office and take two weeks to produce the publication. Objective measures of work performance and incentives based on meeting those criteria are important considerations.

The need to develop future talent in information technology has important implications for everyone. Who will maintain the systems we have today when many baby boomers retire? Who will continue to develop new products and services using new technologies that have not yet been developed? Some schools are requiring all students to take computer literacy courses. Some individuals and organizations are offering incentives to minorities and women to enter technical fields. For example, some wealthy alumni offer to pay all college expenses for minority students from the alumni's high school or grade school. Several colleges and government agencies have programs to help recruit more women into technical fields. All of these efforts will help to develop the human resources needed for future information technology projects.

WHAT IS PROJECT HUMAN RESOURCE MANAGEMENT?

Project human resource management includes the processes required to make the most effective use of the people involved with a project. Human resource management includes all project stakeholders: sponsors, customers, project team members, support staff, suppliers supporting the project, and so on. Project human resource management involves three processes.

- **Organizational planning** involves identifying, assigning, and documenting project roles, responsibilities, and reporting relationships. Key outputs of this process include roles and responsibility assignments, often shown in a matrix form, and an organizational chart for the project.
- **Staff acquisition** involves getting the needed personnel assigned to and working on the project. Getting personnel is one of the crucial challenges of information technology projects.
- **Team development** involves building individual and group skills to enhance project performance. Building individual and group skills is also a challenge for many information technology projects.

Some topics related to human resource management (understanding organizations, stakeholders, and different organizational structures) were introduced in Chapter 2, The Project Management and Information Technology Context. This chapter will expand on some of those topics and introduce other important concepts in project human resource management, including theories about managing people, resource loading, and resource leveling. You will also learn how to use software to assist in project resource management.

KEYS TO MANAGING PEOPLE

Industrial-organizational psychologists and management theorists have devoted much research and thought to the field of managing people at work. Psychosocial issues that affect how people work and how well they work include motivation, influence and power, and effectiveness. This section will review Abraham Maslow's, Frederick Herzberg's, David McClelland's, and Douglas McGregor's contributions to an understanding of motivation; H. J. Thamhain's and D. L. Wilemon's work on influencing workers and reducing conflict; the effect of power on project teams; and Stephen Covey's work on how people and teams can become more effective. Finally, you will look at some implications and recommendations for project managers.

Motivation Theories

Psychologists, managers, coworkers, teachers, parents, and most people in general still struggle to understand what motivates people, or, why they do what they do. **Intrinsic motivation** causes people to participate in an activity for their own enjoyment. For example, some people love to read, write, or play an instrument because it makes them feel good. **Extrinsic motivation** causes people to do something for a reward or to avoid a penalty. For example, some young children would prefer to *not* play an instrument, but they do because they receive a reward or avoid a punishment for doing so. Why do some people require no external motivation whatsoever to produce high-quality work while others require significant external motivation to perform routine tasks? Why can't you get someone who is extremely productive at work to do simple tasks at home? Mankind will continue to try to answer these types of questions. A basic understanding of motivational theory will help anyone who has to work or live with other people to understand themselves and others.

Maslow's Hierarchy of Needs

Abraham Maslow, a highly respected psychologist who rejected the dehumanizing negativism of psychology in the 1950s, is best known for developing a hierarchy of needs. In the 1950s, proponents of Sigmund Freud's psychoanalytic theory were promoting the idea that human beings were not the masters of their destiny and that all their actions were governed by unconscious processes dominated by primitive sexual urges. During the same period, behavioral psychologists saw human beings as controlled by the environment. Maslow argued that both schools of thought failed to recognize unique qualities of human behavior: love, self-esteem, belonging, self-expression, and creativity. He argued that these unique qualities enable a person to make independent choices, which gives them full control of their destiny.

Figure 9-1 shows the basic pyramid structure of Maslow's **hierarchy of needs**, which states that people's behaviors are guided or motivated by a sequence of needs. At the bottom of the hierarchy are physiological needs. Once physiological needs are satisfied, safety needs guide behavior. Once safety needs are satisfied, social needs come to the forefront, and so on up the hierarchy. The order of these needs and their relative sizes in the pyramid are significant. Maslow suggests that each level of the hierarchy is a prerequisite for the levels above. For example, it is not possible for a person to consider self-actualization if he or she has not addressed basic needs concerning security and safety. People in an emergency situation, such as a flood or hurricane, are not going to worry about personal growth. Personal survival will be their main motivation. Once a particular need is satisfied, however, it no longer serves as a potent motivator of behavior.

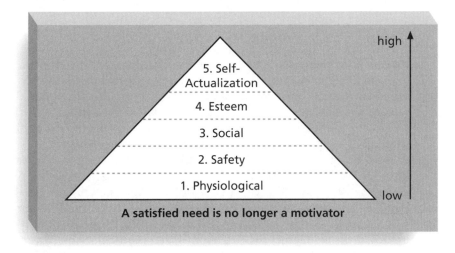

Figure 9-1. Maslow's Hierarchy of Needs

Notice in Figure 9-1 that each layer in the pyramid is smaller than the previous layer. The issues in each level are of greater value than issues in the preceding level, which presumably have been satisfied. Issues higher in the hierarchy that are associated with needs that are currently not satisfied are more important than earlier issues, although, by definition, they are less urgent than issues lower in the hierarchy.

The bottom four needs in Maslow's hierarchy—physiological, safety, social, and esteem needs—are referred to as deficiency needs, and the highest level, self-actualization, is considered a growth need. Only after meeting deficiency needs can individuals act upon growth needs. Self-actualized people are problem-focused, have an appreciation for life, are concerned about personal growth, and have the ability to have peak experiences.

Most people working on an information technology project will probably have their basic physiological and safety needs met. If someone has a sudden medical emergency or is laid off from work, however, physiological and safety needs will move to the forefront. To motivate project team members, the project manager needs to understand each person's motivation, especially with regard to social, esteem, and self-actualization or growth needs. Team members new to a company and city might be motivated by social needs. To address social needs, some companies organize gatherings and social events for new workers in information technology. Other project members may find these events to be an invasion of personal time they would rather spend with their friends and family or working on an advanced degree.

Maslow's hierarchy conveys a message of hope and growth. People can work to control their own destinies and naturally strive to achieve higher and higher needs. Project managers should know something about team members' professional and personal lives so they can provide appropriate motivation.

Herzberg's Motivation-Hygiene Theory

Frederick Herzberg is best known for distinguishing between motivational factors and hygiene factors when considering motivation in work settings. He called factors that cause job satisfaction motivators and factors that cause dissatisfaction hygiene factors. The term "hygiene" is used in the sense that these factors are considered maintenance factors that are necessary to avoid dissatisfaction but by themselves, they do not provide satisfaction.

Head of Case Western University's Psychology Department, Herzberg wrote the book *Work and the Nature of Man* in 1966 and the famous *Harvard Business Review* article, "One More Time: How Do You Motivate Employees?" in 1968. Herzberg analyzed the factors that affected productivity among a sample of 1,685 employees. Popular beliefs at that time were that work output was most improved through larger salaries, more supervision, or a more attractive work environment. According to Herzberg, these hygiene factors would cause dissatisfaction if not present, but would not motivate workers to do more if present. Herzberg found that people were motivated to work mostly by feelings of personal achievement and recognition. Motivators, Herzberg concluded, included achievement, recognition, the work itself, responsibility, advancement, and growth.

In his books and articles, Herzberg explained why attempts to use positive factors such as reducing time spent at work, upward spiraling wages, offering fringe benefits, providing human relations and sensitivity training, and so on did not instill motivation. He argued that people want to actualize themselves. They need stimuli for their growth and advancement needs, in accordance with Maslow's hierarchy of needs. Factors such as achievement, recognition, responsibility, advancement, and growth produce job satisfaction and are work motivators.[10]

McClelland's Acquired-Needs Theory

David McCelland proposed that an individual's specific needs are acquired or learned over time and shaped by life experiences. He wrote several books and articles on motivation, including *The Achieving Society* in 1961. The main categories of acquired needs include achievement, affiliation, and power. Normally one or two of these needs will be dominant in individuals.

- Achievement: People with a high need for achievement (nAch) seek to excel and tend to avoid both low-risk and high-risk situations to improve their chances for achieving something worthwhile. Achievers need regular feedback and often prefer to work alone or with other high achievers. Managers should give high achievers challenging projects with achievable goals. Achievers should receive frequent performance feedback, and although money is not an important motivator to them, it is an effective form of feedback.

- Affiliation: People with a high need for affiliation (nAff) desire harmonious relationships with other people and need to feel accepted by others. They tend to conform to the norms of their work group and prefer work that involves significant personal interaction. Managers should try to create a cooperative work environment to meet the needs of people with a high need for affiliation.

- Power: People with a need for power (nPow) desire either personal power or institutional power. People who need personal power want to direct others and can be seen as bossy. People who need institutional power or social power want to organize others to further the goals of the organization. Management should provide power seekers with the opportunity to manage others, emphasizing the importance of meeting organizational goals.

The Thematic Apperception Test (TAT) is a tool to measure the individual needs of different people using McClelland's categories. The TAT presents subjects with a series of ambiguous pictures and asks them to develop a spontaneous story for each picture, assuming they will project their own needs into the story.

[10] Herzberg, Frederick, "One More Time: How Do You Motivate Employees?" *Harvard Business Review* (February 1968), 51–62.

McGregor 's Theory X and Theory Y

Douglas McGregor was one of the great popularizers of a human relations approach to management, and he is best known for developing Theory X and Theory Y. In his research, documented in his 1960 book *The Human Side of Enterprise,* McGregor found that although many managers spouted the right ideas, they actually followed a set of assumptions about worker motivation that he called Theory X (sometimes referred to as classical systems theory). People who believe in Theory X assume that workers dislike and avoid work if possible, so managers must use coercion, threats, and various control schemes to get workers to make adequate efforts to meet objectives. They assume that the average worker wants to be directed and prefers to avoid responsibility, has little ambition, and wants security above all else. Research seemed to demonstrate clearly that these assumptions were not valid. McGregor suggested a different series of assumptions about human behavior that he called Theory Y (sometimes referred to as human relations theory). Managers who believe in Theory Y assume that individuals do not inherently dislike work, but consider it as natural as play or rest. The most significant rewards are the satisfaction of esteem and self-actualization needs, as described by Maslow. McGregor urged managers to motivate people based on these more valid Theory Y notions.

In 1981, William Ouchi introduced another approach to management in his book *Theory Z: How American Business Can Meet the Japanese Challenge.* Theory Z is based on the Japanese approach to motivating workers, which emphasizes trust, quality, collective decision making, and cultural values. Whereas Theory X and Theory Y emphasize how management views employees, Theory Z also describes how workers perceive management. Theory Z workers, it is assumed, can be trusted to do their jobs to their utmost ability, as long as management can be trusted to support them and look out for their well-being. Theory Z emphasizes things such as job rotation, broadening of skills, generalization versus specialization, and the need for continuous training of workers.

Influence and Power

Many people working on a project do not report directly to project managers, and project managers often do not have control over project staff who report to them. For example, people are free to change jobs. If they are given work assignments they do not like, many workers will simply quit or transfer to other departments or projects. Starting in the 1970s, H. J. Thamhain and D. L. Wilemon investigated the approaches project managers use to deal with workers and how those approaches relate to project success. They identified nine influence bases available to project managers:

1. Authority — the legitimate hierarchical right to issue orders
2. Assignment — the project manager's perceived ability to influence a worker's later work assignments

3. Budget — the project manager's perceived ability to authorize others' use of discretionary funds
4. Promotion — the ability to improve a worker's position
5. Money — the ability to increase a worker's pay and benefits
6. Penalty — the project manager's perceived ability to dispense or cause punishment
7. Work challenge — the ability to assign work that capitalizes on a worker's enjoyment of doing a particular task, which taps an intrinsic motivational factor
8. Expertise — the project manager's perceived special knowledge that others deem important
9. Friendship — the ability to establish friendly personal relationships between the project manager and others

Top management grants authority to the project manager. Assignment, budget, promotion, money, and penalty influence bases may or may not be inherent in a project manager's position. Unlike authority, they are not automatically available to project managers as part of their position. Others' perceptions are important in establishing the usefulness of these influence bases. For example, any manager can influence workers by providing challenging work; the ability to provide challenging work (or take it away) is not a special ability of project managers. In addition, project managers must earn the ability to influence by using expertise and friendship.

Thamhain and Wilemon found that projects were more likely to fail when project managers relied too heavily on using authority, money, or penalty to influence people. When project managers used work challenge and expertise to influence people, projects were more likely to succeed. The effectiveness of work challenge in influencing people is consistent with Maslow's and Herzberg's research on motivation. The importance of expertise as a means of influencing people makes sense on projects that involve special knowledge, as in most information technology projects.

Influence is related to the topic of power. Power is the potential ability to influence behavior to get people to do things they would not otherwise do. Power has a much stronger connotation than influence, especially since it is often used to force people to change their behavior.

There are five main types of power based on French and Raven's 1959 book, *Studies in Social Power*:

■ **Coercive power** involves using punishment, threats, or other negative approaches to get people to do things they do not want to do. This type of power is similar to Thamhain's and Wilemon's influence category called penalty. For example, a project manager can threaten to fire workers or subcontractors to try to get them to change their behavior. If the project manager really has the power to fire people, he or she could follow through on the threat. Recall, however, that influencing using penalties is correlated with unsuccessful projects. However, coercive power can be very effective in stopping negative behavior. For example, if students tend to hand in assignments late, an instructor can have a policy in his or her syllabus stating late penalties, such as a 20 percent grade reduction for each day an assignment is late.

- **Legitimate power** is getting people to do things based on a position of authority. This type of power is similar to the authority basis of influence. If top management gives project managers organizational authority, project managers can use legitimate power in several situations. They can make key decisions without involving the project team, for example. Overemphasis of legitimate power or authority also correlates with project failure.

- **Expert power** involves using personal knowledge and expertise to get people to change their behavior. If people perceive that project managers are experts in certain situations, they will follow their suggestions. For example, if a project manager has expertise in working with a particular information technology supplier and their products, the project team will be more likely to follow the project manager's suggestions on how to work with that vendor and its products.

- **Reward power** involves using incentives to induce people to do things. Rewards can include money, status, recognition, promotions, special work assignments, or other means of rewarding someone for desired behavior. Many motivation theorists suggest that only certain types of rewards, such as work challenge, achievement, and recognition, truly induce people to change their behavior or work hard.

- **Referent power** is based on an individual's personal charisma. People hold someone with referent power in very high regard and will do what they say based on their regard for the person. People such as Martin Luther King, Jr., John F. Kennedy, and Bill Clinton had referent power. Very few people possess the natural charisma that underlies referent power.

It is important for project managers to understand what types of influence and power they can use in different situations. New project managers often overemphasize their position—their legitimate power or authority influence—especially when dealing with project team members or support staff. They also neglect the importance of reward power or work challenge influence. People often respond much better to a project manager who motivates them with challenging work and provides positive reinforcement for doing a good job. It is important for project managers to understand the basic concepts of influence and power, and to practice using them to their own and their project team's advantage.

Improving Effectiveness

Stephen Covey, author of the 1980 book, *The 7 Habits of Highly Effective People*, expanded on the work done by Maslow, Herzberg, and others to develop an approach for helping people and teams become more effective. Covey's first three habits of effective people—be proactive, begin with the end in mind, and put first things first—help people achieve a private victory by becoming independent. After achieving independence, people can then strive for interdependence by developing the next three habits—think win/win, seek first to understand then to be understood, and synergize. (**Synergy** is the concept that the whole is equal to more than the sum of its parts.) Finally, everyone can work on Covey's seventh habit—sharpen the saw—to develop and renew their physical, spiritual, mental, and social/emotional selves.

Project managers can apply Covey's seven habits to improve effectiveness on projects, as follows:

1. *Be proactive.* Covey, like Maslow, believes that people have the ability to be proactive and choose their responses to different situations. Project managers must be proactive and anticipate and plan for problems and inevitable changes on projects. They can also encourage their team members to be proactive in working on their project activities.

2. *Begin with the end in mind.* Covey suggests that people focus on their values, what they really want to accomplish, and how they really want to be remembered in their lives. He suggests writing a mission statement to help achieve this habit. Many organizations and projects have mission statements that help them focus on their main purpose.

3. *Put first things first.* Covey developed a time management system and matrix to help people prioritize their time. He suggests that most people need to spend more time doing things that are important, but not urgent. Important but not urgent activities include planning, reading, and exercising. Project managers need to spend a lot of time working on important and not urgent activities, such as developing the project plan, building relationships with major project stakeholders, and mentoring project team members. They also need to avoid focusing only on important yet urgent activities—putting out fires.

4. *Think win/win.* Covey presents several paradigms of interdependence, with think win/win being the best choice in most situations. When you use a win/win paradigm, parties in potential conflict work together to develop new solutions that make them all winners. Project managers should strive to use a win/win approach in making decisions, but sometimes, especially in competitive situations, they must use a win/lose paradigm.

5. *Seek first to understand, then to be understood.* **Empathic listening** is listening with the intent to understand. It is even more powerful than active listening because you forget your personal interests and focus on truly understanding the other person. To really understand other people, you must learn to focus on others first. When you practice empathic listening, you can begin two-way communication. This habit is critical for project managers so they can really understand their stakeholders' needs and expectations.

6. *Synergize.* In projects, a project team can synergize by creating collaborative products that are much better than a collection of individual efforts. Covey also emphasizes the importance of valuing differences in others to achieve synergy. Synergy is essential to many highly technical projects; in fact, several major breakthroughs in information technology occurred because of synergy. For example, in his Pulitzer Prize–winning book, *The Soul of a New Machine*, Tracy Kidder documented the 1970s synergistic efforts of a team of Data General researchers to create a new 32-bit super-minicomputer.[11]

7. *Sharpen the saw.* When you practice sharpening the saw, you take time to renew yourself physically, spiritually, mentally, and socially. The practice of self-renewal helps people avoid burnout. Project managers must make sure that they and their project team have time to retrain, reenergize, and occasionally even relax to avoid burnout.

Douglas Ross, author of *Applying Covey's Seven Habits to a Project Management Career* (see the Suggested Readings), related Covey's seven habits to project management. Ross suggests that Habit 5—Seek first to understand, then to be understood—differentiates good project managers from average or poor project managers. People have a tendency to focus on their own agendas instead of first trying to understand other people's points of view. Empathic listening can help project managers and team members find out what motivates different people. Understanding what motivates key stakeholders and customers can mean the difference between project success and project failure. Once project managers and team members begin to practice empathic listening, they can communicate and work together to tackle problems more effectively.

Before you can practice empathic listening, you first have to get people to talk to you. In many cases, you must work on developing a rapport with the other person before he or she will really talk to you. **Rapport** is a relation of harmony, conformity, accord, or affinity. Without rapport, people cannot begin to communicate. For example, in the opening case, Ben was not ready to talk to anyone from the Information Technology Department, even if Ed and Sarah were ready to listen. Ben was angry about the lack of support he received from the Information Technology Department and bullied anyone who reminded him of that group. Before Sarah could begin to communicate with Ben, she had to establish rapport.

One technique for establishing rapport is using a process called mirroring. **Mirroring** is the matching of certain behaviors of the other person. You can mirror someone's voice tone and/or tempo, breathing, movements, or body postures. After Ben started yelling at her, Sarah quickly decided to mirror his voice tone, tempo, and body posture. She stood nose to nose with Ben and started yelling right back at him. This action made Ben realize what he was doing, and

---

[11] Kidder, Tracy, *The Soul of a New Machine*. New York: Modern Library, 1997.

it also made him notice and respect this colleague from the Information Technology Department. Once Ben overcame his anger, he could start communicating his needs. In most cases, such extreme measures are not needed.

Recall the importance of getting users involved in information technology projects. For organizations to be truly effective in information technology project management, they must find ways to help users and developers of information systems work together. It is widely accepted that organizations make better decisions about information technology projects when business professionals and information technology staff collaborate. It is also widely accepted that this task is easier said than done. Many companies have been very successful in integrating technology and business departments, but many other companies continue to struggle with this issue.

What Went Right?

Forrester Research, a well-known technology research company in Cambridge, Massachusetts, interviewed twenty-five Chief Information Officers (CIOs) at companies including L.L. Bean, Shell, GE, Taco Bell, Texas Instruments, and Coca-Cola to identify the best practices for ensuring partnerships between people in business and technology areas. Its research study identified three successful practices:

■ Businesspeople, not information technology people, take the lead in determining and justifying investments in new computer systems. At several companies, business managers defined the expected benefits of new systems, with information technology support, and made the pitch for the investment to senior management. At some companies, top business managers are running information technology departments. At GE, for example, the head of the corporate strategy unit assumed responsibility for the information technology division.

■ CIOs continually push their staff to recognize that the needs of the business must drive all technology decisions. The CIOs at Textron and Texas Instruments, for example, go so far as to insist that everyone express all presentations, reports, and even discussions about information technology projects in business terms, not in the technical language used by information technology insiders.

■ Several companies that are successful at creating partnerships between people in business and technology areas are reshaping their information technology units to look and perform like consulting firms. For instance, Texas Instruments borrowed an idea from McKinsey & Company by assigning central information technology employees to one of seventeen centers of excellence for specific technology disciplines. When a business challenge arises, managers pull together ad hoc teams of people drawn from those centers and the business itself to work on the problem. When the problem is solved, managers reassign team members to other problems. This approach allows managers to use resources more flexibly and keeps employees' skills fresh.

> The Forrester Research study also found that successful CIOs downplayed the "T" (technology) of IT, and played up the "I" (information). The CIOs believed that choosing the right technology was important, but success rested more on having the right people involved in the right project, putting the right processes in place, and having a business-results focus at all times.[12]

It is important to understand and pay attention to concepts of motivation, influence, power, and improving effectiveness in all project processes. It is likewise important to remember that projects operate within an organizational environment. The challenge comes from applying these theories to the many unique individuals involved in particular projects in particular organizations.

As you can see, there are many important topics related to motivation, influence, power, and effectiveness that are relevant to project management. Projects are done by and for people, so it is important for project managers and team members to understand and practice key concepts related to these topics. It is important to understand the specific processes of project human resource management: organizational planning, project staff acquisition, and team development.

ORGANIZATIONAL PLANNING

Organizational planning for a project involves identifying, documenting, and assigning project roles, responsibilities, and reporting relationships. This process generates an organizational chart for the project, roles, and responsibility assignments, often shown in a matrix form called a responsibility assignment matrix (RAM), and a staffing management plan.

Before creating an organizational chart for a project, top management and the project manager must identify what types of people the project really needs to ensure project success. If the key to success lies in having the best Java programmers you can find, the organizational planning should reflect that need. If the real key to success is having a top-notch project manager and team leaders whom people respect in the company, that need should drive organizational planning.

What Went Wrong?

The organizational culture of a company has a tremendous influence on how effectively human resources are utilized. For example, one author describes a well-run software development organization where programmers received top pay and status. As a result, the ratio of programmers to managers was about ten to one. In that organization, the company developed and installed a customized publication system for less than $1 million. The author then described the next software development organization where he worked.

[12] Kiely, Thomas, "Making Collaboration Work," *Harvard Business Review* (January–February 1997), 10–11.

Programming was considered a job of underlings. The typical software project had a manager, a project leader, a project manager, a marketing manager, an architect, a system engineer, and one or two programmers. In that environment, the company spent over $30 million in personnel costs to try to develop an almost identical publication system, without a single line of usable code ever being written.[13]

It is also crucial not to ignore training needs of users when developing project plans. At the new $2.2 billion Kuala Lumpur International Airport in Malaysia, flights were delayed for hours when the new total airport management system was unable to cope with incorrect commands entered by improperly trained employees. Many information technology project managers are pressured to meet unrealistic schedule deadlines. Since training often comes at the end of the schedule, it is often compressed or neglected altogether before new systems go live. In addition, it takes a special kind of person to train users on a new system, and project managers cannot expect the same people who wrote the code to be also the best people to provide the user training.[14]

Project Organizational Charts

Recall from Chapter 2 that the nature of information technology projects often means that project team members come from different backgrounds and possess a wide variety of skill sets. It can be very difficult to manage such a diverse group of people, so it is important to provide a clear organizational structure for a project. After identifying important skills and the types of people needed to staff a project, the project manager should work with top management and project team members to create an organizational chart for the project. Figure 9-2 provides a sample organizational chart for a large information technology project. Note that the project personnel include a deputy project manager, subproject managers, and teams. **Deputy project managers** fill in for project managers in their absence and assist them as needed, which is similar to the role of a vice president. **Subproject managers** are responsible for managing the subprojects into which a large project might be divided. There are often subproject managers that focus on managing various software (S/W) and hardware (H/W) components of large projects. This structure is typical for large projects. With many people working on a project, clearly defining and allocating project work is essential. (See the companion Web site for this text to see the project organizational chart that Northwest Airlines used for their large ResNet project.) Smaller information technology projects usually do not have deputy project managers or subproject managers. On smaller projects, the project managers might have just team leaders reporting directly to them.

[13] James, Geoffrey, "IT Fiascoes . . . and How to Avoid Them," *Datamation* (November 1997).

[14] Moad, Jeff, "Grounding those high-flying IT projects," *PC Week* (July 17, 1998).

Figure 9-2. Sample Organizational Chart for a Large Information Technology Project

Figure 9-3 provides a framework for defining and assigning work. This process consists of four steps:

1. Finalizing the project requirements
2. Defining how the work will be accomplished
3. Breaking down the work into manageable elements
4. Assigning work responsibilities

The work definition and assignment process is carried out during the proposal and startup phases of a project. Note that the process is iterative, meaning it often takes more than one pass to refine it. A Request for Proposal (RFP) or draft contract often provides the basis for defining and finalizing work requirements, which are then documented in a final contract and technical baseline. If there were not an RFP, then the internal project charter and scope statement would provide the basis for defining and finalizing work requirements, as described in Chapter 5, Project Scope Management. The project team leaders then decide on a technical approach for how to do the work. Should work be broken down using a product-oriented approach or a phased approach? Will the project team outsource some of the work or subcontract to other companies? Once the project team has decided on a technical approach, they develop a work breakdown structure (WBS) to establish manageable elements of work (see Chapter 5, Project Scope Management). They then develop activity definitions to define the work involved in each activity on the WBS further (see Chapter 6, Project Time Management). The last step is assigning the work.

Figure 9-3. Work Definition and Assignment Process

Once the project manager and project team have broken down the work into manageable elements, the project manager assigns work to organizational units. The project manager often bases these work assignments on where the work fits in the organization and uses an organizational breakdown structure to conceptualize the process. An **organizational breakdown structure (OBS)** is a specific type of organizational chart that shows which organizational units are responsible for which work items. The OBS can be based on a general organizational chart and then broken down into more detail, based on specific units within departments in the company or units in any subcontracted companies.

Responsibility Assignment Matrices

After developing an OBS, the project manager is in a position to develop a responsibility assignment matrix (RAM). A **responsibility assignment matrix (RAM)** is a matrix that maps the work of the project as described in the WBS to the people responsible for performing the work as described in the OBS. Figure 9-4 shows an example of a RAM. The RAM allocates work to responsible and performing organizations, teams, or individuals, depending on the desired level of detail. For smaller projects, it would be best to assign individual people to WBS activities. For very large projects, it is more effective to assign the work to organizational units or teams.

Figure 9-4. Sample Responsibility Assignment Matrix (RAM)

In addition to using a RAM to assign detailed work activities, you can also use a RAM to define general roles and responsibilities on projects. This type of RAM can include the stakeholders in the project. Figure 9-5 provides a RAM that shows whether stakeholders are accountable or just participants in part of a project and whether they are required to provide input, review, or sign off on parts of a project. This simple tool can be a very effective way for the project manager to communicate roles and expectations of important stakeholders on projects.

Figure 9-5. RAM Showing Stakeholder Roles

Some organizations use **RACI charts** to show Responsibility, Accountability, Consultation, and Informed roles for project stakeholders. As shown in Table 9-1, a RACI chart lists tasks vertically, individuals or groups horizontally, and each intersecting cell contains an R, A, C, or I. Each task may have multiple A, C, or I entries, but there can be only one R entry to clarify which particular individual or group has responsibility for each task.

Table 9-1: Sample RACI Chart

	GROUP A	GROUP B	GROUP C	GROUP D	GROUP E
Task 1	R	A	C	C	I
Task 2	C	I	R	A	I
Task 3	A	R	I	C	C
Task 4	I	C	A	I	R
Task 5	A	I	C	R	A

Staffing Management Plans and Resource Histograms

Another output of organizational planning is a staffing management plan. A **staffing management plan** describes when and how people will be added to and taken off the project team. It can be a formal or informal plan, and the level of detail may vary based on the type of project. For example, if an information technology project is projected to need one hundred people on average over a year, the staffing management plan would describe the types of people needed to work on the project, such as Java programmers, business analysts, technical writers, and so on, and the number of each type of person needed each month.

The staffing management plan often includes a **resource histogram**, which is a column chart that shows the number of resources assigned to a project over time. Figure 9-6 provides an example of a histogram that might be used for a six-month information technology project. Notice that the columns represent the number of people needed in each area—managers, business analysts, programmers, and technical writers. By stacking the columns, you can see the total number of people needed each month. After determining the project staffing needs, the next steps in project human resource management are to acquire the necessary staff and then develop the project team.

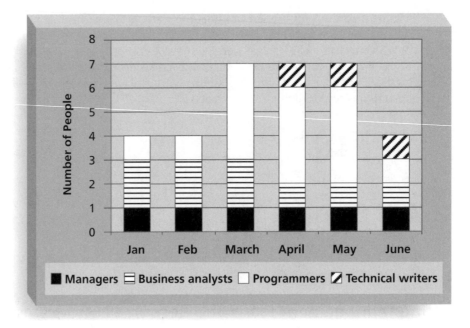

Figure 9-6. Sample Resource Histogram

PROJECT STAFF ACQUISITION

During the late 1990s, the information technology job market became extremely competitive. It was a seller's market with corporations competing fiercely for a shrinking pool of qualified, experienced information technology professionals. In the early 2000s, the market declined tremendously, but, as stated earlier in this chapter, things started turning around in late 2002. Regardless of the current job market, finding qualified information technology professionals—staff acquisition—is critical. It is also important to assign the appropriate type and number of resources to work on projects at the appropriate times. This section addresses important topics related to project staff acquisition: resource assignment, resource loading, and resource leveling.

Resource Assignments

After developing a staffing management plan, project managers must work with other people in their organizations to assign particular personnel to their projects or to acquire additional human resources needed to staff the project. Project managers with strong influencing and negotiating skills are often good at getting internal people to work on their projects. However, the organization must ensure that people are assigned to the projects that best fit their skills and the needs of the organization. The main outputs of the staff acquisition process are project staff assignments and a project team directory.

Organizations that do a good job of staff acquisition have good staffing plans. These plans describe the number and type of people who are currently in the organization and the number and type of people anticipated to be needed for the project based on current and upcoming activities. An important component of staffing plans is maintaining a complete and accurate inventory of employees' skills. If there is a mismatch between the current mix of people's skills and needs of the organization, it is the project manager's job to work with top management, human resource managers, and other people in the organization to address staffing and training needs.

It is also important to have good procedures in place for hiring subcontractors and recruiting new employees. Since the Human Resource Department is normally responsible for hiring people, project managers must work with their human resource managers to address any problems in recruiting appropriate people. It is also a priority to address retention issues, especially for information technology professionals.

One innovative approach to hiring and retaining information technology staff is to offer existing employees incentives for helping recruit and retain personnel. For example, several consulting companies give their employees one dollar for every hour a new person they helped hire works. This provides an incentive for current employees to help attract new people and to keep both them and the person they recruited working at that company. Another approach that several companies are taking to attract and retain information technology professionals is to provide benefits based on their personal needs. For example, some people might want to work only four days a week or have the option of working a couple of days a week from home. As it gets more difficult to find good information technology professionals, organizations must become more innovative and proactive in addressing this issue.

Several organizations, publications, and Web sites address the need for good staff acquisition and retention. For example, ICEX, Inc. conducts several research studies to identify best practices. They have found that world-class information technology organizations have groups dedicated to managing information technology staff recruiting. They have also found that the most effective recruiting message emphasizes that "this is a great place to work and you'll have a chance to work on innovative information technology projects."[15] William C. Taylor, cofounder of *Fast Company* magazine and a public speaker, also believes that people today are more demanding and have

---

[15] Quillard, Judith, "Leading Practices for Recruiting IT Staff," ICEX Web site (1998), *www.icex.com*.

higher expectations of their jobs than just earning a paycheck. His company's research has found that people leave their jobs because:

- They feel they do not make a difference.
- They do not get proper recognition.
- They are not learning anything new or growing as a person.
- They do not like their coworkers.
- They want to earn more money.[16]

It is very important to consider the needs of individuals and the organization when making recruiting and retention decisions and to study the best practices of leading companies in these areas.

Resource Loading

Chapter 6, Project Time Management, described using network diagrams to help manage a project's schedule. One of the problems or dangers inherent in scheduling processes is that they often do not address the issues of resource utilization and availability. (Hence the development of critical chain scheduling, as described in Chapter 6.) Schedules tend to focus primarily on time rather than on both time and resources, which includes people. An important measure of a project manager's success is how well he or she balances the trade-offs among performance, time, and cost. During a period of crises, it is occasionally possible to add additional resources—such as additional staff—to a project at little or no cost. Most of the time, however, resolving performance, time, and cost trade-offs entails additional costs to the organization. The project manager's goal must be to achieve project success without increasing the costs or time required to complete the project. The key to accomplishing this goal is effectively managing human resources on the project.

Once people are assigned to projects, there are two techniques available to project managers that help them use project staff most effectively: resource loading and resource leveling. **Resource loading** refers to the amount of individual resources an existing schedule requires during specific time periods. Resource loading helps project managers develop a general understanding of the demands a project will make on the organization's resources, as well as on individual people's

---

[16] Taylor, William C., Keynote address at the Project Management Institute's 2000 Annual Seminars & Symposium (September 2000).

schedules. Project managers often use histograms, as described in Figure 9-6, to depict period-by-period variations in resource loading. A histogram can be very helpful in determining staffing needs or in identifying staffing problems.

A resource histogram can also show when work is being overallocated to a certain person or group. **Overallocation** means more resources than are available are assigned to perform work at a given time. For example, Figure 9-7 provides a sample resource histogram created in Microsoft Project 2002. This histogram illustrates how much one individual, Joe Franklin, is assigned to work on the project each week. The percentage numbers on the vertical axis represent the percentage of Joe's available time that is allocated for him to work on the project. The top horizontal axis represents time in weeks. Note that Joe Franklin is overallocated most of the time. For example, for most of March and April and part of May, Joe's work allocation is 300 percent of his available time. If Joe is normally available eight hours per day, this means he would have to work twenty-four hours a day to meet this staffing projection!

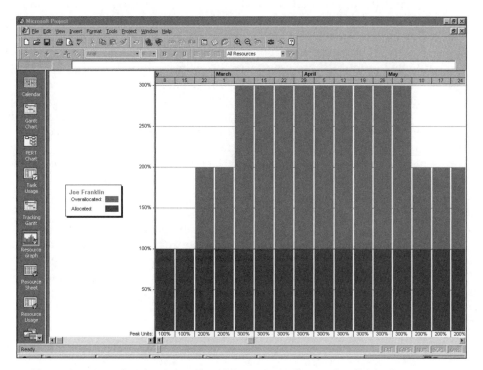

Figure 9-7. Sample Histogram Showing an Overallocated Individual

Resource Leveling

Resource leveling is a technique for resolving resource conflicts by delaying tasks. It is a form of network analysis in which resource management concerns drive scheduling decisions (start and finish dates). The main purpose of resource leveling is to create a smoother distribution of resource usage. Project managers examine the network diagram for areas of slack or float, and to identify resource conflicts. For example, you can sometimes remove overallocations by delaying noncritical tasks, which does not result in an overall schedule delay. Other times you will need to delay the project completion date to reduce or remove overallocations. See Appendix A for information on using Project 2002 to level resources using both of these approaches. You can also view resource leveling as addressing the resource constraints described in critical chain scheduling (see Chapter 6, Project Time Management).

Overallocation is one type of resource conflict. If a certain resource is overallocated, the project manager can change the schedule to remove resource overallocation. If a certain resource is underallocated, the project manager can change the schedule to try to improve the use of the resource. Resource leveling, therefore, aims to minimize period-by-period variations in resource loading by shifting tasks within their slack allowances.

Figure 9-8 illustrates a simple example of resource leveling. The network diagram at the top of this figure shows that Activities A, B, and C can all start at the same time. Activity A has a duration of two days and will take two people to complete; Activity B has a duration of five days and will take four people to complete; and Activity C has a duration of three days and will take two people to complete. The histogram on the lower-left of this figure shows the resource usage if all activities start on day one. The histogram on the lower-right of Figure 9-8 shows the resource usage if Activity C is delayed two days, its total slack allowance. Notice that the lower-right histogram is flat or leveled; that is, its pieces (activities) are arranged to take up the least space (saving days and numbers of workers). You may recognize this strategy from the computer game Tetris, in which you earn points for keeping the falling shapes as level as possible. The player with the most points (most level shape allocation) wins. Resources are also used best when they are leveled.

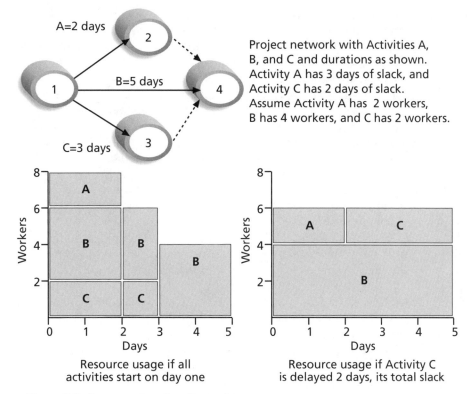

Project network with Activities A, B, and C and durations as shown. Activity A has 3 days of slack, and Activity C has 2 days of slack. Assume Activity A has 2 workers, B has 4 workers, and C has 2 workers.

Resource usage if all activities start on day one

Resource usage if Activity C is delayed 2 days, its total slack

Figure 9-8. Resource Leveling Example

Resource leveling has several benefits. First, when resources are used on a more constant basis, they require less management. For example, it is much easier to manage a part-time project member who is scheduled to work twenty hours per week on a project for the next three months than it is to manage the same person who is scheduled to work ten hours one week, forty the next, five the next, and so on.

Second, resource leveling may enable project managers to use a just-in-time inventory type of policy for using subcontractors or other expensive resources. For example, a project manager might want to level resources related to work that must be done by particular subcontractors such as testing consultants. This leveling might allow the project to use four outside consultants full-time to do testing for four months instead of spreading out the work over more time or needing to use more than four people. The latter approach is usually more expensive.

Third, resource leveling results in fewer problems for project personnel and accounting departments. Increasing and decreasing labor levels and particular human resources often produce additional work and confusion. For example, if a person with expertise in a particular area is only assigned to a project two days a week and another person they need to work with is not assigned to the project those same days, they cannot work well together. The Accounting Department might complain when subcontractors charge a higher rate for billing less than twenty hours a week on a project. The accountants will remind project managers to strive for getting the lowest rates possible.

Finally, resource leveling often improves morale. People like to have some stability in their jobs. It is very stressful for people to not know from week to week or even day to day what projects they will be working on and with whom they will be working.

Project management software can automatically level resources. However, the project manager must be careful in using the results without making adjustments. Automatic leveling often pushes out the project's completion date. Resources may also be reallocated to work at times that are inappropriate with other constraints. A wise project manager would have one of his or her team members who is proficient in using the project management software ensure that the leveling is done appropriately.

TEAM DEVELOPMENT

Even if a project manager has successfully recruited enough skilled people to work on a project, he or she must ensure that people can work together as a team to achieve project goals. Many information technology projects have had very talented individuals working on them. However, it takes teamwork to complete most projects successfully. The main goal of team development is to help people work together more effectively to improve project performance. There is an extensive body of literature on team development. This section will highlight a few important tools and techniques for team development, including training, team-building activities, and reward and recognition systems. It will also provide general advice on the effective use of teams.

Training

Project managers often recommend that people take specific training courses to improve individual and team development. For example, Sarah from the opening case had gone through training in emotional intelligence and dealing with difficult people. She was familiar with the mirroring technique and felt comfortable using that approach with Ben. Many other people would not have reacted so quickly and effectively in the same situation. If Ben and Sarah did reach agreement on what actions they could take to resolve the F-44 aircraft program's information technology problems, it might result in a new project to develop and deliver a new system for Ben's group. If Sarah became the project manager for this new project, she would understand the need for special training in interpersonal skills for specific people in her and Ben's departments. Individuals could take special training classes to improve their personal skills. If Sarah thought the whole project team could benefit from taking training together to learn to work as a team, she could arrange for a special team-building session for the entire project team and key stakeholders.

It is very important to provide training in a just-in-time fashion. For example, if Sarah was preparing for a technical assignment where she would need to learn a new programming language, training to deal with difficult people would not help her much. However, the training was very timely for her new consulting position. Many organizations provide e-learning opportunities for their employees so they can learn specific skills at any time and any place. They have also found e-learning to sometimes be more cost-effective than traditional instructor-led training courses. It is important to make sure that the timing and delivery method for the training is appropriate for specific situations and individuals. Organizations have also found that it is often more economical to train current employees in particular areas than it is to hire new people who already possess those skills.

Several organizations that have successfully implemented Six Sigma principles have taken a unique and effective approach to training. They only let their high-potential employees attend Six Sigma Black Belt training, which is a substantial investment in terms of time and money. In addition, they do not let employees into a particular Black Belt course until they have had a potential Six Sigma project approved that relates to their current job. Attendees can then apply the new concepts and techniques they learn in the classes to their work settings. High-potential employees feel rewarded by being picked to take this training, and the organization benefits by having these employees implement high-payoff projects because of the training.

Team-Building Activities

Many organizations provide in-house team-building training activities, and many also use services provided by external companies that specialize in this area. Two common approaches to team-building activities include using physical challenges and psychological preference indicator tools.

Several organizations have teams of people go through certain physically challenging activities to help them develop as a team. Military basic training or boot camps provide one example. Men and women who wish to join the military must first make it through basic training, which often involves several strenuous physical activities such as rappelling off towers, running and marching in full military gear, going through obstacle courses, passing marksmanship training, and mastering survival training. Many organizations use a similar approach by sending teams of people to special locations where they work as a team to navigate white water rapids, climb mountains or rocks, participate in paintball exercises, and so on.

Even more organizations have teams participate in mental team-building activities in which they learn about themselves, each other, and how to work as a group most effectively. It is important for people to understand and value each other's differences in order to work effectively as a team. Two common exercises used in mental team building include the Myers-Briggs Type Indicator and the Wilson Learning Social Styles Profile.

The **Myers-Briggs Type Indicator (MBTI)** is a popular tool for determining personality preferences. During World War II, Isabel B. Myers and Katherine C. Briggs developed the first version of the MBTI based on psychologist Carl Jung's theory of psychological types. The four dimensions of personality types in the MBTI include:

- Extrovert/Introvert (E/I): This first dimension determines if you are generally extroverted or introverted. The dimension also signifies whether people draw their energy from other people (extroverts) or from inside themselves (introverts).

- Sensation/Intuition (S/N): This second dimension relates to the manner in which you gather information. Sensation type people take in facts, details, and reality and describe themselves as practical. Intuitive type people are imaginative, ingenious, and attentive to hunches or intuition. They describe themselves as innovative and conceptual.

- Thinking/Feeling (T/F): This third dimension represents thinking judgment and feeling judgment. Thinking judgment is objective and logical, and feeling judgment is subjective and personal.

- Judgment/Perception (J/P): This fourth dimension concerns people's attitude toward structure. Judgment type people like closure and task completion. They tend to establish deadlines and take them seriously, expecting others to do the same. Perceiving types prefer to keep things open and flexible. They regard deadlines more as a signal to start rather than complete a project and do not feel that work must be done before play or rest begins.[17]

There is much more involved in personality types, and many books are available on this topic. In 1998, David Keirsey and Ray Choiniere published *Please Understand Me II: Temperament Character Intelligence*. This book includes an easy-to-take and interpret test called The Keirsey Temperament Sorter, which is a personality type preference test based on the work done by Jung, Myers, and Briggs.

In 1985, an interesting study of the MBTI types of the general population in the United States and information systems (IS) developers revealed some significant contrasts.[18] Both groups of people were most similar in the judgment/perception dimension, with slightly more than half of each group preferring the judgment type (J). There were significant differences, however, in the other three dimensions. Most people would not be surprised to hear that most information systems developers are introverts. This study found that 75 percent of IS developers were introverts (I), and only 25 percent of the general population were introverts. This personality type difference might help explain some of the problems users have communicating with developers. Another sharp contrast found in the study was that almost 80 percent of IS developers were thinking types (T) compared to 50 percent of the general population. IS developers were also much more likely to be intuitive (N) (about 55 percent) than the general population (about 25 percent). These results fit with Keirsey's classification of NT (Intuitive/Thinking types) people as rationals. Educationally, they tend to study the sciences, enjoy technology as a hobby, and pursue systems work. Keirsey also suggests that no more than 7 percent of the general population are NTs. Would you be surprised to know that Bill Gates is classified as a rational?[19]

Project managers can often benefit from knowing their team members' MBTI profiles by adjusting their management styles for each individual. For example, if the project manager is a strong N and one of the team members is a strong S, the project manager should take the time to provide more concrete, detailed explanations when discussing that person's task assignments. Project managers may also want to make sure they have a variety of personality types on their teams. For example, if all team members are strong introverts, it may be difficult for them to work well with users and other important stakeholders who are often extroverts.

---

[17] Briggs, Isabel Myers, with Peter Myers, *Gifts Differing: Understanding Personality Type*. Palo Alto, CA: Consulting Psychologists Press, 1995.

[18] Lyons, Michael L., "The DP Psyche," *Datamation* (August 15, 1985).

[19] The Web Site for the Keirsey Temperament Sorter and Keirsey Temperament Theory, *http://keirsey.com/personality/nt.html*.

Many organizations also use the Social Styles Profile in team-building activities. Psychologist David Merril, who helped develop the Wilson Learning Social Styles Profile, describes people as falling into four approximate behavioral profiles, or zones. People are perceived as behaving primarily in one of four zones, based on their assertiveness and responsiveness:

- "Drivers" are proactive and task-oriented. They are firmly rooted in the present, and they strive for action. Adjectives to describe drivers include pushy, severe, tough, dominating, harsh, strong-willed, independent, practical, decisive, and efficient.

- "Expressives" are proactive and people-oriented. They are future-oriented and use their intuition to look for fresh perspectives on the world around them. Adjectives to describe expressives include manipulating, excitable, undisciplined, reacting, egotistical, ambitious, stimulating, wacky, enthusiastic, dramatic, and friendly.

- "Analyticals" are reactive and task-oriented. They are past-oriented and strong thinkers. Adjectives to describe analyticals include critical, indecisive, stuffy, picky, moralistic, industrious, persistent, serious, expecting, and orderly.

- "Amiables" are reactive and people-oriented. Their time orientation varies depending on whom they are with at the time, and they strongly value relationships. Adjectives to describe amiables include conforming, unsure, ingratiating, dependent, awkward, supportive, respectful, willing, dependable, and agreeable.[20]

Figure 9-9 shows these four social styles and how they relate to assertiveness and responsiveness. Note that the main determinants of the social style are your levels of assertiveness—if you are more likely to tell people what to do or ask what should be done—and how you respond to tasks—by focusing on the task itself or on the people involved in performing the task.

---

[20] Robbins, Harvey A. and Michael Finley, *The New Why Teams Don't Work: What Goes Wrong and How to Make It Right.* San Francisco, CA: Berrett-Koehler Publishers, 1999.

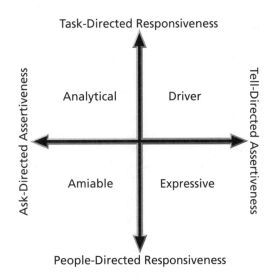

Figure 9-9. Social Styles

Knowing the social styles of project stakeholders can help project managers understand why certain people may have problems working together. For example, drivers are often very impatient working with amiables, and analyticals often have difficulties understanding expressives. Project managers can use their facilitation skills to help all types of people to communicate better with each other and to focus on meeting project goals.

Reward and Recognition Systems

Another important tool for promoting team development is the use of team-based reward and recognition systems. If management rewards teamwork, they will promote or reinforce people to work more effectively in teams. Some organizations offer bonuses, trips, or other rewards to workers who meet or exceed company or project goals. In a project setting, project managers can recognize and reward people who willingly work overtime to meet an aggressive schedule objective or go out of their way to help a teammate. Project managers should not reward people who work overtime just to get extra pay or because of their own poor work or planning.

General Advice on Teams

Effective project managers must be good team builders. Suggestions for ensuring that teams are productive include the following:

- Be patient and kind with your team, and assume the best about people. Do not assume that your team members are lazy and careless.
- Fix the problem instead of blaming people. Help people work out problems by focusing on behaviors.
- Establish regular, effective meetings. Focus on meeting project objectives and producing positive results.
- Allow time for teams to go through the basic team-building stages of "forming, storming, norming, and performing." Don't expect teams to work at the highest performance level right away.
- Limit the size of work teams to three to seven members.
- Plan some social activities to help project team members and other stakeholders get to know each other better. Make the social events fun and not mandatory.
- Stress team identity. Create traditions that team members enjoy.
- Nurture team members and encourage them to help each other. Identify and provide training that will help individuals and the team as a whole become more effective.
- Acknowledge individual and group accomplishments.

As you can imagine, staffing and team development are critical concerns on many information technology projects. Many information technology project managers must break out of their rational/NT preference and focus on empathically listening to other people to address their concerns and create an environment in which individuals and teams can grow and prosper.

USING SOFTWARE TO ASSIST IN HUMAN RESOURCE MANAGEMENT

Earlier in this chapter, you read that a simple responsibility assignment matrix (Figures 9-4 and 9-5) or histograms (Figures 9-6 and 9-7) are useful tools that can help you effectively manage human resources on projects. You can use several different software packages, including spreadsheets or project management software such as Microsoft Project 2002, to create matrixes and histograms. Many people do not realize that Project 2002 provides a variety of human resource management tools, some of which include assigning and tracking resources, resource leveling, resource usage reports, overallocated resource reports, and to-do lists. You will learn how to use many of these functions and features in Appendix A. (Also, note that collaborative software to help people communicate is described in Chapter 10, Project Communications Management.)

You can use Project 2002 to assign resources—including equipment, materials, facilities, or people—to tasks. Project 2002 enables you to allocate individual resources to individual projects or to pool resources and share them across multiple projects. By defining and assigning resources in Project 2002, you can:

- Keep track of the whereabouts of resources through stored information and reports on resource assignments.

- Identify potential resource shortages that could force a project to miss scheduled deadlines and possibly extend the duration of a project.

- Identify underutilized resources and reassign them, which may enable you to shorten a project's schedule and possibly reduce costs.

- Use automated leveling to make level resources easier to manage.

Two particularly helpful features of Project 2002 are the resource usage view and the resource usage report. Both of these features help you see who is scheduled to work how many hours on which project activities. Figure 9-10 shows an example of a resource usage view, and Figure 9-11 shows an example of a resource usage report. The resource usage view shows the names of the people or resources working on the project, the total number of hours they are scheduled to work, and a calendar indicating how many hours they are scheduled to work each month for each task. If someone is scheduled to work more than his or her allowable number of hours, Project 2002 automatically puts a warning symbol (an exclamation point) in the column to the left of that person's name. Note the exclamation point next to Joe Franklin's name in the first row of Figure 9-10. Upon reviewing the number of hours more closely, you can see that Joe Franklin is overallocated February through May. These numbers are displayed in red in Project 2002 to help them stand out. You can adjust the calendar by clicking an icon to see hours scheduled by months, days, and so on.

	ⓘ	Resource Name	Work	Details	Qtr 1, 2003 Jan	Feb	Mar	Qtr 2, 2003 Apr	May	Jun
1	◈	⊟ Joe Franklin	1,440 hrs	Work		112h	440h	528h	360h	
		Packaging	480 hrs	Work		96h	168h	176h	40h	
		Datasheets	520 hrs	Work		16h	168h	176h	160h	
		Reseller kits	440 hrs	Work			104h	176h	160h	
2		⊟ Rich Anderson	240 hrs	Work			64h	176h		
		Working Model	240 hrs	Work			64h	176h		
3		⊟ Mark Smith	280 hrs	Work	56h	160h	64h			
		Develop creative briefs	200 hrs	Work	56h	144h				
		Develop concepts	80 hrs	Work		16h	64h			
4		⊟ Lisa Adams	144 hrs	Work	80h	64h				
		Launch planning	144 hrs	Work	80h	64h				
5		⊟ Intern	800 hrs	Work			82h	176h	176h	168h
		Working model	800 hrs	Work			82h	176h	176h	168h

Figure 9-10. Resource Usage View from Microsoft Project

Figure 9-11 is a resource usage report based on the same software launch file. This standard report shows similar information, but the data is automatically formatted to show the number of hours each person is scheduled to work each week during the project. In this figure, you can see that Joe Franklin is scheduled to work 120 hours the weeks of March 16, March 23, and every week in April. He is also scheduled to work 104 hours the week of May 4. If a standard work week is 40 or even 60 hours per week, the project manager can easily see from this report that Joe Franklin cannot do all of the work assigned to him.

	Mar 16	Mar 23	Mar 30	Apr 6	Apr 13	Apr 20	Apr 27	May 4	May
Joe Franklin	120 hrs	120 hrs	120 hrs	120 hrs	120 hrs	120 hrs	120 hrs	104 hrs	
Packaging	40 hrs	40 hrs	40 hrs	40 hrs	40 hrs	40 hrs	40 hrs	24 hrs	
Datasheets	40 hrs	40 hrs	40 hrs	40 hrs	40 hrs	40 hrs	40 hrs	40 hrs	
Reseller kits	40 hrs	40 hrs	40 hrs	40 hrs	40 hrs	40 hrs	40 hrs	40 hrs	
Rich Anderson	16 hrs	40 hrs	40 hrs	40 hrs	40 hrs	40 hrs	24 hrs		
Working Model	16 hrs	40 hrs	40 hrs	40 hrs	40 hrs	40 hrs	24 hrs		
Mark Smith									
Develop creative briefs									
Develop concepts									
Lisa Adams									
Launch planning									
Intern	34 hrs	40 hrs	40 hrs	40 hrs	40 hrs	40 hrs	40 hrs	40 hrs	
Working model	34 hrs	40 hrs	40 hrs	40 hrs	40 hrs	40 hrs	40 hrs	40 hrs	
Total	170 hrs	200 hrs	200 hrs	200 hrs	200 hrs	200 hrs	184 hrs	144 hrs	

Figure 9-11. Resource Usage Report from Microsoft Project

Just as many project management professionals are not aware of the powerful cost-management features of Microsoft Project 2002, many are also unaware of its powerful human resource management features. Several companies also provide add-in software to perform even more human resource management functions. With the aid of this type of software, project managers can have more information available in useful formats to help them decide how to manage human resources most effectively.

Project resource management involves much more than using software to assess and track resource loading, level resources, and so on. People are the most important asset on most projects, and human resources are very different from other resources. You cannot simply replace people the same way that you would replace a piece of equipment. People need far more than a tune-up now and then to keep them performing well. It is essential to treat people with consideration and respect, to understand what motivates them, and to communicate carefully with them. What makes good project managers great is not their use of tools, but rather their ability to enable project team members to deliver the best work they possibly can on a project.

CASE WRAP-UP

After Sarah yelled right back at Ben, he said, "You're the first person who's had the guts to stand up to me." After that brief introduction,

Sarah, Ben, and the other meeting participants had a good discussion about what was really happening on the F-44 upgrade project. Sarah was able to write a justification to get Ben's group special software and support to download key information from the old system so they could manage their project better. When Sarah stood nose to nose with Ben and yelled at him, she used a technique for establishing rapport called mirroring. Although Sarah was by no means a loud and obnoxious person, she saw that Ben was and decided to mirror his behavior and attitude. She put herself in his shoes for a while, and doing so helped break the ice so Sarah, Ben, and the other people at the meeting could really start communicating and working together as a team to solve their problems.

CHAPTER SUMMARY

People are the most important assets in organizations and on projects. Therefore, it is essential for project managers to be good human resource managers.

Psychosocial issues that affect how people work and how well they work include motivation, influence and power, and effectiveness.

Maslow developed a hierarchy of needs that suggests physiological, safety, social, esteem, and self-actualization needs motivate behavior. Once a need is satisfied, it no longer serves as a motivator.

Herzberg distinguished between motivators and hygiene factors. Hygiene factors such as larger salaries or a more attractive work environment will cause dissatisfaction if not present, but do not motivate workers to do more if present. Achievement, recognition, the work itself, responsibility, and growth are factors that contribute to work satisfaction and motivate workers.

McClellan proposed the acquired-needs theory, suggesting that an individual's needs are acquired or learned over time and shaped by their life experiences. The three types of acquired needs are a need for achievement, a need for affiliation, and a need for power.

McGregor developed Theory X and Theory Y to describe different approaches to managing workers, based on assumptions of worker motivation. Research supports the use of Theory Y, which assumes that people see work as natural and indicates that the most significant rewards are the satisfaction of esteem and self-actualization needs that work can provide. According to Ouchi's Theory Z, workers can be trusted to do their jobs to their utmost ability, as long as management can be trusted to support them and look out for their well-being. Theory Z emphasizes things such as job rotation, broadening of skills, generalization versus specialization, and the need for continuous training of workers.

Thamhain and Wilemon identified nine influence bases available to project managers: authority, assignment, budget, promotion, money, penalty, work challenge, expertise, and friendship. Their research found that project success is associated with project managers who use work challenge and expertise to influence workers. Project failure is associated with using too much influence by authority, money, or penalty.

Power is the potential ability to influence behavior to get people to do things they would not otherwise do. The five main types of power are coercive power, legitimate power, expert power, reward power, and referent power.

Project managers can use Steven Covey's seven habits of highly effective people to help themselves and project teams become more effective. The seven habits include being proactive; beginning with the end in mind; putting first things first; thinking win/win; seeking first to understand, then to be understood; achieving synergy; and sharpening the saw. Using empathic listening is a key skill of good project managers.

The major processes involved in project human resource management include organizational planning, staff acquisition, and team development. Organizational planning involves identifying, assigning, and documenting project roles, responsibilities, and reporting relationships. A responsibility assignment matrix (RAM) and RACI charts are key tools for defining roles and responsibilities on projects.

Staff acquisition is getting the appropriate staff assigned to and working on the project. This is an important issue in today's competitive environment. Companies must use innovative approaches to find and retain good information technology staff.

Resource loading shows the amount of individual resources an existing schedule requires during specific time frames. Histograms show resource loading and identify overallocation of resources.

Resource leveling is a technique for resolving resource conflicts, such as overallocated resources, by delaying tasks. Leveled resources require less management, lower costs, produce fewer personnel and accounting problems, and often improve morale.

One of the most important skills of a good project manager is team development. Teamwork helps people work more effectively to achieve project goals. Project managers can recommend individual training to improve skills related to teamwork, organize team-building activities for the entire project team and key stakeholders, and provide reward and recognition systems that encourage teamwork.

Spreadsheets and project management software such as Microsoft Project 2002 can help project managers in project human resource management. Software makes it easy to produce responsibility assignment matrixes, create resource histograms, identify overallocated resources, level resources, and provide various views and reports related to project human resource management.

Project human resource management involves much more than using software to facilitate organizational planning and assign resources. What makes good project managers great is their ability to enable project team members to deliver the best work they possibly can on a project.

DISCUSSION QUESTIONS

1. Discuss the changes in the job market for information technology workers. How does the job market and current state of the economy affect human resource management?
2. Summarize the processes involved in project human resource management.
3. Briefly summarize the work done by Maslow, Herzberg, McClellan, McGregor, Ouchi, Thamhain and Wilemon, and Covey. How do their theories relate to project management?
4. Describe situations where it would be appropriate to create a project organizational chart, a responsibility assignment matrix, a RACI chart, and a resource histogram. Describe what these charts or matrices look like.
5. Discuss the difference between resource loading and resource leveling and provide an example of when you would use each technique.
6. Explain two different types of team-building activities described in this chapter.
7. How can you use Project 2002 to assist in project human resource management?

EXERCISES

1. Develop a responsibility assignment matrix for building a new state-of-the-art multimedia classroom for your university. Assume the faculty in the Computer Science/MIS Department received a grant to lead this project. The grant includes funds for all of the hardware and software plus some money for faculty and student labor. Also assume the university must match the funds provided by the granting agency. Key stakeholders are faculty who might use the facility, students, the Dean, the head of facilities, the head of finance, and the head of information technology. Major activities might include deciding where to put the classroom, defining the classroom configuration (number and type of computers, projections system software, layout, and so on), writing a request for proposal, evaluating proposals, overseeing the winning contractor, obtaining matching funds from the college, surveying faculty and students on the use of the classroom, providing training on using the classroom, and scheduling use of the classroom. Use the responsibility assignment matrix in Figures 9-4 and 9-5 as models.

2. Take the MBTI test and research information on this tool. There are several Web sites that have different versions of the test, such as *www.humanmetrics.com*, *www.personalitytype.com*, and *www.keirsey.com*. Then write a one- to two-page paper describing your MBTI type and what you think about this test as a team-building tool.

3. Read Douglas Ross's article "Applying Covey's Seven Habits to a Project Management Career" (see Suggested Readings) or another article that describes the seven habits. Summarize three of Covey's habits in your own words and give examples of how these habits would apply to project management.

4. Research recruiting and retention strategies at three different companies. What distinguishes one company from another in this area? Are strategies such as signing bonuses, tuition reimbursement, and business casual dress codes standard for new information technology workers? What strategies appeal most to you?

5. Write a paper summarizing the main features of Microsoft Project 2002 that can assist project managers in human resource management. Also interview someone who uses Project 2002. Ask him or her if their organization uses any project human resource management features as described in this chapter and Appendix A and document their reasons for using or not using certain features.

MINICASE

Your company is planning to launch an important new project starting January 1. You estimate that you will need one full-time project manager for this one-year project, two business analysts full-time for the first six months, two senior programmers full-time for the whole year, four junior programmers full-time for the months of July, August, and September, and one technical writer full-time for the last three months. For simplicity's sake, assume that the main tasks these people will perform are as follows, respectively:

■ Project management
■ Systems analysis
■ Systems design
■ Coding and testing
■ Documentation

Part 1: Use Microsoft Excel to create a stacked column chart showing a resource histogram for this project, similar to the one shown in Figure 9-6. Be sure to include a legend to label the types of resources needed. Use appropriate titles and axis labels.

Part 2: Enter the five project tasks—project management, systems analysis, systems design, coding and testing, and documentation—and their given durations in Microsoft Project 2002. Assume that one year is fifty-two weeks, six months is twenty-six weeks, and so on. Assume the first three tasks will all start at the same time, that coding and testing can start when systems design is 50 percent completed, and that documentation starts when the coding and testing is completed. Then enter the resource information, as provided in the first paragraph of the Minicase. Instead of entering names for each resource, put the job title. For two people with the same job title, enter business analyst 1 and business analyst 2, for example. Then assign the resources to the appropriate tasks. Print the Gantt chart view of your work on one page when you are finished.

RUNNING CASES

Several of the running cases in Appendix C provide questions related to human resource management. Appendix A also includes a section on using Project 2002 to assist in human resource management.

SUGGESTED READINGS

1. Brooks, Frederick P., Jr. *The Mythical Man-Month: Essays on Software Engineering*. Reading, MA: Addison-Wesley, 1995.

    The Mythical Man-Month is one of the classics in the field of software program management. Brooks wrote the first version of this book in 1975. He draws on his experience as the head of operating systems development for IBM's famous 360 mainframe and some of his most recent experiences.

2. Covey, Stephen. *The 7 Habits of Highly Effective People*. New York: Simon & Schuster, 1990.

    Stephen Covey is famous for his books, audiotapes, videotapes, and seminars on improving effectiveness. This book continues to be a business bestseller. Covey also has written related books based on these seven habits. Visit his Web site at www.franklincovey.com.

3. Goleman, Daniel. *Emotional Intelligence*. New York: Bantam Books, 1997.

    Another important topic related to human resource development is emotional intelligence. Daniel Goleman has written this book and several others to help people understand what emotional intelligence is and how to help people, especially young children, develop all aspects of intelligence.

4. Herzberg, Frederick. "One More Time: How Do You Motivate Employees?" *Harvard Business Review* (February 1968): 51–62.

    This classic article describes the results of Herzberg's studies on what motivates people. He distinguishes between hygiene factors and motivating factors in a work setting and offers suggestions on how to motivate workers effectively.

5. Keirsey, David, and Ray Choiniere. *Please Understand Me II: Temperament, Character, Intelligence*. Del Mar, CA: Prometheus Nemesis Book Co., May 1998.

> *Keirsey and Bates's* Please Understand Me, *first published in 1978, sold nearly two million copies in its first twenty years, becoming a bestseller all over the world. Keirsey has continued to investigate personality differences, and recently published this sequel to refine his theory of the four temperaments and to define the facets of character that distinguish one person from another. You can get more information on this book and take an online version of the Keirsey Temperament Sorter test at www.keirsey.com.*

6. Ross, Douglas. "Applying Covey's Seven Habits to a Project Management Career." *PM Network*, Project Management Institute (April 1996): 26–30.

> *Ross summarizes key points from Covey's seven habits and relates them to improving effectiveness in project management. This article helps project managers translate Covey's principles to their own teams, projects, and personal lives.*

7. Verma, Vijay K. *Organizing Projects for Success: The Human Aspects of Project Management*. Upper Darby, PA: Project Management Institute, 1995.

> *This book focuses specifically on managing human resources in a project environment. There are separate chapters on interfacing with major stakeholders, designing organizational structures, addressing important issues in project organization design, designing a project organizational structure, and making matrix structures work. Verma has also written several other books related to project human resource management.*

KEY TERMS

- **coercive power** — using punishment, threats, or other negative approaches to get people to do things they do not want to do
- **deputy project managers** — people who fill in for project managers in their absence and assist them as needed, similar to the role of a vice president
- **empathic listening** — listening with the intent to understand
- **expert power** — using one's personal knowledge and expertise to get people to change their behavior
- **extrinsic motivation** — causes people to do something for a reward or to avoid a penalty
- **hierarchy of needs** — a pyramid structure illustrating Maslow's theory that people's behaviors are guided or motivated by a sequence of needs
- **intrinsic motivation** — causes people to participate in an activity for our own enjoyment

- **legitimate power** — getting people to do things based on a position of authority
- **mirroring** — the matching of certain behaviors of the other person
- **Myers-Briggs Type Indicator (MBTI)** — a popular tool for determining personality preferences
- **organizational breakdown structure (OBS)** — a specific type of organizational chart that shows which organizational units are responsible for which work items
- **organizational planning** — identifying, assigning, and documenting project roles, responsibilities, and reporting relationships
- **overallocation** — when more resources than are available are assigned to perform work at a given time
- **power** — the potential ability to influence behavior to get people to do things they would not otherwise do
- **RACI charts** — charts that show Responsibility, Accountability, Consultation, and Informed roles for project stakeholders
- **rapport** — a relation of harmony, conformity, accord, or affinity
- **referent power** — getting people to do things based on an individual's personal charisma
- **resource histogram** — a column chart that shows the number of resources assigned to a project over time
- **resource leveling** — a technique for resolving resource conflicts by delaying tasks
- **resource loading** — the amount of individual resources an existing schedule requires during specific time periods
- **responsibility assignment matrix (RAM)** — a matrix that maps the work of the project as described in the WBS to the people responsible for performing the work as described in the OBS
- **reward power** — using incentives to induce people to do things
- **staff acquisition** — getting the needed personnel assigned to and working on the project
- **staffing management plan** — a document that describes when and how people will be added to and taken off the project team
- **subproject managers** — people responsible for managing the subprojects that a large project might be broken into
- **synergy** — an approach where the whole is greater than the sum of the parts
- **team development** — building individual and group skills to enhance project performance

10

Project Communications Management

Objectives

After reading this chapter, you will be able to:

1. *Understand the importance of good communication on projects and describe the major components of a communications management plan*
2. *Discuss the elements of project communications planning, including information distribution, performance reporting, and administrative closure*
3. *Discuss various methods for project information distribution and the advantages and disadvantages of each*
4. *Understand individual communication needs and how to determine the number of communications channels needed for a project*
5. *Understand how the main outputs of performance reporting help stakeholders stay informed about project resources*
6. *Recognize how the main outputs of administrative closure are used to formally end a project*
7. *List various methods for improving project communications, such as managing conflicts, running effective meetings, using e-mail effectively, and using templates*
8. *Describe how software can enhance project communications*

OPENING CASE

Peter Gumpert worked his way up the corporate ladder in a large telecommunications company. He was intelligent, competent, and a strong leader, but a new fiber-optic undersea telecommunications program was much larger and more complicated than anything he had worked on, let alone managed. The undersea telecommunications program consisted of several distinct projects, and Peter was the program manager in charge of overseeing them all. The changing marketplace for undersea telecommunications systems and the large number of projects involved made communications and flexibility critical concerns for Peter. For missing milestone and completion dates, his company would suffer huge financial penalties ranging from thousands of dollars per day for smaller projects to more than $250,000 per day for larger projects. Many projects depended on the success of other projects, so Peter had to understand and actively manage those critical interfaces.

Peter held several informal and formal discussions with the project managers reporting to him on this program. He worked with them and his project executive assistant, Christine Braun, to develop a communications plan for the program. He was still unsure, however, of the best way to distribute information and manage all of the inevitable changes that would occur. He also wanted to develop consistent ways for all of the project managers to develop their plans and track performance without stifling their creativity and autonomy. Christine suggested that they consider using some new communications technologies to keep important project information up to date and synchronized. Although Peter knew a lot about telecommunications and laying fiber-optic lines, he was not an expert in using information technology to improve the communication process. In fact, that was part of the reason he asked Christine to be his assistant. Could they really develop a process for communicating that would be flexible and easy to use? Time was of the essence as more projects were being added to the undersea telecommunications program every week.

THE IMPORTANCE OF PROJECT COMMUNICATIONS MANAGEMENT

Many experts agree that the greatest threat to the success of any project, especially information technology projects, is a failure to communicate. Recall that the 1995 Standish Group study found the three major factors related to information technology project success were user involvement, executive management support, and a clear statement of requirements.[1] All of these factors depend on having good communications skills, especially with non-information technology personnel.

Our culture often portrays computer professionals as nerds who like to sit in dark corners hacking away on computers. When computer professionals have to communicate with non-computer professionals, it can be as if they were talking to someone from another planet. The information technology field is constantly changing, and these changes bring with them a great deal of technical jargon. Even though more and more people use computers, the gap between users and developers increases as technology advances. Of course, not every computer professional is a poor communicator, but most people in any field can improve their communication skills.

Compounding the problem, most educational systems for information technology graduates promote strong technical skills over strong communication and social skills. Most information technology-related degree programs have many technical requirements, but few require courses in communication (speaking, writing, listening), psychology, sociology, and the humanities. People often assume it is easy to pick up these soft skills, but they *are* important skills and, as such, people must learn and develop them.

Many studies have shown that information technology professionals need these soft skills just as much or even more than other skills. You cannot totally separate technical skills and soft skills when working on information technology projects. For projects to succeed, every project team member needs both types of skills and needs to develop them continuously through formal education and on-the-job training.

An article in the *Journal of Information Systems Education* on the importance of communications skills for information technology professionals presented the following conclusion:

> Based on the results of this research we can draw some general conclusions. First, it is evident that IS professionals engage in numerous verbal communication activities that are informal in nature, brief in duration, and with a small number of people at a time. Second, we can infer that most of the communication is indeed verbal in nature but sometimes it is supported by notes or graphs on a board or a handout and also by computer output. Third, it is clear that people expect their peers to listen carefully during a conversation and respond correctly to the issues at hand. Fourth, all IS professionals must be aware of the fact that they will have to engage in some form of informal public speaking. Fifth, it is evident that IS professionals must be able to communicate effectively in order to be successful in their current position but they must also be able to do so in order to move to higher positions. Since our respondents, on average, seem to have moved

---

[1] The Standish Group, "CHAOS," *www.standishgroup.com* (1995).

throughout their IS career, from lower to higher positions, and they ranked verbal skills more important for their advancement than for their current job, the ability to communicate verbally seems to be a key factor in career advancement.[2]

This chapter will highlight key aspects of project communications management, provide some suggestions for improving communications management, and describe how software can assist in project communications management.

The goal of project communications management is to ensure timely and appropriate generation, collection, dissemination, storage, and disposition of project information. There are four main processes in project communications management:

- **Communications planning** involves determining the information and communications needs of the stakeholders: who needs what information, when will they need it, and how will the information be given to them.
- **Information distribution** involves making needed information available to project stakeholders in a timely manner.
- **Performance reporting** involves collecting and disseminating performance information, including status reports, progress measurement, and forecasting.
- **Administrative closure** involves generating, gathering, and disseminating information to formalize phase or project completion.

COMMUNICATIONS PLANNING

Because communication is so important on projects, every project should include a **communications management plan**—a document that guides project communications. This plan should be part of the overall project plan (described in Chapter 4, Project Integration Management). The type of communications management plan will vary with the needs of the project, but some type of written plan should always be prepared. For example, for small projects, such as the Project Management Intranet Site project described in Chapter 3, the communications management plan can be part of a team contract. The main parts of a communications management plan include:

- *A description of a collection and filing structure for gathering and storing various types of information.* For example, if a project team member attends a conference and brings back valuable information, where should he or she file and store it? If a supplier sends a new product brochure, where should it be filed and stored? If the same supplier sends a new product brochure six

---

[2] Sivitanides, Marcos P., James R. Cook, Roy B. Martin, and Beverly A. Chiodo, "Verbal Communication Skills Requirements for Information Systems Professionals," *Journal of Information Systems Education* (Spring 1995), 7(1).

months later, how should the new brochure and the old brochure be handled? If the new product is used on the project and might affect other areas such as accounting or engineering, how will that new product information be communicated to those areas? Everyone knows it can be difficult to organize his or her individual work, so it is imperative to develop and follow a system for filing important project-related work for many people. In addition, several government agencies require a detailed filing system and hold inspections to ensure their instructions for filing are followed.

■ *A distribution structure describing what information goes to whom, when, and how.* For example, are all status reports written or are some oral? Does every stakeholder receive every master schedule update? Does top management receive different formats of status reports?

■ *A format for communicating key project information.* Is there a template for project team members to follow in preparing written and oral status reports? Is there a master list of all acronyms and definitions, or do they need to be repeated on different project documentation? Many hours of confusion can be avoided by providing templates and examples of key project reports. You will find several examples later in this chapter.

■ *A production schedule for producing the information.* Have resources been assigned to create, assemble, and disseminate key project information? Do stakeholders know when to expect different information, when they need to attend key meetings, and so on? Has time been allowed for review and approval of key project documentation? Many people procrastinate when it comes to documenting their work. It is important to allow time for creating key project information and ensuring its quality.

■ *Access methods for obtaining the information.* Who can see a draft document? Can everyone access all project documentation? Which information is online and which is in hard copy or other formats? Can anyone check out hard copies of documents? Who can attend what meetings?

■ *A method for updating the communications management plans as the project progresses and develops.* Who will update the communications management plan as changes occur? How will the new plan be distributed?

■ *A stakeholder communications analysis.* It is important to know what kinds of information will be distributed to which stakeholders. By analyzing stakeholder communications, you can avoid wasting time or money on creating or disseminating unnecessary information. The project's organizational chart is a starting point for identifying internal stakeholders. You must also include key stakeholders outside the project organization such as the customer, the customer's top management, and subcontractors.

Table 10-1 provides a sample stakeholder communications analysis that shows which stakeholders should get which written communications. Note that the stakeholder communications analysis includes information such as the contact person for the information, when the information is due, and the preferred format for the information. You can create a similar table to show which stakeholders should attend which project meetings. It is always a good idea to include comment sections with these types of tables to record special considerations or details related to each stakeholder, document, meeting, and so on. Having stakeholders review and approve all stakeholder analysis materials will ensure that the information is correct and useful.

Table 10-1: Sample Stakeholder Communications Analysis

STAKEHOLDERS	DOCUMENT NAME	DOCUMENT FORMAT	CONTACT PERSON	DUE
Customer Management	Monthly status report	Hard copy	Tina Erndt, Tony Silva	First of month
Customer Business Staff	Monthly status report	Hard copy	Julie Grant, Jeff Martin	First of month
Customer Technical Staff	Monthly status report	E-mail	Evan Dodge, Nancy Michaels	First of month
Internal Management	Monthly status report	Hard copy	Bob Thomson	First of month
Internal Business and Technical Staff	Monthly status report	Intranet	Angie Liu	First of month
Training Subcontractor	Training plan	Hard copy	Jonathan Kraus	11/1/2004
Software Subcontractor	Software implementation plan	E-mail	Barbara Gates	6/1/2004

Many projects do not include enough initial information on communications. Project managers, top management, and project team members assume using existing communications channels to relay project information is sufficient. The problem with using existing communication channels is that each of these groups, as well as other stakeholders, has different communication needs. Creating some sort of communications management plan and reviewing it with project stakeholders early in a project helps prevent or reduce later communication problems. If organizations work on many projects, developing some consistency in handling project communications helps the organization run smoothly.

Consistent communication helps organizations improve project communications, especially for programs composed of multiple projects. For example, Peter Gumpert, the undersea telecommunications program manager in the opening case, would benefit greatly from having a communications management plan that all of the project managers under him helped develop and follow. Since several of the projects probably have some of the same stakeholders, it is even more important to develop a coordinated communications management plan. For example, if customers receive status reports from Peter's company that have totally different formats and do not coordinate information from related projects within the same company, they will question the ability of Peter's company to manage large programs.

Information regarding the content of essential project communications comes from the work breakdown structure (WBS). In fact, many WBSs include a section for project communications to ensure that reporting key information is a project deliverable. If reporting essential information is an activity defined in the WBS, it becomes even more important to develop a clear understanding of what project information to report, when to report it, how to report it, who is responsible for generating the report, and so on.

INFORMATION DISTRIBUTION

Getting project information to the right people at the right time and in a useful format is just as important as developing the information in the first place. The stakeholder communications analysis serves as a good starting point for information distribution. Project managers and their teams must decide who receives what information, but they must also decide the best way to distribute the information. Is it sufficient to send written reports for project information? Are meetings alone effective in distributing project information? Are meetings and written communications both required for project information?

After answering these questions, project managers and their teams must decide the best way to distribute the information. Important considerations for information distribution include the use of technology, formal and information communications, and the complexity of communications.

Using Technology to Enhance Information Distribution

Technology can facilitate the process of distributing information. Using an internal project management information system, you can organize project documents, meeting minutes, customer requests, requests to change status, and so on, and make them available in an electronic format. You can store this information in local software or make it available on an intranet, an extranet, or the Internet, if the information is not sensitive. Storing templates and samples of project documents electronically can make accessing standard forms easier, thus making the information distribution process easier. You will learn more about using the Internet and Microsoft Project 2002 to assist in project communications management later in this chapter.

Formal and Informal Methods for Distributing Information

It is not enough for project team members to submit status reports to their project managers and other stakeholders and assume that everyone who needs to know that information will read the reports. Many technical professionals assume that submitting the appropriate status reports is sufficient. Informal verbal communications are equally effective ways to distribute information, but technical professionals tend to neglect techniques that are more informal. In contrast to many technical professionals, many non-technical professionals—from colleagues to managers—prefer to hear important project information informally and have a two-way conversation about it. They prefer not to read stacks of paperwork, e-mails, or Web pages to try to find pertinent information.

Instead of focusing on getting information by reading technical documents, many colleagues and managers want to know the people working on their projects and develop a trusting relationship with them. They use informal discussions about the project to develop these relationships. Therefore, project managers must be good at nurturing relationships through good communication. Many experts believe that the difference between good project managers and excellent project managers is their ability to nurture relationships and use empathic listening skills, as described in Chapter 9, Project Human Resource Management.

Effective distribution of information depends on project managers and project team members having good communication skills. Communicating includes many different dimensions such as writing, speaking, and listening, and project personnel need to use all of these dimensions in their daily routines. In addition, different people respond positively to different levels or types of communication. For example, a project sponsor may prefer to stay informed through informal discussions held once a week over coffee. The project manager needs to be aware of and take advantage of this special communication need. The project sponsor will give better feedback about the project during these informal talks than he or she could give through some other form of communication. Informal conversations allow the project sponsor to exercise his or her role of leadership and provide insights and information that are critical to the success of the project and the organization as a whole. Short face-to-face meetings are often more effective than electronic communications, particularly for sensitive information.

┌─────────────────────────┐
│ *What Went Wrong?* │
└─────────────────────────┘

A well-publicized example of misuse of e-mail comes from the 1998 Justice Department's high-profile, antitrust suit against Microsoft. E-mail emerged as a star witness in the case. Many executives sent messages that they should have never put in writing. The court used e-mail as evidence, even though the senders of the notes said the information was being interpreted out of context.[3]

Some companies, such as Amazon.com, have established policies to encourage employees to watch their use of e-mail and delete it often. Their "Sweep and Clean" program instructed employees to purge e-mails that were no longer required for business or legal purposes. They even offered free café lattes to employees who complied with the program immediately.[4]

Another, more amusing, example of miscommunication comes from a director of communications at a large firm:

"I was asked to prepare a memo reviewing our company's training programs and materials. In the body of the memo in one of the sentences, I mentioned the "pedagogical approach" used by one of the training manuals. The day after I routed the memo to the executive committee, I was called into the HR director's office, and told that the executive vice president wanted me out of the building by lunch. When I asked why, I was told that she wouldn't stand for perverts (pedophiles?) working in her company. Finally, he showed me her copy of the memo, with her demand that I be fired—and the word "pedagogical" circled in red. The HR manager was fairly reasonable, and once he looked the word up in his dictionary and made a copy of the definition to send back to her, he told me not to worry. He would take care of it. Two days later, a memo to the entire staff came out directing us that no words which could not be found in the local Sunday newspaper could be used in company memos. A month later, I resigned. In accordance with company policy, I created my resignation memo by pasting words together from the Sunday paper."[5]

Many written reports neglect to provide the important information that good managers and technical people have a knack for asking about. For example, it is important to include detailed technical information that will affect critical performance features of products the company is producing as part of a project. It is even more important to document any changes in technical

---

[3] Harmon, Amy, "E-mail Comes Back to Haunt Companies," *Minneapolis Star Tribune* (from the *New York Times*), (November 29, 1998).

[4] *Ibid.*

[5] Projectzone, "Humor.," *www.projectzone.com/memos.html* (2001).

specifications that might affect product performance. For example, if the undersea telecommunications program included a project to purchase and provide special diving gear, and the supplier who provided the oxygen tanks enhanced the tanks so divers could stay under water longer, it would be very important to let other people know about this new capability. The information should not be buried in an attachment with the supplier's new product brochure. People also have a tendency to not want to report bad information. If the oxygen tank vendor was behind on production, the person in charge of the project to purchase the tanks might wait until the last minute to report this critical information. Oral communication via meetings and informal talks helps bring important information—positive or negative—out into the open.

Oral communication also helps build stronger relationships among project personnel and project stakeholders. People make or break projects, and people like to interact with each other to get a true feeling for how a project is going. Studies show that less than 10 percent of communications consist of the actual content or words communicated. A person's tone of voice and body language say a lot about how they really feel.

Since information technology projects often require a lot of coordination, it is a good idea to have short, frequent meetings. For example, some information technology project managers require all project personnel to attend a "stand-up" meeting every week or even every morning, depending on the project needs. Stand-up meetings have no chairs, and the lack of chairs forces people to focus on what they really need to communicate.

To encourage face-to-face, informal communications, some companies have instituted policies that workers cannot use e-mail between certain hours of the business day. Sending frequent e-mails to a large group of people has a much different effect than sending frequent e-mails within a small team of people working on a particular task. Many people also waste a lot of time by not using e-mail effectively. You will learn more about improving the effectiveness of e-mail later in this chapter.

Table 10-2 provides guidelines from Practical Communications, Inc., a communications consulting firm, about how well different types of media, such as hard copy, telephone calls, voice mail, e-mail, meetings, and Web sites, are suited to different communication needs. For example, if you were trying to assess commitment of project stakeholders, a meeting would be the most appropriate medium to use. A telephone call would be adequate, but the other media would not be appropriate. Project managers must assess the needs of the organization, the project, and individuals in determining which communication medium to use, and when.

Table 10-2: Media Choice Table

KEY: 1 = EXCELLENT		2 = ADEQUATE			3 = INAPPROPRIATE	
How Well Medium is Suited to:	Hard Copy	Telephone Call	Voice Mail	E-Mail	Meeting	Web Site
Assessing commitment	3	2	3	3	1	3
Building consensus	3	2	3	3	1	3
Mediating a conflict	3	2	3	3	1	3
Resolving a misunderstanding	3	1	3	3	2	3
Addressing negative behavior	3	2	3	2	1	3
Expressing support/appreciation	1	2	2	1	2	3
Encouraging creative thinking	2	3	3	1	3	3
Making an ironic statement	3	2	2	3	1	3
Conveying a reference document	1	3	3	3	3	2
Reinforcing one's authority	1	2	3	3	1	1
Providing a permanent record	1	3	3	1	3	3
Maintaining confidentiality	2	1	2	3	1	3
Conveying simple information	3	1	1	1	2	3
Asking an informational question	3	1	1	1	3	3
Making a simple request	3	1	1	1	3	3
Giving complex instructions	3	3	2	2	1	2
Addressing many people	2	3 or 1*	2	2	3	1

*Depends on system functionality
Galati, Tess. *Email Composition and Communication (EmC2)*. Practical Communications, Inc., (*www.praccom.com*) (2001).

Understanding Human and Individual Communication Needs

Many top managers think they can just add more people to a project that is falling behind schedule. Unfortunately, this approach often causes more setbacks because of the increased complexity of communications. In his popular book, *The Mythical Man-month*, Frederick Brooks illustrates this concept very clearly. People are not interchangeable parts. You cannot assume that a task originally scheduled to take two months of one person's time can be done in one month by two people. A popular analogy is that you cannot take nine women and produce a baby in one month!

In addition to understanding that people are not interchangeable, it is also important to understand individuals' personal preferences for communications.

As described in Chapter 9, people have different personality traits, which often affect their communication preferences. For example, if you want to praise a project team member for doing a good job, an introvert would be more comfortable receiving that praise in private while an extrovert would like everyone to hear about his or her good work. An intuitive person would want to understand how something fits into the big picture, while a sensing person would prefer to have more focused, step-by-step details. A strong thinker would want to know the logic behind information, while a feeling person would want to know how the information affects him or her personally as well as other people. Someone who is a judging would be very driven to meet deadlines with few reminders while a perceiving person would need more assistance in developing and following plans.

Rarely does the receiver interpret a message exactly as the sender intended. Therefore, it is important to provide many methods of communication and an environment that promotes open dialogue. It is important for project managers and their teams to be aware of their own communication styles and preferences and those of other project stakeholders. As described in the previous chapter, many information technology professionals have different personality traits than the general population, such as being more introverted, intuitive, and oriented to thinking (as opposed to feeling). These personality differences can lead to miscommunication with people who are extroverted, sensation-oriented, and feeling types. For example, information technology professionals are notorious for writing poor documentation. Their version of a user guide might not provide the detailed steps most users need. Many users also prefer face-to-face meetings to learn how to use a new system instead of trying to follow a cryptic user guide.

Geographic location and cultural background also affect the complexity of project communications. For example, if project stakeholders are in different countries, it is often difficult or impossible to schedule times for two-way communication during normal working hours. Language barriers can also cause communication problems. The same word may have very different meanings in different languages. Times, dates, and other units of measure are also interpreted differently. People from some cultures also prefer to communicate in ways that may be uncomfortable to others. For example, managers in some countries still do not allow workers of lower ranks or women to give formal presentations. Some cultures also reserve written documents for binding commitments.

Determining the Number of Communication Channels

Another important aspect of information distribution is the number of people involved in a project. As the number of people involved increases, the complexity of communications increases because there are more communications channels or pathways through which people can communicate. There is a simple formula for determining the number of communications channels as the number of people involved in a project increases. You can calculate the number of communications channels as follows:

$$\text{number of communications channels} = \frac{n(n\text{-}1)}{2}$$

where *n* is the number of people involved.

For example, two people have one communications channel: $(2(2-1))/2 = 1$. Three people have three channels: $(3(3-1))/2 = 3$. Four people have six channels, five people have ten, and so on. Figure 10-1 illustrates this concept. You can see that as the number of people communicating increases above three, the number of communications channels increases rapidly. The lesson is a simple one: If you want to enhance communications, you must consider the interactions among different project team members and stakeholders. If you raise an issue by sending e-mail to 100 people, the result will be quite different than if you bring up the issue at a large review meeting or at smaller team meetings. Sending an e-mail to 100 people might result in more problems than would arise if you discussed the same issue in a meeting.

number of communications channels = $\frac{n(n-1)}{2}$

2 people, 1 communications channel

4 people, 6 communications channels, etc.

3 people, 3 communications channels

Figure 10-1. The Impact of the Number of People on Communications Channels

When asked why you cannot always send an e-mail to a team of 100 people, just as you would to a team of five, one CIO answered, "As a group increases in size, you have a whole slew of management challenges. Communicating badly exponentially increases the possibility of making fatal mistakes. A large-scale project has a lot of moving parts, which makes it that much easier to break down. Communication is the oil that keeps everything working properly. It's much easier to address an atmosphere of distrust among a group of five team members than it is with a team of 500 members."[6]

However, there are situations in which you cannot have face-to-face meetings and must e-mail a large group of people. Many information technology professionals work on virtual projects where they never meet their project sponsors, other team members, or other project stakeholders. In a virtual project environment, it is crucial for project managers to develop clear communication procedures. They can and must use e-mail, discussion threads, project Web sites, and other technologies to communicate most information. They might be able to use phone calls or other media occasionally, but in general, they must rely on good written communications.

As you can see, information distribution involves more than creating and sending status reports or holding periodic meetings. Many good project managers know their personal strengths and weaknesses in this area and surround themselves with people who complement their skills, just as Peter Gumpert in the opening case did in asking Christine to be his assistant. It is good practice to share the responsibility for project communications management with the entire project team.

PERFORMANCE REPORTING

Performance reporting keeps stakeholders informed about how resources are being used to achieve project objectives. The project plan and work results are important inputs to performance reporting. The main outputs of performance reporting include status reports, progress reports, forecasts, and change requests.

- **Status reports** describe where the project stands at a specific point in time. Recall the importance of the triple constraint. Status reports address where the project stands in terms of meeting scope, time, and cost goals. How much money has been spent to date? How long did it take to do certain tasks? Is work being accomplished as planned? Status reports can take various formats depending on the stakeholders' needs. Chapter 6, Project Cost Management, provides detailed information on earned value management, a technique used for reporting project performance that integrates scope, time, and cost data.

- **Progress reports** describe what the project team has accomplished during a certain period. Many projects have each team member prepare a monthly or sometimes weekly progress report. Team leaders often create consolidated progress reports based on the information received from

6 Hildenbrand, Carol, "Loud and Clear," *CIO Magazine* (April 15, 1996), *http://www.cio.com/archive/041596_qa.html.*

team members. A sample template for a monthly progress report is provided later in this chapter.

- **Project forecasting** predicts future project status and progress based on past information and trends. How long will it take to finish the project based on how things are going? How much more money will be needed to complete the project? Project managers can also use earned value management to answer these questions by estimating the budget at completion based on how the project is progressing.

Another important technique for performance reporting is the status review meeting. Status review meetings, as described in Chapter 4, Project Integration Management, are a good way to highlight information provided in important project documents, empower people to be accountable for their work, and have face-to-face discussions about important project issues. Many program and project managers hold monthly status review meetings to exchange important project information and motivate people to make progress on their parts of the project. Likewise, many top managers hold monthly or quarterly status review meetings where program and project managers must report overall status information.

Status review meetings sometimes become battlegrounds where conflicts between different parties come to a head. Project managers or higher-level top managers should set ground rules for status review meetings to control the amount of conflict and should work to resolve any potential problems. It is important to remember that project stakeholders should work together to address performance problems.

ADMINISTRATIVE CLOSURE

A project or phase of a project (concept, development, implementation, or close-out) requires closure. Administrative closure consists of verifying and documenting project results. This process formalizes sponsor or customer acceptance of the project's products. Administrative closure also allows time to collect project records, ensure those records reflect final specifications, analyze project effectiveness, and archive information for future use.

The main outputs of administrative closure are project archives, formal acceptance, and lessons learned. **Project archives** include a complete set of organized project records that provide an accurate history of the project. **Formal acceptance** is documentation that the project's sponsor or customer signs to show they have accepted the products of the project. **Lessons learned** are reflective statements written by project managers and their team members. Many projects are done under contract, and contracts usually specify what each of these outputs should consist of. If projects are not done under contract, it is still very important to prepare these items.

Project archives often come in handy many years after the project is completed. For example, a new project manager might want to know more details on some of the tools or techniques used in a certain aspect of an older project. Documentation contained in the project archives could save time and money on the current project. There might be an audit of the organization, and good project archives could provide valuable information very quickly.

Formal acceptance should be provided on internal as well as external projects. This process helps to end the project formally and avoids dragging out project termination. In a contract situation, the buyer must legally accept the products produced as part of the contract so the seller can receive payment. There are usually added costs if the contract does not end as planned. In a non-contract situation, all parties must agree on completion of the work so that people and other resources can be reassigned.

The project manager and project team members should all prepare a lessons-learned report after completing a project. Some items discussed in lessons-learned reports include the causes of variances on the project, the reasoning behind corrective actions chosen, the use of different project management tools and techniques, and personal words of wisdom based on team members' experiences. Some projects require all project members to write a brief lessons-learned report. These reports provide valuable reflections by people who know what really worked or did not work on the project. Everyone learns in different ways and has different insights into a project. These reports can be an excellent resource and help future projects run more smoothly.

To reinforce the benefits of lessons-learned reports, some companies require new project managers to read several past project managers' lessons-learned reports and discuss how they will incorporate some of their ideas into their own projects. Other companies actually charge a small fee for purchasing lessons-learned reports, so the authors can make a little money for documenting useful lessons-learned reports and the buyers are more motivated to read them.

SUGGESTIONS FOR IMPROVING PROJECT COMMUNICATIONS

You have seen that good communication is vital to the management and success of information technology projects; you have also learned that project communications management can ensure that essential information reaches the right people at the right time, that feedback and reports are appropriate and useful, and that there is a formalized process of administrative closure. This section highlights a few areas that all project managers and project team members should consider in their quests to improve project communications. The following sections provide guidelines for managing conflict, developing better communication skills, running effective meetings, using e-mail effectively, using templates for project communications, and developing a communications infrastructure.

Using Communication Skills to Manage Conflict

Most large information technology projects are high-stake endeavors that are highly visible within organizations. They require tremendous effort from team members, are expensive, commandeer significant resources, and can have an extensive impact on the way work is done in an organization. When the stakes are high, conflict is never far away; when the potential for conflict is high, good communication is a necessity.

Chapter 6, Project Time Management, explained that schedule issues cause the most conflicts over the project life cycle and provided suggestions for improving project scheduling. Other common conflicts occur over project priorities, staffing, technical issues, administrative procedures, personalities, and cost. It is crucial for project managers to develop and use their human resources and communication skills to help identify and manage conflict on projects. Project managers should lead their teams in developing norms for dealing with various types of conflicts that might arise on their projects. For example, team members should know that disrespectful behavior toward any project stakeholder is inappropriate, and that team members are expected to try to work out small conflicts themselves before elevating them to higher levels.

Blake and Mouton (1964) delineated five basic modes for handling conflicts: confrontation, compromise, smoothing, forcing, and withdrawal.

1. *Confrontation.* When using the **confrontation mode**, project managers directly face a conflict using a problem-solving approach that allows affected parties to work through their disagreements. This approach is also called the problem-solving mode.

2. *Compromise.* With the **compromise mode**, project managers use a give-and-take approach to resolving conflicts. They bargain and search for solutions that bring some degree of satisfaction to all the parties in a dispute.

3. *Smoothing.* When using the **smoothing mode**, the project manager deemphasizes or avoids areas of differences and emphasizes areas of agreement.

4. *Forcing.* The **forcing mode** can be viewed as the win-lose approach to conflict resolution. Project managers exert their viewpoint at the potential expense of another viewpoint. Managers who are very competitive or autocratic in their management style might favor this approach.

5. *Withdrawal.* When using the **withdrawal mode**, project managers retreat or withdraw from an actual or potential disagreement. This approach is the least desirable conflict-handling mode.

Research indicates that project managers favor using confrontation for conflict resolution over the other four modes. The term confrontation may be misleading. This mode really focuses on addressing conflicts using a problem-solving approach. Using Stephen Covey's paradigms of interdependence, this mode focuses on a win-win approach. All parties work together to find the best way to solve the conflict. The next most favored approach to conflict resolution is compromise. Successful project managers are less likely to use smoothing, forcing, or withdrawal than they are to use confrontation or compromise.

Project managers must also realize that not all conflict is bad. In fact, conflict can often be good. Conflict often produces important results, such as new ideas, better alternatives, and motivation to work harder and more collaboratively. Project team members may become stagnant or develop **groupthink**— conformance to the values or ethical standards of a group—if there are no conflicting viewpoints on various aspects of a project. Research by Karen Jehn, Professor of Management at Wharton, suggests that task-related conflict, which is derived from differences over team objectives and how to achieve them, often improves team performance. Emotional conflict, however, which stems from personality clashes and misunderstandings, often depresses team performance.[7] Project managers should create an environment that encourages and maintains the positive and productive aspects of conflict.

Several organizations are emphasizing the importance of conflict management. For example, an innovative program at California State University, Monterey Bay, has incorporated conflict resolution as one of the program's eleven major learning objectives that students must know and understand in order to graduate. The academic program, based at the Institute for Community Collaborative Studies, focuses on preparing students for careers in the field of health and human services. The program's belief is that collaboration is an essential aspect of success for workers in the modern health and human services delivery field. Core competencies for developing collaboration skills include conflict resolution, negotiation, and mediation. These skills are also important for project managers in any field.

Developing Better Communication Skills

Some people seem to be born with great communication skills. Others seem to have a knack for picking up technical skills. It is rare to find someone with a natural ability for both. Both communication and technical skills, however, can be developed. Most information technology professionals enter the field because of their technical skills. Most find, however, that communication skills are the key to advancing in their careers, especially if they want to become good project managers.

Most companies spend a lot of money on technical training for their employees, even when employees might benefit more from communication training. Individual employees are also more likely to enroll voluntarily in classes on the latest technology than those on developing their soft skills. Communication skills training usually includes role-playing activities in which participants learn concepts such as building rapport, as described in Chapter 9, Project Human Resource Management. Training sessions also give participants a chance to develop specific skills in small groups. Training sessions that focus

[7] *Wharton Leadership Digest*, "Constructive Team Conflict" (March, 1997), 1(6).

on presentation skills usually videotape the participants. Most people are surprised to see some of their mannerisms on tape and enjoy the challenge of improving their skills. A minimal investment in communication and presentation training can have a tremendous payback to individuals, their projects, and their organizations. These skills also have a much longer shelf life than many of the skills learned in technical training courses.

As organizations become more global, they realize that they must also invest in ways to improve communication with people from different countries and cultures. For example, many Americans are raised to speak their minds, while in some other cultures people are offended by outspokenness. Not understanding how to communicate effectively with other cultures and people of diverse backgrounds hurts projects and businesses. Many training courses are available to educate people in cultural awareness, international business, and international team building.

It takes leadership to help improve communication. If top management lets employees give poor presentations, write sloppy reports, offend people from different cultures, or behave poorly at meetings, the employees will not want to improve their communication skills. Top management must set high expectations and lead by example. Some organizations send all information technology professionals to training that includes development of technical *and* communication skills. Some organizations allocate time in project schedules for preparing drafts of important reports and presentations and incorporating feedback on the drafts. Some include time for informal meetings with customers to help develop relationships. Some organizations even provide staff to assist in relationship management. As with any other goal, improving communication can be achieved with proper planning, support, and leadership from top management.

Running Effective Meetings

A well-run meeting can be a vehicle for fostering team building and reinforcing expectations, roles, relationships, and commitment to the project. However, a poorly run meeting can have a detrimental effect on a project. For example, a terrible **kickoff meeting**—a meeting held at the beginning of a project or project phase where all major project stakeholders discuss project objectives, plans, and so on—may cause some important stakeholders to decide not to support the project further. Many people complain about the time they waste in unnecessary or poorly planned and poorly executed meetings. Following are some guidelines to help improve time spent at meetings:

- *Determine if a meeting can be avoided*. Do not have a meeting if there is a better way of achieving the objective at hand. For example, a project manager might know that he or she needs approval from a top manager to hire another person for the project team. It could take a week or longer to schedule even a ten-minute meeting on the top manager's calendar.

Instead, an e-mail or phone call describing the situation and justifying the request is a faster, more effective approach than having a meeting.

■ *Define the purpose and intended outcome of the meeting.* Be specific about what should happen as a result of the meeting. Is the purpose to brainstorm ideas, provide status information, or solve a problem? Make the purpose of a meeting very clear to all meeting planners and participants. For example, if a project manager calls a meeting of all project team members without knowing the true purpose of the meeting, everyone will start focusing on their own agendas and very little will be accomplished. All meetings should have a purpose and intended outcome.

■ *Determine who should attend the meeting.* Do certain stakeholders have to be at a meeting to make it effective? Should only the project team leaders attend a meeting, or should the entire project team be involved? Many meetings are most effective with the minimum number of participants possible, especially if decisions must be made. Other meetings require many attendees. It is important to determine who should attend a meeting based on the purpose and intended outcome of the meeting.

■ *Provide an agenda to participants before the meeting.* Meetings are most effective when the participants come prepared. Did they read reports before the meeting? Did they collect necessary information? Some professionals refuse to attend meetings if they do not have an agenda ahead of time. Insisting on an agenda forces meeting organizers to plan the meeting and gives potential attendees the chance to decide whether they really need to attend the meeting.

■ *Prepare handouts, visual aids, and make logistical arrangements ahead of time.* By creating handouts and visual aids, the meeting organizers must organize their thoughts and ideas. This usually helps the entire meeting run more effectively. It is also important to make logistical arrangements by booking an appropriate room, having necessary equipment available, and providing refreshments or entire meals, if appropriate. It takes time to plan for effective meetings. Project managers and their team members must take time to prepare for meetings, especially important ones with key stakeholders.

■ *Run the meeting professionally.* Introduce people, restate the purpose of the meeting, and state any ground rules that attendees should follow. Have someone facilitate the meeting to make sure important items are discussed, watch the time, encourage participation, summarize key issues, and clarify decisions and action items. Designate someone to take minutes and send the minutes out soon after the meeting. Minutes should be short and focus on the crucial decisions and action items from the meeting.

■ *Build relationships.* Depending on the culture of the organization and project, it may help to build relationships by making meetings fun experiences. For example, it may be appropriate to use humor, refreshments, or prizes for good ideas to keep meeting participants actively involved. If used effectively, meetings are a good way to build relationships.

Using E-Mail Effectively

As more and more people use e-mail, communications should improve, right? Not necessarily. In fact, few people have received any training or guidelines on when or how to use e-mail. As discussed earlier in this chapter, e-mail is not an appropriate medium for several types of communications. The media choice table (Table 10-2) suggests that e-mail is not appropriate for assessing commitment, building consensus, mediating a conflict, resolving a misunderstanding, making an ironic statement, conveying a reference document, reinforcing one's authority, or maintaining confidentiality.

Even if people do know when to use e-mail for project communications, they also need to know how to use it. New features are added to e-mail systems with every new release, but often users are unaware of these features and do not receive any training on how to use them. Do you know how to file your e-mail messages, or do you have hundreds of e-mail messages sitting in your inbox? Do you know how to use your address book or how to create distribution lists? Have you ever used sorting features to find e-mail messages by date or author or key words?

Even if the user knows how to use all the features of an e-mail system, he or she will likely need to learn how to put ideas into words clearly. For example, meaningful subjects for all e-mail messages to clearly state the intention of the e-mail should be created. A business professional that is not a very good writer may rather talk to people than send them written communications in an e-mail message. When they do send e-mail, it often results in misunderstandings or confusion due to poor writing.

Unwanted e-mail also increases as advertisers use e-mail as an inexpensive way to reach potential customers. Most e-mail systems can block out the receipt of unwanted or junk e-mail (known as spam), but few users know how to implement this feature. Some people are so frustrated with unwanted e-mail that lawmakers are proposing a bill to address this and related electronic communication problems. In 2003, Minnesota Senator Mark Dayton introduced a Computer Owner's Bill of Rights to address issues such as computer spam. Among other things, the bill would order the Federal Trade Commission to set up a registry similar to the national "do not call" list used for telemarketing, and companies would be required to buy the list and be fined $10,000 for sending an e-mail to someone who did not want it.[8] Project managers should do whatever they can to help their project stakeholders use e-mail effectively and not waste time with unwanted e-mails.

Following are some guidelines for using e-mail more effectively:

■ Information sent via e-mail should be appropriate for that medium, versus other media. If you can communicate the information better with a phone call or meeting, for example, then do so.

[8] Hotakainen, Rob and Tom Ford, "Dayton bill tackles bad computer service," *Minneapolis Star Tribune* (March 7, 2003).

- Be sure to send the e-mail to the right people. Do not automatically "reply to all" if you do not need to.

- Use meaningful subjects so readers can quickly see what information the message will contain. If the entire message can be put in the subject line, put it there. For example, if a meeting is cancelled, just type that in the subject. Also, do not continue replying to e-mail messages without changing the subject. The subject should always relate to the latest correspondence.

- Limit the content of the e-mail to one main subject. Send a second or third e-mail if it relates to a different subject.

- The body of the e-mail should be as clear and concise as possible, and you should always re-read your e-mail before you send it. Also, be sure to check your spelling using the spell check function. If you have three questions you need answered, number them as question 1, 2, and 3.

- Limit the number and size of e-mail attachments. If you can include a link to an online version of a document instead of attaching a file, do so.

- Delete e-mail that you do not need to save or respond to. Do not even open e-mail that you know is not important, such as spam. Use the e-mail blocking feature of the software, if available, to block unwanted junk mail.

- Make sure your virus protection software is up to date. Never open e-mail attachments if you do not trust the source.

- Respond to e-mail quickly, if possible. It will take you longer to open and read it again later. In addition, if you send an e-mail that does not require a response, make that clear as well.

- If you need to keep e-mail, file each message appropriately. Create folders with meaningful names to file the e-mail messages you want to keep. File them as soon as possible.

- Learn how to use important features of your e-mail software such as mailing lists, sorting, and so on.

Using Templates for Project Communications

Many intelligent people have a hard time writing a performance report or preparing a ten-minute technical presentation for a customer review. Some people in these situations are too embarrassed to ask for help. To make preparing project communications easier, project managers need to provide examples and templates for common project communications items such as project descriptions, project charters, monthly performance reports, oral status reports, and so on. Good documentation from past projects can be an ample source of examples. Samples and templates of both written and oral reports are particularly helpful for people who have never before had to write project documents or give project presentations. Finding, developing, and sharing relevant templates is an important task for many project managers. Several examples of project documentation such as a business case, project charter, scope statement, stakeholder analysis,

WBS, Gantt chart, cost estimate, and so on are provided throughout this text. Appendix D includes these and other templates as well as resource information for finding additional templates. The companion Web site for this text includes the actual files used in creating the templates. A few of these templates and guidelines for preparing them are provided in this section.

Figure 10-2 shows a sample template for a one-page project description. This form could be used to show a "snapshot" of an entire project on one page. For example, top managers might require all project managers to provide a brief project description as part of a quarterly management review meeting. Peter Gumpert, the program manager in the opening case, might request this type of document from all of the project managers working for him to get an overall picture of what each project involves. According to Figure 10-2, a project description should include the project objective, scope, assumptions, cost information, and schedule information. This template suggests including information from the project's Gantt chart to highlight key deliverables and other milestones.

Project X Descripton

Objective: Describe the objective of the project in one or two sentences. Focus on the business benefits of doing the project.

Scope: Briefly describe the scope of the project. What business functions are involved, and what are the main products the project will produce?

Assumptions: Summarize the most critical assumptions for the project.

Cost: Provide the total estimated cost of the project. If desired, list the total cost each year.

Schedule: Provide summary information from the project's Gantt chart, as shown. Focus on summary tasks and milestones.

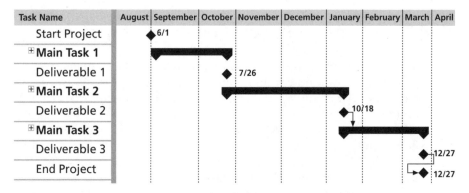

Figure 10-2. Sample Template for a Project Description

Table 10-3 shows a template for a monthly progress report. Sections of the progress report include accomplishments from the current period, plans for the next period, issues, and project changes. Table 3-13 in Chapter 3 provided a sample of a weekly status report that includes similar information. Recall that progress reports focus on accomplishments during a specific time period while status reports focus on where the project stands at a certain point in time. Because progress and status reports are important ways to communicate project information, it is important for project teams to tailor them to meet their specific needs. Some organizations, such as JWD Consulting from Chapter 3, combine both progress and status information on the same template.

Table 10-3: Sample Template for a Monthly Progress Report

I. Accomplishments for Month of January (or appropriate month):

- Describe most important accomplishments. Relate them to project's Gantt chart.
- Describe other important accomplishments, one bullet for each. If any issues were resolved from the previous month, list them as accomplishments.

II. Plans for February (or following month):

- Describe most important items to accomplish in the next month. Again, relate them to the project's Gantt chart.
- Describe other important items to accomplish, one bullet for each.

III. Issues: Briefly list important issues that surfaced or are still important. Managers hate surprises and want to help the project succeed, so be sure to list issues.

IV. Project Changes (Dates and Description): List any approved or requested changes to the project. Include the date of the change and a brief description.

Table 10-4 provides a template for writing a letter of agreement for a class project. Many projects, such as class projects, do not have a contract or official project charter. Instead, they have a letter of agreement, which provides similar information in a friendlier and less formal manner than a contract or project charter. A letter of agreement can also be used to document and clarify goals and expectations for projects. Table 10-4 suggests that a letter of agreement for a class project include a project description (see Figure 10-2), the sponsoring organization's goals and expectations, the students' goals and expectations, meeting information, contact information and communications plans, and signatures of the main sponsor, student project manager, and all students on the project team.

Table 10-4: Sample Template for a Letter of Agreement for a Class Project

I.	**Project Description:** Describe the project's objective, scope, assumptions, cost information, and schedule information, as shown in Figure 10-2. Be sure to include important dates that the project's sponsor needs to be aware of.
II.	**Organizational Goals and Expectations:** Have the main sponsor from the organization briefly state his or her goals and expectations for the project.
III.	**Student Goals and Expectations:** Students on the project team should briefly state their goals and expectations for the project.
IV.	**Meeting Information:** The project's sponsor and all students on the project team should agree to meet for at least one hour per week when all parties can work on this project. The meeting place should be a convenient location without distractions. Virtual meetings may be an option for some people and projects.
V.	**Contact Information and Communications Plan:** List the sponsor's and students' names, phone numbers, e-mail addresses, and important procedures for communications. It is a good idea to set up a Web site for all project information.
VI.	**Signatures:** Have the main sponsor and students on the project team sign the letter of agreement. Designate which student is the project manager. This student should be the main contact for all project information.

Table 10-5 provides an outline for a final project report that could be used for a class project or other projects. It is important to include a cover page and table of contents. For a long report, it is also a good idea to include a one-page executive summary that highlights the most important information in the report. Other major sections of a final project report include a description of the need for the project, the original project description and letter of agreement (see Figure 10-2 and Table 10-4), the overall outcome of the project and reasons for success or failure, project management tools and techniques used and an assessment of them, project team recommendations and future considerations, the final project Gantt chart, and attachments with all project deliverables.

Table 10-5: Outline for a Final Project Report

I.	Cover page
II.	Table of contents and executive summary (for a long report)
III.	Need for the project
IV.	Project description and letter of agreement
V.	Overall outcome of the project and reasons for success or failure
VI.	Project management tools and techniques used and assessment of them
VII.	Project team recommendations and future considerations
VIII.	Final project Gantt chart
IX.	Attachments with all deliverables

For very large projects, the contents of a final project report might be similar to those listed in Table 10-5, but the report would probably include many more items, depending on the nature of the project. It is important to write a final report and archive all important project documents. Table 10-6 provides an exhaustive list of all of the documentation that should be organized and filed at the end of a major project. From this list, you can see that a large project can generate a lot of documentation. In fact, some project professionals have observed that documentation for designing an airplane usually weighs more than the airplane itself. (Smaller projects usually generate much less documentation!)

Table 10-6: Final Project Documentation Items

I.	Project description
II.	Project proposal and backup data (request for proposal, statement of work, proposal correspondence, and so on)
III.	Original and revised contract information and client acceptance documents
IV.	Original and revised project plans and schedules (WBS, Gantt charts and network diagrams, cost estimates, communications management plan, etc.)
V.	Design documents
VI.	Final project report
VII.	Deliverables, as appropriate
VIII.	Audit reports
IX.	Lessons-learned reports
X.	Copies of all status reports, meeting minutes, change notices, and other written and electronic communications

Figure 10-3 provides a template file created in Microsoft Project 2002 that students can use for class projects. People working in companies can develop and follow a similar template for their projects. Notice the main tasks in this template follow the project management process of project initiating, planning, executing, controlling, and closing. Milestones are included to highlight significant events, such as having the letter of agreement signed, or the completion of major deliverables. In addition, notice the column labeled Hyperlink Address. One of the biggest problems on projects has often been providing the most recent project plans, Gantt charts, specifications, meeting information, change requests, and so on to all or selective stakeholders in a timely fashion. Most project management software products allow users to insert hyperlinks to other project-related files. In Project 2002, you can insert a hyperlink into a Word file that contains the project charter or letter of agreement, as shown in this figure. You could also link to an Excel file that contains a staffing management plan or cost estimate, or link to a PowerPoint file with an important presentation. The

Project 2002 file and all associated hyperlinked files could then be placed on a local area network server or Web server so all project stakeholders can easily access important project information. In Figure 10-3, hyperlinks are included to template files for creating a letter of agreement, progress report, and final report. Project 2002 also comes with several template files for various types of projects, including a commercial construction project, an engineering project, an infrastructure deployment project, a project office project, a software development project, as well as several others. Many other Web sites provide sample files in Project 2002. Links to several of these sites are provided on the companion Web site for this text.

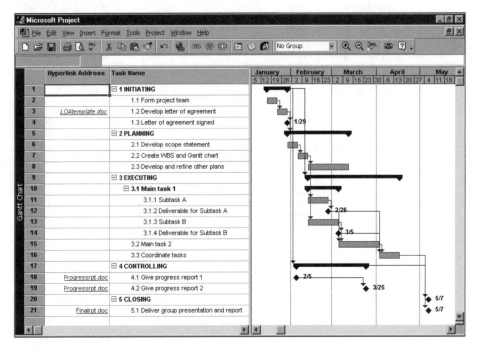

Figure 10-3. Gantt Chart Template for a Class Project

Table 10-7 provides guidance for writing a lessons-learned report. These particular guidelines apply to a project for students taking a class in project management. The lessons-learned report used by JWD Consulting in Chapter 3 had a different format. As you can see, templates, outlines, and guidelines for project documentation often need to be tailored to the particular organization and project.

Table 10-7: Guidance for Students' Lessons-Learned Report

Every two weeks or after a major event in your group project, write a brief journal entry describing what happened and how you felt about it. At the end of the term, write a 2- to 3-page paper describing your lessons learned based on your group project. Answer the following questions:

- What were your roles and responsibilities on the team? How were they decided?

- What did you like or dislike about the project?

- What did you learn about project management and yourself by doing the group project?

- What did you learn about teamwork and yourself by doing the group project?

- What would you have done differently? What will you remember to do on the next project you work on after this experience?

In the past few years, more and more project teams have started putting all or part of their project information, including various templates, on project Web sites. Project Web sites provide a centralized way of delivering project documents and other communications. Project teams can develop project Web sites using Web-authoring tools, such as Microsoft FrontPage or Macromedia Dreamweaver. Table 10-8 provides a sample template for a project Web site. The home page for the project Web site should include summary information about the project and other pertinent information. The home page should also include contact information, such as names and e-mail addresses for the project manager and Webmaster. Links should be provided to items such as the project description, team member list, project documents, templates, a discussions area, and other materials related to the project. If the project involves creating research reports, software, design documents, or other items that can also be accessed via the Web site, links can be provided to those files as well. The project team should also address other issues in creating and using a project Web site, such as security, access, and type of content that should be included on the site.

Table 10-8: Sample Template for a Project Web Site

PROJECT X WEB SITE	
Project Description	Welcome to the home page for Project X.
Team Members	

Project Documents
   Charter
   Gantt Chart
   Progress Reports
   Product Deliverables
   Final Report
	Provide contact information, such as the names, e-mail addresses, and telephone numbers of the project manager and Webmaster for the project Web site.
	
	Provide summary information about the project and other pertinent information.

Templates	
Discussions	
Related Links	

For more sophisticated Web sites, project teams can also use one of the many software products created specifically to assist in project communications through the Web. These products vary considerably in price and functionality, as described in Chapter 1.

When the project team develops their communications management plan, they should also determine what templates could be used for key documentation. To make it even more convenient to use templates, the organization should make project templates readily available online for all projects. The project team should also be certain to understand top management's and customers' documentation expectations for each particular project. For example, if a project sponsor or customer wants a one-page monthly progress report for a specific project, but the project team delivers a twenty-page report, there are communication problems. In addition, if particular customers or top managers want specific items in all final project reports, they should make sure the project team is aware of those expectations and modifies any templates for those reports to take these requirements into account.

Developing a Communications Infrastructure

A **communications infrastructure** is a set of tools, techniques, and principles that provides a foundation for the effective transfer of information among people. Tools include e-mail, project management software, groupware, fax machines, telephones, teleconferencing systems, document management systems, and word processing software. Techniques include reporting guidelines and templates, meeting ground rules and procedures, decision-making processes, problem-solving approaches, conflict resolution and negotiation techniques, and the like. Principles include providing an environment for open dialogue using "straight talk" and following an agreed-upon work ethic.

Since the defense industry has been involved in project management for a long time, many government and defense contractors have formal communications infrastructures already in place. For example, in the early 1980s—before most people ever used personal computers—the U.S. Air Force had standard forms for reporting project progress information, outlines for developing project final reports, regulations describing the progression of major projects, inspections of project archives, and forms and procedures for creating project Gantt charts. There were even special stickers you could place on Gantt charts to show changes, such as a white diamond for a slipped milestone. (No project manager wanted to ask for that particular sticker!)

Today, several organizations have designed customized systems for collecting and reporting project information, and most use various forms of information technology as part of their project communications infrastructure. Some organizations have intranets to keep track of all project information. Some use enterprise project management software while others use Lotus Notes or other groupware to maintain consistent and complete project information. Still other industries are developing communications infrastructures specifically to help in managing projects.

In his 1999 book, *Business @ the Speed of Thought: Using a Digital Nervous System*, Microsoft CEO Bill Gates suggests that organizations must develop a communications infrastructure or "digital nervous system" that allows for rapid movement of information inside a company as well as with customers, suppliers, and other business partners. Gates suggests that how organizations gather, manage, and use information to empower people will determine whether they win or lose in a competitive business environment.[9] In an interview, Gates mentioned systems to aid project management as an example of a great use of technology that really makes a difference in the companies that implement such systems.[10] In 2002, Microsoft released their enterprise version of project management software to address the need for this type of software. See Appendix A for more information describing the differences between the enterprise and standard versions of Microsoft Project 2002.

---

[9] Gates, Bill, *Business @ the Speed of Thought: Using a Digital Nervous System.* New York: Warner Books, 1999.

[10] Schlender, Brent, "E-Business According to Gates," *Fortune* (April 12, 1999), 73.

USING SOFTWARE TO ASSIST IN PROJECT COMMUNICATIONS

Even though organizations routinely use many types of hardware and software to enhance communications, they need to take advantage of new technologies and adjust existing systems to serve the special communications needs of customers and project teams. In addition to diverse customer and project needs, they also have to address the changing expectations of consumers and the workforce. For example, several television shows have harnessed communications technology to engage their audiences by letting them vote online or via telephone for their favorite singers, e-mail celebrities, or access information on their Web sites. If television shows can use advanced communications technologies, why can't their employers?

Employers have made changes to meet changing expectations and needs. Today, more than 37 percent of people telecommute or work remotely at least part-time, according to Gartner Inc.[11] On several information technology projects, project managers have found that their team members can be more productive when they are allowed to work from home. Other project managers have no choice in the matter when some or all of their project team members work remotely. As described in Chapter 9, Project Human Resource Management, studies show that providing a quiet work environment and a dedicated workspace increases programmer productivity. Most people who work from home have well-equipped, comfortable home offices with fewer distractions and more space than corporate offices. Workers also appreciate the added bonus of avoiding traffic and having more flexible work schedules. However, it is important to make sure work is well defined and communications are in place to allow for remote workers to work effectively.

What Went Right?

Frito-Lay, Inc. has taken advantage of various types of communication technology for decades. Its sales force used handheld computers in the mid-1980s. Today, Frito Lay is taking advantage of new technologies to manage its software development projects. "We do all our own development, and many of our developers work from home full- or part-time—some out of state," said Anthony King, director of software and engineering systems for Frito Lay in Tucson, Arizona. Frito Lay's applications were so large that using its virtual private network (VPN) was very slow, almost to the point of not being functional. In 2001, the company installed GoToMyPC from Expertcity Inc. in Santa Barbara, California. This Web-based technology allows developers to work much faster from a remote environment

---

[11] Melymuka, Kathleen, "Far from the mother ship: Managing remote workers," *Computerworld* (December 9, 2002).

and tracks users' online time. This feature came in handy when auditors wanted to know how King monitors remote workers.[12]

Everyday citizens use new communications technologies to bring people together from around the world to support projects, especially when it helps their loved ones. As service men and women were called to duty in the Persian Gulf in 2003, people used technology in innovative ways to show their support. NBC's "*Today*" show interviewed the wife of a serviceman who set up a Web site called Hugs to Kuwait to facilitate communications with her husband's unit and help their family members at home. She got so many people interested in what she was doing that groups from around the world decided to show their support. Girl Scout troops adopted units and sent soldiers letters of support as well as much appreciated Girl Scout cookies.

Several products are now available to assist individual consumers and organizations with communications. Many products have been developed or enhanced in the early 2000s to address the issue of providing fast, convenient, consistent, and up-to-date project information. Webcasts are now a common tool for presenting video, graphics, sound, voice, and participant feedback live over the Web. Most individuals have cell phones, and today it is not uncommon to see someone take and send a picture or send and receive text messages with a cell phone.

These same technologies can enhance project communications. For even more powerful and integrated communications, enterprise project management software provides many workgroup functions that allow a team of people at different locations to work together on projects and share project information. Workgroup functions allow the exchange of messages through e-mail, an intranet, or the Web. For example, you can also use Project 2002 to alert members about new or changed task assignments, and members can return status information and notify other workgroup members about changes in the schedule or other project parameters.

Microsoft's Enterprise Project Management solution also provides the following tools to enhance communications:

- Portfolio management—By providing a centralized and consolidated view of programs and projects, the user can evaluate and prioritize activities across the organization. This feature makes it possible to maximize productivity, minimize costs, and keep activities aligned with strategic objectives.

- Resource management—Maximizing human resources is often the key to minimizing project costs. This feature enables the user to maximize resource use across the organization to help effectively plan and manage the workforce.

[12] *Ibid.*

■ Project collaboration—Sharing project information is often a haphazard endeavor. Project collaboration enables an organization to share knowledge immediately and consistently to improve communications and decision making, eliminate redundancies, and take advantage of best practices for project management.[13]

Communication is among the more important factors for success in project management. Whereas technology can aid in the communications process and can be the easiest aspect of the process to address, it is not the most important aspect of the communications process. Far more important is improving an organization's ability to communicate. Improving the ability to communicate often requires a cultural change in an organization that takes a lot of time, hard work, and patience. Information technology personnel, in particular, often need special coaching to improve their communications skills. The project manager's chief role in the communications process is that of facilitator. Project managers must educate all stakeholders—management, team members, and customers—on the importance of good project communications and ensure that the project has a communications plan to help make good communication happen.

CASE WRAP-UP

Christine Braun worked closely with Peter Gumpert and his project managers to develop a communications plan for all of the fiber-optic undersea telecommunications systems projects. Peter was very skilled at running effective meetings, so everyone focused on meeting specific objectives. Peter stressed the importance of keeping himself, the project managers, and other major stakeholders informed about the status of all projects. He emphasized that the project managers were in charge of their projects, and that he did not intend to tell them how to do their jobs. He just wanted to have accurate and consistent information to help coordinate all of the projects and make everyone's jobs easier. When some of the project managers balked at the additional work of providing more project information in different formats, Peter openly discussed the issues with them in more detail. He then authorized each project manager to use additional staff to help develop and follow standards for all project communications.

[13] Microsoft, Enterprise Project Management (*www.microsoft.com/solutions/EPM/ evaluation/benefits*), (June 6, 2002).

Christine used her strong technical and communications skills to create a Web site that included samples of important project documents, presentations, and templates for other people to download and use on their own projects. After determining the need for more remote communications and collaboration between projects, Christine and other staff members researched the latest hardware and software products available. Peter authorized funds for a new project led by Christine to evaluate and then purchase several wireless handheld devices and enterprise project management software that could be accessed via the Web. Any project stakeholder could check out one of these handheld devices and get one-on-one training on how to use it with the new Web-based software. Even Peter learned how to use one and took it along on all of his business trips.

CHAPTER SUMMARY

A failure to communicate is often the greatest threat to the success of any project, especially information technology projects. Communication is the oil that keeps a project running smoothly.

Communication planning involves information distribution, performance reporting, and administrative closure. It requires determining the information and communications needs of the project stakeholders. A communications management plan should be created for all projects. A stakeholder analysis for project communications helps determine communications needs for different people involved in a project.

The various methods for distributing project information include formal, informal, written, and verbal. It is important to determine the most appropriate means for distributing different types of project information. Project managers and their teams should focus on the importance of building relationships as they communicate project information. As the number of people that need to communicate increases, the number of communications channels also increases.

Performance reporting involves collecting and disseminating information about how well a project is moving toward meeting its goals. Project teams can use earned value charts and other forms of progress information to communicate and assess project performance. Status review meetings are an important part of communicating, monitoring, and controlling projects.

Administrative closure involves generating, gathering, and disseminating information to formalize phase or project completion. The main customer for the project should formally accept the product or products produced by the project. Project archives and lessons-learned reports should be generated as part of administrative closure.

To improve project communications, project managers and their teams must develop good conflict management skills, as well as other communication skills. Conflict resolution is an important part of project communications management. The main causes of conflict during a project are schedules, priorities, staffing, technical opinions, procedures, cost, and personalities. A problem-solving approach to managing conflict is often the best approach. Other suggestions for improving project communications include learning how to run more effective meetings, how to use e-mail more effectively, how to use templates for project communications, and how to develop a communications infrastructure.

New hardware and software continues to become available to help improve communications. As more people work remotely, it is important to make sure they have the necessary tools to be productive. Enterprise project management software provides many features to enhance communications across the organization.

DISCUSSION QUESTIONS

1. Think of examples in the media that poke fun at the communications skills of technical professionals. How does poking fun at technical professionals' communications skills influence the industry and educational programs?

2. What are the key components of a project communications plan? How can a stakeholder analysis assist in preparing and implementing parts of this plan?

3. Discuss the advantages and disadvantages of different ways of distributing project information.

4. What are some of the ways to create and distribute project performance information?

5. Discuss the use of project archives, formal acceptance, and lessons learned as part of administrative closure. Give examples of how each of these items could be used in closing a course on project management.

6. Do you agree with some of the suggestions provided in this chapter for improving project communications? What other suggestions do you have?

7. How can software assist in project communications? How can it hurt project communications?

EXERCISES

1. Create a stakeholder analysis for project communications. Assume your organization has a project to determine employees' training needs and then provide in-house and external sources for courses in developing communications skills for employees. Stakeholders could be various levels and

types of employees, vendors, the Human Resources department in charge of the project, and so on. Determine specific stakeholders for the project. List at least one type of project communication for each stakeholder and the format for disseminating information to each. Use Table 10-1 as an example.

2. How many different communications channels does a project team with six people have? How many more communications channels would there be if the team grew to ten people?

3. Review the templates for various project documents provided in this chapter. Pick one of them and apply it to a project of your choice. Make suggestions for improving the template.

4. Write a lessons-learned report for a project of your choice using Table 10-7 as a guide. Do you think it is important for all project managers and team members to write lessons-learned reports? Would you take the time to read them if they were available in your organization?

5. Research software products that assist in communications management for large projects. Write a one- to two-page paper summarizing your findings. Include Web sites for software vendors and your opinion of some of the products.

6. Write a one- to two-page paper summarizing one of the suggested readings.

MINICASE

You are a member of a project team with twelve members. Several team members work at remote locations, so good virtual communication is crucial. Your team has decided to use a project Web site to store all project documentation and facilitate other communications. You have the task of determining the requirements for the project Web site, researching alternate solutions, making a recommendation, getting team consensus on what to implement, and then implementing the project Web site and acting as the Webmaster.

Part 1: Create a survey that you could use to solicit inputs on requirements for the project Web site from your team members. First, research information on how to design and administer surveys. One potential source of information is *www.surveysystem.com*. Then create the survey, either using word-processing software or Web-based survey software. Be sure to address a wide range of requirements, such as functionality, cost, security, and maintenance.

Part 2: Your project team has decided to use Web-authoring software to create its project Web site. Using the template provided in Table 10-8, create a project Web site, using your choice of Web-authoring software (for example, Microsoft FrontPage, Macromedia Dreamweaver, Allaire Homesite, or even Microsoft Word). Create the home page and pages for all of the links mentioned in the template. Test the Web pages to make sure they work, and then post them on the project Web site.

RUNNING CASES

All of the running cases in Appendix C include questions related to project communications management, including the use of templates provided in this chapter and Appendix D. Appendix A includes a section on using Microsoft Project 2002 to enhance project communications.

SUGGESTED READINGS

1. Bohner, Shawn. "PMOs: Projects in Harmony." META Group, *www.gantthead.com* (June 2000).

   As companies face increasingly complex initiatives involving multiple information technology projects, executives are turning to project management offices (PMOs) to improve communication, manage multiple risks, and target their vital resources. This article describes some of the patterns of evolution and maturity that are emerging.

2. Gates, Bill. *Business @ The Speed of Thought: Using a Digital Nervous System.* New York: Warner Books, 1999.

   The main premise of the Microsoft CEO's 1999 book is that the speed of business is accelerating at an ever-increasing rate, and to survive, organizations must develop a "digital nervous system" that allows for rapid movement of information. Gates suggests that how organizations gather, manage, and use information to empower people will determine whether they win or lose in the competitive business environment.

3. Kirchof, Nicki S. and John R. Adams. *Conflict Management for Project Managers.* Upper Darby, PA: Project Management Institute, 1989.

   This 53-page booklet contains information on the theory of conflict management, conflict in project organizations, conflict management and the project manager, and two-party conflict management. It also provides more information on Blake and Mouton's five basic modes for handling conflict and many references on the subject of conflict management.

4. Sivitanides, Marcos P., James R. Cook, Roy B. Martin, and Beverly A. Chiodo. "Verbal Communication Skills Requirements for Information Systems Professionals." *Journal of Information Systems Education*, *http://www.gise.org/JISE/Vol7/v71_7.htm* (Spring 1995), 7(1).

   This research investigates the verbal communication skills requirements for practicing professionals and new university graduates in information technology. Organizations can use the findings to identify areas of weakness in their information technology personnel's communication skills and improve them with continuing education. Universities can use the findings to identify and make adjustments for skills that they may not be currently emphasizing in their curricula.

5. Van Slyke, Erik J. "Resolving Team Conflict." *PMNetwork* (June 2000).

This article states that conflict is a regular part of managing projects. Project managers should encourage the positive and productive aspects of conflict and approach all conflict by seeking constructive resolution.

6. Wetherbe, Bond and James C. Wetherbe. *So, What's Your Point?: A Practical Guide to Learning and Applying Effective Techniques for Interpersonal Communication.* Houston, TX: Mead Publishing, 1996.

James Wetherbe is well known in the information technology profession for several of his books and entertaining speeches at various conferences. This short book uses a compelling and humorous approach to provide practical techniques for improving communication skills.

KEY TERMS

- **administrative closure** — generating, gathering, and disseminating information to formalize phase or project completion
- **communications infrastructure** — a set of tools, techniques, and principles that provide a foundation for the effective transfer of information among people
- **communications management plan** — a document that guides project communications
- **communications planning** — determining the information and communications needs of the stakeholders: who needs what information, when will they need it, and how will the information be given to them
- **compromise mode** — using a give-and-take approach to resolving conflicts; bargaining and searching for solutions that bring some degree of satisfaction to all the parties in a dispute
- **confrontation mode** — directly facing a conflict using a problem-solving approach that allows affected parties to work through their disagreements
- **forcing mode** — using a win-lose approach to conflict resolution to get one's way
- **formal acceptance** — documentation that the project's sponsor or customer signs to show they have accepted the products of the project
- **groupthink** — conformance to the values or ethical standards of a group
- **information distribution** — making needed information available to project stakeholders in a timely manner
- **kickoff meeting** — a meeting held at the beginning of a project or project phase where all major project stakeholders discuss project objectives, plans, and so on
- **lessons learned** — reflective statements written by project managers and their team members

- **performance reporting** — collecting and disseminating performance information, which includes status reports, progress measurement, and forecasting
- **progress reports** — reports that describe what the project team has accomplished during a certain period of time
- **project archives** — a complete set of organized project records that provide an accurate history of the project
- **project forecasting** — predicting future project status and progress based on past information and trends
- **smoothing mode** — deemphasizing or avoiding areas of differences and emphasizing areas of agreements
- **status reports** — reports that describe where the project stands at a specific point in time
- **withdrawal mode** — retreating or withdrawing from an actual or potential disagreement

11

Project Risk Management

Objectives

After reading this chapter, you will be able to:

1. *Understand what risk is and the importance of good project risk management*
2. *Discuss the elements involved in risk management planning*
3. *List common sources of risks on information technology projects*
4. *Describe the risk identification process and tools and techniques to help iden-tify project risks*
5. *Discuss the qualitative risk analysis process and explain how to calculate risk factors, use probability/impact matrixes, the Top Ten Risk Item Tracking tech-nique, and expert judgment to rank risks*
6. *Explain the quantify risk analysis process and how to use decision trees and simulation to quantitative risks*
7. *Provide examples of using different risk response planning strategies such as risk avoidance, acceptance, transference, and mitigation*
8. *Discuss what is involved in risk monitoring and control*
9. *Describe how software can assist in project risk management*
10. *Explain the results of good project risk management*

OPENING CASE

Cliff Branch was the president of a small information technology consulting firm that specialized in developing Internet applications and providing full-service support. The staff consisted of program-mers, business analysts, database specialists, Web designers, project managers, and so on. The firm had fifty people and planned to hire at least ten more in the next year. The company had done very well the past few years, but was recently having difficulty winning con-tracts. Spending time and resources to respond to various requests for proposals from prospective clients was becoming expensive. Many clients were starting to require presentations and even some prototype development before awarding a contract.

Cliff knew he had a fairly aggressive approach to risk and liked to bid on the projects with the highest payoff. He did not use a systematic approach to evaluating the risks involved in various projects before bidding on them. He focused on the profit potentials and on how challenging the projects were. His strategy was now causing problems for the company because it was investing heavily in the preparation of proposals, yet winning few contracts. Several consultants were not currently assigned to projects but were still on the payroll. What could Cliff and his company do to get a better understanding of project risks? Should Cliff adjust his strategy for deciding what projects to pursue? How?

THE IMPORTANCE OF PROJECT RISK MANAGEMENT

Project risk management is the art and science of identifying, analyzing, and responding to risk throughout the life of a project and in the best interests of meeting project objectives. A frequently overlooked aspect of project management, risk management can often result in significant improvements in the ultimate success of projects. Risk management can have a positive impact on selecting projects, determining the scope of projects, and developing realistic schedules and cost estimates. It helps project stakeholders understand the nature of the project, involves team members in defining strengths and weaknesses, and helps to integrate the other project management knowledge areas.

All industries, especially the software development industry, tend to neglect the importance of project risk management. As mentioned in Chapter 8, Project Quality Management, William Ibbs and Young H. Kwak performed a study to assess project management maturity. The thirty-eight organizations participating in the study were divided into four industry groups: engineering and construction, telecommunications, information systems/software development, and high-tech manufacturing. Survey participants answered 148 multiple-choice questions to assess how mature their organization was in the project management knowledge areas of scope, time, cost, quality, human resources, communications, risk, and procurement. The rating scale ranged from 1 to 5, with 5 being the highest maturity rating. Table 11-1 shows the results of the survey. Notice that risk management was the only knowledge area for which all ratings were less than 3. This study shows that all organizations should put more effort into project risk management, especially the information systems industry, which had the lowest rating of 2.75.[1]

Table 11-1: Project Management Maturity by Industry Group and Knowledge Area

KEY: 1 = LOWEST MATURITY RATING   5 = HIGHEST MATURITY RATING

KNOWLEDGE AREA	ENGINEERING/ CONSTRUCTION	TELECOMMUNICATIONS	INFORMATION SYSTEMS	HI-TECH MANUFACTURING
Scope	3.52	3.45	3.25	3.37
Time	3.55	3.41	3.03	3.50
Cost	3.74	3.22	3.20	3.97
Human Resource	3.18	3.20	2.93	3.18
Communications	3.53	3.53	3.21	3.48
Risk	**2.93**	**2.87**	**2.75**	**2.76**
Procurement	3.33	3.01	2.91	3.33

Ibbs, C. William and Young Hoon Kwak, "Assessing Project Management Maturity," *Project Management Journal* (March 2000).

Other studies also note the lack of attention paid to project risk management. KPMG, a large consulting firm, did a study that found that 55 percent of **runaway projects**—projects that have significant cost or schedule overruns—did no risk management at all, 38 percent did some (but half did not use their risk findings after the project was underway), and 7 percent did not know whether they did risk management or not.[2] This study suggests that performing risk management is important to improving the likelihood of project success and preventing runaway projects.

Before you can improve project risk management, you must understand what risk is. A basic dictionary definition says that **risk** is "the possibility of loss or injury." This definition highlights the negativity often associated with risk and suggests that uncertainty is involved. Project risk involves understanding potential problems that might occur on the project and how they might impede project success.

In many respects, risk management is like a form of insurance. It is an activity undertaken to lessen the impact of potentially adverse events on a project. It is important to note that risk management *is* an investment—there are costs associated with it. The investment a project is willing to make in risk management activities depends on the nature of the project, the experience of the

---

[1] Ibbs, C. William and Young Hoon Kwak, "Assessing Project Management Maturity," *Project Management Journal* (March 2000).

[2] McConnell, Steve, "10 Keys to Successful Software Projects Presentation," *Construx* (2000).

project team, and the constraints imposed on both. In any case, the cost for risk management should not exceed the potential benefits.

Why do organizations continue to introduce new systems, then, if they often entail the possibility of loss or injury? If the information technology project failure rate is so high, why do companies continue to start new information technology projects? Many companies are in business today because they took risks that created great opportunities. Organizations survive over the long term when they pursue opportunities. Information technology is often a key part of a business's strategy; without it, many businesses might not survive. Given that all projects involve risks and opportunities, the question is how to decide which projects to pursue and how to manage project risk throughout a project's life cycle.

Several risk experts suggest that organizations and individuals strive to find a balance between risks and opportunities in all aspects of projects and their personal lives. The idea of striving to balance risks and opportunities suggests that different organizations and people have different tolerances for risk. Some organizations or people have a neutral tolerance for risk, some have an aversion to risk, and others are risk-seeking. These three preferences for risk are part of the utility theory of risk.

Risk utility or **risk tolerance** is the amount of satisfaction or pleasure received from a potential payoff. Figure 11-1 shows the basic difference between risk-averse, risk-neutral, and risk-seeking preferences. The y-axis represents utility, or the amount of pleasure received from taking a risk. The x-axis shows the amount of potential payoff, opportunity, or dollar value of the opportunity at stake. Utility rises at a decreasing rate for a **risk-averse** person. In other words, when more payoff or money is at stake, a person or organization that is risk-averse gains less satisfaction from the risk, or has lower tolerance for the risk. Those who are **risk-seeking** have a higher tolerance for risk, and their satisfaction increases when more payoff is at stake. A risk-seeking person prefers outcomes that are more uncertain and is often willing to pay a penalty to take risks. A **risk-neutral** person achieves a balance between risk and payoff.

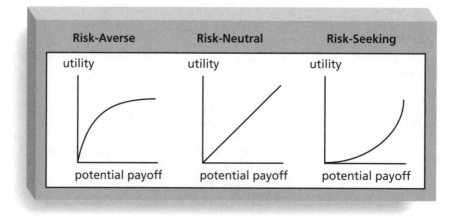

Figure 11-1. Risk Utility Function and Risk Preference

The goal of project risk management can be viewed as minimizing potential risks while maximizing potential opportunities or payoffs. There are six major processes involved in risk management:

- **Risk management planning** involves deciding how to approach and plan the risk management activities for the project. By reviewing the project charter, work breakdown structure (WBS), roles and responsibilities, stakeholder risk tolerances, and the organization's risk management policies, the project team can formulate a risk management plan.
- **Risk identification** involves determining which risks are likely to affect a project and documenting the characteristics of each. The risk management plan, outputs from project planning, risk categories, and historical information are key inputs to the process of identifying risks.
- **Qualitative risk analysis** involves characterizing and analyzing risks and prioritizing their effects on project objectives. After identifying risks, project teams can use various tools and techniques to develop an overall risk ranking for the project.
- **Quantitative risk analysis** involves measuring the probability and consequences of risks and estimating their effects on project objectives. After identifying risks, project teams can use additional tools and techniques to prioritize quantified risks and estimate probabilities of achieving project objectives.
- **Risk response planning** involves taking steps to enhance opportunities and reduce threats to meeting project objectives. Using outputs from the preceding risk management processes, project teams can develop a risk response plan.
- **Risk monitoring and control** involves monitoring known risks, identifying new risks, reducing risks, and evaluating the effectiveness of risk reduction throughout the life of the project. The main outputs of this process include corrective actions in response to risks and updates to the risk response plan.

The first step in project risk management is deciding how to address this knowledge area for your particular project by performing risk management planning.

RISK MANAGEMENT PLANNING

Risk management planning is the process of deciding how to approach and plan for risk management activities for a project, and the main output of this process is a risk management plan. A **risk management plan** documents the procedures for managing risk throughout the project. Project teams should hold several planning meetings to help develop the risk management plan. The project team should review project documents such as the project charter, WBS, and definitions of roles and responsibilities, as well as organizational documents such as corporate risk management policies and templates for creating a risk management plan. It is also important to review the risk tolerances of various stakeholders. For example, if the project sponsor is risk-averse, the project might require a different approach to risk management than if the project sponsor were a risk seeker.

A risk management plan summarizes the results of the risk identification, qualitative analysis, quantitative analysis, response planning, and monitoring and control processes. Table 11-2 lists the questions that a risk management plan should address. It is important to define specific deliverables for the project related to risk, assign people to work on those deliverables, and evaluate milestones associated with the risk mitigation approach. **Risk mitigation** is reducing the impact of a risk event by reducing the probability of its occurrence. The level of detail included in the risk management plan will vary with the needs of the project.

Table 11-2: Questions Addressed in a Risk Management Plan

- Why is it important to take/not take this risk in relation to the project objectives?
- What is the specific risk, and what are the risk mitigation deliverables?
- How is the risk going to be mitigated? (What risk mitigation approach is to be used?)
- Who are the individuals responsible for implementing the risk management plan?
- When will the milestones associated with the risk mitigation approach occur?
- How much is required in terms of resources to mitigate risk?

The risk management plan can include a methodology for risk management, roles and responsibilities for activities involved in risk management, budgets and schedules for the risk management activities, descriptions of scoring and interpretation methods used for the qualitative and quantitative analyses of risk, threshold criteria for risks, reporting formats for risk management activities, and a description of how the project team will track and document risk activities.

In addition to a risk management plan, many projects also include contingency plans, fallback plans, and contingency reserves. **Contingency plans** are predefined actions that the project team will take if an identified risk event occurs. For example, if the project team knows that a new release of a software package may not be available in time for them to use it for their project, they might have a contingency plan to use the existing, older version of the software. **Fallback plans** are developed for risks that have a high impact on meeting project objectives, and are put into effect if attempts to reduce the risk are not effective. For example, a new college graduate might have a main plan and several contingency plans on where to live after graduation, but if none of those plans works out, a fallback plan might be to live at home for a while. Sometimes the terms contingency plan and fallback plan are used interchangeably. **Contingency reserves** or **contingency allowances** are provisions held by the project sponsor that can be used to mitigate cost or schedule risk if changes in project scope or quality occur. For example, if a project appears to be off course because the staff is inexperienced with some new technology, the project sponsor may provide additional funds from contingency reserves to hire an outside consultant to train and advise the project staff in using the new technology.

Before you can begin to understand and use the other project risk management processes on information technology projects, you must be able to recognize and understand the common sources of risk.

COMMON SOURCES OF RISK ON INFORMATION TECHNOLOGY PROJECTS

Several studies have shown that information technology projects share some common sources of risk. For example, the Standish Group did a follow-up study to the 1995 CHAOS research, which they called Unfinished Voyages. This study brought together sixty information technology professionals to elaborate on how to evaluate a project's overall likelihood of being successful. Table 11-3 shows the Standish Group's success potential scoring sheet and shows the relative importance of the project success criteria factors.[3]

Table 11-3: Information Technology Success Potential Scoring Sheet

SUCCESS CRITERION	RELATIVE IMPORTANCE
User Involvement	19
Executive Management Support	16
Clear Statement of Requirements	15
Proper Planning	11
Realistic Expectations	10
Smaller Project Milestones	9
Competent Staff	8
Ownership	6
Clear Visions and Objectives	3
Hardworking, Focused Staff	3
Total	**100**

[3] The Standish Group, "Unfinished Voyages" (1996), *www.standishgroup.com/voyages.html*.

The Standish Group provides specific questions for each success criterion to help decide the number of points to assign to a project. For example, the five questions related to user involvement include the following:

- Do I have the right user(s)?
- Did I involve the user(s) early and often?
- Do I have a quality user(s) relationship?
- Do I make involvement easy?
- Did I find out what the user(s) need(s)?

The number of questions corresponding to each success criterion determines the number of points each positive response is assigned. For example, in the case of user involvement there are five questions. For each positive reply, you would get 3.8 (19/5) points; 19 represents the weight of the criterion, and 5 represents the number of questions. Therefore, you would assign a value to the user involvement criterion by adding 3.8 points to the score for each question you can answer positively.

Another risk questionnaire developed by F. W. McFarlan and the Dayton Tire Co. can be used to identify the major sources of risk in the categories of people, structure, and technology.[4] Risks in the people category include inadequate skills (both technical and managerial), inexperience in general, and inexperience in a specific application area or technology. Structural risk includes the degree of change a new project will introduce into user areas and business procedures, the number of distinct user groups the project must satisfy, and the number of other systems with which the new project must interact. The experience of the organization with the project's technology and with other similar projects also affects structural risk. Technological risk involves using new or untried technology.

Table 11-4 provides samples of questions from McFarlan's risk questionnaire. You can select more than one answer for a question. After you answer, multiply the total points by the question's weight (not shown here). McFarlan suggests comparing your scores in each risk area—people, structure, and technology—to the norms for the organization. High scores should warn you that high risk is involved. Review your answers to the questions in the high-risk areas to determine the causes for increased risk. For example, perhaps the project could be restructured to use less new technology or to reduce the size of the project or number of user areas involved.

---

[4] McFarlan, F. W., "Portfolio Approach to Information Systems," *Harvard Business Review* (September–October 1981): 142–150.

Table 11-4: McFarlan's Risk Questionnaire

WHAT IS THE PROJECT ESTIMATE IN CALENDAR (ELAPSED) TIME?

( ) 12 months or less	Low = 1 point
( ) 13 months to 24 months	Medium = 2 points
( ) Over 24 months	High = 3 points

WHAT IS THE ESTIMATED NUMBER OF PERSON DAYS FOR THE SYSTEM?

( ) 12 to 375	Low = 1 point
( ) 375 to 1875	Medium = 2 points
( ) 1875 to 3750	Medium = 3 points
( ) Over 3750	High = 4 points

NUMBER OF DEPARTMENTS INVOLVED (EXCLUDING INFORMATION TECHNOLOGY)

( ) One	Low = 1 point
( ) Two	Medium = 2 points
( ) Three	High = 3 points

IS ADDITIONAL HARDWARE REQUIRED FOR THE PROJECT?

( ) None	Low = 0 points
( ) Central processor type change	Low = 1 point
( ) Peripheral/storage device changes	Low = 1 point
( ) Terminals	Med = 2 points
( ) Change of platform, for example, PCs replacing mainframes	High = 3 points

Other broad categories of risk include:

- *Market risk:* If the information technology project is to produce a new product or service, will it be useful to the organization or marketable to others? Will users accept and use the product or service? Will someone else create a better product or service faster, making the project a waste of time and money?

- *Financial risk:* Can the organization afford to undertake the project? How confident are stakeholders in the financial projections? Will the project meet NPV, ROI, and payback estimates? If not, can the organization afford to continue the project? Is this project the best way to use the organization's financial resources?

- *Technology risk:* Is the project technically feasible? Will hardware, software, and networks function properly? Will the technology be available in time to meet project objectives? Could the technology be obsolete before a useful product can be produced?

What Went Wrong?

Many information technology projects fail because of technology risk. One project manager learned an important lesson on a large information technology project: focus on business needs first, not technology. David Anderson, a project manager for Kaman Sciences Corp., shared his experience with a project failure in an article for *CIO Enterprise Magazine*. After spending two years and hundreds of thousands of dollars on a project to provide new client/server-based financial and human resources information systems for their company, Anderson and his team finally admitted they had a failure on their hands. Anderson revealed that he had been too enamored of the use of cutting-edge technology and had taken a high-risk approach on the project. He "ramrodded through" what the project team was going to do and then admitted that he was wrong. The company finally decided to switch to a more stable technology to meet its business needs.[5]

Reviewing a proposed project in terms of the Standish Group's success criteria, McFarlan's risk questionnaire, or any other similar tool is a good method for understanding common sources of risk on information technology projects. Understanding common sources of risk also helps in risk identification, the next step in project risk management.

RISK IDENTIFICATION

Identifying risks is the process of understanding what potential unsatisfactory outcomes are associated with a particular project. By reviewing a project's risk management plan, other planning documents, and the broad categories of risk, project managers and their teams can often identify several potential risks. A review of historical information related to risks on similar projects is also an important input to the risk identification process. Understanding historical risks can help project managers learn from the past and thereby identify and reduce risks on current and future projects.

In addition to identifying risk based on the nature of the project or products produced, it is also important to identify potential risks according to project management knowledge areas, such as scope, time, cost, and quality. Table 11-5 lists potential risk conditions that can exist within each knowledge area.[6]

[5] Hildebrand, Carol, "If At First You Don't Succeed," *CIO Enterprise Magazine* (April 15, 1998).
[6] Wideman, R. Max, *Project and Program Risk Management: A Guide to Managing Project Risks and Opportunities*, Upper Darby, PA: Project Management Institute, 1992, II–4.

Table 11-5: Potential Risk Conditions Associated with Each Knowledge Area

KNOWLEDGE AREA	RISK CONDITIONS
Integration	Inadequate planning; poor resource allocation; poor integration management; lack of post-project review
Scope	Poor definition of scope or work packages; incomplete definition
Time	Errors in estimating time or resource availability; errors in determining the critical path; poor allocation and management of float; early release of competitive products
Cost	Estimating errors; inadequate productivity, cost, change, or contingency
Quality	Poor attitude toward quality; substandard design/materials/ workmanship; inadequate quality assurance program
Human Resources	Poor conflict management; poor project organization and definition of responsibilities; absence of leadership
Communications	Carelessness in planning or communicating; lack of consultation with key stakeholders
Risk	Ignoring risk; unclear analysis of risk; poor insurance management
Procurement	Unenforceable conditions or contract clauses; adversarial relations

There are several tools and techniques for identifying risks. Project teams often begin the risk identification process by reviewing project documentation, analyzing project assumptions, and gathering risk-related information. For example, after reviewing project documents and assumptions, project team members and outside experts often hold meetings to discuss these documents and assumptions and ask important questions about them as they relate to risk. After identifying potential risks at this initial meeting, the project team might then use different information-gathering techniques to further identify risks. Four common information-gathering techniques include brainstorming, the Delphi Technique, interviewing, and SWOT analysis.

Brainstorming is a technique by which a group attempts to generate ideas or find a solution for a specific problem by amassing ideas spontaneously and without judgment. This approach could be used to create a comprehensive list of risks that could be addressed later in the qualitative and quantitative risk analysis processes. Care must be taken, however, not to overuse or misuse brainstorming. Although businesses use brainstorming widely to generate new ideas, the psychology literature shows that individuals, working alone, produce a greater number of ideas than the same individuals produce through brainstorming in small face-to-face groups. Group effects, such as fear of social disapproval, the effects of authority hierarchy, and domination of the session by one or two very vocal people often inhibit idea generation for many participants.[7]

[7] Couger, J. Daniel, *Creative Problem Solving and Opportunity Finding*, Boston: Boyd & Fraser Publishing Company, 1995.

One common approach to gathering information from experts is the Delphi Technique. The basic concept of the **Delphi Technique** is to derive a consensus among a panel of experts who make predictions about future developments. Developed by the Rand Corporation for the U.S. Air Force in the late 1960s, the Delphi Technique is a systematic, interactive forecasting procedure based on independent and anonymous input regarding future events. The Delphi Technique uses repeated rounds of questioning and written responses, including feedback to earlier-round responses, to take advantage of group input, while avoiding the biasing effects possible in oral panel deliberations. To use the Delphi Technique, you must select a panel of experts for the particular area in question. For example, Cliff Branch from the opening case could use the Delphi Technique to help him understand why his company is no longer winning many contracts. Cliff could assemble a panel of people with knowledge in his business area. Each expert would answer questions related to Cliff's scenario, and then Cliff or a facilitator would evaluate their responses, together with opinions and justifications, and provide that feedback to each expert in the next iteration. Cliff would continue this process until the group responses converge to a specific solution. If the responses diverge, the facilitator of the Delphi Technique needs to determine if there is a problem with the process.

Interviewing is a fact-finding technique for collecting information in face-to-face, telephone, e-mail, or instant messaging discussions. Interviewing people with similar project experience is an important tool for identifying potential risks. For example, if a new project involves using a particular type of hardware or software, someone with recent experience with that hardware or software could describe problems he or she had on a past project. If someone has worked with a particular customer, he or she might provide insight into potential risks involved in working for that group again.

As described in Chapter 5, Project Scope Management, strengths, weaknesses, opportunities, and threats (SWOT) analysis is often used in strategic planning. It can also assist in risk identification by having project teams focus on the broad perspectives of potential risks for particular projects. For example, before writing a particular proposal, Cliff Branch could have a group of his employees discuss in detail what their company's strengths are, what their weaknesses are related to that project, and what opportunities and threats exist. Do they know that several competing firms are much more likely to win a certain contract? Do they know that winning a particular contract will likely lead to future contracts and help expand their business? Applying SWOT to specific potential projects can help identify the broad risks and opportunities that apply in that scenario.

Two other techniques for risk identification include the use of checklists and diagrams. Checklists based on risks that have been encountered in previous projects provide a meaningful template for understanding risks in a current project. You can use checklists similar to those developed by the Standish Group, McFarlan, or other groups to help identify risks on information technology projects. Diagramming techniques include using cause-and-effect diagrams or Fishbone diagrams, flow charts, and influence diagrams. Recall from Chapter 8, Project Quality Management, that Fishbone diagrams help you trace problems back to their root cause. System or process **flow charts** are diagrams that show how different parts of a system interrelate. For example, many programmers create flow charts to show programming logic. Another type of diagram, an **influence diagram**, represents decision problems by displaying essential elements, including decisions, uncertainties, and objectives, and how they influence each other.[8]

The main outputs of the risk identification process are identified risk events for the project, triggers or risk symptoms, and inputs to other processes, such as updates to the WBS or schedule based on identified risks. **Risk events** are specific things that may occur to the detriment of the project. Examples of risk events are significant changes in scope, the performance failure of products produced as part of a project, specific delays in the project due to rejection of work or labor unavailability, supply shortages, litigation against the company, strikes, and so on. **Triggers**, or **risk symptoms**, are indicators of actual risk events. For example, cost overruns on early activities may be symptoms of poor cost estimates. Defective products may be symptoms of a low-quality supplier. Documenting potential risk symptoms for projects also helps the project team identify potential risk events and determine what response to take. In the opening case, the fact that Cliff Branch's firm was not winning more contracts may have been a symptom that there was more competition in their market or that they were not writing competitive proposals.

QUALITATIVE RISK ANALYSIS

Qualitative risk analysis involves assessing the likelihood and impact of identified risks, to determine their magnitude and priority. This section describes examples of using a probability/impact matrix to produce a prioritized list of risks and examples of using the Top Ten Risk Item Tracking technique to produce an overall ranking for project risks and to track trends in qualitative risk analysis. It also discusses the importance of expert judgment in performing risk analysis.

[8] Lumina Analytical, *Influence Diagrams* (2001), *www.lumina.com/software/ influencediagrams.html.*

Using Probability/Impact Matrixes to Calculate Risk Factors

People often describe a risk probability or consequence as being high, medium or moderate, or low. For example, a meteorologist might predict that there is a high probability, or likelihood, of severe rain showers on a certain day. If that day happens to be your wedding day and you are planning a large outdoor wedding, the consequences or impact of severe showers might also be high.

A project manager can chart the probability and impact of risks on a **probability/impact matrix** or **chart**. A probability/impact matrix or chart lists the relative probability of a risk occurring on one side of a matrix or axis on a chart and the relative impact of the risk occurring on the other. Many project teams would benefit from using this simple technique to help them identify risks that they need to pay attention to. To use this approach, project stakeholders list the risks they think might occur on their projects. They then label each risk as being high, medium, or low in terms of its probability of occurrence and its impact if it did occur.

The project manager then summarizes the results in a probability/impact matrix or chart, as shown in Figure 11-2. For example, Cliff Branch and some of his project managers in the opening case could each identify three potential risks for their company or a particular project. They could then label each risk as being high, medium, or low in terms of probability and impact. For example, one project manager might identify a risk of a severe market downturn. He or she might think that particular risk is low in probability and high in impact. Cliff may have listed the risk of a severe market downturn as being medium in both probability and impact. The team could then plot all of the risks on a matrix or chart using sticky notes. As a group, they could then combine any common risks and decide where those risks should be on the matrix or chart, if there were differences of opinion. After agreeing on the risks and their placement on the chart, the team should then focus on any risks that fall in the high sections of the probability/impact matrix or chart. For example, Risks 1 and 4 are listed as high in both categories of probability and impact. Risk 6 is high in the probability category but low in the impact category. Risk 9 is high in the probability category and medium in the impact category. Risk 12 is low in the probability category and high in the impact category. The team should then discuss how they plan to respond to those risks if they occur, as discussed later in this chapter in the section on risk response planning.

Figure 11-2. Sample Probability/Impact Matrix

A more sophisticated approach to using probability/impact information is to calculate risk factors. To quantify risk probability and consequence, the Defense Systems Management College (DSMC) developed a technique for calculating **risk factors**—numbers that represent the overall risk of specific events, based on their probability of occurring and the consequences to the project if they do occur. The technique makes use of a probability/impact matrix that shows the probability of risks occurring and the impact or consequences of the risks.

Probabilities of a risk occurring can be estimated based on several factors, as determined by the unique nature of each project. For example, factors evaluated for potential hardware or software technology risks could include the technology not being mature, the technology being too complex, and an inadequate support base for developing the technology. The impact of a risk occurring could include factors such as the availability of fallback solutions or the consequences of not meeting performance, cost, and schedule estimates.

Table 11-6 provides an example of two probability/impact matrixes that were used in assessing the risk of various technologies that could make aircraft more reliable. The U.S. Air Force sponsored a multimillion-dollar research project in the mid-1980s, the High Reliability Fighter study, to evaluate potential technologies for improving the reliability of fighter aircraft. Many aircraft were too often unfit for flying because of reliability problems with various components of the planes and problems with their maintenance. The purpose of the High Reliability Fighter study was to help the U.S. Air Force decide what technologies to invest in to make planes more reliable and maintainable, thereby decreasing down time. The matrixes in Table 11-6 were used to assess the risk of proposed technologies, such as providing radial tires for planes, using a more efficient fuel system, or developing a sophisticated onboard computer system to monitor and adjust various systems in the aircraft. The top matrix in Table 11-6 was used to assign a Probability of Failure (Pf) value to each proposed technology, and the bottom matrix was used to assign a Consequence of Failure (Cf) value. Experts used their judgment to assign a value for both the probability and impact of each proposed technology. For example, an expert assigned a Pf value of .1 for the radial tires technology since the hardware existed, had a simple design, and had multiple programs and services that used it. Likewise, the expert assigned a Cf value of .9 to a risk that had no acceptable alternatives, would increase project life cycle costs, was unlikely to meet schedule dates, and might not reduce the aircraft down time factor. These values were then used in a formula to calculate an overall risk factor. A risk factor is defined as the probability of failure (Pf) plus the consequence of failure (Cf) minus the product of the two.[9] For example, a technology with a Pf of .1 and a Cf of .9 would have a risk factor of .91, or (.1 + .9) - (.1 * .9) = .91.

[9] Defense Systems Management College, *Systems Engineering Management Guide*, Washington, D.C., 1989.

Table 11-6: Sample Probability/Impact Matrixes for Qualitative Risk Assessment

PROBABILITY OF FAILURE (PF) ATTRIBUTES OF SUGGESTED TECHNOLOGY

VALUE	MATURITY HARDWARE/SOFTWARE	COMPLEXITY HARDWARE/SOFTWARE	SUPPORT BASE
0.1	Existing	Simple design	Multiple programs and services
0.3	Minor redesign	Somewhat complex	Multiple programs
0.5	Major change feasible	Fairly complex	Several parallel programs
0.7	Complex hardware design/ new software similar to existing	Very complex	At least one other program
0.9	Some research completed/ never done before	Extremely complex	No additional programs

CONSEQUENCE OF FAILURE (CF) ATTRIBUTES OF SUGGESTED TECHNOLOGY

VALUE	FALLBACK SOLUTIONS	LIFE CYCLE COST (LCC) FACTOR	SCHEDULE FACTOR (INITIAL OPERATIONAL CAPABILITY = IOC)	DOWNTIME (DT) FACTOR
0.1	Several acceptable alternatives	Highly confident will reduce LCC	90–100% confident will meet IOC significantly	Highly confident will reduce DT
0.3	A few known alternatives	Fairly confident will reduce LCC	75–90% confident will meet IOC	Fairly confident will reduce DT significantly
0.5	Single acceptable alternative	LCC will not change much	50–75% confident will meet IOC	Highly confident will reduce DT somewhat
0.7	Some possible alternatives	Fairly confident will increase LCC	25–50% confident will meet IOC	Fairly confident will reduce DT somewhat
0.9	No acceptable alternatives	Highly confident will increase LCC	0–25% confident will meet IOC	DT may not be reduced much

Figure 11-3 provides an example of how the risk factors were used to graph the probability of failure and consequence of failure for proposed technologies in this study. The figure classifies potential technologies (dots on the chart) as high-, medium-, or low-risk, based on the probability of failure and consequences of failure. The researchers for this study highly recommended that the U.S. Air Force invest in the low- to medium-risk technologies and suggested that it not pursue the high-risk technologies.[10] You can see that the rigor behind using the probability/impact matrix and risk factors provides a much stronger argument than simply stating that risk probabilities or consequences are high, medium, or low.

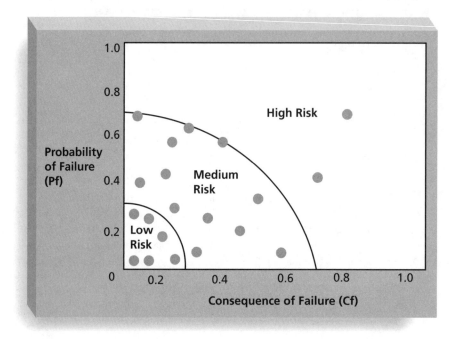

Figure 11-3. Chart Showing High-, Medium-, and Low-Risk Technologies

Top Ten Risk Item Tracking

Top Ten Risk Item Tracking is a qualitative risk analysis tool, and in addition to identifying risks, it maintains an awareness of risks throughout the life of a project. It involves establishing a periodic review of the project's most significant risk items with management and, optionally, with the customer. The review begins with a summary of the status of the top ten sources of risk on the project. The summary includes each item's current ranking, previous ranking, number of times it appears on the list over a period of time, and a summary of progress made in resolving the risk item since the previous review. The Microsoft Solution Framework (MSF) includes a risk management model that includes developing and monitoring a top ten master list of risks. MSF is the

[10] McDonnell Douglas Corporation, "Hi-Rel Fighter Concept," Report MDC B0642, 1988.

methodology Microsoft uses for managing projects. It combines aspects of software design and development and building and deploying infrastructure into a single project lifecycle for guiding technology solutions of all kinds. Consult the Microsoft Web site for more information on MSF (*www.microsoft.com/msf*).

Table 11-7 provides an example of a Top Ten Risk Item Tracking chart that could be used at a management review meeting for a project. This particular example includes only the top five risk items. Notice that each risk item is ranked on the basis of the current month, previous month, and how many months it has been in the top ten. The last column briefly describes the progress for resolving each particular risk item.

Table 11-7: Example of Top Ten Risk Item Tracking

	MONTHLY RANKING			
RISK ITEM	THIS MONTH	LAST MONTH	NUMBER OF MONTHS	RISK RESOLUTION PROGRESS
Inadequate planning	1	2	4	Working on revising the entire project plan
Poor definition	2	3	3	Holding meetings with project customer and sponsor to clarify scope
Absence of leadership	3	1	2	After previous project manager quit, assigned a new one to lead the project
Poor cost estimates	4	4	3	Revising cost estimates
Poor time estimates	5	5	3	Revising schedule estimates

A risk management review accomplishes several objectives. First, it keeps management and the customer (if included) aware of the major influences that could prevent the project from being a success. Second, by involving the customer, the project team may be able to consider alternatives that could mitigate the risk, such as reducing the scope of a project by postponing some work for a later project in order to meet cost and schedule goals. Third, it is a means of promoting confidence in the project team by demonstrating to management and the customer that the team is aware of the significant risks, has a mitigation strategy in place, and is effectively carrying out that strategy.

Expert Judgment

Many organizations rely on the intuitive feelings and experience of experts in performing qualitative risk analyses. They may use expert judgment in lieu of or in addition to other techniques for analyzing risks. For example, experts can categorize risks as being high, medium, or low with or without sophisticated techniques such as calculating risk factors, as described earlier.

Using sophisticated risk analysis techniques has a number of disadvantages. For example, the outputs are only as good as the inputs, and the people using the techniques may be using poor assumptions. Many people get confused when someone tries to explain the math and statistics behind the various techniques. Because of these disadvantages, it is important to include expert opinions when using both qualitative and quantitative risk assessment techniques.

QUANTITATIVE RISK ANALYSIS

Quantitative risk analysis often follows qualitative risk analysis, yet both processes can be done together or separately. On some projects, the team may only perform qualitative risk analysis. The nature of the project and availability of time and money affect the type of risk analysis techniques used. Large, complex projects involving leading-edge technologies often require extensive quantitative risk analysis. The main techniques for quantitative risk analysis include decision tree analysis and simulation, as described in the following sections. Project teams can also use simpler approaches, such as interviewing and sensitivity analysis, to assist in quantitative risk analysis.

Decisions Trees and Expected Monetary Value

A **decision tree** is a diagramming analysis technique used to help select the best course of action in situations in which future outcomes are uncertain. A common application of decision tree analysis involves calculating expected monetary value. **Expected monetary value (EMV)** is the product of a risk event probability and the risk event's monetary value. Figure 11-4 uses the decision of which project(s) an organization might pursue to illustrate this concept. Suppose Cliff Branch's firm was trying to decide if it should submit a proposal for Project 1, Project 2, both projects, or neither project. The team could draw a decision tree with two branches, one for Project 1 and one for Project 2. The firm could then calculate the expected monetary value to help make this decision.

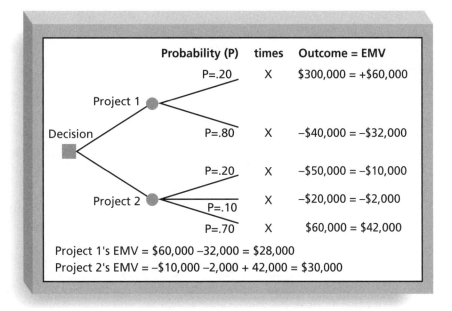

Figure 11-4. Expected Monetary Value (EMV) Example

In order to create a decision tree, and to calculate expected monetary value specifically, you must estimate the probabilities, or chances, of certain events occurring. For example, in Figure 11-4 there is a 20 percent probability or chance (P = .20) that Cliff's firm will win the contract for Project 1, which is estimated to be worth $300,000 in profits, the outcome of the top branch in the figure. There is an 80 percent probability (P = .80) that it will not win the contract for Project 1, and the outcome is estimated to be –$40,000, meaning that the firm will have to invest $40,000 into Project 1 with no reimbursement if it is not awarded the contract. The sum of the probabilities for outcomes for each project must equal one (for Project 1, 20 percent plus 80 percent). Probabilities are normally determined based on expert judgment. Cliff or other people in his firm should have some sense of their likelihood of winning certain projects.

Figure 11-4 also shows probabilities and outcomes for Project 2. Suppose there is a 20 percent probability that Cliff's firm will lose $50,000 on Project 2, a 10 percent probability that it will lose $20,000, and a 70 percent probability that it will earn $60,000. Again, experts would need to estimate these dollar amounts and probabilities.

To calculate the expected monetary value for each project, multiply the probability by the outcome value for each potential outcome for each project and sum the results. To calculate expected monetary value (EMV) for Project 1, going from left to right, multiply the probability by the outcome for each branch and sum the results. In this example, the EMV for Project 1 is $28,000.

.2($300,000) + .8(–$40,000) = $60,000 – $32,000 = $28,000

The EMV for Project 2 is $30,000.

.2(–$50,000) + .1(–$20,000) + .7($60,000) = –$10,000 – $2,000 + $42,000 = $30,000

Because the EMV provides an estimate for the total dollar value of a decision, you want to have a positive number; the higher the EMV, the better. Since the EMV is positive for both Projects 1 and 2, Cliff's firm would expect a positive outcome from each and could bid on both projects. If it had to choose between the two projects, perhaps because of limited resources, Cliff's firm should bid on Project 2 because it has a higher EMV.

Also notice in Figure 11-4 that if you just looked at the potential outcome of the two projects, Project 1 looks more appealing. You could earn $300,000 in profits from Project 1, but you can only earn $60,000 for Project 2. If Cliff were a risk seeker, he would naturally want to bid on Project 1. However, there is only a 20 percent chance of getting that $300,000 on Project 1, and there is a 70 percent chance of earning $60,000 on Project 2. Using EMV helps account for all possible outcomes and their probabilities of occurrence, thereby reducing the tendency to pursue overly aggressive or conservative risk strategies.

Simulation

Another sophisticated quantitative risk analysis technique is simulation. Simulation uses a representation or model of a system to analyze the expected behavior or performance of the system. Most simulations are based on some form of Monte Carlo analysis. **Monte Carlo analysis** simulates a model's outcome many times to provide a statistical distribution of the calculated results. For example, a Monte Carlo analysis can determine if a project will meet its schedule or cost goals given a 10 percent, 50 percent, or 90 percent probability. The basic steps of a Monte Carlo analysis are:

1. Assess the range for the variables being considered and determine the probability distribution for each. In other words, collect the most likely, optimistic, and pessimistic estimates for the variables in the model and determine the probability of each variable falling between the optimistic and most likely estimates.

2. For each variable, select a random value based on the probability distribution for the occurrence of the variable. For example, suppose an optimistic

estimate is 10 (units can be days, dollars, labor hours, or whatever unit is in the model). Also, suppose the most likely estimate is 20, and the pessimistic estimate is 50. If there is a 30 percent probability of being between 10 and 20 (the optimistic and most likely estimates), then 30 percent of the time, select a random number between 10 and 20, and 70 percent of the time, select a number between 20 and 50 (the pessimistic estimate).

3. Run a deterministic analysis or one pass through the model using the combination of values selected for each one of the variables.

4. Repeat Steps 2 and 3 many times to obtain the probability distribution of the results. The number of iterations depends on the number of variables and the degree of confidence required in the results, but it typically lies between 100 and 1,000.

Another technique for quantifying risk, described in Chapter 6, Project Time Management, is Program Evaluation and Review Technique (PERT) analysis. Recall that PERT analysis involves making three estimates of each activity's duration. The three estimates represent a pessimistic or worst-case estimate, an optimistic or best-case estimate, and a most-likely estimate, similar to those described in a Monte Carlo analysis. However, the PERT formula simply weights the most-likely estimate four times more than the pessimistic or optimistic estimates, instead of assigning a probability for the estimate falling between the optimistic and most-likely estimates and then running a simulation of the model. Although this approach can be better than using one discrete estimate, it does not provide the flexibility or accuracy of a Monte Carlo analysis.

As you can imagine, people now use software to perform the steps required for a Monte Carlo analysis. Several PC-based software packages are available that perform Monte Carlo simulations. You will learn more about using simulation software later in this chapter.

What Went Right?

A large aerospace company used Monte Carlo analysis to help quantify risks on several advanced-design engineering projects. The National Aerospace Plan (NASP) project involved many risks. The purpose of this multibillion-dollar project was to design and develop a vehicle that could fly into space using a single-stage-to-orbit approach. A single-stage-to-orbit approach meant the vehicle would have to achieve a speed of Mach 25 (25 times the speed of sound) without a rocket booster. A team of engineers and business professionals worked together in the mid-1980s to develop a software model for estimating the time and cost of developing the NASP project. This model was then linked with simulation software to determine the sources of cost and schedule risk for the project. The results of the Monte Carlo analysis were then used to determine how the company would invest its internal research and development funds. Although the NASP project was terminated, the resulting research has helped develop more advanced materials and propulsion systems used on many modern aircraft.

RISK RESPONSE PLANNING

After an organization identifies and quantifies risks, it must develop a response to them. Developing a response to risks involves defining steps for enhancing opportunities and developing plans for handling risks or threats to project success. The four basic response strategies are avoidance, acceptance, transference, and mitigation. Important outputs of the risk response development process include a risk management plan, contingency plans, and reserves.

Risk avoidance involves eliminating a specific threat or risk, usually by eliminating its causes. Of course, not all risks can be eliminated, but specific risk events can be. For example, a project team may decide to continue using a specific piece of hardware or software on a project because they know it works. Other products that could be used on the project may be available, but if the project team is unfamiliar with them, they could cause significant risk. Using familiar hardware or software eliminates this risk.

Risk acceptance means accepting the consequences should a risk occur. For example, a project team planning a big project review meeting could take an active approach to risk by having a contingency or backup plan and contingency reserves if they cannot get approval for a specific site for the meeting. On the other hand, they could take a passive approach and accept whatever facility their organization provides them.

Risk transference is shifting the consequence of a risk and responsibility for its management to a third party. For example, risk transference is often used in dealing with financial risk exposure. A project team may purchase special insurance or warranty protection for specific hardware needed for a project. If the hardware fails, the insurer must replace it within an agreed-upon period of time.

Risk mitigation, as discussed earlier in the chapter, involves reducing the impact of a risk event by reducing the probability of its occurrence. Suggestions for reducing common sources of risk on information technology projects were provided at the beginning of this chapter. Other examples of risk mitigation include using proven technology, having competent project personnel, using various analysis and validation techniques, and buying maintenance or service agreements from subcontractors.

Table 11-8 provides general mitigation strategies for technical, cost, and schedule risks on projects.[11] Note that increasing the frequency of project monitoring and using a work breakdown structure (WBS) and Critical Path Method (CPM) are strategies for all three areas. Increasing the project manager's authority is a strategy for mitigating technical and cost risks, and selecting the most experienced project manager is recommended for reducing schedule risks. Improving communication is also an effective strategy for mitigating risks.

---

[11] Couillard, Jean, "The Role of Project Risk in Determining Project Management Approach," *Project Management Journal*, Project Management Institute (December 1995).

Table 11-8: General Risk Mitigation Strategies for Technical, Cost, and Schedule Risks

TECHNICAL RISKS	COST RISKS	SCHEDULE RISKS
Emphasize team support and avoid stand-alone project structure	Increase the frequency of project monitoring	Increase the frequency of project monitoring
Increase project manager authority	Use WBS and CPM	Use WBS and CPM
Improve problem handling and communication	Improve communication, project goals understanding, and team support	Select the most experienced project manager
Increase the frequency of project monitoring	Increase project manager authority	
Use WBS and CPM		

Important outputs from risk response planning include development of a risk response plan, analysis of residual risks, and analysis of secondary risks. The risk response plan describes identified risks, people assigned responsibilities for managing those risks, results from risk analyses, response strategies, budget and schedule estimates for responses, and contingency and fallback plans. **Residual risks** are risks that remain after all of the response strategies have been implemented. For example, even though a more stable hardware product may have been used on a project, there may still be some risk of it failing to function properly. **Secondary risks** are a direct result of implementing a risk response. For example, using the more stable hardware may have caused a risk of peripheral devices failing to function properly. Other outputs of risk response planning include contractual agreements, estimates of needed contingency reserve, and inputs to other processes and the project plan.

RISK MONITORING AND CONTROL

Risk monitoring and control involves executing the risk management processes and the risk management plan to respond to risk events. Executing the risk management processes means ensuring that risk awareness is an ongoing activity performed by the entire project team throughout the entire project. Project risk management does not stop with the initial risk analysis. Identified risks may not materialize, or their probabilities of occurrence or loss may diminish. Previously identified risks may be determined to have a greater probability of occurrence or a higher estimated loss value. Similarly, new risks will be identified as the project progresses. Newly identified risks need to go through the same process as those identified during the initial risk assessment. A redistribution of resources devoted to risk management may be necessary because of relative changes in risk exposure.

Carrying out individual risk management plans involves monitoring risks on the basis of defined milestones and making decisions regarding risks and mitigation strategies. It may be necessary to alter a mitigation strategy if it becomes ineffective, implement a planned contingency activity, or eliminate a risk from the list of potential risks when it no longer exists. Project teams sometimes use **workarounds**—unplanned responses to risk events—when they do not have contingency plans in place.

Project risk audits, periodic risk reviews such as the Top Ten Risk Item Tracking method, earned value management, technical performance measurement, and additional risk response planning are all tools and techniques for performing risk monitoring and control. Outputs of this process are corrective action, project change requests, and updates to other plans.

USING SOFTWARE TO ASSIST IN PROJECT RISK MANAGEMENT

As mentioned earlier, you can use software tools to enhance various risk management processes. Databases can keep track of risks, spreadsheets can aid in tracking and quantifying risks, and more sophisticated risk management software can help you develop models and use simulations to analyze and respond to various risks.

You can use Microsoft Project 2002 to perform PERT analysis, and you can use add-on software to perform Monte Carlo analysis. For example, Risk + (by C/S Solutions, Inc.) is a comprehensive risk analysis tool that integrates with Project 2002 to quantify the cost and schedule uncertainty associated with projects. It uses Monte Carlo-based simulation techniques to answer questions such as: What are the chances of completing the project by December 1, 2005? How confident are we that costs will remain below $10 million? What are the chances that this task will end up on the critical path?

To use a Monte Carlo-based simulation to estimate the probability of meeting specific schedule goals, you would collect optimistic, pessimistic, and most-likely duration estimates for project tasks on a network diagram, which is similar to PERT analysis. You must also collect estimates for the probability of completing each task between the optimistic and most-likely times. The same approach can be used for cost estimates. You would collect optimistic, pessimistic, and most likely estimates for factors that determine project costs and the probability of the cost factors being between the optimistic and most likely values.

For example, an expert might estimate that a certain task will most likely take three months to complete, but it could take as little as one month or up to nine months. When asked the probability of completing the task between one and three months, the expert might honestly reply that the probability is only 20 percent. Another expert might estimate that another project task will take five months to complete, but it could take as little as two months, or it could take up to seven months. This expert might estimate that the probability of completing that task between two and five months is 80 percent. Estimating the probability of completing tasks between the optimistic and most likely times helps to account for estimating bias. Compared with a PERT calculation, the simulation shows various probability distributions for each estimate instead of applying the same simple PERT variation for all estimated durations. Unlike PERT analysis, which focuses on schedule estimates, Monte Carlo analysis can also be used to estimate cost risks.

Figure 11-5 illustrates the results from a Monte Carlo-based simulation of a project schedule. The simulation was done using Microsoft Project 2002 and Risk + software. On the left side of Figure 11-5 is a chart displaying columns and an S-shaped curve. The height of each column, read by the scale on the left of the chart, indicates how many times the project was completed in a given time interval during the simulation run, which is the sample count. In this example, the time interval is two working days, and the simulation was run 250 times. The first column shows that the project was completed by 1/29/05 only two times during the simulation. The S-shaped curve, read from the scale on the right of the chart, shows the cumulative probability of completing the project on or before a given date. The information is also shown in tabular form on the right side of Figure 11-5. For example, there is a 10 percent probability that the project will be completed by 2/8/05, a 50 percent chance of completion by 2/17/05, and a 90 percent chance of completion by 2/25/05.

Figure 11-5. Sample Monte Carlo-Based Simulation Results for Project Schedule

You can also use Monte Carlo analysis to help estimate project costs. First, you would develop a model for estimating the total project cost. Suppose project costs could be estimated based on the number of pounds of a certain material, the cost per pound of the material, the number of hours of specific workers, and the cost per hour for each category of worker (such as managers, programmers, and electrical engineers). You could run a Monte Carlo-based simulation of the total project cost based on estimates of the optimistic, pessimistic, and most likely number of pounds of material, costs per pound of the material, number of hours of specific workers, and the cost per hour for the various workers.

Figure 11-6 shows the results of a Monte Carlo-based simulation to estimate total project cost using Microsoft Project 2002 and Risk + software. These simulation results show that there is a 20 percent chance of the project costing less than $175,693, a 65 percent chance of it costing less than $180,015, and a 95 percent chance of the total project cost being under $184,528. For example, you can use this information to decide how much to bid on a project, if you are the seller, or how much to budget for the project, if you are the buyer, given your risk tolerance. For example, if you are risk-averse, you might want to bid $185,000 to be extremely confident that you will not go over budget.

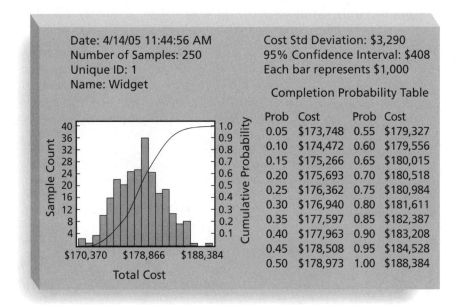

Figure 11-6. Sample Monte Carlo-Based Simulation Results for Project Cost

You can also use Microsoft Excel with an add-on software such as Crystal Ball (by Decisioneering, Inc.) to perform Monte Carlo-based simulations. For example, you could enter costs as described above into a model developed in Excel and then run the simulation. You would get similar results that show a probability distribution of the potential cost outcomes based on the data in the model.

In addition to estimating overall probabilities of project goals such as completion dates or cost estimates, you can also find top sources of risk (risk drivers). For example, a cost simulation might show that the number of labor hours for the electrical engineers was the main source of cost risk for a project.

Simulations are very powerful tools, and it is important that people who use them understand all of the variables, inputs, and outputs involved. As with any software product, the information that comes out is only as good as the information that goes in. It is important to collect data from people who understand the project or specific tasks involved. It is also important to test the model used in simulations to ensure that it provides realistic results.

RESULTS OF GOOD PROJECT RISK MANAGEMENT

Good project risk management often goes unnoticed, unlike crisis management. With crisis management, there is an obvious danger to the success of a project. The crisis, in turn, receives the intense interest of the entire project team. Resolving a crisis has much greater visibility, often accompanied by rewards from management, than successful risk management. In contrast, when risk management is effective, it results in fewer problems, and for the few problems that exist, it results in more expeditious resolutions. It may be difficult for outside observers to tell whether risk management or luck was responsible for the smooth development of a new system, but project teams will always know that their projects worked out better because of good risk management.

Well-run projects, like a master violinist's performance, an Olympic athlete's gold medal win, or a Pulitzer-Prize–winning book, appear to be almost effortless. Those on the outside—whether audiences, customers, or managers—cannot observe the effort that goes into a superb performance. They cannot see the hours of practice, the edited drafts, or the planning, management, and foresight that create the appearance of ease. To improve information technology project management, project managers should strive to make their jobs look easy—it reflects the results of a well-run project.

CASE WRAP-UP

Cliff Branch and two of his senior people attended a seminar on project risk management where the speaker discussed several techniques such as estimating the expected monetary value of projects, Monte Carlo simulations, and so on. Cliff asked the speaker how he might use these techniques to help his company decide which projects to bid on, since bidding on projects often required up-front investments with the possibility of no payback. The speaker walked through an example of EMV and then ran a quick Monte Carlo simulation. Cliff did not have a strong math background and had a hard time understanding the EMV calculations. He thought the simulation was much too confusing to have any practical use for him. He believed in his gut instincts much more than any math calculation or computer output.

The speaker finally sensed that Cliff was not impressed, so she explained the importance of looking at the odds of winning project awards and not just at the potential profits. She suggested using a risk-neutral strategy by bidding on projects that the firm had a good chance of winning (50 percent or so) and that had a good profit potential, instead of focusing on projects that they had a small chance of winning and that had a larger profit potential. Cliff disagreed with this advice, and he continued to bid on high-risk projects. The two other managers who attended the seminar now understood why the firm was having problems—their leader loved taking risks, even if it hurt the firm. They soon found jobs with competing firms, as did several others.

CHAPTER SUMMARY

Risk is the possibility of loss or injury. Projects, by virtue of their unique nature, involve risk.

Risk management is an investment; that is, there are costs associated with identifying risks, analyzing those risks, and establishing plans to mitigate those risks. Those costs must be included in cost, schedule, and resource planning. Organizations take risks to benefit from potential opportunities.

Risk utility or risk tolerance is the amount of satisfaction or pleasure received from a potential payoff. Risk seekers enjoy high risks, risk-averse people do not like to take risks, and risk-neutral people seek to balance risks and potential payoff.

Project risk management is an ethic in which the project team continually assesses what may negatively impact the project, determines the probability of such events occurring, and determines the impact should such events occur. It also involves analyzing and determining alternate strategies to deal with risks. The six main processes involved in risk management are risk management planning, risk identification, qualitative risk analysis, quantitative risk analysis, risk response planning, and risk monitoring and control.

Risk management planning is the process of deciding how to approach and plan for risk management activities for a particular project. A risk management plan is a key output of risk management. Contingency plans are predefined actions that a project team will take if an identified risk event occurs. Fallback plans are developed for risks that have a high impact on meeting project objectives, and are implemented if attempts to reduce the risk are not effective. Contingency reserves or allowances are provisions held by the project sponsor that can be used to mitigate cost or schedule risk, if changes in project scope or quality occur.

Information technology projects often involve several risks: lack of user involvement, lack of executive management support, unclear requirements, poor planning, and so on. Lists developed by the Standish Group, McFarlan, and other organizations can help you identify potential risks on information technology projects. Lists of common risk conditions in project management knowledge areas can also be helpful in identifying risks, as can information-gathering techniques such as brainstorming, the Delphi Technique, interviewing, and SWOT analysis.

Risks can be assessed qualitatively and quantitatively. Tools for qualitative risk analysis include a probability/impact matrix, the Top Ten Risk Item Tracking technique, and expert judgment. Tools for quantitative risk analysis include decision trees and Monte Carlo simulation. Expected monetary value uses decision trees to evaluate potential projects based on their expected value. Simulations are a more sophisticated method for creating estimates to help you determine the likelihood of meeting specific project schedule or cost goals.

The four basic responses to risk are avoidance, acceptance, transference, and mitigation. Risk avoidance involves eliminating a specific threat or risk. Risk acceptance means accepting the consequences of a risk, should it occur. Risk transference is shifting the consequence of a risk and responsibility for its management to a third party. Risk mitigation is reducing the impact of a risk event by reducing the probability of its occurrence.

Risk monitoring and control involves executing the risk management processes and the risk management plan to respond to risks. Outputs of this process include corrective action, project change requests, and updates to other plans.

Several types of software can assist in project risk management. Monte Carlo-based simulation software is a particularly useful tool for helping you get a better idea of project risks and top sources of risk or risk drivers.

DISCUSSION QUESTIONS

1. Discuss the risk utility function and risk preference chart in Figure 11-1. Would you rate yourself as being risk-averse, risk-neutral, or risk-seeking? Give examples of each approach from different aspects of your life, such as your current job, your personal finances, romances, and eating habits.

2. What are some questions that should be addressed in a risk management plan?

3. Discuss the common sources of risk on information technology projects and suggestions for managing them. Which suggestions do you find most useful? Which do you feel would not work in your organization? Why?

4. What is the difference between using brainstorming and the Delphi technique for risk identification? What are some of the advantages and disadvantages of each approach?

5. Describe how to use a probability/impact matrix, risk factors, and the Top Ten Risk Item Tracking approaches for performing qualitative risk analysis. How could you use each technique in your organization?

6. Explain how to use decision trees and Monte Carlo analysis for quantifying risk. Give an example of how you could use each technique on an information technology project.

7. Provide examples of risk avoidance, risk acceptance, risk transference, and risk mitigation as responses to risks.

8. List the tools and techniques for performing risk monitoring and control.

9. How can you use Microsoft Project and Excel to assist in project risk management?

10. Why should project managers strive to make their jobs look easy?

EXERCISES

1. Suppose your college or organization is considering a new project that would involve developing an information system that would allow all employees and students/customers to access and maintain their own human resources-related information, such as address, marital status, tax information, and so on. The main benefits of the system would be a reduction in human resources personnel and more accurate information. For example, if an employee, student, or customer had a new telephone number or e-mail address, he or she would be responsible for entering the new data in the new system. The new system would also allow employees to change their tax withholdings or pension plan contributions. Identify five potential risks for this new project. Provide a detailed description of each risk and propose strategies for mitigating each risk.

2. Review a project risk management plan. Did the organization do a good job of answering the questions that should be addressed in a risk management plan? If you cannot find a risk management plan, write one of your own for an information technology project your organization is considering.

3. Research risk management software. Are many products available? What are the main advantages of using them? What are the main disadvantages? Write a paper discussing your findings, and include at least three references.

4. Suppose your organization is deciding which of four projects to bid on. Information on each is in the Table below. Assume that all up-front investments are not recovered, so they are shown as negative profits. Draw a diagram and calculate the EMV for each project. Write a few paragraphs explaining which projects you would bid on. Be sure to use the EMV information and your personal risk tolerance to justify your answer.

PROJECT	CHANCE OF WINNING	ESTIMATED PROFITS
Project 1	50 percent	$120,000
	50 percent	−$50,000
Project 2	30 percent	$100,000
	40 percent	$50,000
	30 percent	−$60,000
Project 3	70 percent	$20,000
	30 percent	−$5,000
Project 4	30 percent	$40,000
	30 percent	$30,000
	20 percent	$20,000
	20 percent	−$50,000

5. Find an example of a company that took a big risk on an information technology project and succeeded. In addition, find an example of a company that took a big risk and failed. Summarize each project and situation. Did anything besides luck make a difference between success and failure?

MINICASE

Your college or organization implemented a project, described in Exercise 1, to provide an information system for all employees, students, and customers, allowing them to access and maintain their own human resources-related information, such as address, marital status, and tax information. Many problems arose after implementation. You knew there was a potential risk that people would not use the system, and you have discovered that less than half of all potential users are using it. This low usage has caused bottlenecks in the human resources department, because the staff was reduced to pay for the new system. You also anticipated some security problems if users shared their passwords with friends or family members, but now your organization is facing potential litigation.

Part 1: List four alternatives to address the problem of eligible users not using the system. Describe each option in detail. Include estimates of how long it would take to implement, how much it would cost, and how effective it would be. Also, list at least one potential secondary risk that might arise after implementing each option.

Part 2: You have confirmed the fact that the new system has several security problems, including users sharing their passwords with friends or family members. This behavior has resulted in several pranks, such as entries of bogus names and addresses. More serious compromises of security have caused financial and legal problems. One woman is threatening to sue the organization because someone gathered a list of e-mail addresses from the system and sent a mass mailing of a multilevel marketing spam. Write a one- to two-page policy statement that your organization could implement to transfer the risk of these inappropriate uses of the system to the users of the system or to a third party.

RUNNING CASES

Review the running cases provided in Appendix C for several questions related to project risk management as well as related templates in Appendix D.

SUGGESTED READINGS

1. Boehm, Barry W. "Software Risk Management: Principles and Practices." *IEEE Software* (January 1991).

    Boehm describes detailed approaches for several risk management techniques, including risk identification checklists, risk prioritization, risk management planning, and risk monitoring. This article includes an example of Top Ten Risk Item Tracking.

2. Microsoft, White Papers on Microsoft Solutions Framework (MSF), *www.microsoft.com/enterpriseservices* (2001).

    Microsoft has written several white papers describing their approach to managing risks. The Microsoft Solutions Framework includes three core disciplines, one being risk management. Their "Process of Risk Management" white paper describes techniques that project managers can use to create an environment in which teams successfully identify and manage risks.

3. Project Management Institute. *PM Network* (February 1998).

    This entire issue of the magazine focuses on managing risk in projects. Topics of articles include analyzing risk decisions, using risk-based scheduling and analysis, and encouraging team-based risk assessment.

4. Risk Driver. *www.Riskdriver.com*.

    This Web site provides information regarding best practices in project risk management, risk assessment tools, and mailing lists to provide the latest news and events related to risk management. The site also links to other Web sites, such as PMI's Risk Management Specific Interest Group and Carnegie Mellon's SEI Web site.

5. Van Scoy, Roger L. "Software Development Risk: Opportunity, Not Problem." *Technical Report CMU/SEI-92-TR-30*, Software Engineering Institute, Carnegie Mellon University (September 1992).

This report summarizes risk management for software projects. The Software Engineering Institute (SEI) provides this and several other publications on the topic of risk management and other subjects related to software development. More recent articles are available on their Web site at www.sei.cmu.edu.

6. Wideman, R. Max (editor) and Rodney J. Dawson. *Project & Program Risk Management: A Guide to Managing Project Risks and Opportunities.* Upper Darby, PA: Project Management Institute, 1998.

Wideman has written several books and articles on project risk management. This textbook includes practical advice for managing project risks and opportunities.

7. Williams, Ray C., Julie A. Walker, and Audrey J. Dorofee. "Putting Risk Management into Practice." *IEEE Software* (May/June 1997): 75–82.

The authors of this article use a road map provided by the Software Engineering Institute (SEI) to discuss effective and ineffective risk management methods based on six years' experience with software-intensive Department of Defense programs. These programs followed the SEI approach of continuous risk management, selecting processes and methods that would work best for their work cultures.

KEY TERMS

- **brainstorming** — a technique by which a group attempts to generate ideas or find a solution for a specific problem by amassing ideas spontaneously and without judgment

- **contingency plans** — predefined actions that the project team will take if an identified risk event occurs

- **contingency reserves** or **allowances** — provisions held by the project sponsor that can be used to mitigate cost and/or schedule risk, should possible changes in project scope or quality occur

- **Delphi technique** — an approach used to derive a consensus among a panel of experts, to make predictions about future developments

- **expected monetary value (EMV)** — the product of the risk event probability and the risk event's monetary value

- **fallback plans** — plans developed for risks that have a high impact on meeting project objectives, to be implemented if attempts to reduce the risk are not effective

- **flowcharts** — diagrams that show how various elements of a system relate to each other

- **influence diagrams** — diagrams that represent decision problems by displaying essential elements, including decisions, uncertainties, and objectives, and how they influence each other
- **interviewing** — a fact-finding technique that is normally done face-to-face or via telephone, but can also occur through e-mail or instant messaging
- **Monte Carlo analysis** — a risk quantification technique that simulates a model's outcome many times, to provide a statistical distribution of the calculated results
- **probability/impact matrix or chart** — a matrix or chart that lists the relative probability of a risk occurring on one side of a matrix or axis on a chart and the relative impact of the risk occurring on the other
- **qualitative risk analysis** — qualitatively analyzing risks and prioritizing their effects on project objectives
- **quantitative risk analysis** — measuring the probability and consequences of risks and estimating their effects on project objectives
- **residual risks** — risks that remain after all of the response strategies have been implemented
- **risk** — the possibility of loss or injury
- **risk acceptance** — accepting the consequences should a risk occur
- **risk-averse** — having a low tolerance for risk
- **risk avoidance** — eliminating a specific threat or risk, usually by eliminating its causes
- **risk events** — specific circumstances that may occur to the detriment of the project
- **risk factors** — numbers that represent overall risk of specific events, given their probability of occurring and the consequence to the project if they do
- **risk identification** — determining which risks are likely to affect a project and documenting the characteristics of each
- **risk management plan** — a plan that documents the procedures for managing risk throughout the project
- **risk management planning** — deciding how to approach and plan the risk management activities for a project, by reviewing the project charter, WBS, roles and responsibilities, stakeholder risk tolerances, and the organization's risk management policies and plan templates
- **risk mitigation** — reducing the impact of a risk event by reducing the probability of its occurrence
- **risk monitoring and control** — monitoring known risks, identifying new risks, reducing risks, and evaluating the effectiveness of risk reduction throughout the life of the project
- **risk-neutral** — a balance between risk and payoff
- **risk response planning** — taking steps to enhance opportunities and reduce threats to meeting project objectives

- **risk-seeking** — having a high tolerance for risk
- **risk symptoms** or **triggers** — indications for actual risk events
- **risk transference** — shifting the consequence of a risk and responsibility for its management to a third party
- **risk utility** or **risk tolerance** — the amount of satisfaction or pleasure received from a potential payoff
- **runaway projects** — projects that have significant cost or schedule overruns
- **secondary risks** — risks that are a direct result of implementing a risk response
- **Top Ten Risk Item Tracking** — a qualitative risk analysis tool for identifying risks and maintaining an awareness of risks throughout the life of a project
- **triggers** — indications for actual risk events
- **workarounds** — unplanned responses to risk events that project teams use when there is no contingency plan in place

12

Project Procurement Management

Objectives

After reading this chapter, you will be able to:

1. *Understand the importance of project procurement management and the increasing use of outsourcing for information technology projects*
2. *Describe the procurement planning process, procurement planning tools and techniques, types of contracts, and statements of work*
3. *Discuss what is involved in solicitation planning and the difference between a request for proposal and a request for quote*
4. *Explain what occurs during the solicitation process*
5. *Describe the source selection process and different approaches for evaluating proposals or selecting suppliers*
6. *Discuss the importance of good contract administration*
7. *Describe the contract close-out process*
8. *Discuss types of software available to assist in project procurement management*

OPENING CASE

Marie McBride could not believe how much money her company was paying for outside consultants to help the company finish an important operating system conversion project. The consulting company's proposal said it would provide experienced professionals who had completed similar conversions and that the job would be finished in six months or less with four consultants working full-time. Nine months later her company was still paying high consulting fees, and half of the original consultants on the project had been replaced with new people. One new consultant had graduated from college only two months before. Marie's internal staff complained that they were wasting time training some of these "experienced professionals." Marie talked to her company's purchasing manager about the contract, fees, and special clauses that might be relevant to the problems they were experiencing.

Marie was dismayed at how difficult it was to interpret the contract. It was very long and obviously written by someone with a legal background. When she asked what her company could do since the consulting firm was not following their proposal, the purchasing manager stated that the proposal was not part of the official contract. Marie's company was paying for time and materials, not specific deliverables. There was no clause stating the minimum experience level required for the consultants, nor were there penalty clauses for not completing the work on time. There was a termination clause, however, meaning the company could terminate the contract. Marie wondered why her company had signed such a poor contract. Was there a better way to deal with procuring services from outside the company?

IMPORTANCE OF PROJECT PROCUREMENT MANAGEMENT

Procurement means acquiring goods and/or services from an outside source. The term procurement is widely used in the government; many private companies use the term *purchasing*; and information technology professionals use the term *outsourcing*. Organizations or individuals who provide procurement services are referred to as suppliers, vendors, subcontractors, or sellers, with suppliers being the most widely used term. Many information technology projects involve the use of goods and services from outside the organization. In fact, the U.S. market for information technology outsourcing was projected to reach more than $110 billion by 2003.[1] U.S. federal spending on information technology outsourcing services is projected to increase from $6.6 billion in 2002 to nearly $15 billion by 2007, due to the emphasis on e-government, the role information technology plays in homeland security, and the shortage of information technology workers in government.[2] Because outsourcing is a growing area, it is important for project managers to understand project procurement management.

[1] Input.com, "Internet-related Spending to Propel the U.S. Outsourcing Market to $110 Billion by 2003," *www.input.com/public/article38.cfm* (2001).

[2] Wait, Patience, "IT outsourcing market to reach $15 billion by 2007," *Washington Technology,* (January 8, 2003) (*www.washintontechnology.com*).

Many organizations are turning to outsourcing to:

- Reduce both fixed and recurrent costs. Outsourcing suppliers are often able to use economies of scale that may not be available to the client alone. For example, clients can save on information technology hardware and software costs through consolidation of operations. Many foreign companies can also pay talented workers less than U.S. companies can. The February 3, 2002 edition of *Business Week* magazine's cover story was titled: "Is your job next?" and described how globalization is exporting U.S. technical jobs to companies in China, India, Russia, and other countries.

- Allow the client organization to focus on its core business. Most organizations are not in business to provide information technology services, yet many have spent valuable time and resources on information technology functions when they should have focused on core competencies such as marketing, customer service, and new product design. By outsourcing many information technology functions, employees can focus on jobs that are critical to the success of the organization.

- Access skills and technologies. Organizations can gain access to specific skills and technologies when they are required by using outside resources. For example, a project may require an expert in a particular field or use of expensive hardware or software for one particular month on a project. Planning for this procurement will ensure that the needed skills or technology will be available for the project.

- Provide flexibility. Outsourcing to provide extra staff during periods of peak workloads can be much more economical than trying to staff entire projects with internal resources.

- Increase accountability. A well-written **contract**—a mutually binding agreement that obligates the supplier to provide the specified products or services and obligates the buyer to pay for them—can clarify responsibilities and sharpen focus on key deliverables of a project. Because contracts are legally binding, there is more accountability for delivering the work as stated in the contract.

Outsourcing of important information technology functions is growing tremendously. For example, many companies outsource data operations or help desk support to other companies that specialize in those areas. In addition, many companies are outsourcing some information technology projects. For example, a company might hire a large consulting firm to facilitate development of a strategic information technology plan. Another company might hire an outside firm to develop and manage its e-commerce applications. These unique needs in an organization can be filled by an outside organization in a finite period of time.

A recent trend in outsourcing is the increasing size of contracts. The $6.9 billion Navy-Marine Corps Intranet contract awarded in October 2000 was the biggest technology outsourcing contract ever for the U.S. government.[3] Procter & Gamble recently opened its information technology and business operations outsourcing work for bids, and analysts have estimated the potential value of the contract at anywhere from $4 billion to $10 billion over ten years.[4] With so much money at stake, project managers must be well versed in procurement management.

Organizations must also consider reasons they might not want to outsource. When an organization outsources work, it often does not have as much control over those aspects of projects that suppliers carry out. In addition, an organization could become too dependent on particular suppliers. If those suppliers went out of business or lost key personnel, it could cause great damage to a project. Organizations must also be careful to protect strategic information that could become vulnerable in the hands of suppliers. According to Scott McNeally, CEO of Sun Microsystems, Inc., "What you want to handle in-house is the stuff that gives you an edge over your competition—your core competencies. I call it your 'secret sauce.' If you're on Wall Street and you have your own program for tracking and analyzing the market, you'll hang onto that. At Sun, we have a complex program for testing microprocessor designs, and we'll keep it."[5] Project teams must think carefully about procurement issues and make wise decisions based on the unique needs of their projects and organizations.

The success of many information technology projects that use outside resources is often due to good project procurement management. **Project procurement management** includes the processes required to acquire goods and services for a project from outside the performing organization. There are six main processes of project procurement management:

- **Procurement planning** involves determining what to procure and when. In procurement planning, one must decide what to outsource, determine the type of contract, and create a statement of work. The project team also creates a procurement management plan as part of the procurement planning process.

- **Solicitation planning** involves documenting product requirements and identifying potential sources. This process entails writing procurement documents, such as a Request for Proposal (RFP), and developing evaluation criteria. At the end of the solicitation planning process, the organization often issues an RFP.

[3] Chen, Yu-Che and James L. Perry, "IT Outsourcing: A Primer for Public Managers, *IBM Endowment for The Business of Government* (February 2002).

[4] Weiss, Todd R. and Juan Carlos Perez, "EDS Renews P&G Outsourcing Talks," *Computerworld* (September 23, 2002).

[5] McNeally, Scott, "The Future of the Net: Why We Don't Want You to Buy Our Software," Executive Perspectives, Sun Microsystems, Inc., *www.sun.com/dot-com/perspectives/stop.html* (1999): 1.

- **Solicitation** involves obtaining quotations, bids, offers, or proposals as appropriate. In the solicitation process, one usually needs to finalize procurement documents, advertise, hold a bidders' conference, and receive proposals or bids for the work. Occasionally, work is outsourced without a formal solicitation.

- **Source selection** involves choosing from among potential suppliers through a process of evaluating prospective suppliers, negotiating the contract, and awarding the contract.

- **Contract administration** involves managing the relationship with the supplier. This process entails monitoring contract performance, making payments, and awarding contract modifications. By the end of the contract administration process, project teams expect that a substantial amount of the contracted work has been completed.

- **Contract close-out** involves completion and settlement of the contract, including resolution of any open items. This process usually includes product verification, formal acceptance and closure, and a contract audit.

Figure 12-1 summarizes the major processes involved in project procurement management and lists important milestones that often occur after completing each stage of the process.[6] For example, an important milestone after procurement planning is the make-or-buy decision. **A make-or-buy decision** is one in which an organization decides if it is in their best interests to make certain products or perform certain services inside the organization, or if it is better to buy them from an outside organization. If there is no need to buy any products or services from outside the organization, then there is no need to perform any of the other procurement management processes. At the end of solicitation planning, a key milestone is often the issuance of an RFP. After the solicitation process, the organization receives proposals. At the end of source selection, a contract is awarded. After the contract administration process, the supplier completes a substantial amount of the work described in the contract. At the end of contract close-out, there is a formal acceptance and closure of the contract.

Figure 12-1. Project Procurement Management Processes and Key Outputs

[6] Modified figure based on Rita Mulcahy's *Mastering Contracts for Project Managers* (Training Materials), 1998.

PROCUREMENT PLANNING

Procurement planning is the process of identifying which project needs can best be met by using products or services outside the organization. It involves deciding whether to procure, how to procure, what to procure, how much to procure, and when to procure. For many projects, properly outsourcing some information technology functions can be a great investment, as shown in the following examples of *What Went Right*.

What Went Right?

The Boots Company PLC, a pharmacy and health care company in Nottingham, England, outsourced its information technology systems to IBM in October 2002. The Boots Company signed a ten-year contract worth about $1.1 billion and expected to save $203.9 million over that period compared with the cost of running the systems iself. IBM managed and developed The Boots Company's systems infrastructure "from the mainframes to the tills in our 1,400 stores, to the computer on my desk," said spokesman Francis Thomas. More than 400 Boots employees were transferred to IBM's payroll but continued to work at Boots' head office, with extra IBM staff brought in as needed. Thomas added, "The great thing about this is that if IBM has an expert in Singapore and [if] we need that expertise, we can tap into it for three months. It keeps our costs on an even keel."[7]

Properly planning procurement can also save organizations millions of dollars. Many companies centralize purchasing for products such as personal computers, software, and printers to earn special pricing discounts. For example, when the U.S. Air Force decided to allow for a minimum contract award and flexible pricing strategy for its local online network system for automating fifteen Air Force Systems Command bases in the mid-1980s, the winning supplier lowered its bid by over $40 million.[8]

Inputs needed for procurement planning include the project scope statement, product descriptions, market conditions, constraints, and assumptions. For example, a large clothing company might consider outsourcing the delivery of, maintenance of, and basic user training and support for laptops supplied to its international sales and marketing force. This decision would be made according to the procurement planning inputs. If there were suppliers who could provide this service well at a reasonable price, it would make sense to outsource, because this could reduce fixed and recurring costs for the clothing company and let them focus on their core business of selling clothes.

It is important to understand why a company would want to procure goods or services and what inputs are needed to do procurement planning. In the opening case, Marie's company hired outside consultants to help complete an

---

[7] Law, Gillian, "IBM wins $1.1B outsourcing deal in England," *Computerworld* (October 1, 2002).
[8] Schwalbe, Kathy, Air Force Commendation Medal Citation, 1986.

operating system conversion project because it needed people with specialized skills for a short period of time. This is a common occurrence in many information technology projects. It can be more effective to hire skilled consultants to perform specific tasks for a short period of time than to keep employees on staff full-time.

However, it is also important to define clearly the scope of the project, the products, market conditions, and constraints and assumptions. In Marie's case, the scope of the project and products was relatively clear, but her company may not have adequately discussed or documented the market conditions or constraints and assumptions involved in using the outside consultants. Were there many companies that provided consultants to do operating conversion projects similar to theirs? Did the project team investigate the background of the company that provided their consultants? Did they list important constraints and assumptions for using the consultants, such as limiting the time that the consultants had to complete the conversion project or the minimum years of experience for any consultant assigned to their project?

Procurement Planning Tools and Techniques

Tools and techniques of procurement management include performing make-or-buy analysis and consulting with experts.

Make-or-buy analysis is a general management technique used to determine whether an organization should make or perform a particular product or service inside the organization or buy from someone else. This form of analysis involves estimating the internal costs of providing a product or service and comparing that estimate to the cost of outsourcing. Consider a company that has 1000 international salespeople who require laptops with maintenance, training, and user support. Using make-or-buy analysis, the company would estimate the cost to provide this product and service using only internal resources. The estimate would need to include costs for hardware and software, travel, shipping, and technical support. Suppose the company estimated an initial investment of $3 million and annual support costs of $2 million. Note that a make-or-buy analysis must include the life cycle cost, as described in Chapter 7, Project Cost Management. If the company received supplier quotes that were less than its internal estimates, it should definitely consider outsourcing the project.

Many organzations also use make-or-buy analysis to decide if they should either purchase or lease items for a particular project. For example, suppose you need a piece of equipment for a project that has a purchase price of $12,000. Assume it also had a daily operational cost of $400. Suppose you could lease the same piece of equipment for $800 per day, including the operational costs. You can set up an equation in which the purchase cost equals the lease cost to determine when it makes sense financially to lease or buy the equipment. In this example, d = the number of days you need the piece of equipment. The equation would then be:

$12,000 + $400d = $800d

Subtracting $400d from both sides, you get:

$12,000 = $400d

Dividing both sides by $400, you get:

30 = d

which means that the purchase cost equals the lease cost in 30 days. So, if you need the equipment for less than 30 days, it would be more economical to lease it. If you need the equipment for more than 30 days, you should purchase it.

Internal experts should be consulted as part of procurement planning. They might suggest that the company mentioned earlier could not provide quality maintenance, training, and service for the 1000 laptops since the service involves so many people with different skill levels in so many different locations. Experts in the company might also know that most of their competitors outsource this type of work and know who the qualified outside suppliers are. Experts outside the company, including potential suppliers themselves, can also provide expert judgment. For example, suppliers might suggest an option for salespeople to purchase the laptops themselves at a reduced cost. This option would solve problems of employee turnover—exiting employees would own their laptops and new employees would purchase a laptop through the program. An internal expert might then suggest that employees receive a technology bonus to help offset what they might view as an added expense. Expert judgment, both internal and external, is an asset in making procurement decisions.

Types of Contracts

Contract type is an important consideration. Different types of contracts can be used in different situations. Four broad categories of contracts are fixed price or lump sum, cost reimbursable, time and material, and unit price. Descriptions of each contract type and examples for information technology projects follow.

Fixed-price or **lump-sum contracts** involve a fixed total price for a well-defined product or service. The buyer incurs little risk in this situation. For example, a company could award a fixed-price contract to purchase 100 laser printers with a certain print resolution and print speed to be delivered to one location within two months. In this example, the product and delivery date are well defined. Fixed-price contracts may also include incentives for meeting or exceeding selected project objectives. For example, the contract could include an incentive fee paid if the laser printers are delivered within one month. A firm-fixed price (FFP) contract has the least amount of risk for the buyer, followed by a fixed-price incentive (FPI) contract.

Cost-reimbursable contracts involve payment to the supplier for direct and indirect actual costs. Recall from Chapter 7 that direct costs are costs that are related to a project and can be traced back in a cost-effective way. Indirect costs are costs related to the project that cannot be traced back in a cost-effective way. For example, the salaries for people working directly on a project and hardware or software purchased for a specific project are direct costs, while the cost of providing a work space with electricity, a cafeteria, and so on, are indirect costs. Indirect costs are often calculated as a percentage of direct costs. Cost reimbursable contracts often include fees such as a profit percentage or incentives for meeting or exceeding selected project objectives.

These contracts are often used for projects that include providing goods and services that involve new technologies. The buyer absorbs more of the risk with cost-reimbursable contracts than they do with fixed-price contracts. Three types of cost-reimbursable contracts, in order of lowest to highest risk to the buyer, include cost plus incentive fee, cost plus fixed fee, and cost plus percentage of costs.

■ With a **cost plus incentive fee (CPIF) contract**, the buyer pays the supplier for allowable performance costs along with a predetermined fee and an incentive bonus. If the final cost is less than the expected cost, both the buyer and the supplier benefit from the cost savings, according to a prenegotiated share formula. For example, suppose the expected cost of a project is $100,000, the fee to the supplier is $10,000, and the share formula is 85/15, meaning that the buyer absorbs 85 percent of the uncertainty and the supplier absorbs 15 percent. If the final price is $80,000, the cost savings are $20,000. The supplier would be paid the final cost and the fee plus an incentive of $3,000 (15 percent of $20,000), for a total reimbursement of $93,000.

■ With a **cost plus fixed fee (CPFF) contract**, the buyer pays the supplier for allowable performance costs plus a fixed fee payment usually based on a percentage of estimated costs. This fee does not vary, however, unless the scope of the contract changes. For example, suppose the expected cost of a project is $100,000 and the fixed fee is $10,000. If the actual cost of the contract rises to $120,000 and the scope of the contract remains the same, the contractor will still receive the fee of $10,000.

■ With a **cost plus percentage of costs (CPPC) contract**, the buyer pays the supplier for allowable performance costs along with a predetermined percentage based on total costs. From the buyer's perspective, this is the least desirable type of contract because the supplier has no incentive to decrease costs. In fact, the supplier may be motivated to increase costs, since doing so will automatically increase profits based on the percentage of costs. This type of contract is prohibited for federal government use, but is sometimes used in private industry, particularly in the construction industry. All of the risk is borne by the buyer.

Time and material contracts are a hybrid of both fixed-price and cost-reimbursable contracts. For example, an independent computer consultant might have a contract with a company based on a fee of $80 per hour for his or her services plus a fixed price of $10,000 for providing specific materials for the project. The materials fee might also be based on approved receipts for purchasing items, with a ceiling of $10,000. The consultant would send an invoice to the company each week or month, listing the materials fee, the number of hours worked, and a description of the work produced. This type of contract is often used for services that are needed when the work cannot be clearly specified and total costs cannot be estimated in a contract. Many contract programmers and consultants, such as those Marie's company hired in the opening case, prefer time and material contracts.

Unit price contracts require the buyer to pay the supplier a predetermined amount per unit of service, and the total value of the contract is a function of the quantities needed to complete the work. Consider an information technology department that might have a unit price contract for purchasing computer hardware. If the company purchases only one unit, the cost might be $1000. If it purchases ten units, the cost would be $10,000. This type of contract often involves volume discounts. For example, if the company purchases between ten and fifty units, the contracted cost might be $900 per unit. If it purchases over fifty units, the cost might go down to $800 per unit.

Any type of contract should include specific clauses that take into account issues unique to the project. For example, if a company uses a time and material contract for consulting services, the contract should stipulate different hourly rates based on the level of experience of each individual contractor. The services of a junior programmer with no Bachelor's degree and less than three years' experience might be billed at $40 per hour, whereas the services of a senior programmer with at least a Bachelor's degree and more than ten years of experience might be billed at $80 per hour.

Figure 12-2 summarizes the spectrum of risk to the buyer and supplier for different types of contracts. Buyers have the lowest risk with firm-fixed price contracts because they know exactly what they will need to pay the supplier. Buyers have the most risk with cost plus percentage of costs contracts because they do not know what the supplier's costs will be in advance, and the suppliers may be motivated to keep increasing costs. From the supplier's perspective, there is the least risk with a cost plus percentage of costs contract and the most risk with the firm-fixed price contract.

Figure 12-2. Contract Types Versus Risk

Time and material and unit price contracts can be viewed as high- or low-risk, depending on the nature of the project and other contract clauses. For example, if an organization is unclear on what work needs to be done, it cannot expect a supplier to sign a firm-fixed price contract. However, the buyer could find a consultant or group of consultants to work on specific tasks based on a predetermined hourly rate. The buying organization could evaluate the work produced every day or week to decide if it wants to continue using the consultants. In this case the contract would include a **termination clause**—a contract clause that allows the buyer or supplier to end the contract. Some termination clauses state that the buyer can terminate a contract for any reason and give the supplier only 24 hours' notice. Suppliers must often give a one-week notice to terminate a contract and must have sufficient reasons for the termination. The buyer could also include a contract clause specifying hourly rates that are based on education and experience of consultants. These contract clauses reduce the risk incurred by the buyer while providing flexibility for accomplishing the work.

Statement of Work

Many contracts often include a statement of work. The **statement of work (SOW)** is a description of the work required for the procurement. The SOW is a type of scope statement that describes the work in sufficient detail to allow prospective suppliers to determine if they are capable of providing the goods and services required and to determine an appropriate price. A SOW should be clear, concise, and as complete as possible. It should describe all services required and include performance reporting. It is important to use appropriate words in a SOW such as *must* instead of *may*. For example, *must* implies that something has to be done; *may* implies that there is a choice involved in doing something or not. The SOW should specify the product of the project, use industry terms, and refer to industry standards.

Many organizations use samples and templates to generate SOWs. Figure 12-3 provides a template for a SOW that Marie's organization could use when they hire outside consultants or purchase other goods or services. For example, for the operating system conversion project, Marie's company should specify the specific manufacturer and model number for the hardware involved, the former operating systems and new ones for the conversion, the number of pieces of each type of hardware involved (mainframes, midrange computers, or PCs), and so on. The SOW should also specify the location of the work, the expected period of performance, specific deliverables and when they are due, applicable standards, acceptance criteria, and special requirements. A good SOW gives bidders a better understanding of the buyer's expectations. A SOW should become part of the official contract to ensure that the buyer gets what the supplier bid on. See Appendix D for templates and resources for procurement-related documents.

Statement of Work (SOW)

I. **Scope of Work:** Describe the work to be done in detail. Specify the hardware and software involved and the exact nature of the work.

II. **Location of Work:** Describe where the work must be performed. Specify the location of hardware and software and where the people must perform the work.

III. **Period of Performance:** Specify when the work is expected to start and end, working hours, number of hours that can be billed per week, where the work must be performed, and related schedule information.

IV. **Deliverables Schedule:** List specific deliverables, describe them in detail, and specify when they are due.

V. **Applicable Standards:** Specify any company or industry-specific standards that are relevant to performing the work.

VI. **Acceptance Criteria:** Describe how the buyer organization will determine if the work is acceptable.

VII. **Special Requirements:** Specify any special requirements such as hardware or software certifications, minimum degree or experience level of personnel, travel requirements, and so on.

Figure 12-3. Statement of Work (SOW) Template

SOLICITATION PLANNING

Solicitation planning involves preparing the documents needed for solicitation and determining the evaluation criteria for the contract award. Two common examples of solicitation documents include a Request for Proposal (RFP) and a Request for Quotes (RFQ) document. A **Request for Proposal (RFP)** is a document used to solicit proposals from prospective suppliers. Many organizations issue RFPs to potential suppliers. For example, if a government department wants to automate its work practices, it writes and issues an RFP so suppliers can respond with proposals. Suppliers might propose various hardware, software, and networking solutions to meet the government's need. A **Request for Quotes (RFQ)** is a document used to solicit quotes or bids from prospective suppliers. Organizations often use an RFQ for solicitations that involve specific items. For example, if the government wanted to purchase 100 personal computers with specific features, it might issue an RFQ to potential suppliers. RFQs usually do not take nearly as long to prepare as RFPs, nor do responses to them.

Writing a good RFP is a critical part of project procurement management. Many people have never had to write or respond to an RFP. To generate a good RFP, expertise is invaluable. Many examples of RFPs are available within different companies, from potential contractors, and from government agencies. There are often legal requirements involved in issuing RFPs and reviewing proposals, especially for government projects. It is important to consult with experts familiar with the solicitation planning process for particular organizations. To make sure the RFP has enough information to provide the basis for a good proposal, the buying organization should try to put themselves in the suppliers' shoes. Could you develop a good proposal based on the information in the RFP? Could you determine detailed pricing and schedule information based on the RFP? Developing a good RFP is difficult, as is writing a good proposal.

Figure 12-4 provides a basic outline for an RFP. The main sections of an RFP usually include a statement of the purpose of the RFP, background information on the organization issuing the RFP, the basic requirements for the products and/or services being proposed, the hardware and software environment (usually important information for information technology-related proposals), a description of the RFP process, the statement of work and schedule information, and possible appendices. A simple RFP might be three to five pages long, while an RFP for a larger, more complicated procurement might be hundreds of pages long.

Request for Proposal Outline

I.   Purpose of RFP

II.   Organization's Background

III. Basic Requirements

IV.   Hardware and Software Environment

V.   Description of RFP Process

VI.   Statement of Work and Schedule Information

VII. Possible Appendices

     A. Current System Overview

     B. System Requirements

     C. Volume and Size Data

     D. Required Contents of Vendor's Response to RFP

     E. Sample Contract

Figure 12-4. Outline for a Request for Proposal (RFP)

Other documents often used in solicitation planning include invitations for bid, invitations for negotiation, and initial contractor responses. All solicitation documents should be written to facilitate accurate and complete responses from prospective suppliers. They should include background information on the organization and project, the relevant statement of work, a schedule, a description of the desired form of response, evaluation criteria, pricing forms, and any required contractual provisions. They should also be rigorous enough to ensure consistent, comparable responses, but flexible enough to allow consideration of supplier suggestions for better ways to satisfy the requirements.

It is very important for organizations to prepare some form of evaluation criteria, preferably before they issue a formal RFP or RFQ. Organizations use criteria to rate or score proposals, and they often assign a weight to each criterion to indicate how important it is. Some examples of criteria include the technical approach (30 percent weight), management approach (30 percent weight), past performance (20 percent weight), and price (20 percent weight). The criteria should be specific and objective. For example, if the buyer wants the supplier's project manager to be a certified Project Management Professional (PMP), that requirement should be stated clearly in the procurement documents and followed during the award process. Losing bidders may pursue legal recourse if the buyer does not follow a fair and consistent evaluation process.

INPUT, a Web-based information technology market research and marketing services firm, provides market reports and buyer guides to assist organizations in information technology outsourcing. One buyer guide offers suggestions on selection criteria, which should include reputation and past performance, industry knowledge, strategic partnership, and ability to meet needs. INPUT's research shows that most contracts include provisions to safeguard against unsatisfactory partnering, but most companies hesitate to exercise the provisions. Therefore, buyers should look for a supplier with a solid record of excellence and a proven reputation for quality.[9]

Organizations should heed the saying, "Let the buyer beware." It is critical to evaluate proposals based on more than the appearance of the paperwork submitted. A key factor in evaluating bids, particularly for projects involving information technology, is the past performance record of the bidder. The RFP should require bidders to list other similar projects they have worked on and provide customer references for those projects. Reviewing performance records and references helps to reduce the risk of selecting a supplier with a poor track record. Suppliers should also demonstrate their understanding of the buyer's need, their technical and financial capabilities, their management approach to the project, and their price for delivering the desired goods and services.

Some information technology projects also require potential suppliers to deliver a technical presentation as part of their proposal. The proposed project manager should lead the potential supplier's presentation team. When the outside project manager leads the proposal presentation, the organization can begin building a relationship with the potential provider from the beginning. Visits to contractor sites can also help the buyer get a better feeling for the supplier's capabilities and management style.

SOLICITATION

Solicitation involves obtaining proposals or bids from prospective suppliers. Prospective suppliers do most of the work in this process, normally at no cost to the buyer or project. The buying organization is responsible for advertising

---

[9] INPUT, "Executive Guide to IT Outsourcing in the U.S.," *www.input.com* (March 2000).

the solicitation, and often holds some sort of bidders' conference to answer questions about the solicitation. The main output of this process is receipt of proposals or bids.

Organizations can advertise to procure outside goods and services in many different ways. Sometimes a specific supplier might be the number-one choice for the buyer. In this case, the buyer gives solicitation information to just that company. If the preferred supplier responds favorably, both organizations proceed to work together. Many organizations have formed good working relationships with certain suppliers, so they want to continue working with them.

In many cases, however, there may be more than one supplier qualified to provide the goods and services. Providing information and receiving bids from multiple sources often takes advantage of the competitive business environment. As a result of pursuing a competitive bidding strategy, the buyer receives better goods and services than expected at a lower price.

A bidders' conference, also called a supplier conference or pre-bid conference, is a meeting with prospective suppliers prior to preparation of a proposal. These conferences help ensure that everyone has a clear, common understanding of the buyer's desired products or services. Before, during, or after the conference, the buyer may incorporate responses to questions into the procurement documents as amendments.

SOURCE SELECTION

Once buyers receive proposals, they must select a supplier or decide to cancel the procurement. Source selection involves evaluating bidders' proposals, choosing the best one, negotiating the contract, and awarding the contract. It is often a long, tedious process. Stakeholders in the procurement process should be involved in selecting the best supplier for the project. Often, teams of people are responsible for evaluating various sections of the proposals. There might be a technical team, a management team, and a cost team to focus on each of those major areas. Often buyers develop a short list of the top three to five suppliers to reduce the work involved in selecting a source.

Experts in source selection highly recommend that buyers use formal proposal evaluation sheets during source selection. Figure 12-5 provides a sample proposal evaluation sheet that the project team might use to help create a short list of the best three to five proposals. Notice that this example is a form of a weighted scoring model as described in Chapter 5, Project Scope Management. Experts also recommend that technical criteria should not be given more weight than management or cost criteria. Many organizations have suffered the consequences of paying too much attention to the technical aspects of proposals. For example, the project might cost much more than expected or take longer to complete because the source selection team focused only on technical aspects of proposals. Paying too much attention to technical aspects of proposals is especially likely to occur on information technology projects. However, it is often the supplier's management team—not the technical team—that makes procurement successful.

Criteria	Weight	Proposal 1		Proposal 2		Proposal 3	
		Rating	Score	Rating	Score	Rating	Score
Technical approach	30%						
Management approach	30%						
Past performance	20%						
Price	20%						
Total score	100%						

Figure 12-5. Sample Proposal Evaluation Sheet

After developing a short list of possible suppliers, organizations often follow a more detailed proposal evaluation process. Figure 12-6 lists items that might be part of an evaluation of the top three suppliers for a large information technology project. This particular list focuses on the project management capabilities of each supplier. Notice that the criteria include the project manager's educational background and PMP certification, his or her presentation (meaning the suppliers had to give a formal presentation as part of the evaluation process), and the organization's project management methodologies. All of these criteria are assigned a certain number of possible points, and the project team members and other stakeholders performing the evaluation would then assign points to each of the supplier finalists for each criterion. In the example in Figure 12-6, Supplier 3 has the highest score (28 out of 30 points) based on the management approach criteria for the award. Similar scores could be assigned for the selection criteria. The supplier with the most points based on all of the criteria for each category, therefore, should be offered the award.

Criteria	Possible Points	Supplier 1 Points	Supplier 2 Points	Supplier 3 Points
Project manager's educational background and experience	10	8	6	9
Project manager is PMP certified	5	5	0	5
Presentation on management approach	5	4	3	5
Organization's project management methodology	10	7	4	9
Total score	30	24	13	28

Figure 12-6. Detailed Criteria for Selecting Suppliers

It is customary to have contract negotiations during the source selection process. Suppliers on the short list are often asked to prepare a best and final offer (BAFO). People who negotiate contracts for a living often conduct these negotiations for contracts involving large amounts of money. In addition, top managers from both the buying and selling organizations often meet before making final decisions. The final output from the source selection process is a contract that obligates the supplier to provide the specified products or services and obligates the buyer to pay for them.

What Went Wrong?

The growth in information technology outsourcing may be one of the reasons for the creation and subsequent failure of several dot-com companies. Hundreds of dot-coms started in the late 1990s and early 2000s to provide products and services such as Internet hosting, e-commerce applications, and online publications. However, more than 130 Internet companies and more than 26 business-to-business companies shut down in 2000.[10] Several Web sites, such as

[10] Davis, Jessica, "Dot-com Failures Abound," InfoWorld, *www.infoworld.com* (November 16, 2000).

www.whytheyfailed.com, attempt to track dot-com failures. Kevin Scully, President and CEO of a successful dot-com, Webcritical Technologies, suggests that most dot-com failures were due to poor business planning, lack of top management operations experience, lack of leadership, and lack of vision.[11] Buyers should definitely consider the business stability of suppliers when making source selection decisions, and many dot-coms failed because buyers were not willing to take the risk of doing business with unstable dot-coms.

Selecting a stable company does not guarantee procurement success, however. Department 56, Inc., a giftware company in Minnesota, filed a $6 billion lawsuit against Arthur Andersen Worldwide in March 2001. Department 56, Inc. claimed that the well-known consulting firm badly managed the overhaul of their computer systems in 1999. Arthur Andersen assured Department 56 officials that the new system would not cost more than $3.3 million, but costs ultimately exceeded $14 million. Computer system problems allegedly required Department 56 to spend an additional $4 million to hire other consultants to fix the problems. In addition, Department 56 allegedly lost more than $8 million in revenues, took extraordinary charges to earnings, had $18 million in debt, and spent almost two years dealing with computer problems instead of focusing on company growth. Andersen claims the lawsuit was retaliation for a lawsuit they had filed five months earlier against Department 56 for nonpayment.[12]

CONTRACT ADMINISTRATION

Contract administration ensures that the supplier's performance meets contractual requirements. The contractual relationship is a legal relationship and as such is subject to state and federal contract laws. It is very important that appropriate legal and contracting professionals be involved in writing and administering contracts. Good contracts and contract administration can help organizations avoid problems such as those experienced by Department 56, Inc. and Arthur Andersen Worldwide, described in *What Went Wrong*.

Many project managers know very little about contract administration. Many technical professionals would prefer not to look at contracts at all. However, many project managers and technical people do not understand or trust their own purchasing departments, if their organization even has one, because of problems they may have encountered in the past. Ideally, the project manager and his or her team should be actively involved in writing and administering the contract so that everyone understands the importance of good procurement management. The project team should also seek expert advice in working with contractual issues.

Project team members must be aware of potential legal problems they might cause by not understanding a contract. For example, most projects involve

[11] Scully, Kevin, "There Are Solid Reasons for Dot.com Failures," Washington Techway, *www.washtech.com* (October 23, 2000).

[12] Moore, Janet, "Department 56 Sues Arthur Andersen for $6 Billion Over Computer System," *Minneapolis Star Tribune* (March 2, 2001).

changes, and these changes must be handled properly for items under contract. Without understanding the provisions of the contract, a project manager may not realize he or she is authorizing the contractor to do additional work at additional costs. Therefore, change control is an important part of the contract administration process.

Following are suggestions to ensure adequate change control on projects that involve outside contracts:

- Changes to any part of the project need to be reviewed, approved, and documented by the same people in the same way that the original part of the plan was approved.

- Evaluation of any change should include an impact analysis. How will the change affect the scope, time, cost, and quality of the goods or services being provided? There must also be a baseline to understand and analyze changes.

- Changes must be documented in writing. Project team members should document all important meetings and telephone calls.

When a project involves outside contracts, it is critical that project managers and team members watch for constructive change orders. **Constructive change orders** are oral or written acts or omissions by someone with actual or apparent authority that can be construed to have the same effect as a written change order. For example, if a member of the buyer's project team has met with the contractor on a weekly basis for three months to provide guidelines for performing work, he or she can be viewed as an apparent authority. If he or she tells the contractor to redo part of a report that has already been delivered and accepted by the project manager, that action can be viewed as a constructive change order and the contractor can legally bill the buyer for the additional work. Likewise, if this apparent authority tells the contractor to skip parts of a critical review meeting in the interests of time, the omission of that information is not the contractor's fault.

CONTRACT CLOSE-OUT

The final process in project procurement management is contract close-out. Contract close-out includes product verification to determine if all work was completed correctly and satisfactorily. Close-out also includes administrative activities to update records to reflect final results and archiving information for future use. Procurement audits are often done during contract close-out to identify lessons learned in the procurement process. Outputs from contract close-out include a contract file and formal acceptance and closure. The organization responsible for contract administration should provide the supplier with formal written notice that the contract has been completed. The contract itself should include requirements for formal acceptance and closure.

USING SOFTWARE TO ASSIST IN PROJECT PROCUREMENT MANAGEMENT

Over the years, organizations have used various types of productivity software to assist in project procurement management. For example, most organizations use word-processing software to write proposals or contracts, spreadsheet software to create proposal evaluation worksheets, databases to track suppliers, and presentation software to present procurement-related information.

It was not until the late 1990s and early 2000s, however, that many companies started using more advanced software to assist in procurement management. In fact, the term "e-procurement" often describes various procurement functions that are now done electronically. "The purchasing function was neglected for many, many years and has only recently begun to be viewed as strategic," said Hank Adamany, of PricewaterhouseCoopers. "Purchasing was a place where you had people who spent 20 to 25 years in positions below the senior team, and as a result the department never got much attention or investment. That's odd when you consider that those departments can account for up to 70 percent of a company's total spending."[13]

Many different Web sites and software tools are now available to assist in various procurement functions. For example, most college students have discovered Web sites to help them purchase books, airline tickets, or even automobiles. Likewise, many organizations can now access free Web-based portals for purchasing numerous items, or they can buy specialized software to help streamline their procurement activities. Companies such as Commerce One, Ariba, Concur Technologies, and others started providing corporate procurement services over the Internet in the past few years. Other established companies, such as Oracle, SAS, and Baan, have developed new software products to assist in procurement management. Traditional procurement methods were very inefficient and costly, and new e-procurement services have proved to be very effective in reducing the costs and burdens of indirect procurement. As Hank Adamany of PricewaterhouseCoopers said, "In fact, most user organizations that we surveyed were able to realize more than a 300 percent return on investment in Internet procurement automation within the first year of deployment."[14]

Organizations can also take advantage of information available on the Web, in industry publications, or in various discussion groups offering advice on selecting suppliers. For example, many organizations invest millions of dollars in enterprise resource planning (ERP) systems. ERP systems attempt to integrate several business functions, such as manufacturing, finance, human resources, sales, and distribution across an enterprise into a single computer system that can serve all those different departments' particular needs. Before deciding which ERP to use, find information describing specific products

[13] *Fortune*, "What Tools These Portals Be," *www.fortune.com/fortune/sections/eprocurement* (January 16, 2001).
[14] Ibid.

provided by various suppliers, prices, and current customer information to assist in making procurement decisions.

As with any information or software tool, organizations must focus on using the information and tools to meet project and organizational needs. Many non-technical issues are often involved in getting the most value out of new technologies, especially new e-procurement software. For example, organizations must often develop partnerships and strategic alliances with other organizations to take advantage of potential cost savings. Organizations should practice good procurement management in selecting new software tools and managing relationships with the chosen suppliers.

The processes involved in project procurement management follow a clear, logical sequence. However, many project managers are not familiar with the many issues involved in purchasing goods and services from other organizations. If projects will benefit by procuring goods or services, then project managers and their teams must follow good project procurement management. As outsourcing for information technology projects increases, it is important for all project managers to have a fundamental understanding of this knowledge area.

CASE WRAP-UP

After reading the contract for their consultants carefully, Marie McBride found a clause giving her company the right to terminate the contract with a one-week notice. She met with her project team to get their suggestions. They still needed help completing the operating system conversion project. One team member had a friend who worked for a competing consulting firm. The competing consulting firm had experienced people available, and their fees were lower than the fees in the current contract. Marie asked this team member to help her research other consulting firms in the area who could work on the operating system conversion project. Marie then requested bids from these companies. She personally interviewed people from the top three suppliers' management teams and checked their references for similar projects.

Marie worked with the purchasing department to terminate the original contract and issue a new one with a new consulting firm that had a much better reputation and lower hourly rates. This time, she made certain the contract included a statement of work, specific deliverables, and requirements stating the minimum experience level of consultants provided. The contract also included incentive fees for completing the conversion work within a certain time period. Marie learned the importance of good project procurement management.

CHAPTER SUMMARY

Procurement, purchasing, or outsourcing is acquiring goods and/or services from an outside source. Experts predicted that information technology outsourcing would grow to more than $110 billion by the year 2003. U.S. federal spending on information technology outsourcing services is projected to increase from $6.6 billion in 2002 to nearly $15 billion by 2007. Organizations outsource to reduce costs, focus on their core business, access skills and technologies, provide flexibility, and increase accountability. It is becoming increasingly important for information technology professionals to understand project procurement management.

Project procurement management processes include procurement planning, solicitation planning, solicitation, source selection, contract administration, and contract close-out.

Procurement planning involves deciding what to procure or outsource, what type of contract to use, and how to describe the effort in a statement of work. A make-or-buy analysis helps an organization determine whether it can cost-effectively procure a product or service. Project managers should consult internal and external experts to assist them with procurement planning because many legal, organizational, and financial issues are often involved.

The basic types of contracts are fixed price, cost reimbursable, time and material, and unit price contracts. Fixed-price contracts involve a fixed total price for a well-defined product and entail the least risk to buyers. Cost-reimbursable contracts involve payments to suppliers for direct and indirect actual costs and require buyers to absorb some of the risk. Time and material contracts are a hybrid of fixed price and cost reimbursable contracts and are commonly used by consultants. Unit price contracts involve paying suppliers a predetermined amount per unit of service and impose different levels of risk on buyers, depending on how the contract is written. It is important to decide which contract type is most appropriate for a particular procurement. All contracts should include specific clauses that address unique aspects of a project and that describe termination requirements.

A statement of work (SOW) describes the work required for the procurement in enough detail to allow prospective suppliers to determine if they are capable of providing the goods and services and to determine an appropriate price.

Solicitation planning involves writing procurement documents such as a Request for Proposal (RFP) and developing source selection evaluation criteria.

Solicitation involves finalizing procurement documents, advertising the work, holding a bidders' conference, and receiving proposals or bids.

Source selection is the process used to evaluate prospective suppliers and negotiate a contract. Organizations should use a formal proposal evaluation form when evaluating suppliers. Technical criteria should not be given more weight than management or cost criteria during evaluation.

Contract administration involves finalizing and awarding a contract, monitoring performance, and awarding contract modifications. The project manager and key team members should be involved in writing and administering the contract. Project managers must be aware of potential legal problems they might cause when they do not understand a contract. Project managers and teams should use change control procedures when working with outside contracts and should be especially careful about constructive change orders.

Contract close-out includes product verification and administrative close-out, and outputs include a contract file, formal acceptance, and closure. Procurement audits identify lessons learned during the procurement process.

Several types of software can assist in project procurement management. A new category of e-procurement software has been developed in the past few years to help organizations save money in procuring various goods and services. Organizations can also use the Web, industry publications, and discussion groups to research and compare various suppliers.

DISCUSSION QUESTIONS

1. List five reasons why organizations outsource. Why is there a growing trend in outsourcing, especially in the government?

2. Explain the make-or-buy process and how to perform the financial calculations involved in the process. What are the main types of contracts if you do decide to outsource? What are the advantages and disadvantages of each?

3. Do you think many information technology professionals have experience writing RFPs and evaluating proposals for information technology projects? What skills would be useful for these tasks?

4. What is involved in the solicitation process? How do organizations decide whom to send RFPs or RFQs?

5. How can organizations use a weighted decision matrix to evaluate proposals as part of source selection?

6. What are two suggestions for ensuring adequate change control on projects that involve outside contracts?

7. How can software assist in procuring goods and services? What is e-procurement software?

EXERCISES

1. Research information on information technology outsourcing. Find at least two articles and summarize them. Answer the following questions:

   a. What are the main types of goods and services being outsourced?

   b. What are some of the largest companies that provide information technology outsourcing services?

c. Why is outsourcing growing so rapidly?

d. Have most organizations benefited from outsourcing? Why or why not?

2. Interview someone who was involved in an information technology procurement process and have him or her explain the process that was followed. Write a paper describing the procurement and any lessons learned by the organization.

3. Suppose your company is trying to decide whether it should buy special equipment to prepare some of its high-quality publications itself or lease the equipment from another company. Suppose leasing the equipment costs $240 per day. If you decide to purchase the equipment, the initial investment is $6,800, and operations will cost $70 per day. After how many days will the lease cost be the same as the purchase cost for the equipment? Assume your company would only use this equipment for thirty days. Should your company buy the equipment or lease it?

4. Find an example of a contract for information technology services. Analyze the key features of the contract. What type of contract was used and why? Review the language and clauses in the contract. What are some of the key clauses? List questions you have about the contract and try to get answers from someone familiar with the contract.

5. Draft an RFP for an information technology project to provide laptops for all students, faculty, and staff at your college or university. Use the outline provided in Figure 12-4. List the assumptions you made in preparing the RFP.

6. Draft the source selection criteria that might be used for evaluating proposals for providing laptops for all students, faculty, and staff at your college or university. Use Figure 12-6 as a guide. Include at least five criteria, and make the total scores add up to 100. Write a paper justifying the criteria and scores.

MINICASE

Your company is reevaluating its policies concerning cellular telephones, and its approach to providing cellular telephone services to its employees. All 1500 consultants and salespeople are authorized to use their choice of cellular telephone services with unlimited calling privileges, and these expenses have increased dramatically. Several employees are using new features such as Internet access via their cell phones. To provide for new capabilities and to benefit from a corporate discount, your company is issuing a request for proposals to provide cellular telephone services for the entire company.

Part 1: Research at least three companies that provide cellular telephone services. Write a paragraph describing the purpose of the RFP that your company plans to issue and another paragraph describing the basic requirements. Write these paragraphs so that at least three potential suppliers would meet the basic requirements. Then develop a list of criteria that your company could use for evaluating a cellular telephone services supplier. Include at least ten criteria, and assign weights to each criterion to total 100 percent. Create a spreadsheet to accompany your proposal evaluation sheet, using Figure 12-5 as a guide.

Part 2: You want to include special clauses in your contract for cellular telephone services to provide for the following:

- To ensure that your company is getting the best rates possible. Assume that you will commit to a one-year renewable contract and a minimum of 1000 cell phones.
- To be able to make special features, such as Internet access, available to only certain employees
- To receive special reports identifying employees who have excessive charges

Write a paragraph for each of these clauses, using appropriate contractual language.

RUNNING CASES

Review the running cases provided in Appendix C for several questions related to project procurement management. In addition, review the related templates in Appendix D.

SUGGESTED READINGS

1. Chen, Yu-Che and James L. Perry, "IT Outsourcing: A Primer for Public Managers," IBM Endowment for The Business of Government (*www.businessofgovernment.org/pdfs/ChenReport.pdf*), (February 2003)

    This forty-page report describes recent trends in outsourcing and provides a five-phase process model for government executives to follow in making and then implementing information technology outsourcing decisions. Chen emphasizes the importance of government to move toward a partnership model for procuring information technology equipment and services.

2. Community B2B *(www.communityb2b.com)* (March 17, 2003).

    This Web site provides several articles related to e-procurement and using business-to-business (B2B) electronic commerce. There are links to several articles and white papers describing technologies and case studies of how companies use e-procurement.

3. Klepper, Robert, and Wendell Jones. *Outsourcing Information Technology, Systems, and Services*, Upper Saddle River, NJ: Prentice Hall, 1997.

    Klepper and Jones based this book on real-world experience and lessons learned from information technology executives, service providers, outsourcing consultants, attorneys, and others who have successfully and unsuccessfully outsourced. The book includes many lists, charts, facts, figures, and sample documents such as RFPs and contracts to guide you through outsourcing decisions and other processes.

4. Project Management Institute. *PM Network,* Special Topic Issue: Service and Outsourcing Projects (October 1998).

The special topic of this magazine is outsourcing. Articles include the growth in outsourcing, using tiger teams (an integrated team of subject matter experts) on outsourcing projects, developing a statement of work, enhancing supplier relationships, and financial constraints on outsourcing projects.

5. Ripin, Kathy M. and Leonard R. Sayles. *Insider Strategies for Outsourcing Information Systems: How to Build Productive Partnerships and Avoid Seductive Traps.* New York: Oxford University Press, 1999.

This book provides advice for negotiating with potential suppliers and facilitating information technology development projects. The authors provide several case studies to illustrate how client managers obtain new skills through project participation that enable them to make effective use of new technologies.

KEY TERMS

- **constructive change orders** — oral or written acts or omissions by someone with actual or apparent authority that can be construed to have the same effect as a written change order
- **contract** — a mutually binding agreement that obligates the supplier to provide the specified products or services, and obligates the buyer to pay for them
- **contract administration** — managing the relationship with the supplier
- **contract close-out** — completion and settlement of the contract, including resolution of any open items
- **cost plus fixed fee (CPFF) contract** — a contract in which the buyer pays the supplier for allowable performance costs plus a fixed fee payment usually based on a percentage of estimated costs

- **cost plus incentive fee (CPIF) contract** — a contract in which the buyer pays the supplier for allowable performance costs along with a predetermined fee and an incentive bonus

- **cost plus percentage of costs (CPPC) contract** — a contract in which the buyer pays the supplier for allowable performance costs along with a predetermined percentage based on total costs

- **cost-reimbursable contracts** — contracts involving payment to the supplier for direct and indirect actual costs

- **fixed-price** or **lump-sum contracts** — contracts with a fixed total price for a well-defined product or service

- **make-or-buy decision** — when an organization decides if it is in its best interests to make certain products or perform certain services inside the organization, or if it is better to buy them from an outside organization

- **procurement** — acquiring goods and/or services from an outside source

- **procurement planning** — determining what to procure and when

- **project procurement management** — the processes required to acquire goods and services for a project from outside the performing organization

- **Request for Proposal (RFP)** — a document used to solicit proposals from prospective suppliers

- **Request for Quote (RFQ)** — a document used to solicit quotes or bids from prospective suppliers

- **solicitation** — obtaining quotes, bids, offers, or proposals as appropriate

- **solicitation planning** — documenting product requirements and identifying potential sources

- **source selection** — choosing from among potential suppliers

- **statement of work (SOW)** — a description of the work required for the procurement

- **termination clause** — a contract clause that allows the buyer or supplier to end the contract

- **time and material contracts** — a hybrid of both fixed-price and cost-reimbursable contracts

- **unit price contract** — a contract where the buyer pays the supplier a predetermined amount per unit of service, and the total value of the contract is a function of the quantities needed to complete the work

Appendix A

Guide to Using Microsoft Project 2002

OPENING CASE

Kathy Nelson was managing a project to develop a Project Tracking database for all of her firm's projects. She asked one of her assistants, Terry Dunlay, to work with her in using Microsoft Project 2002 to help manage the project. Kathy was familiar with using previous versions of Microsoft Project, and Terry had never used any type of project management software before. Terry had been out of the job market for a few years, but she did use word-processing software and e-mail almost daily. She had also built some very simple spreadsheets to keep track of her family's finances.

Kathy suggested that Terry familiarize herself with Project 2002 by reading information on Microsoft's Web site, working through the online tutorials and documentation that came with the software, opening several template files, and just playing around with the software. Kathy also suggested that Terry read a good textbook on project management so she would understand some of the concepts behind the software. Kathy knew from experience that project management software was very confusing if you did not know what it meant to create a work breakdown structure (WBS), link tasks, assign resources, view Gantt charts and network diagrams, set a baseline, enter actuals, and so on. Once Terry was ready, Kathy would help her enter information regarding the project and use this powerful software to help manage the Project Tracking Database project and future projects.

INTRODUCTION

This appendix provides a concise guide to using Microsoft Project 2002 to assist in performing project management. The CD-ROM in the back of this text includes a 120-day trial version of Project 2002 Standard. You can also download evaluation copies of various Microsoft Project products from *www.microsoft.com/project*. Project 2002 can help users manage different aspects of all nine project management knowledge areas. Most users, however, focus on using Project 2002 to assist with scope, time, cost, human resource, and communications management. This appendix uses these project management knowledge areas as the context for learning how to use Project 2002.

As described in Chapter 1, Introduction to Project Management, hundreds of project management software products are available today. Microsoft Project is the clear market leader among midrange applications. A 1998 survey of 1,000 project managers showed that Microsoft Project was by far the most widely used computerized project management tool. When asked to list three project management tools they were currently using or had used within the past three years, 48 percent of the 159 survey respondents listed Microsoft Project. Only 13.8 percent of respondents mentioned the second most-listed tool, Primavera Project Planner. The third most-listed tool was Microsoft Excel, the only one of the ten tools listed that is not categorized as a project management software tool. Survey respondents used Microsoft Project for project control and tracking, detailed scheduling, early project planning, communication, reporting, high-level planning, Gantt chart and Critical Path Method (CPM) development, and Program Evaluation and Review Technique (PERT). Respondents used Excel for budgeting, cost analysis, variance analysis, tracking and reporting, and creating WBSs.[1] In a December 2001 press release, the pci group and the American Productivity and Quality Council (APQC) reported that 88 percent of the companies in their study reported using scheduling software, with more than 65 percent using Microsoft Project.[2] An August 2002 software review stated that Microsoft Project has dominated the standalone desktop market for years, but up until 2002 Microsoft had ceded the enterprise market to other vendors like Welcom and Primavera. Enterprise project management software, as mentioned in Chapter 1, includes high-end tools that provide robust capabilities to handle very large projects, dispersed workgroups, and enterprise functions that summarize and combine individual project information to provide an enterprise view of all projects. Microsoft entered the enterprise market with its Solution for Enterprise Project Management, which consists of Project 2002 combined with Project Server 2002.[3]

[1] Fox, Terry L. and J. Wayne Spence, "Tools of the Trade: A Survey of Project Management Tools," *Project Management Journal* (September 1998): 20–27.

[2] Trent, Theresa, "The pci group Presents Project Management Best Practices Findings," the pci group (December 10, 2001).

[3] Miller, Ron, "Aim For the Enterprise," *New Architect* (August 2002).

Even with its popularity, however, Microsoft Project is not nearly as widespread in its use as other Microsoft products such as Word or Excel. At Inmark Communications' Project World Conference in December 1998, Microsoft issued a press release stating that it had surpassed the three-millionth-customer mark for Microsoft Project products.[4] Customers of Word or Excel are much higher—well over 100 million. In January 2000, Microsoft announced that revenues from Microsoft Project had grown to more than $100 million and that it planned to develop and enhance applications.[5]

Before being able to use Project 2002 effectively, you must understand the fundamental concepts of project management. Therefore, to master the proper use of Project 2002, use this guide in conjunction with the main text. Each of the chapters on the project management knowledge areas (Chapters 4 through 12) includes a section describing how software, including Project 2002, can assist in managing that area.

NEW FEATURES OF MICROSOFT PROJECT 2002

Because there have been several previous versions of Microsoft Project, it is useful to understand some of the new capabilities of Project 2002, especially if you are working with people who are upgrading from a previous version.

One of the most significant differences between Project 2002 and Project 2000 is that Microsoft now provides two different versions of Project 2002—Microsoft Project Standard 2002 and Microsoft Solution for Enterprise Project Management.

1. Microsoft Project Standard 2002 is most similar to previous versions of Project and is intended for individual desktop usage, although it does include one Client Access License (CAL) for connection to Project Server 2002 via the Web, if available. This appendix focuses on using Microsoft Project Standard 2002.

2. The Microsoft Solution for Enterprise Project Management version includes several components, such as Microsoft Project Professional 2002, Microsoft Project Server 2002 (which includes Microsoft Project Web Access), and Microsoft Project Server Client Access. Project Server 2002 also requires SQL Server. Microsoft SQL Server is a complete database and analysis software package that helps organizations develop enterprise applications. Organizations should develop and apply many standards, templates, codes, and procedures before using the enterprise version of Project 2002 to make the best use of its capabilities.

[4] Microsoft Corporation, *Microsoft Project Surpasses 3 Millionth-customer Mark: Customers Help Make Microsoft Project a Category Leader*, San Jose, CA: Microsoft (December 7, 1998).
[5] Business Wire, "Pacific Edge Software Appoints Jim Dunnigan as Vice President of Marketing" (1/19/00).

Microsoft's Web site lists details on these different versions (*www.microsoft. com/project*). Figure A-1 provides an overview of using either Project Standard 2002 or Solution for Enterprise Project Management.

Microsoft Project Standard 2002

Project 2002 Standard

Desktop application
For individual business and project managers
Most similar to Project 2000
Includes one Client Access License (CAL)

Microsoft Solution for Enterprise Project Management

Project team members and executives use CALs to access project information

Project managers use on their desktops, provides full functionality

Project Web Access

Project Professional 2002

Project Server 2002

Server software, central location for project-related information, comes with five CALs

Microsoft SQL Server

Required for enterprise project management functionality

Figure A-1. Microsoft Project Standard 2002 Versus Microsoft Solution for Enterprise Project Management

There are also several new features available in Project 2002. Both versions of Project 2002 include the following new features:

- New interactive help tool called the Project Guide
- Microsoft Office XP interface
- Smart tags
- Improved integration with Microsoft Excel and Outlook
- Resource availability graphs
- XML file format

The Microsoft Solution for Enterprise Project Management also includes the following features:

- Enterprise resource pool
- Skill-based resource assignment and replacement
- Improved e-mail notifications and Web-based timesheets and status reports
- Document library using Microsoft SharePoint Team Services
- Issue tracking database
- Project portfolio views
- Modeling and scenario analysis tools
- Real-time reporting
- Enterprise codes, custom fields, and templates
- Project data security layer
- Load balancing and clustering

The steps in this appendix assume you are using Project 2002 Standard, running Windows 2000 or XP, and are familiar with other Windows-based applications. The screen shots were taken using Windows 2000 and Project 2002 Standard, and they will help you check your work. You can also check your work against the solution files that are available for download from the companion Web site for this text (*www.course.com/mis/schwalbe*) or from your instructor.

Project 2002 supports many features that help you to perform as an effective project manager. You will learn how to take advantage of these features as you work through this appendix, which uses a fictitious information technology project—the Project Tracking Database—to illustrate how to use the software. Each section of the appendix includes hands-on activities for you to work through. To complete the hands-on activities in the appendix, you will need to create a Data Disk or copy files to a location on your hard drive or network drive. If you are making your own Data Disk, you will need a blank, formatted high-density disk. You will need to copy a set of files from a file server or from the companion Web site for this text (*www.course.com/mis/schwalbe*). When you begin each set of steps, make sure you are using the correct file from your Data Disk, hard drive, or network drive. You should have the following files before you begin your work:

- Software Development.mpt
- resource.mpp
- kickoffmeeting.doc

In addition, you will create the following files from scratch as you work through the steps. After you create the following files, you will need to save each file:

- scope.mpp
- time.mpp
- tracking.mpp
- baseline.mpp
- level.mpp
- taskinfo.html

For further practice, complete the exercises at the end of this appendix. The running cases in Appendix C and some of the exercises and Minicases in other chapters also provide questions involving Project 2002.

Now that you understand project management concepts and the basic project management terminology, you will learn how to start Project 2002, review the Help facility and a template file, and begin to plan the Project Tracking Database project.

OVERVIEW OF MICROSOFT PROJECT 2002

The first step to mastering Project 2002 is to become familiar with the Help facility and online tutorials, major screen elements, views, and filters. This section describes each of these features.

Starting Project 2002 and the Project Help Window

To start Microsoft Project 2002:

1. *Open Project 2002*. Click **Start** on the Windows taskbar, select **Programs** (or **All Programs** if using Windows XP), and then click **Microsoft Project**. Alternatively, a shortcut or icon might be available on the desktop; in this case, double-click the icon to start the software.
2. *Close the New Project Task pane*. Click the **Close** button ✖ in the upper-right corner of the New Project Task pane. The Project Guide window will display along with a blank project screen, as shown in Figure A-2.

Figure A-2. Project Guide and Project 2002 Main Screen

 If the View bar is not open, select View from the Menu bar, and then select View bar.

3. *Maximize Project 2002.* If the Project 2002 program window does not fill the entire screen as shown in Figure A-2, click the Maximize button in the upper-right corner of the program window.

 Do not worry if your toolbars or other screen elements look slightly different from Figure A-2. Focus on the Project 2002 screen and the Project Guide window.

Project 2002 is now running and ready to use. Notice some of the elements of the Project 2002 screen. The default view is the Gantt Chart view. You can access other views by clicking on each icon in the View bar. When Project 2002 opens, the Project Guide window automatically opens, as shown in Figure A-2.

You can change this option and other options by selecting Options from the Tools menu, as shown in Figure A-3. The Project Guide window helps you learn Project 2002 by providing instructions for completing various steps in building a Project 2002 file.

Display Project Guide →

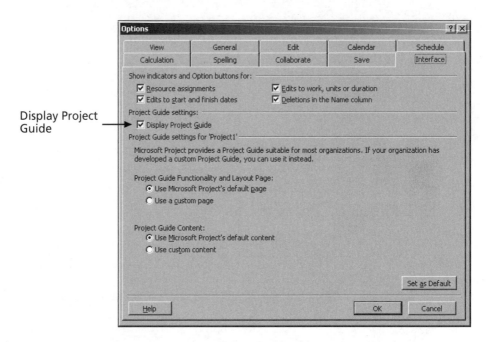

Figure A-3. Displaying the Project Guide in the Options Dialog Box

You can also display a Project Guide toolbar or other toolbars, as desired, by selecting them under Toolbars from the View menu, as shown in Figure A-4. To close the Project Guide window, click the Close button in the Project Guide window.

Figure A-4. Displaying the Project Guide and Other Toolbars from the View Menu

In addition to using the Project Guide window, you can access other information to help you learn how to use Project 2002 under the Help menu. Figure A-5 displays the Help menu options available, and Figure A-6 provides a description of each option.

Figure A-5. Project 2002 Help Menu Options

MENU OPTION	DESCRIPTION
Microsoft Project Help	Displays the Office Assistant, a context-sensitive, interactive guide to finding information on Microsoft Project 2002 and that sometimes opens automatically to help you with a task. You can also type a direct question to search the Help tool for an answer.
Show the Office Assistant	Acts as a toggle to display and hide the Office Assistant.
Contents and Index	Enables you to look up information in the Project 2002 Help tool as you would in an encyclopedia. This option bypasses the Office Assistant and goes directly into the Help manual.
Reference	Provides information on Microsoft Project features, reference content in the Help tool, and links to additional resources.
Getting Started	Accesses the Project Map, tutorial, and What's New features to help you learn how to use Project 2002.
What's This?	Provides context-sensitive Help information. When you choose this command, the pointer changes to the Help pointer, which you can then use to click any object or option on the screen, including menu commands, to see a description of the item.
Office on the Web	When you are connected to the Internet, provides up-to-the-minute information, news, and answers to common Project 2002 questions on the Microsoft Web site.
Detect and Repair	Automatically finds and fixes errors in all Office files.
About Microsoft Project	Displays the version of Microsoft Project, licensed use, and other information about Project 2002.

Figure A-6. Descriptions of Project 2002 Help Menu Options

Microsoft realizes that Project 2002 can take some time to learn and provides a number of resources on its Web site. Select Office on the Web from the Project 2002 Help menu to go to Microsoft's Web site for Office products. Microsoft's Web site for Project 2002 (*www.microsoft.com/project*) provides files for users to download, case studies, articles, and other useful materials.

Within Project 2002, the Office Assistant helps answer questions as they arise. You can access the Office Assistant at any time by selecting Microsoft Project Help from the Help menu. You can also press F1 or click the Office Assistant icon on the Standard toolbar. As in other Microsoft applications, the Office Assistant allows you to type a question and search for related information. You can also hide the Office Assistant from the Help menu if you no longer want it to appear on your screen. Feel free to experiment with the various Help tools.

Main Screen Elements

Figure A-7 shows the Project 2002 main screen after closing the Project Guide window. The Project 2002 default main screen, called the Gantt Chart view, has three parts: a Gantt chart, an Entry table, and the View bar. At the top of the main screen, the menu bar and toolbar are similar to those in other Windows applications. Note that the order and appearance of icons on your toolbar may vary from those shown in Figure A-7, depending on the features you are using. You can display the toolbar in one row or two, and icons often change according to which ones you use. Figure A-7 shows several commonly used icons, such as the Link Tasks icon, Zoom In icon, and Indent icon. Below the toolbar is the Entry bar, which displays entries you make in the Entry table, located right below the Entry text box. The Gantt chart appears on the right of the split bar, which separates the Entry table and the Gantt chart. Use the Entry table to enter WBS task information. The column to the left of the Task Name column is the Indicators column. The Indicators column displays indicators or symbols related to items associated with each task, such as task notes or hyper-links to other files.

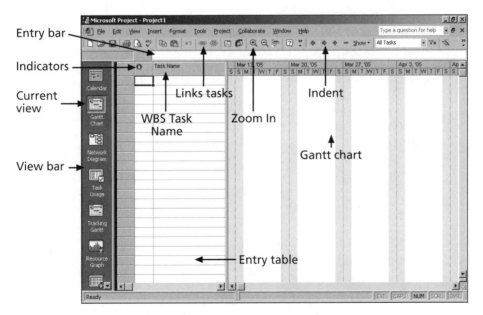

Figure A-7. Project 2002 Main Screen

Many features in Project 2002 are similar to features in other Windows applications. For example, to collapse or expand tasks, click the appropriate symbols to the left of the task name. To access shortcut items, right-click in either the Entry table area or the Gantt chart. Many of the Entry table operations in Project 2002 are very similar to operations in Excel. For example, to adjust a column width, click and drag or double-click between the column heading titles.

If you select another view and want to return to the Gantt Chart view, select Gantt Chart from the expanded View bar on the left of the screen or select View from the menu bar, and click Gantt Chart, as shown in Figure A-8. If the table on the left appears to be different, select View, point to Table: Entry, and then click Entry to return to the default Entry table view.

Figure A-8. Project 2002 View Menu Options

Notice the split bar that separates the Entry table from the Gantt chart. When you move the mouse over the split bar, your cursor changes to the resize pointer. Clicking and dragging the split bar to the right reveals other task information in the Entry table, including duration, start and finish dates, predecessors, and resource names.

To adjust Project 2002 screen elements:

1. *Open a new template file.* Click **File** on the menu bar, and then click **New**. Click the **General Templates** link below **New from Template** in the New Project Task pane, as shown in Figure A-9.

New Project task pane

General Templates

Figure A-9. Opening a Template File

? **Help** After you open template files, those filenames will appear on your screen under the New from Template section. For example, Figure A-9 shows that the template files for Project Office and Home Move were recently accessed on this user's screen.

2. *Open the Software Development template file.* Click the **Project Templates** tab in the Templates dialog box, scroll down to find the file named Software Development, and then double-click that file's icon.

Help

You may be prompted to install the template file. If you cannot find the Software Development file, you can download it from the companion Web site for this text.

3. *Close the New Project Task pane.* Click the **Close** button to close the New Project Task pane. Your screen should resemble Figure A-10.

Figure A-10. Software Development Template File

4. *Adjust the timescale.* Click the **Zoom Out** button ⊖ on the toolbar *twice* to display the timescale in years. If you cannot find the Zoom Out button, you can add it to the Standard toolbar by clicking the **Toolbar options** button.

5. *Select Outline Level 1 to display WBS level 1 tasks.* Click the **Show** button's drop-down arrow on the toolbar, and then click **Outline Level 1**. Your screen should resemble Figure A-11 as you select Outline Level 1. Notice that only the level 1 WBS items display in the Entry table after you select Outline Level 1, and the timescale shows years. The black bars on the Gantt chart represent the summary tasks.

Task Name column   Show list arrow   Summary tasks

Figure A-11. Viewing Summary Tasks Using the Show Button

6. *Widen the Task Name column.* To make all the text display in the Task Name column, move the cursor over the right-column gridline in the Task Name header until you see the resize pointer, then double-click the **left mouse** button to automatically resize the column width to show all the text.

7. *Move the split bar to reveal more Entry table columns.* Move the split bar to the right to reveal the entire Duration column.

8. *Expand a task.* Click the expand symbol ⊞ to the left of Task 7, Analysis/Software Requirements. Your screen should resemble Figure A-12.

Expand and collapse symbols   Split bar

Figure A-12. Adjusting Screen Elements

Project 2002 Views

In Project 2002, there are many ways to display project information. These displays are called views. Several views discussed in the main text are on the default View bar: Gantt Chart, Network Diagram, Tracking Gantt Chart, and Resource Graph. Other views include Calendar, Task Usage, Resource Sheet, Resource Usage, and an option to display More Views. These different views allow you to examine project information in different ways, which helps you analyze and understand what is happening on your project.

To the left of the Entry table is the View bar. Instead of using commands on the View menu to change views, you can click icons on the View bar. To save screen space, you can hide the View bar by selecting View from the menu bar and deselecting View Bar. When the View bar is not visible, a blue line appears to the far left of the main screen. When you right-click this blue line, a shortcut menu appears, which gives you quick access to the other views.

The View menu also provides access to different tables and reports that display information in a variety of ways. Some tables that you can access from the View menu include Schedule, Cost, Tracking, and Earned Value. Some reports you can access from the View menu include Overview, Current Activities, Costs, Assignments, and Workload.

Some Project 2002 views, such as the Chart views, present a broad look at the entire project, whereas others, such as the Form views, focus on specific pieces of information about each task. Three main categories of views are available:

- Graphical: A chart or graphical representation of data using bars, boxes, lines, and images.
- Task Sheet or Table: A spreadsheet-like representation of data in which each task appears as a new row, and each piece of information about the task is represented by a column. Different tables are applied to a task sheet to display different kinds of information.
- Form: A specific view of information for one task. Forms are used to focus on the details of one task.

Table A-1 describes some of the predesigned views within each category that Project 2002 provides to help you display the project or task information that you need.

Table A-1: Common Project 2002 Views

CATEGORY OF VIEW	VIEW NAME	DESCRIPTION OF VIEW
Graphical	Gantt Chart	Standard format for displaying project schedule information that lists project tasks and their corresponding start and finish dates in a calendar format. Shows each task as a horizontal bar with the length and position corresponding to the timescale at the top of the Gantt chart.
	Network Diagram	Schematic display of the logical relationships or sequencing of project activities. Shows each task as a box with linking lines between tasks to show sequencing. Critical tasks appear in red.
Task Sheet or Table	Entry Table	Default table view showing columns for the Task Name and Duration. By revealing more of the Entry table, you can enter start and end dates, predecessors, and resource names.
	Schedule Table	Displays columns for Task Name, Start, Finish, Late Start, Late Finish, Free Slack, and Total Slack.
	Cost Table	Displays columns for Task Name, Fixed Cost, Fixed Cost Accrual, Total Cost, Baseline, Variance, Actual, and Remaining.
	Tracking Table	Displays columns for Task Name, Actual Start, Actual Finish, % Complete, Actual Duration, Remaining Duration, Actual Cost, and Actual Work.
	Earned Value	Displays columns for Task Name, BCWS, BCWP, ACWP, SV, CV, EAC, BAC, and VAC. See Chapter 6, Project Cost Management, for descriptions of these earned value terms.

Table A-1: Common Project 2002 Views (continued)

CATEGORY OF VIEW	VIEW NAME	DESCRIPTION OF VIEW
Form	Task Details Form	Displays detailed information about a single task in one window.
	Task Name Form	Displays columns for the Task Name, Resources, and Predecessors for a single task.

Next, you will use the Software Development template file to access and explore some of the views available in Project 2002. As mentioned earlier, you can configure the toolbars and icons in different ways, as with other Office XP products. When you first start Project 2002, the menus and toolbars display basic commands and buttons. As you work with Project 2002, the commands and buttons that you use most often are stored as personalized settings and displayed on menus and toolbars. Your toolbar might be displayed on two rows instead of one. To make the toolbar appear on one line, as shown in this text, click the left side of the second toolbar row and drag it to the first toolbar row. You can also select Customize from the Tools menu, then Toolbars, and select the option for the Standard and Formatting toolbars to be shown on two rows.

To access and explore different views:

1. *Explore the Network Diagram view.* Click the **Network Diagram** button on the View bar, and then click the **Zoom Out** button Ｑ on the toolbar *twice*. Your screen should resemble Figure A-13.

Figure A-13. Network Diagram View of Software Development File

2. *Explore the Calendar view.* Click the **Calendar** button [▦] on the View bar. Your screen should resemble Figure A-14.

Figure A-14. Calendar View of Software Development File

3. *Examine columns in the Entry table.* Click the **Gantt Chart** button [▦] on the View bar, and then move the **split bar** by moving your mouse between the Entry table and Gantt chart until you see the resize pointer. Click and drag your mouse to the right almost to the end of the screen to reveal other columns in the Entry table.

4. *Examine the Table: Schedule view.* Click **View** on the menu bar, move your mouse to **Table: Entry**, and then click **Schedule** in the cascading menu to the right. Notice that the columns to the left of the Gantt chart now display more detailed schedule information.

5. *Right-click the Select All button to access different table views.* Move your mouse to the **Select All** button to the left of the Indicator column symbol, and then **right-click** with your mouse. A shortcut menu to different table views displays, as shown in Figure A-15. Experiment with other table views, then return to the Table: Entry view.

Select All button

Figure A-15. Changing Table Views

6. *Explore the Reports feature.* Click **View** on the menu bar, click **Reports**, and then double-click **Overview** from the Reports dialog box. Double-click **Project Summary** in the Overview Reports dialog box. Notice that the insertion point now resembles a magnifying glass. Click inside the report to Zoom In or Zoom Out. Click the **Close** button to close this report, and then experiment with viewing other reports. You will use several reports and other views throughout this appendix.

7. *Close the Reports feature.* Click **Close** to close the Reports dialog box and return to the Gantt Chart view.

Project 2002 Filters

Project 2002 uses an underlying relational database to filter, sort, store, and display information. This database is compatible with Microsoft Access, and it is relatively easy to import information from and export information to other applications. Project 2002 is an Object Linking and Embedding for Databases (OLE-DB) provider, meaning that it has low-level connectivity between multiple types of data sources. For example, you can import or export data to other applications such as Excel, Access, Oracle, and so on.

You can filter information by clicking the Filter text box list arrow on the toolbar. Figure A-16 shows the resulting list of filters. You can access more filter options by using the scroll bar.

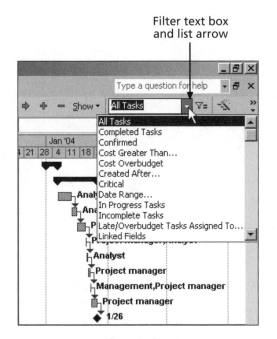

Figure A-16. Filter List Options

Filtering project information is very useful. For example, if a project includes hundreds of tasks, you might want to view only summary or milestone tasks to get an overview of the project. To get this type of overview of a project, select the Milestones or Summary Tasks filter from the Filter list. If you are concerned about the schedule, you can select a filter that shows only tasks on the critical path. Other filters include Completed Tasks, Late/Overbudget Tasks, and Date Range, which displays tasks based on dates you provide. As shown earlier, you can also click the Show list arrow on the toolbar to quickly display different levels in the WBS. For example, Outline Level 1 shows the highest-level items in the WBS, Outline Level 2 shows the next level of detail in the WBS, and so on.

To explore Project 2002 filters:

1. *Filter to show specific tasks.* From the Table: Entry view in the Software Development file, apply a filter to see only summary tasks. Click the **Filter** list arrow, scroll down until you see Summary Tasks, and then click **Summary Tasks**. Click **Zoom Out** on the toolbar once, or as needed, so the timescale displays quarters. Your screen should resemble Figure A-17.

 If the Filter text box does not appear on your toolbar, click **Toolbar options** and add the Filter button to the Formatting toolbar.

Figure A-17. Summary Tasks Filter for Software Development File

2. *Show outline levels.* Select **All Tasks** from the Filter list box to reveal all the tasks in the WBS again. Click the **Show** list arrow, and then click **Outline Level 2**. Now the level 2 items and above appear in the WBS. This file only goes down to WBS level 2. Experiment with other outline levels and filters.

3. *Examine other template files.* When you are finished reviewing the Software Development file, click **Close** from the File menu or click the **Close** button. A Save Changes dialog box appears. Click **No**. Open and examine other template files as you desire.

4. *Exit Project 2002.* Select **Exit** from the File menu or click the **Close** button for Project 2002.

Now that you are familiar with Project 2002 main screen elements, views, and filters, you will learn how to use Project 2002 to assist in project scope management by creating a new project file, developing a WBS, and setting a baseline.

PROJECT SCOPE MANAGEMENT

Project scope management involves defining the work to be done to carry out the project. To use Project 2002, you must first determine the scope of the project. To begin determining the project's scope, create a new file with the project name and start date. Develop a list of tasks that need to be done to carry out the project. This list of tasks becomes the work breakdown structure (WBS; see Chapter 5). If you intend to track actual project information against the initial plan, you must set a baseline. In this section, you will learn how to create a new project file, develop a WBS, and set a baseline. Assume you are just like Terry Dunlay from the opening case, and you are learning how to use Project 2002 to help plan and manage the Project Tracking Database project. To start, you will enter the scope-related information.

 In this section, you will go through several steps to create a Project 2002 file named scope.mpp. If you want to download the completed file to check your work or continue to the next section, a copy of scope.mpp is available on the companion Web site for this text or from your instructor. You should try to complete an entire section of this appendix (project scope management, project time management, and so on) in one sitting to create the complete file.

Creating a New Project File

To create a new project file:

1. *Create a blank project.* Open Project 2002, and close the New Project Task pane and the Project Guide window. Project 2002 automatically opens a blank project file when you start the application. If Project 2002 is already open and you want to open a new file, click the **New** icon on the toolbar, similarly to other Office applications.

2. *Open the Project Information dialog box.* Click **Project** on the menu bar, and then click **Project Information** to display the Project Information dialog box, as shown in Figure A-18. The Project Information dialog box enables you to set dates for the project, select the calendar to use, and view project statistics. The project start date will default to today's date. The default filenames are Project1, Project2, and so on.

Start date
text box →

Project Information for 'Project1'	? X
Start date: Sun 3/20/05 ▼	Current date: Sun 3/20/05 ▼
Finish date: Sun 3/20/05 ▼	Status date: NA ▼
Schedule from: Project Start Date ▼	Calendar: Standard ▼
All tasks begin as soon as possible.	Priority: 500 ▼
Help Statistics...	OK Cancel

Figure A-18. Project Information Dialog Box

3. *Enter the project start date.* In the Start date text box, enter **4/04/05**. Setting your project start date to 4/04/05 will ensure that your work matches the results that appear in this appendix. Leave the Finish date, Current date, and other information at the default settings. Click **OK**.

4. *Enter project properties.* Click **File** on the menu bar, and then click **Properties**. You may need to click the arrows to expand the menu to show Properties.

 Any command that you click in the expanded menu is added immediately to the personalized (short) version of the menu. If you stop using a command for a while, Project 2002 stops showing it on the short version of the menu.

5. *Enter project information.* Type **Project Tracking Database** in the Title text box, type **Terry Dunlay** in the Author text box, as shown in Figure A-19, and then click **OK**.

Figure A-19. Project Properties Dialog Box

You may have some default information entered in the Project Properties dialog box, such as your company's name. Also, as with other Office applications, you can change the default toolbar and menu options. Select Tools, Customize, and then Toolbars from the menu bar. On the Options tab, you can personalize menus and toolbars. To display the Project 2002 toolbars on one row, uncheck the first check box for Standard and Formatting toolbars to be on two rows. To always display the full set of menu options, make sure the check box for the second item—Menus show recently used commands first—is unchecked.

Developing a Work Breakdown Structure

Before using Project 2002, you must develop a work breakdown structure (WBS) for your project. Developing a good WBS takes time, and it will make entering tasks into the Entry table easier if you develop the WBS first. It is also a good idea to establish milestones before entering tasks in Project 2002. See

Chapter 5, Project Scope Management, for more information on creating a WBS and determining milestones. You will use the information in Table A-2 to enter tasks for the Project Tracking Database project. Be aware that this example is much shorter and simpler than most WBSs.

To develop a work breakdown structure and enter milestones for the Project Tracking Database project:

1. *Enter task names.* Enter the twenty tasks in Table A-2 into the Task Name column in the order shown. Do not worry about durations or any other information at this time. Type the name of each task into the Task Name column of the Entry table, beginning with the first row. Press **Enter** or the **down arrow** key on your keyboard to move to the next row. If you accidentally skip a task, highlight the task row and select **Insert** from the menu bar, then select **New Task**, to insert a blank row. To edit a task entry, click the text for that task, click the entry bar under the formatting toolbar, and either type over the old text or edit the existing text.

Table A-2: Project Tracking Database Tasks

ORDER	TASKS	ORDER	TASKS
1	Initiating	11	Design
2	Kickoff meeting	12	Implementation
3	Develop project charter	13	System implemented
4	Charter signed	14	Controlling
5	Planning	15	Report performance
6	Develop project plans	16	Control changes
7	Review project plans	17	Closing
8	Project plans approved	18	Prepare final project report
9	Executing	19	Present final project
10	Analysis	20	Project completed

Entering tasks into Project 2002 and editing the information is similar to entering and editing data in an Excel spreadsheet. The 2002 version of Project 2002 has a new feature called a SmartTag, which appears when you delete a row to clarify whether you want to delete the entire task or only clear the contents of a cell.

2. *Move the split bar to reveal more columns.* If necessary, move the split bar ↔|↦ to the right to reveal the entire Task Name and Duration columns.

3. *Adjust the Task Name column width, as needed.* To make all the text display in the Task Name column, move the mouse over the right-column gridline in the Task Name header until you see the resize pointer, then click the **left mouse** button and drag the line to the right to make the column wider or double-click to automatically adjust the column width.

Notice that this WBS separates tasks according to the project management process groups of initiating, planning, executing, controlling, and closing. These tasks will be the level 1 items in the WBS for this project. It is a good idea to include all of these process group tasks, rather than focus only on the executing tasks of a project, as many people do. Recall that the WBS should include all of the work required for the project. In the Project Tracking Database WBS, the tasks will purposefully be left at a high WBS level (level 2). You will create these levels, or the WBS hierarchy, next when you create summary tasks. For a real project, you would usually break the WBS into even more levels to provide more details to describe all the work involved in the project. For example, analysis tasks for a database project might be broken down further to include preparing entity relationship diagrams for the database and developing guidelines for the user interface. Design tasks might be broken down to include preparing prototypes, incorporating user feedback, entering data, and testing the database. Implementation tasks might include more levels, such as installing new hardware or software, training the users, fully documenting the system, and so on.

Creating Summary Tasks

After entering the WBS tasks listed in Table A-2 into the Entry table, the next step is to show the WBS levels by creating summary tasks. The summary tasks in this example are Tasks 1 (initiating), 5 (planning), 9 (executing), 14 (controlling), and 17 (closing). You create summary tasks by highlighting and indenting their respective subtasks.

To create the summary tasks:

1. *Select lower-level or subtasks.* Highlight **Tasks 2** through **4** by left-clicking the text for Task 2 and dragging the mouse through the text for Task 4.

2. *Indent subtasks.* Click the **Indent** icon ➡ on the Formatting toolbar, so your screen resembles Figure A-20. After the subtasks (Tasks 2 through 4) are indented, notice that Task 1 automatically becomes boldface, which indicates that it is a summary task. A minus sign ⊟ appears to the left of the new summary task name. Clicking the minus sign will collapse the summary task and hide the subtasks beneath it. When subtasks

are hidden, a plus sign ⊞ appears to the left of the summary task name. Clicking the plus sign will expand the summary task. Also, notice that the symbol for the summary task on the Gantt chart has changed from a blue line to a black line with arrows indicating the start and end dates.

Expand or collapse symbols by summary tasks   Indent   Summary task symbol

		ⓘ	Task Name	Duration	
	1		⊟ **Initiating**	**1 day?**	
	2		Kickoff meeting	1 day?	
	3		Develop project charter	1 day?	
	4		Charter signed	1 day?	
	5		Planning	1 day?	
	6		Develop project plans	1 day?	
	7		Review project plans	1 day?	
	8		Project plans approved	1 day?	
	9		Executing	1 day?	
	10		Analysis	1 day?	
	11		Design	1 day?	
	12		Implementation	1 day?	
	13		System implemented	1 day?	
	14		Controlling	1 day?	
	15		Report performance	1 day?	
	16		Control changes	1 day?	
	17		Closing	1 day?	
	18		Prepare final project report	1 day?	
	19		Present final project	1 day?	
	20		Project completed	1 day?	

Figure A-20. Indenting Tasks to Create the WBS Hierarchy

3. *Create other summary tasks and subtasks.* Create subtasks and summary tasks for Tasks 5, 9, 14, and 17 by following the same steps. Indent **Tasks 6** through **8** to make Task 5 a summary task. Indent **Tasks 10** through **13** to make Task 9 a summary task. Indent **Tasks 15** through **16** to make Task 14 a summary task. Indent **Tasks 18** through **20** to make Task 17 a summary task.

To change a task from a subtask to a summary task or to change its level in the WBS, you can "outdent" the task. Highlight the task or tasks you wish to change, and click the Outdent icon ⬅ on the Formatting toolbar. Remember that the tasks in Project 2002 should be entered in an appropriate WBS format with several levels in the hierarchy.

Numbering Tasks

Depending on how Project 2002 is set up on your computer, you may or may not see numbers associated with tasks as you enter and indent them.

To display automatic numbering of tasks, using the standard tabular numbering system for a WBS:

1. *Display the Options dialog box.* Click **Tools** on the menu bar, and then click **Options**. The Options dialog box opens.
2. *Show outline numbers.* In the Options dialog box, click the **View** tab, if necessary. In the Outline options section of the View options for Project 1 section (at the bottom of the dialog box), click **Show Outline Number** so that a check mark appears in the check box. Figure A-21 shows the Options dialog box and resulting outline numbers applied to the WBS tasks. Click **OK** to close the dialog box.

Outline numbers show WBS Hierarchy

Show outline number check box

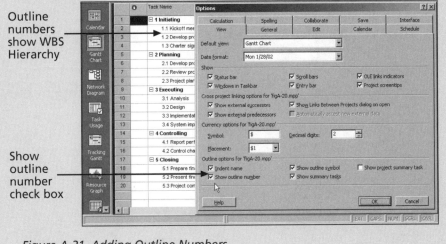

Figure A-21. Adding Outline Numbers

Saving Project Files With or Without a Baseline

An important part of project management is tracking performance against a baseline. Project 2002, unlike previous versions, does not prompt you to save a file with or without a baseline each time you save it. The default is to save without a baseline. It is important to wait until you are ready to save your file with a baseline, because Project 2002 will show changes against a baseline. Because you are still developing your project file for the Project Tracking Database project, you want to save the file without a baseline. Later in this appendix, you will save the file with a baseline by selecting Tools, Tracking, and then Set Baseline. You will then enter actual information to compare planned and actual performance data.

To save a file without a baseline:

1. *Save your file.* Save your file by clicking **File** on the menu bar and then clicking **Save**, or by clicking the **Save** icon 🖫.

2. *Enter a filename.* In the Save As dialog box, type **scope** in the File name text box. Save the file in your personal folder, on a floppy disk, or in the C:\temp folder, depending on how your computer is set up. Click **Save**. Your Project 2002 file should look like Figure A-22. (You can adjust the timescale to show months by clicking Zoom In or Zoom Out.)

 If you wish to download the Project 2002 file to check your work or continue to the next section, a copy of scope.mpp is available on the companion Web site for this text or from your instructor.

	ⓘ	Task Name	Duration
1		⊟ **1 Initiating**	**1 day?**
2		1.1 Kickoff meeting	1 day?
3		1.2 Develop project charte	1 day?
4		1.3 Charter signed	1 day?
5		⊟ **2 Planning**	**1 day?**
6		2.1 Develop project plans	1 day?
7		2.2 Review project plans	1 day?
8		2.3 Project plans approve	1 day?
9		⊟ **3 Executing**	**1 day?**
10		3.1 Analysis	1 day?
11		3.2 Design	1 day?
12		3.3 Implementation	1 day?
13		3.4 System implemented	1 day?
14		⊟ **4 Controlling**	**1 day?**
15		4.1 Report performance	1 day?
16		4.2 Control changes	1 day?
17		⊟ **5 Closing**	**1 day?**
18		5.1 Prepare final project r	1 day?
19		5.2 Present final project	1 day?
20		5.3 Project completed	1 day?

Figure A-22. Project 2002 Scope File

Microsoft provides a free download for Project 2002 called the Visio WBS Chart Wizard from *http://office.microsoft.com/downloads/2002/prjcomvi.aspx*. This tool makes it easy to create a WBS in chart form based on the hierarchy entered in Project 2002. Figure A-23 shows the WBS chart created with the Visio WBS Chart Wizard. Contact Microsoft Product Support for assistance with this tool.

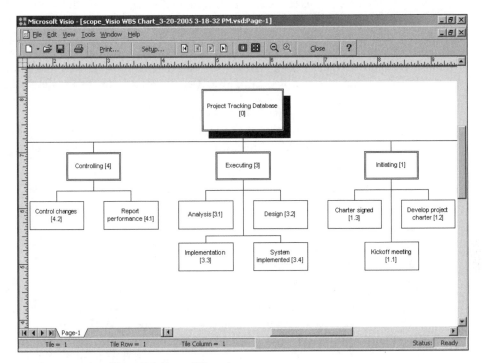

Figure A-23. Visio WBS Chart Wizard

Now that you have finished entering all twenty tasks, created the summary tasks and subtasks, set the options to show the standard WBS tabular numbering system, and saved your file, you will learn how to use the Project 2002 time management features.

PROJECT TIME MANAGEMENT

Many people use Project 2002 for its time management features. The first step in using these features, after entering the WBS for the project, is to enter durations for tasks or specific dates when tasks will occur. Entering durations or specific dates will automatically update the Gantt chart. If you plan to do critical path analysis, you must also enter task dependencies. After entering durations and task dependencies, you can view the network diagram and critical path information. This section describes how to use each of these time management features.

Entering Task Durations

When you enter a task, Project 2002 automatically assigns to it a default duration of one day, followed by a question mark. To change the default duration, type a task's estimated duration in the Duration column. If you are unsure of an estimate and want to review it again later, enter a question mark after it. For example, you could enter 5d? for a task with an estimated duration of five days that you want to review later. You can then use the Tasks With Estimated Durations filter to quickly see the tasks for which you need to review duration estimates.

To indicate the length of a task's duration, you must type both a number and an appropriate duration symbol. If you type only a number, Project 2002 automatically enters days as the duration unit. Duration unit symbols include:

- d = days (default)
- w = weeks
- m = minutes
- h = hours
- mon = months
- ed = elapsed days
- ew = elapsed weeks

For example, to enter one week for a task duration, type 1w in the Duration column. To enter two days for a task duration, type 2d in the Duration column. The default unit is days, so if you enter 2 for the duration, it will be entered as 2 days. You can also enter elapsed times in the Duration column. For example, 2ed means two elapsed days, and 2ew means two elapsed weeks. You would use an elapsed duration for a task like "Allow paint to dry." The paint will dry in exactly the same amount of time regardless of whether it is a workday, a weekend, or a holiday.

 If the Duration column is not visible, move the mouse over the split bar until the resize pointer is visible. Click and drag the split bar to the right until the Duration column is in view.

Entering time estimates or durations might seem like a straightforward process. However, you must follow a few important procedures:

- As you saw in the previous section, when you enter a task in the Task Name column, 1 day? appears in the Duration column. The question mark means that you are unsure of the duration or want to go back to it. Be sure to enter the durations you want to appear.

- Do not enter durations for summary tasks. Summary task durations are calculated automatically, based on the subtasks. If you enter a duration for a task and then make it a summary task, its duration will automatically change to match the durations of its subtasks. Project 2002 will not allow you to enter or change the duration of a summary task. Think of it as entering durations for work packages, the lowest-level items in your WBS. The other items are really the WBS hierarchy. Let Project 2002 do the duration calculations for you.

- To mark a task as a milestone, enter 0 for the duration. You can also make tasks that have a non-zero duration as milestones by checking the Mark as milestone option in the Task Information dialog box on the Advanced tab. The milestone symbol for them will appear at their start date. Double-click on a task to access the Task Information dialog box.

- To enter recurring tasks, such as weekly meetings or monthly status reports, select Recurring Task from the Insert menu. Enter the task name, the duration, and when the task occurs. Project 2002 will automatically insert appropriate subtasks based on the length of the project and the number of tasks required for the recurring task. For example, if you enter a recurring task for monthly review meetings that occur on the first day of every month for a twelve-month project, Project 2002 will enter a summary task for monthly review meetings and twelve subtasks—one meeting for each month.

- You can enter the exact start and finish dates for activities, instead of entering durations. To enter start and finish dates, move the split bar to the right to reveal the start and finish columns. Enter start and finish dates only when those dates are certain. If you want task dates to adjust according to any other task dates, do not enter exact start and finish dates. Instead, you should enter a duration and then establish a dependency to related tasks. *The real scheduling power of Project 2002 comes from setting up dependencies or relationships among tasks, as described in the next section.*

- Project 2002 uses a default calendar with standard workdays and hours. Duration estimates will vary according to the calendar used. For example, entering 5d in the standard calendar may result in more than 5 days on the Gantt chart if the time period includes Saturday or Sunday. You can change specific working and nonworking days or the entire project calendar by selecting Change Working Time from the Tools menu.

- You often need to adjust the timescale on the Gantt chart to see your project's schedule in different time frames, such as weeks, months, quarters, or years. You expand the timescale by clicking the Zoom Out icon. The Zoom In icon collapses the timescale. You can also change the format of the timescale by having three tiers instead of two or by formatting dates differently by double-clicking on the timescale section of the Gantt chart.

Next, you will set task durations in your Project Tracking Database file, created in the previous section. If you did not create the file named scope.mpp, you can download it from the companion Web site for this text. You will create a new recurring task and enter its duration, and then you will enter other task durations. First, create a new recurring task called Status Reports above Task 15, Report Performance.

To create a new recurring task:

1. *Insert a recurring task above Task 15, Report performance.* Open scope.mpp, if necessary, and then click **Report performance** (Task 15) in the Task Name column to select that task. Click **Insert** on the menu bar, then click **Recurring Task**. The Recurring Task Information dialog box opens.

2. *Enter task and duration information for the recurring task.* Type **Status Reports** as the task title in the Task Name text box. Type **1h** in the Duration text box. Select **Weekly** from the Recurrence Pattern radio buttons. Select **Every** from the Week On drop-down list. On the list of days, check **Wednesday**. In the Range of Recurrence section, type **4/6/05** in the Start text box, click the **End by**: radio button to select it, and type **9/7/05** in the End By text box. Select the drop-down list in the End By text box to see the calendar, as shown in Figure A-24. You can use the calendar to enter the Start: and End by: dates. You can also enter a number of occurrences instead of a To date for a recurring task. You might need to adjust the End by date after you enter all of your task durations and dependencies. Your screen should resemble Figure A-24. The new recurring task will appear above Task 15, Report Performance, when you are finished.

Figure A-24. Recurring Task Information Dialog Box

3. *View the new summary task and its subtasks.* Click **OK**. Project 2002 inserts a new Status Reports subtask in the Task Name column. Note that this new subtask is boldface and has a plus sign ⊞ to the left of the task name. Expand the new subtask by clicking ⊞ to the left of the words Status Reports. To collapse the recurring task, click the minus sign ⊟.

4. *Adjust the Duration column width and Gantt chart timescale.* Notice the # signs that appear in the Duration column for the Status Reports task. Like Excel, Project 2002 uses these symbols to denote that the column width needs to be increased. Increase the Duration column's width by moving your mouse to the right of the Duration column heading until you see a resize pointer ╬. Double-click ╬ to adjust the column width automatically to display the information. Adjust the time scale of the Gantt chart to display quarters by clicking the Zoom Out ⊝ icon on the toolbar three times. Your screen should resemble Figure A-25. Notice that the recurring task appears on the appropriate dates on the Gantt chart.

14		⊟ 4 Controlling	112.13 days?	
15	↻	⊟ 4.1 Status Reports	110.13 days	
16		4.1.1 Status Reports 1	1 hr	
17		4.1.2 Status Reports 2	1 hr	
18		4.1.3 Status Reports 3	1 hr	
19		4.1.4 Status Reports 4	1 hr	
20		4.1.5 Status Reports 5	1 hr	
21		4.1.6 Status Reports 6	1 hr	
22		4.1.7 Status Reports 7	1 hr	

Figure A-25. Expanding a Recurring Task

Use the information in Table A-3 to enter durations for the other tasks for the Project Tracking Database project. The Project 2002 row number is shown to the left of each task name in the table. Remember that you already entered a duration for the recurring task. Also, remember that you should *not* enter durations for summary tasks. Durations for summary tasks are automatically calculated to match the durations and dependencies of subtasks, as described further in the next section, "Establishing Task Dependencies."

Table A-3: Durations for Project Tracking Database Tasks

TASK NUMBER/ROW	TASK NAME	DURATION
2	Kickoff meeting	2h
3	Develop project charter	10d
4	Charter signed	0d
6	Develop project plans	3w
7	Review project plans	4mon
8	Project plans approved	0d
10	Analysis	1mon
11	Design	2mon
12	Implementation	1mon
13	System implemented	0d
39	Report performance	5mon
40	Control changes	5mon
42	Prepare final project report	2w
43	Present final project	1w
44	Project completed	0d

To enter task durations for the other tasks:

1. *Enter the duration for Task 2.* Click the **Duration** column for row 2, Kickoff meeting, type **2h**, and then press **Enter**.

2. *Enter the duration for Task 3.* In the **Duration** column for row 3, Develop project charter, type **10d**, then press **Enter**.

3. *Enter remaining task durations.* Continue to enter the durations, using information in Table A-3.

4. *Save your file and name it.* Click **File** on the menu bar, and then click **Save As**. Enter **time** for the filename, and save the file in the desired location on your computer or network.

Establishing Task Dependencies

To use Project 2002 to adjust schedules automatically and to do critical path analysis, you *must* determine the dependencies or relationships among tasks (see Chapter 6, Project Time Management). Project 2002 provides three methods for creating task dependencies: using the Link Tasks icon; using the Predecessors column of the Entry table; or clicking and dragging the Gantt chart symbols for tasks with dependencies.

To create dependencies using the Link tasks icon, highlight tasks that are related and click the Link Tasks icon (⊂⊃) on the toolbar. For example, to create a finish-to-start dependency between Task 1 and Task 2, click any cell in row 1, drag down to row 2, and then click the Link Tasks icon. The default type of link is finish-to-start. In the Project Tracking Database example, all the tasks use this default relationship. You will learn about other types of dependencies later in this appendix.

Selecting tasks is similar to selecting cells in Microsoft Excel. To select adjacent tasks, you can click and drag the mouse. You can also click the first task, hold down the Shift key, and then click the last task. To select nonadjacent tasks, hold down the Control (Ctrl) key as you click tasks in order of their dependencies.

When you use the Predecessors column of the Entry table to create dependencies, you must manually enter the information. To create dependencies manually, type the task row number of the preceding task in the Predecessors column of the Entry table. For example, Task 3 in Table A-3 has Task 2 as a predecessor, as entered in the Predecessors column, meaning that Task 3 cannot start until Task 2 is finished. To see the Predecessors column of the Entry table, move the split bar to the right.

You can also create task dependencies by clicking the Gantt chart symbol for a task and then dragging to the Gantt chart symbol for a task that succeeds it. For example, you could click the Milestone symbol ◆ for Task 4, hold down the left mouse button, and drag to the Task Bar symbol ▨▨▨ for Task 6 to create a dependency, as shown in the Gantt chart in Figure A-26. Note the Finish-to-Start Link dialog box that appears when you use this method.

Link tasks

Predecessors column

Gantt chart symbols

Figure A-26. Creating Task Dependencies

Next, you will use information from Figure A-27 to enter the predecessors for tasks as indicated. You will create some dependencies by manually typing the predecessors in the Predecessors column, some by using the Link Tasks icon, some by using the Gantt chart symbols, and the remaining dependencies by using whichever method you prefer.

To link tasks or establish dependencies for the Project Tracking Database project:

1. *Display the Predecessors column in the Entry table.* Move the Split bar ◀∥▶ to the right to reveal the Predecessors column in the time.mpp file. Widen the Task Name or other columns, if needed.

2. *Highlight the cell where you want to enter a predecessor, and then type the task number for the preceding task.* Click the **Predecessors cell** for Task 3, type **2**, and press **Enter**.

3. *Enter predecessors for Task 4.* In the Predecessors cell for Task 4, type **3**, and press **Enter**. Notice that as you enter task dependencies, the Gantt chart changes to reflect the new schedule.

4. *Establish dependencies using the Link Tasks icon.* To link Tasks 10 through 13, click the task name for Task 10 in the Task Name column and drag down through Task 13. Then, click the **Link Tasks** icon ⬚⬚⬚ on the toolbar.

5. *Create a dependency using Gantt chart symbols.* Click the **Milestone** symbol ◆ for Task 4 on the Gantt chart, hold down the **left mouse** button, and drag to the **Task Bar** symbol for Task 6. (See Figure A-26 to see this step in progress.)

6. *Enter remaining dependencies.* Link the other tasks by manually entering the predecessors into the Predecessors column, or by using the Link Tasks icon, or clicking and dragging the Gantt chart symbols. You can view the dependencies in Figure A-27. For example, Task 8 has Task 6 as a predecessor, Task 10 has Task 8 as a predecessor, and so on. If you have entered all data correctly, the project should end on 9/19/05.

7. *Adjust screen elements.* Select **View Bar** from the **View** menu to collapse the View bar, and then drag the resize line between the Indicator column and the Task Name columns to the left to remove that column. When you finish, your screen should look like Figure A-27. Double-check your screen to make sure you entered the dependencies correctly.

Figure A-27. Project Tracking Database File with Durations and Dependencies Entered

8. *Preview and save your file.* Select **Print Preview** from the File menu or click the **Print Preview** icon on the toolbar. If your previewed file does not resemble Figure A-27, you may need to adjust the location of the split bar between the Entry table and the Gantt chart. The entire Predecessors column must be visible in the Gantt chart view for it to show up in the Print Preview. When you are finished, close the Print Preview and save your file again by clicking the **Save** icon. If you desire, print your file by clicking the **Print** icon or by selecting Print from the File menu.

> If information in a column on the Entry table does not appear in the Print Preview, close the Print Preview and move the split bar to fully reveal the column, then select Print Preview again. Also, check your timescale to make sure you are not wasting paper by having too-detailed a timescale. You should not print your file until the Print Preview displays the desired information.

Changing Task Dependency Types and Adding Lead or Lag Time

A task dependency or relationship describes how a task is related to the start or finish of another task. Project 2002 allows for four task dependencies: finish-to-start (FS), start-to-start (SS), finish-to-finish (FF), and start-to-finish (SF). You can find detailed descriptions of these dependencies in Chapter 6 and in the Project 2002 Help facility under the keyword "dependency." By using these dependencies effectively, you can modify the critical path and shorten your project schedule. (See Chapter 6 for a detailed discussion of critical path analysis.) The most common type of dependency is finish-to-start (FS). All of the dependencies in the Project Tracking Database example are FS dependencies. However, sometimes you need to establish other types of dependencies. This section describes how to change task dependency types. It also explains how to add lead or lag times between tasks. You will shorten the duration of the Project Tracking Database project by adding lead time between some tasks.

To change a dependency type, open the Task Information dialog box for that task by double-clicking the task name. On the Predecessors tab of the Task Information dialog box, select a new dependency type from the Type column list arrow.

The Predecessor tab also allows you to add lead or lag time to a dependency. You can enter both lead and lag time using the Lag column on the Predecessor tab. Lead time reflects an overlap between tasks that have a dependency. For example, if Task B can start when its predecessor, Task A, is half-finished, you can specify a finish-to-start dependency with a lead time of 50 percent for the successor task. Enter lead times as negative numbers. In this example, enter –50% in the first cell of the Lag column. Adding lead times is also called fast tracking and is one way to compress a project's schedule (see Chapter 6).

Lag time is the opposite of lead time—it is a time gap between tasks that have a dependency. If you need a 2-day delay between the finish of Task C and the start of Task D, establish a finish-to-start dependency between Tasks C and D and specify a 2-day lag time. Enter lag time as a positive value. In this example, type 2d in the Lag column.

In the Project Tracking Database example, notice that work on design tasks does not begin until all the work on the analysis tasks has been completed (see Rows 10 and 11 on the WBS), and work on implementation tasks does not begin until all the work on the design tasks has been completed (see Rows 11 and 12 on the WBS). In reality, it is rare to wait until all of the analysis work is complete before starting any design work, or to wait until all of the design work is finished before starting any implementation work. It is also a good idea to add some additional time, or a buffer, before crucial milestones, such as a system being implemented. To create a more realistic schedule, add lead times to the design and implementation tasks and lag time before the system implemented milestone.

To add lead and lag times:

1. *Open the Task Information dialog box for Task 11, Design.* In the Task Name column, double-click the text for Task 11, **Design**. The Task Information dialog box opens. Click the **Predecessors** tab.

2. *Enter lead time for Task 11.* Type **–10%** in the Lag column, as shown in Figure A-28. Click **OK**. You could also type a value such as –5d to indicate a 5-day overlap. In the resulting Gantt chart, notice that the bar for this task has moved slightly to the left. Also, notice that the project completion date has moved from 9/19 to 9/15.

Figure A-28. Adding Lead or Lag Time to Task Dependencies

3. *Enter lead time for Task 12.* Double-click the text for Task 12, **Implementation**, type **–3d** in the Lag column, and click **OK**. Notice that the project now ends on 9/12 instead of 9/15.

4. *Enter lag time for Task 13.* Double-click the text for Task 13, **System Implemented**, type **5d** in the Lag column for this task, and click **OK**. Move the split bar to the right, if necessary, to reveal the Predecessors column. When you are finished, your screen should resemble Figure A-29. Notice the slight overlap in the taskbars for Tasks 10 and 11 and the short gap between the taskbars for Tasks 12 and 13. You can zoom in to see the gap more clearly. The project completion date should be 9/19.

	Task Name	Duration	Start	Finish	Predecessors
1	⊟ 1 Initiating	10.25 days	Mon 4/4/05	Mon 4/18/05	
2	1.1 Kickoff meeting	2 hrs	Mon 4/4/05	Mon 4/4/05	
3	1.2 Develop project charter	10 days	Mon 4/4/05	Mon 4/18/05	2
4	1.3 Charter signed	0 days	Mon 4/18/05	Mon 4/18/05	3
5	⊟ 2 Planning	95 days	Mon 4/18/05	Mon 8/29/05	
6	2.1 Develop project plans	3 wks	Mon 4/18/05	Mon 5/9/05	4
7	2.2 Review project plans	4 mons	Mon 5/9/05	Mon 8/29/05	6
8	2.3 Project plans approved	0 days	Mon 5/9/05	Mon 5/9/05	6
9	⊟ 3 Executing	80 days	Mon 5/9/05	Mon 8/29/05	
10	3.1 Analysis	1 mon	Mon 5/9/05	Mon 6/6/05	8
11	3.2 Design	2 mons	Thu 6/2/05	Thu 7/28/05	10FS-10%
12	3.3 Implementation	1 mon	Mon 7/25/05	Mon 8/22/05	11FS-3 days
13	3.4 System implemented	0 days	Mon 8/29/05	Mon 8/29/05	12FS+5 days
14	⊟ 4 Controlling	112.13 days	Mon 4/4/05	Wed 9/7/05	
15	⊞ 4.1 Status Reports	110.13 days	Wed 4/6/05	Wed 9/7/05	
39	4.2 Report performance	5 mons	Mon 4/4/05	Fri 8/19/05	
40	4.3 Control changes	5 mons	Mon 4/4/05	Fri 8/19/05	
41	⊟ 5 Closing	15 days	Mon 8/29/05	Mon 9/19/05	
42	5.1 Prepare final project report	2 wks	Mon 8/29/05	Mon 9/12/05	13
43	5.2 Present final project	1 wk	Mon 9/12/05	Mon 9/19/05	42
44	5.3 Project completed	0 days	Mon 9/19/05	Mon 9/19/05	43

Figure A-29. Schedule for Project Tracking Database File with Lead and Lag Times

5. *Save your file.* Click **File** on the menu bar, and then **Save**. Or, click the **Save** icon 🖫 on the toolbar.

You can enter or modify lead or lag times directly in the Predecessors column of the Entry table. Notice how the predecessors appear for Tasks 11, 12, and 13. For example, the Predecessor column for Task 11 shows 10FS-10%. This notation means that Task 11 has a finish-to-start (FS) dependency with Task 10 and a 10 percent lead. You can enter lead or lag times to task dependencies directly into the Predecessors column of the Entry table by using this same format: task row number, followed by type of dependency, followed by the amount of lead or lag.

Gantt Charts

Project 2002 shows a Gantt chart as the default view along with the Entry table. Gantt charts show the timescale for a project and all of its activities. In Project 2002, dependencies between tasks are shown on the Gantt chart by the arrows between tasks. Many Gantt charts, however, do not show any dependencies. Instead, as you might recall, network diagrams or PERT charts are used to show task dependencies. See Chapter 6 for more information about Gantt charts and network diagrams. This section explains important information about Gantt charts and describes how to make critical path information more visible in the Gantt Chart view.

You must follow a few important procedures when working with Gantt charts:

- To adjust the timescale, click the Zoom Out icon ⊖ or the Zoom In icon ⊕. Clicking these icons automatically makes the dates on the Gantt chart show more or less information. For example, if the timescale for the Gantt chart is showing months and you click the Zoom Out icon, the timescale will adjust to show quarters. Clicking Zoom Out again will display the timescale in years. Similarly, each time you click the Zoom In icon, the timescale changes to display more detailed time information—from years to quarters, quarters to months, and months to weeks.

- You can also adjust the timescale and access more formatting options by selecting Timescale from the Format menu. Adjusting the timescale enables you to see the entire Gantt chart on one screen and in the time increments that you desire.

- A Gantt Chart Wizard is available on the Format menu. This wizard helps you adjust the format of the Gantt chart. For example, you can select an option to display critical path information on the Gantt chart, and those items on the critical path will automatically display using a red bar.

- You can view a tracking Gantt chart by setting a baseline for your entire project or for selected tasks and then entering actual durations for tasks. The tracking Gantt chart view displays two taskbars, one above the other, for each task. One taskbar shows planned or baseline start and finish dates, and the other taskbar shows actual start and finish dates. You will find a sample tracking Gantt chart later in this appendix, after you enter actual information for the Project Tracking Database example.

Because you have already created task dependencies, you can now find the critical path for the Project Tracking Database project. You can view the critical tasks by changing the color of those items on the critical path in the Task Name column. You can also change the color of bars on the Gantt chart. Tasks on the critical path will automatically be red in the Network Diagram view, as described in the following section.

To make the text for the critical path tasks appear in red in the Entry table and on the Gantt chart:

1. *Open the Text Styles dialog box.* Open time.mpp, if necessary. Click **Format** on the menu bar, and then click **Text Styles**. The Text Styles dialog box opens.

2. *Change the critical tasks color option to red.* Click the **list arrow** next to the Item to Change option, and select **Critical Tasks**. In the color section, select **Red**, as shown in Figure A-30. Click **OK** to accept the changes to the text styles.

Figure A-30. Text Styles Dialog Box Settings to Display Critical Tasks in Red

3. *Open the Gantt Chart Wizard.* Select **Gantt Chart Wizard** from the **Format** menu or click the Gantt Chart Wizard icon on the toolbar. The Gantt Chart Wizard opens. Click **Next**, and the next step of the Wizard appears.

4. *Select the Critical Path option.* Click the **Critical path** radio button, as shown in Figure A-31. Notice that the sample Gantt chart in the Gantt Chart Wizard displays some bars in red, representing tasks on the critical path. Click **Finish** to proceed to the next step in the Gantt Chart Wizard.

Figure A-31. Gantt Chart Wizard Showing Critical Path Format

5. *Format the Gantt chart.* Click **Format It**, and then click **Exit Wizard**. Critical task bars on the Gantt chart will now be red.

6. *Save your file without a baseline.* Click the **Save** icon 💾 on the toolbar to save your file.

Network Diagrams

The network diagrams in Project 2002 use the precedence diagramming method, with tasks displayed in rectangular boxes and relationships shown by lines connecting the boxes. In the Network Diagram view, tasks on the critical path are automatically shown in red.

To view the network diagram for the Project Tracking Database project:

1. *View the Network Diagram.* Click **View Bar** on the View menu, and then click the **Network Diagram** icon 🔲 on the View Bar or select **Network Diagram** from the View menu.

2. *Adjust the Network Diagram view.* To see more tasks in the Network Diagram view, click the **Zoom Out** icon on the toolbar twice. You can also use the scroll bars to see different parts of the network diagram. Figure A-32 shows several of the tasks in the Project Tracking

Database network diagram. Note that milestone tasks, such as Charter signed, appear as pointed rectangular boxes, while other tasks appear as rectangles. Tasks on the critical path automatically appear in red, while noncritical tasks appear in blue. Each task in the network diagram also shows information such as the start and finish dates, task ID, and duration. Move your mouse over any of the boxes to see them in a larger view. A dashed line on a network diagram represents a page break. You often need to change some of the default settings for the Network Diagram view before printing it.

Figure A-32. Network Diagram View

3. *View the Help topic on rearranging network diagram boxes.* Click the **Microsoft Project Help** icon [?] on the toolbar or press **F1** to display the Office Assistant. Type **How do I rearrange the network diagram?** in the text box and then click **Search**. Click the resulting Help topic called **Rearrange Network Diagram boxes**. Figure A-33 shows the resulting Project 2002 Help window. Note that you can change several layout options for the network diagram, hide fields, and manually position boxes. Close the Help window.

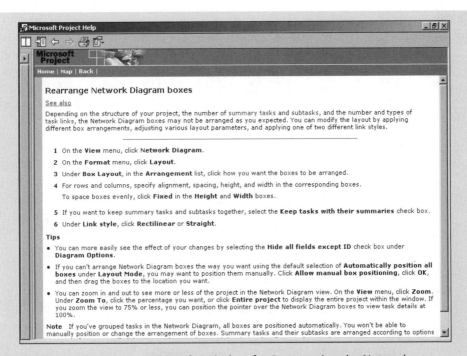

Figure A-33. Project 2002 Help Window for Rearranging the Network Diagram Boxes

4. *Return to Gantt Chart view.* Return to the Gantt Chart view by clicking the **Gantt Chart** icon in the View bar or by selecting **Gantt Chart** from the View menu on the menu bar.

Some people create new Project 2002 files or add tasks to existing files in the Network Diagram view instead of the Gantt Chart view. To add a new task or node in the Network Diagram view, select New Task from the Insert menu, or press the Insert key on your keyboard. Double-click the new node to add a task name and other information. Create finish-to-start dependencies between tasks in Network Diagram view by clicking the preceding node and dragging to the succeeding node. To modify the type of dependency and add lead or lag time, double-click the arrow between the dependent nodes.

Critical Path Analysis

Recall that the critical path is the path through the network diagram with no slack or the least amount of slack; it represents the shortest possible time to complete the project (see Chapter 6). If a task on the critical path takes longer than planned, the project schedule will slip unless time is reduced on a task later on the critical path. Sometimes you can shift resources between tasks to help keep a project on schedule. Project 2002 has several views and reports to help analyze critical path information.

Two particularly useful features are the Schedule Table view and the Critical Tasks report. The Schedule Table view shows the early and late start dates for each task, the early and late finish dates for each task, and the free and total slack for each task. This information shows how flexible the schedule is and helps in making schedule compression decisions. The Critical Tasks report lists only tasks that are on the critical path for the project. If meeting schedule deadlines is essential for a project, project managers want to closely monitor tasks on the critical path.

To access the Schedule Table view and view the Critical Tasks report for a file:

1. *View the Schedule table.* Right-click the **Select All** button to the left of the Task Name column heading and select **Schedule**. Alternatively, you can click **View** on the menu bar, point to **Table: Entry**, and then click **Schedule**. The Schedule table replaces the Entry table to the left of the Gantt chart.

2. *Reveal all columns in the Schedule table.* Move the split bar to the right until you see the entire Schedule table. Your screen should resemble Figure A-34. This view shows the start and finish (meaning the early start and early finish) and late start and late finish dates for each task, as well as free and total slack. Right-click the **Select all** button and select **Entry** to return to the Entry table view.

Select All button

	Task Name	Start	Finish	Late Start	Late Finish	Free Slack	Total Slack
1	⊟ 1 Initiating	**Mon 4/4/05**	**Mon 4/18/05**	**Mon 4/4/05**	**Mon 4/18/05**	**0 days**	**0 days**
2	1.1 Kickoff meeting	Mon 4/4/05	Mon 4/4/05	Mon 4/4/05	Mon 4/4/05	0 hrs	0 hrs
3	1.2 Develop projec	Mon 4/4/05	Mon 4/18/05	Mon 4/4/05	Mon 4/18/05	0 days	0 days
4	1.3 Charter signed	Mon 4/18/05	Mon 4/18/05	Mon 4/18/05	Mon 4/18/05	0 days	0 days
5	⊟ 2 Planning	**Mon 4/18/05**	**Mon 8/29/05**	**Mon 4/18/05**	**Mon 9/19/05**	**0 days**	**0 days**
6	2.1 Develop projec	Mon 4/18/05	Mon 5/9/05	Mon 4/18/05	Mon 5/9/05	0 wks	0 wks
7	2.2 Review projec	Mon 5/9/05	Mon 8/29/05	Mon 5/30/05	Mon 9/19/05	0.75 mons	0.75 mons
8	2.3 Project plans a	Mon 5/9/05	Mon 5/9/05	Mon 5/9/05	Mon 5/9/05	0 days	0 days
9	⊟ 3 Executing	**Mon 5/9/05**	**Mon 8/29/05**	**Mon 5/9/05**	**Mon 8/29/05**	**0 days**	**0 days**
10	3.1 Analysis	Mon 5/9/05	Mon 6/6/05	Mon 5/9/05	Mon 6/6/05	0 mons	0 mons
11	3.2 Design	Thu 6/2/05	Thu 7/28/05	Thu 6/2/05	Thu 7/28/05	0 mons	0 mons
12	3.3 Implementation	Mon 7/25/05	Mon 8/22/05	Mon 7/25/05	Mon 8/22/05	0 mons	0 mons
13	3.4 System implem	Mon 8/29/05	Mon 8/29/05	Mon 8/29/05	Mon 8/29/05	0 days	0 days
14	⊟ 4 Controlling	**Mon 4/4/05**	**Wed 9/7/05**	**Mon 5/2/05**	**Mon 9/19/05**	**8.13 days**	**8.13 days**
15	⊞ 4.1 Status Repor	**Wed 4/6/05**	**Wed 9/7/05**	**Mon 9/19/05**	**Mon 9/19/05**	**8.13 days**	**8.13 days**
39	4.2 Report perform	Mon 4/4/05	Fri 8/19/05	Mon 5/2/05	Mon 9/19/05	1.01 mons	1.01 mons
40	4.3 Control change	Mon 4/4/05	Fri 8/19/05	Mon 5/2/05	Mon 9/19/05	1.01 mons	1.01 mons
41	⊟ 5 Closing	**Mon 8/29/05**	**Mon 9/19/05**	**Mon 8/29/05**	**Mon 9/19/05**	**0 days**	**0 days**
42	5.1 Prepare final p	Mon 8/29/05	Mon 9/12/05	Mon 8/29/05	Mon 9/12/05	0 wks	0 wks
43	5.2 Present final pi	Mon 9/12/05	Mon 9/19/05	Mon 9/12/05	Mon 9/19/05	0 wks	0 wks
44	5.3 Project comple	Mon 9/19/05	Mon 9/19/05	Mon 9/19/05	Mon 9/19/05	0 days	0 days

Figure A-34. Schedule Table View

3. *Open the Reports dialog box.* Click **View** on the menu bar, and then click **Reports**. Double-click **Overview** to open the Overview Reports dialog box. Your screen should resemble Figure A-35.

Figure A-35. Accessing the Critical Tasks Report

4. *Display the Critical Tasks report.* Double-click **Critical Tasks**. A Critical Tasks report as of today's date is displayed.

5. *Close the report and save your file.* When you are finished examining the Critical Tasks report, click **Close**. Click **Close** on the Reports dialog box. Click the **Save** icon on the toolbar to save your file.

If you want to download the completed time.mpp file to check your work or continue to the next section, a copy is available on the companion Web site for this text or from your instructor.

Now that you have entered task durations, established task dependencies, and reviewed the network diagram and critical path information, you are ready to explore some of the Project 2002 cost management features.

PROJECT COST MANAGEMENT

Many people do not use Project 2002 for cost management. Most organizations have more established cost management software products and procedures in place, and many people simply do not know how to use the cost features. Using the cost features of Project 2002, however, makes it possible to integrate total project information more easily. This section offers brief instructions for entering fixed and variable cost estimates and actual cost and time information after establishing a baseline plan. It also explains how to use Project 2002 for earned value management. More details on these features are available in Project 2002 Help, online tutorials, or other books.

To complete the hands-on steps in this section, you need to download the Project 2002 file resource.mpp from the companion Web site for this book and save it on a floppy disk or with your other project files on your hard drive.

Fixed and Variable Cost Estimates

The first step to using the cost features of Project 2002 is entering cost-related information. You enter costs as fixed or variable based on per-use material costs, or variable based on the type and amount of resources used. Costs related to personnel are often a significant part of project costs.

Entering Fixed Costs in the Cost Table

The Cost table allows you to enter fixed costs related to each task. To access the Cost table, right-click the Select All button and select Cost or select Table: Cost from the View menu. Figure A-36 shows the resulting view for the Project Tracking Database project using time.mpp. You can also assign a per-use cost to a resource that represents materials or supplies and use it as a base for calculating the total materials or supplies cost of a task. See Project 2002 Help for details on this feature.

Figure A-36. Cost Table View

Entering Human Resource Costs

Human resources represent a major part of the costs on many projects. By defining and then assigning human resources and their related costs to tasks in Project 2002, you can calculate human resource costs, track how people are used, identify potential resource shortages that could force you to miss deadlines, and identify underutilized resources. It is often possible to shorten a project's schedule by reassigning underutilized resources. This section focuses on entering human resource costs and assigning resources to tasks. The following section describes other features of Project 2002 related to human resource management.

There are several methods for entering resource information in Project 2002. One of the easiest is to enter basic resource information in the Resource Sheet. Access the Resource Sheet from the View bar, or by selecting Resource Sheet from the View menu. This Resource Sheet allows you to enter the resource name, initials, resource group, maximum units, standard rate, overtime rate, cost/use, accrual method, base calendar, and code. Entering data into the Resource Sheet is similar to entering data into an Excel spreadsheet, and you can easily sort items by selecting Sort from the Project menu. In addition, you can use the Filter list on the toolbar to filter resources. Once you have established resources in the Resource Sheet, you can assign those resources to tasks in the Entry table with the list arrow that appears when you click a cell in the Resource Names column. The Resource Names column is the last column of the Entry table. You can also use other methods for assigning resources, as described below.

Next, you will use the Project 2002 file time.mpp, which you saved in the preceding section, to assign resources to tasks. (If you did not save your file, download it from the companion Web site for this text.) Assume that there are four people working on the Project Tracking Database project, and assume that the only costs for this project are for these human resources: Kathy is the project manager; John is the business analyst; Mary is the database analyst; and Chris is an intern.

To enter basic information about each person into the Resource Sheet:

1. *Open the Resource Sheet view.* Open your Project 2002 file, time.mpp, if necessary. Click the **Resource Sheet** icon 🖿 on the View bar (you may need to scroll to see it), or click **View** on the menu bar, and then click **Resource Sheet**.

2. *Enter resource information.* Enter the information from Table A-4 into the Resource Sheet. Type the information as shown and press the **Tab** key to move to the next field. Notice that you are putting abbreviated job titles in the Initials column: PM stands for project manager; BA stands for business analyst; DA stands for database analyst; and IN stands for intern. When you type the standard and overtime rates, you can just type 50, and Project 2002 will automatically enter $50.00/hr. The standard and overtime rates entered are based on hourly rates. You can also enter annual salaries by typing the annual salary number followed by /**y** for "per year." Leave the default values for the other columns in the Resource Sheet as they are. Your screen should resemble Figure A-37 when you are finished entering the resource data.

Table A-4: Project Tracking Database Resource Data

RESOURCE NAME	INITIALS	GROUP	STD. RATE	OVT. RATE
Kathy	PM	1	$50.00/h	$60.00/h
John	BA	1	$40.00/h	$50.00/h
Mary	DA	1	$40.00/h	$50.00/h
Chris	IN	1	$20.00/h	$25.00/h

Figure A-37. Resource Sheet View with Resource Data Entered

If you know that some people will be available for a project only part-time, enter their percentage of availability in the Max Units column of the Resource Sheet. Project 2002 will then automatically assign those people based on their maximum units. For example, if someone can work only 25 percent of their time on a project throughout most of the project, enter 25% in the Max Units column for that person. When you enter that person as a resource for a task, his or her default number of hours will be 25 percent of a standard eight-hour workday, or two hours per day.

Adjusting Resource Costs

To make a resource cost adjustment, such as a raise, for a particular resource, double-click the person's name in the Resource Name column, select the Costs tab in the Resource Information dialog box, and then enter the effective date and raise percentage. You can also adjust other resource cost information, such as standard and overtime rates.

To give the project manager a 10 percent raise starting 6/1/05:

1. *Open the Resource Information dialog box.* From the Resource Sheet view, double-click **Kathy** in the Resource Name column. The Resource Information dialog box opens.

2. *Enter an effective date for a raise.* Select the **Costs** tab, and then select **tab A**, if needed. Type **6/1/05** in the second cell in the Effective Date column and press **Enter**. Alternately, click the **list arrow** in the second cell and use the calendar that appears to enter the effective date.

3. *Enter the raise percentage.* Type **10%** in the second cell for the Standard Rate column, and then press **Enter**. The Resource Information screen should resemble the screen shown in Figure A-38. Notice that Project 2002 calculated the 10% raise to be $55.00/h. Click **OK**.

Resource Information				?|X
General	Working Time	Costs	Notes	Custom Fields

Resource Name: Kathy

Cost rate tables

For rates, enter a value or a percentage increase or decrease from the previous rate. For instance, if a resource's Per Use Cost is reduced by 20%, type -20%.

A (Default)	B	C	D	E

	Effective Date	Standard Rate	Overtime Rate	Per Use Cost
	--	$50.00/h	$60.00/h	$0.00
	Wed 6/1/05	$55.00/h	$60.00/h	$0.00

Cost accrual: Prorated

Help   Details...   OK   Cancel

Figure A-38: Adjusting Resource Costs

Assigning Resources to Tasks

For Project 2002 to calculate resource costs, you must assign the appropriate resources to tasks in your WBS. There are several methods for assigning resources. The Resources column in the Entry table allows you to select a resource using a drop-down list. You can, however, use this method to assign only one resource to a task full-time. You often need to use other methods to assign resources, such as the Assign Resources icon *C* on the toolbar or the Resource Cost window, to ensure that Project 2002 captures resource assignments as you intend them. Next, you will use these three methods for assigning resources to the Project Tracking Database project.

Assigning Resources Using the Entry Table

To assign resources using the Entry table:

1. *Select the task to which you want to assign resources.* Click the **Gantt Chart** icon on the View bar, or click View Table: Entry, if necessary, to return to that view. Make sure that Task 2, **Kickoff meeting**, is selected.

2. *Reveal the Resource Names column of the Entry table.* Move the split bar to the right to reveal the entire Resource Names column in the Entry table.

3. *Select a resource from the Resource Names column.* Click the **drop-down arrow** in the Resource Names column, and then click **Kathy** to assign her to Task 2. Notice that the resource choices are based on information that was entered in the Resource Sheet. If you had not entered any resources, you would not have a drop-down arrow or any choices to select.

4. *Try to select another resource for Task 2.* Again, click the **drop-down arrow** in the Resource Names column for Task 2, click **John,** and then press **Enter**. Notice that only John's name appears in the Resource Names column. You can assign only one resource to a task using this method.

5. *Clear the resource assignment.* Right-click the **Resource Names** column for Task 2, and then select **Clear Contents** to remove the resource assignments.

Assigning Resources Using the Toolbar

To assign resources using the toolbar:

1. *Select the task to which you want to assign resources.* Click the task name for Task 2, **Kickoff meeting**, in the second row of the Task Name column.

2. *Open the Assign Resource dialog box.* Click the **Assign Resources** icon 🗘 on the toolbar. The Assign Resources dialog box, which lists the names of the people assigned to the project, is displayed, as shown in Figure A-39. This dialog box remains open while you move from task to task to assign resources. If your Resource list options button is not expanded, you can click the + sign to expand it. You click the − sign to collapse it.

Figure A-39. Assign Resources Dialog Box

3. *Assign Kathy to Task 2.* Click **Kathy** in the Resource Name column of the Assign Resources dialog box, and then click **Assign**. Notice that the duration estimate for Task 2 remains at 2 hours, and Kathy's name appears on the Gantt chart by the bar for Task 2.

4. *Assign John to Task 2.* Click **John** in the Resource Name column of the Assign Resources dialog box, and then click **Assign**. The duration for Task 2 changes to 1 hour, *but you do not want this change to occur.* Click **Close** in the Assign Resources dialog box.

5. *Clear resource assignments.* Right-click the **Resource Names** column for Task 2, and then select **Clear Contents** to remove the resource assignments.

6. *Reenter the duration for Task 2.* Type **2h** in the duration column for Task 2 and press **Enter**.

Assigning Resources Using the Split Window and Resource Cost View

Even though using the Assign Resources icon seems simpler, it is often better to use a split window and the Resource Cost view when assigning resources. When you assign resources using the split view, the durations of tasks will not automatically change when you assign more resources, and you have more control over how you enter information. Project 2002 Help offers more information on different options for resource assignment.

To assign both Kathy and John to attend the two-hour kickoff meeting:

1. *Split the window to reveal more information.* Select **Window**, and then **Split** from the menu bar. The Gantt Chart view is displayed at the top of the screen and a resource information table is displayed at the bottom of the screen.

2. *Open the Resource Cost window.* Right-click anywhere in the bottom window and select **Resource Cost**.

3. *Assign Kathy to Task 2.* Select **Task 2** and click under **Resource Name** in the Resource Cost window. Click the **down arrow** and select **Kathy**.

4. *Assign John to Task 2.* Click in the cell below Kathy, and select **John**.

5. *Enter the resource assignment and review the Gantt chart.* Click **OK** in the lower window. Your screen should resemble Figure A-40. Notice that the duration for Task 2 is still 2 hours, and both Kathy and John are assigned to that task 100%. You can also enter a different percentage under the Units column if resources will not work 100% of their available time on a task. Note that the default settings are that tasks are Effort driven and the Task type is Fixed Units. You will learn more about these settings later.

Figure A-40. Split Screen View for Entering Resource Information

6. *Open the Resource Schedule window.* Right-click anywhere in the lower window and select **Resource Schedule**. Notice that Kathy and John are assigned to Task 2 for 2 hours, as you intended. You can also enter resource information using the Resource Schedule option.

7. *Close the file and do not save it.* Close the file, but do not save the changes you made. Other resource information has been entered for you in the Project 2002 file named resource.mpp, available on the companion Web site for this text.

A copy of the Project 2002 file resource.mpp is available on the companion Web site for this text or from your instructor. You must use this file to continue the steps in the next section.

As you can see, you must be careful when assigning resources. Project 2002 assumes that the durations of tasks are not fixed, but are effort-driven, and this assumption can create some problems when you are assigning resources. Figure A-41 shows the results of adding resources with different settings. The first four task entries assume the task, Design, is effort-driven. When you assign one resource to the design task, John, the duration is one month. If you later assign two resources, John and Mary, to that same task without changing any settings, Project 2002 automatically adjusts the duration of that task to half its original duration, or .05 months. If you change the task type to be Fixed Duration and assign two resources, the duration remains at one month and each resource is assigned 50 percent to the task. If you change the type to Fixed Work and assign two resources, the duration is cut in half and the resources are assigned full-time. The bottom part of Figure A-41 shows the results when a task is not effort-driven or when you assign the Fixed Work task type. You must be very careful when entering resource information to make sure the task duration and work hours reflect what you want.

Fixed Work forces the
Effort Driven field to "yes"

Note the effects of type
on resource assignments

	Task Name	Effort Driven	Duration	Work	Type	
1	Design	Yes	1 mon	160 hrs	Fixed Units	John
2	Design	Yes	0.5 mons	160 hrs	Fixed Units	John,Mary
3	Design	Yes	1 mon	160 hrs	Fixed Duration	John[50%],Mary[50%]
4	Design	Yes	0.5 mons	160 hrs	Fixed Work	John,Mary
5						
6	Design	No	1 mon	160 hrs	Fixed Units	John
7	Design	No	1 mon	320 hrs	Fixed Units	John,Mary
8	Design	No	1 mon	320 hrs	Fixed Duration	John,Mary
9	Design	Yes	0.5 mons	160 hrs	Fixed Work	John,Mary

Figure A-41. Results of Adding Resources with Various Task Types

Viewing Project Cost Information

Once you enter resource information, Project 2002 automatically calculates resource costs for the project. There are several ways to view project cost information. You can view the Cost table to see cost information, or you can run various cost reports. Next, you will view cost information for the Project Tracking Database project.

To view cost information:

1. *Open resource.mpp.* Download **resource.mpp** from the companion Web site for this text or get it from your instructor. Open the file.

2. *Open the Cost table.* Right click the **Select All** button and click **Cost**, or click **View** on the menu bar, point to **Table**: **Entry**, and then click **Cost**. The Cost table displays cost information and the Gantt chart. Move the split bar to the right to reveal all of the columns. Your screen should resemble Figure A-42. Note that by assigning resources, costs have been automatically calculated for tasks. You could also enter fixed costs for tasks in the cost table, but there are no fixed costs associated with this project.

Figure A-42. Cost Table for the Resource File

3. *Open the Cost Reports dialog box.* Click **View** on the menu bar, and then click **Reports.** Double-click **Costs** to display the Cost Reports dialog box.

4. *Set the time units for the report.* Click **Cash Flow**, if necessary, and then click **Edit**. The Crosstab Report dialog box is displayed. Click the Column **drop-down arrow** and select **Months** instead of weeks in the Crosstab Report dialog box, if needed, as shown in Figure A-43. Click **OK**.

Figure A-43. Crosstab Report Dialog Box

5. *View the Cash Flow report.* Click **Select** in the Cost Reports dialog box. A Cash Flow report for the Project Tracking Database appears, as shown in Figure A-44. Click **Close**.

Microsoft Project - resource.mpp

| Page Setup... | Print... | Close | Help |

Cash Flow as of Mon 3/21/05
Project Tracking Database

	April	May	June	July	August	September	Total
Initiating							
Kickoff meeting	$180.00						$180.00
Develop project charter	$2,320.00						$2,320.00
Charter signed							
Planning							
Develop project plans	$3,510.00	$1,890.00					$5,400.00
Review project plans		$335.00	$484.00	$462.00	$445.50		$1,726.50
Project plans approved							
Executing							
Analysis		$6,566.00	$1,287.00				$7,853.00
Design			$6,225.00	$5,775.00			$12,000.00
Implementation				$2,489.00	$7,991.00		$10,480.00
System implemented							
Controlling							
Status Reports	$590.40	$721.60	$765.60	$730.80	$800.40	$143.55	$3,752.35
Report performance	$160.00	$176.00	$193.60	$184.80	$132.00		$846.40
Control changes	$400.00	$440.00	$484.00	$462.00	$330.00		$2,116.00
Closing							
Prepare final project report					$605.00	$1,595.00	$2,200.00
Present final project						$440.00	$440.00
Project completed							
Total	$7,160.40	$10,128.60	$9,439.20	$10,103.60	$10,303.90	$2,178.55	$49,314.25

Figure A-44. Cash Flow Report

6. *View the Project Summary report.* In the Reports dialog box, double-click **Overview**, and then double-click **Project Summary**. A Project Summary report for the Project Tracking Database is displayed, listing information such as project baseline start and finish dates; actual start and finish dates; summaries of duration, work hours, and costs; as well as variance information. Since you have not yet saved the file as a baseline, much of this information is blank in the report. The Project Summary report provides a high-level overview of a project. Click **Close** to close the report after reviewing it. Click **Close** on the Reports dialog box.

 You can edit many of the report formats in Project 2002. Instead of double-clicking the report, highlight the desired report, and then click **Edit**.

The total estimated cost for this project, based on the information entered, should be $49,314.25, as shown in the Cash Flow and Project Summary reports. In the next section, you will save this data as a baseline plan and enter actual information.

Baseline Plan, Actual Costs, and Actual Times

Once you complete the initial process of creating a plan—entering tasks, establishing dependencies, assigning costs, and so on—you are ready to set a baseline plan. By comparing the information in your baseline plan to an updated plan during the course of the project, you can identify and solve problems. After the project ends, you can use the baseline and actual information to plan similar, future projects more accurately. To use Project 2002 to help control projects, you must establish a baseline plan, enter actual costs, and enter actual durations.

Establishing a Baseline Plan

An important part of project management is setting a baseline plan. If you plan to compare actual information such as durations and costs, you must first save the Project 2002 file as a baseline. The procedure involves selecting Tracking, and then Save Baseline from the Tools menu. Before setting a baseline, you must complete the baseline plan by entering time, cost, and human resources information. Be careful not to save a baseline until you have completed the baseline plan. If you do save a baseline before completing the baseline plan, Project 2002 allows you to save up to ten baselines. You can then clear unwanted baseline plans.

Even though you can clear a baseline plan or save multiple baselines, it is a good idea to save a separate, backup file of the original baseline for your project. Enter actuals and save that information in the main file, but always keep a backup baseline file without actuals. Keeping separate baseline and actual files will allow you to return to the original file in case you ever need to use it again.

To rename resource.mpp and then save it as a baseline plan in Project 2002:

1. *Save resource.mpp as a new file named baseline.mpp.* Open resource.mpp, if necessary, click **File** on the menu bar, and then click **Save As**. Type **baseline** as the filename, and then click **Save**.

2. *Open the Save Baseline dialog box.* Click **Tools** on the menu bar, select **Tracking**, and then click **Save Baseline**. The Save Baseline dialog box is displayed (Figure A-45.)

Save Baseline ? X

○ Save baseline

| Baseline | ▼ |

Baseline
○ Save inte Baseline 1
    Copy: Baseline 2
    Into: Baseline 3
For: Baseline 4
  ○ Entire Baseline 5
  ○ Select Baseline 6
    Roll u   Baseline 7
    ☐ To   Baseline 8
    ☐ En   Baseline 9
              Baseline 10

                          Set as Default

    Help   OK   Cancel

Figure A-45. Save Baseline Dialog Box Showing Ability to Save Multiple Baselines

3. *Review the Save Baseline options, and save the entire project as a baseline.* If necessary, click the **Save baseline** radio button and the **Entire project** radio button. Use the default option of the first Save baseline name of just Baseline. These options should be the default settings in Project 2002. Click **OK**.

Notice that there are several options in the Save Baseline dialog box, as shown in Figure A-45. You can save the file as an interim plan if you anticipate several versions of the plan. You can also select the entire project or selected tasks when saving a baseline or interim plan. You can also save up to ten baselines with Project 2002.

Entering Actual Costs and Times

After you set the baseline plan, it is possible to track information on each of the tasks as the project progresses. You can also adjust planned task information for tasks still in the future. The Tracking Table displays tracking information, and the Tracking Toolbar helps you enter this information. Figure A-46 describes each button on the Tracking Toolbar. The 100% icon and the Update as Schedule icon are the most commonly used icons for entering actual information.

Button	Name	Description
	Project Statistics	Provides summary information about the project baseline, the actual project start and finish dates, and overall project duration, costs, and work
	Update as Scheduled	Updates the selected tasks to indicate that actual dates, costs, and work match the scheduled dates, costs, and work
	Reschedule Work	Schedules the remaining duration for a task that is behind schedule so that it will continue from the status date
	Add Progress Line	Displays a progress line on the Gantt chart from a date that you select on the time scale
	0% Complete	Marks the selected tasks as 0% complete as of the status date (Actual date, work, and duration data is updated)
	25% Complete	Marks the selected tasks as 25% complete as of the status date (Actual date, work, and duration data is updated)
	50% Complete	Marks the selected tasks as 50% complete as of the status date (Actual date, work, and duration data is updated)
	75% Complete	Marks the selected tasks as 75% complete as of the status date (Actual date, work, and duration data is updated)
	100% Complete	Marks the selected tasks as 100% complete as of the status date (Actual date, work, and duration data is updated)
	Update Tasks	Displays the Update Tasks dialog box for the selected tasks so that you can enter their percentages completed, actual durations, remaining durations, or actual start or finish dates
	Collaborate Toolbar	Toggles on and off the Collaborate Toolbar, which displays buttons that enable you to share parts of the project with others

Bunin, Rachel Biheller, Lisa Friedrichsen, and Kathy Schwalbe, (2003) *Microsoft Project 2002, New Perspectives Series Introductory*, Course Technology (pp. 250–251).

Figure A-46. Buttons on the Tracking Toolbar

To practice entering actual information, enter just a few changes to the baseline. Assume that Tasks 1 through 7 were completed as planned, but that Task 10 took longer than planned.

To enter actual information for tasks that were completed as planned:

1. *Display the Tracking toolbar.* Click **View** on the menu bar, select **Toolbars**, and then click **Tracking** from the **Toolbars** submenu to display the Tracking toolbar. Move the toolbar as desired.

2. *Display the Tracking table.* Right-click the **Select All** button, and then click **Tracking** to see more information as you enter actual data. Move the split bar to reveal all the columns in the Tracking table.

3. *Mark Tasks 1 though 8 as 100% complete.* Click the Task Name for Task 1, **Initiating**, and drag down through Task 8 to highlight the first eight tasks. Click the **100 percent icon** |00⁺ on the Tracking toolbar. The columns with dates, durations, and cost information should now contain data instead of the default values, such as NA or 0. The % Comp column should display 100%. Adjust column widths if needed. Your screen should resemble Figure A-47.

	Task Name	Act. Start	Act. Finish	% Comp.	Phys. % Comp.	Act. Dur.	Rem. Dur.	Act. Cost	Act. Work
1	⊟ 1 Initiating	Mon 4/4/05	Mon 4/18/05	100%	0%	10.25 days	0 days	$2,500.00	52 hrs
2	1.1 Kickoff meeting	Mon 4/4/05	Mon 4/4/05	100%	0%	2 hrs	0 hrs	$180.00	4 hrs
3	1.2 Develop project charter	Mon 4/4/05	Mon 4/18/05	100%	0%	10 days	0 days	$2,320.00	48 hrs
4	1.3 Charter signed	Mon 4/18/05	Mon 4/18/05	100%	0%	0 days	0 days	$0.00	0 hrs
5	⊟ 2 Planning	Mon 4/18/05	Mon 8/29/05	100%	0%	95 days	0 days	$7,126.50	164 hrs
6	2.1 Develop project plans	Mon 4/18/05	Mon 5/9/05	100%	0%	3 wks	0 wks	$5,400.00	132 hrs
7	2.2 Review project plans	Mon 5/9/05	Mon 8/29/05	100%	0%	4 mons	0 mons	$1,726.50	32 hrs
8	2.3 Project plans approved	Mon 5/9/05	Mon 5/9/05	100%	0%	0 days	0 days	$0.00	0 hrs
9	⊟ 3 Executing	NA	NA	0%	0%	0 days	80 days	$0.00	0 hrs
10	3.1 Analysis	NA	NA	0%	0%	0 mons	1 mon	$0.00	0 hrs
11	3.2 Design	NA	NA	0%	0%	0 mons	2 mons	$0.00	0 hrs
12	3.3 Implementation	NA	NA	0%	0%	0 mons	1 mon	$0.00	0 hrs
13	3.4 System implemented	NA	NA	0%	0%	0 days	0 days	$0.00	0 hrs
14	⊟ 4 Controlling	NA	NA	0%	0%	0 days	112.13 days	$0.00	0 hrs
15	⊞ 4.1 Status Reports	NA	NA	0%	0%	0 days	110.13 days	$0.00	0 hrs
39	4.2 Report performance	NA	NA	0%	0%	0 mons	5 mons	$0.00	0 hrs
40	4.3 Control changes	NA	NA	0%	0%	0 mons	5 mons	$0.00	0 hrs
41	⊟ 5 Closing	NA	NA	0%	0%	0 days	15 days	$0.00	0 hrs
42	5.1 Prepare final project report	NA	NA	0%	0%	0 wks	2 wks	$0.00	0 hrs
43	5.2 Present final project	NA	NA	0%	0%	0 wks	1 wk	$0.00	0 hrs

Figure A-47. Tracking Table Information

4. *Enter actual completion dates for Task 10.* Click the Task Name for Task 10, **Analysis**, and then click the **Update Tasks** icon on the Tracking toolbar. The Update Tasks dialog box opens. For Task 10, enter the Actual Start date as **5/11/05** and the Actual Finish date as **6/16/05** in the Update Tasks dialog box, as shown in Figure A-48. Click **OK**.

Figure A-48. Update Tasks Dialog Box

5. *Display the Indicators column.* Click **Insert** on the menu bar and click **Column**, click the **Field name** list arrow, select **Indicators**, and then click **OK**. The Indicators column is displayed, showing a check mark by completed tasks.

6. *Review changes in the Gantt chart.* Move the split bar to the left to reveal more of the Gantt chart. Notice that the Gantt chart bars for completed tasks have changed. Gantt chart bars for completed tasks appear with a black line drawn through the middle.

 You can hide the Indicators column by dragging the right border of the column heading to the left.

You can also view actual and baseline schedule information more clearly with the Tracking Gantt Chart view. See Chapter 6 for descriptions of the symbols on a Tracking Gantt chart.

To display the Tracking Gantt chart:

1. *View the Tracking Gantt chart.* Click **View** on the menu bar, and then click **Tracking Gantt**. Alternatively, click the **Tracking Gantt** icon on the View bar. Adjust the horizontal scroll bar at the lower-right corner of the Gantt chart window, if necessary, to see symbols on the Gantt chart.

2. *Display Gantt chart information in months.* Click **Zoom Out** to display information in months. Your screen should resemble Figure A-49. (The Tracking Toolbar was turned off in this figure to display the entire schedule).

Figure A-49. Tracking Gantt Chart View

3. *Save your file as a new file named tracking.mpp.* Click **File** on the menu bar, then click **Save As**. Name the file **tracking**, and then click **Save**.

Notice the additional information available on the Tracking Gantt chart. Completed tasks have 100% by their symbols on the Tracking Gantt chart. Tasks that have not started yet display 0%. Tasks in progress, such as Task 9, show the percentage of the work completed (31% in this example). Also, note that the project completion date has moved to 9/29 since several tasks depended on completion of Task 10, which took longer than planned. Viewing the Tracking Gantt chart allows you to easily see your schedule progress against the baseline plan.

After you have entered some actuals, you can review earned value information for the initiating tasks of this project.

Earned Value Management

Earned value management (see Chapter 7, Project Cost Management) is an important project management technique for measuring project performance. Because you have entered actual information for the initiating tasks in the Project Tracking Database project, you can now view earned value information in Project 2002.

To view earned value information:

1. *View the Earned Value table.* Open **tracking.mpp**, if necessary, and then click **View** on the menu bar. Point to **Table: Entry**. Select **More Tables** to open the More Tables dialog box, and then double-click **Earned Value**.

2. *Display all the Earned Value table columns.* Move the split bar to the right to reveal all of the columns. Note that the Earned Value table includes columns for each earned value term described in Chapter 7. Also note that the EAC (Estimate at Completion) is higher than the BAC (Budget at Completion) starting with Task 9, where the task took longer than planned to complete (Figure A-50). The term VAC stands for Variance at Completion. Task 9 shows a VAC of ($2,779.00), meaning the project is projected to cost $2,779 more than planned at completion at that point in time. Remember that not all of the actual information has been entered yet.

Figure A-50. Earned Value Table

Project 2002 uses earned value terms from the 1996 PMBOK® Guide. Recall that the PMBOK® Guide 2000 uses simpler acronyms for earned value terms. See Chapter 7 of this text for explanations of the 1996 and 2000 terms.

3. *Close and save the file.* Click the **Save** icon and then close the tracking.mpp file.

If you wish to check your work or continue to the next section, copies of the files baseline.mpp and tracking.mpp are available on the companion Web site for this text or from your instructor.

Further information on earned value is available from Project 2002 Help or in Chapter 7, Project Cost Management. To create an earned value chart, you must export data from Project 2002 into Excel. There are also add-in products you can use with Project 2002 to do more with earned value information. Consult Project 2002 Help and Microsoft's Web site for Project 2002 (*www.microsoft.com/project*) for more details.

Now that you have entered and analyzed various project cost information, you will examine some of the human resource management features of Project 2002.

PROJECT HUMAN RESOURCE MANAGEMENT

In the project cost management section, you learned how to enter resource information into Project 2002 and how to assign resources to tasks. Two other helpful human resource features include resource calendars and histograms. In addition, it is important to know how to use Project 2002 to assist in resource leveling.

Resource Calendars

When you created the Project Tracking Database file, you used the standard Project 2002 calendar. This calendar assumes that standard working hours are Monday through Friday, from 8:00 a.m. to 5:00 p.m., with an hour for lunch starting at noon. Rather than using this standard calendar, you can create a different calendar that takes into account each project's unique requirements.

To create a new base calendar:

1. *Open a new file.* Click the New icon 🗋 on the Toolbar to open a new Project 2002 file.

2. *Open the Change Working Time dialog box.* Click **Tools** on the menu bar, and then click **Change Working Time**. The Change Working Time dialog box appears.

3. *Name the new base calendar.* In the Change Working Time dialog box, click **New**. The Create New Base Calendar dialog box appears. Click the **Create new base calendar** radio button, type **Mine** to name the new calendar in the **Name** text box, and then click **OK**. Make adjustments to create your own base calendar. Click **OK**.

You can use this new calendar for the whole project, or you can assign it to specific resources on the project.

To assign the new calendar to the whole project:

1. *Open the Project Information dialog box.* Click **Project** on the menu bar, and then click **Project Information**. The Project Information dialog box appears.

2. *Select a new calendar.* Click the **Calendar list arrow** to display a list of available calendars. Select your new calendar named **Mine** from this list, and then click **OK**.

To assign a specific calendar to a specific resource:

1. *Assign a new calendar.* Click **View** on the menu bar, and then click **Resource Sheet**. Type **Me** in the Resource Name column, press **Enter**, and then select the word **Me**.

2. *Select the calendar.* Click the **Base Calendar** cell for that resource name. Using the left mouse button, click the **list arrow** that appears, drag to select the **Mine** calendar, and release the mouse button.

3. *Block off vacation time.* Double-click a resource name to display the Resource Information dialog box, and then click the **Working Time** tab. You can block off vacation time for people by selecting the appropriate days on the calendar and marking them as nonworking days, as shown in Figure A-51.

4. *Close the file without saving it.* Click **File** on the menu bar, and then click **Close**. Click **No** when you are prompted to save the file.

Figure A-51. Blocking Off Vacation Time

Resource Histograms

A resource histogram is a type of chart that shows the number of resources assigned to a project over time (see Chapter 9, Project Human Resource Management). A histogram by individual shows whether a person is over- or underallocated during certain periods. To view histograms in Project 2002, select Resource Graph from the View bar or select Resource Graph from the View menu. The Resource Graph helps you see which resources are overallocated, by how much, and when. It also shows you the percentage of capacity each resource is scheduled to work, so you can reallocate resources, if necessary, to meet the needs of the project.

To view resource histograms for the Project Tracking Database project:

1. *View the Resource Graph.* Open the **baseline.mpp** file, and open the View Bar, if necessary. Click the **Resource Graph** icon 📊 on the View bar. If you cannot see the Resource Graph icon, you may need to click the up or down arrow on the View bar. Alternatively, you can click **View** on the menu bar, and then click **Resource Graph**. A histogram for Kathy appears, as shown in Figure A-52. Notice that the screen is divided into two sections. The left pane displays a person's name and the right pane displays a resource histogram for that person. Also, notice that Kathy is overallocated slightly in the month of April since the column for that month goes above the 100% line.

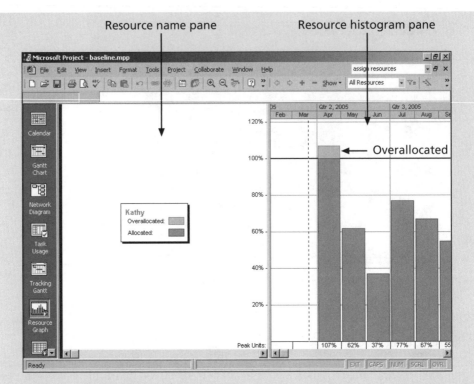

Figure A-52. Resource Histogram View

2. *Adjust the histogram's timescale.* Click **Zoom Out** or **Zoom In**, if necessary, to adjust the timescale for the histogram so that it appears in quarters and then months.

3. *View the next resource's histogram.* Click the **right scroll arrow** at the bottom of the resource name pane. The resource histogram for the next person appears. View the resource histograms for all four people, and then go back to the first one for Kathy.

Notice that Kathy's histogram has a partially red bar in April 2005. This red portion of the bar means that she has been overallocated during that month. The percentages at the bottom of each bar show the percentage each resource is assigned to work. For example, Kathy is scheduled to work 107% of her available time in April, 62% of her available time in May, and so on. Project 2002 has two tools that enable you to see more details about resource overallocation: the Resource Usage view and the Resource Management toolbar.

To see more details about an overallocated resource using the Resource Usage view:

1. *Open the Resource Usage view.* Click the **Resource Usage** icon 🖳 from the View bar or click **View** on the menu bar, and then click **Resource Usage**.

2. *Adjust the information displayed.* On the right side of the screen, click the **right scroll arrow** to display the hours Kathy is assigned to work each day, starting in April. You may also need to click the scroll down arrow to see all of Kathy's hours. If you need to adjust the timescale to display weeks, click Zoom Out or Zoom In. When you are finished, your screen should resemble Figure A-53.

Figure A-53. Resource Usage View

3. *Examine overallocation information.* Notice that Kathy's name appears in red, as does the value 5.57h in the column for Monday. Although Kathy is not scheduled to work more than 8 hours on Monday, some of the tasks were entered as hours, and Project 2002 assumes those hours start as soon as possible, thus causing the scheduling conflict. You'll remove the conflict using the Hour by Hour leveling later in this section.

To learn more about an overallocated resource, you can also use the Resource Management toolbar for resource allocation.

To see more resource allocation information:

1. *View the Resource Management toolbar.* Click **View** on the menu bar, select **Toolbars**, and then click **Resource Management**. The Resource Management toolbar displays below the Formatting toolbar.

2. *Select the Resource Allocation view.* Click **Resource Allocation View** ⊞ on the Resource Management toolbar. The Resource Allocation view appears, showing the Resource Usage view at the top of the screen and the Gantt Chart view at the bottom of the screen. Figure A-54 shows this view with Kathy's information highlighted.

Figure A-54. Resource Allocation View

3. *Close the file, but do not save any changes.* Click **File** on the menu bar, and then click **Close**. Click **No** when you are prompted to save changes.

The Resource Allocation view can help you identify the source of a resource overallocation. It is fairly obvious that Kathy has a slight overallocation the first day of the project because she is assigned to three different tasks, for a total of 8.57 hours. You could fix this problem by reducing Kathy's hours on one of the tasks or by allowing her to work a small amount of overtime. By using the scroll bar, you can view information for other resources. If you scroll down to look at Mary's resource allocation information and scroll to the right to reveal June and July, you can see that overlapping the design and implementation tasks may have caused another overallocation problem. To fix the problem, you can have Mary work overtime, assign another resource to help, or reschedule the implementation task to reduce the overlap. You can also see if resource leveling, as described in the next section, will help solve the problem.

Resource Leveling

Resource leveling is a technique for resolving resource conflicts by delaying tasks. Resource leveling also creates a smoother distribution of resource usage. You can find detailed information on resource leveling in Chapter 9, Project Human Resources Management, and in Project 2002 Help.

To use resource leveling:

1. *Reopen the baseline file.* Click **Open** and select **baseline.mpp**. Right-click the **Select All** button and click **Entry**. Move the split bar to see more of the Gantt chart. Notice that the project completion date is 9/19/05.

2. *Open the Resource Leveling dialog box.* Click **Tools** on the menu bar, and then click **Level Resources**. The Resource Leveling dialog box opens, as shown in Figure A-55. Uncheck the **Level only within available slack** option, if necessary.

Figure A-55. Resource Leveling Dialog Box

3. *Level the file and review the date changes.* Click the **Level Now** button. In this example, resource leveling moved the project completion date from 9/19 to 9/22, as shown on the Gantt chart milestone symbol for Task 44, Project completed.

4. *View Kathy's Resource Graph again.* Click the **Resource Graph** icon on the View bar. Notice that Kathy is still overallocated in April.

5. *Level using the Hour by Hour option.* Click **Tools** on the menu bar, and then click **Level Resources**. Change the basis for leveling from Day by Day to **Hour by Hour**, click the **Level Now** button, keep the default option of Entire pool in the Level Now dialog box, and then click **OK**. Notice that Kathy's overallocation disappears. Use the scroll arrows to reveal information about Mary. Remember that the left pane displays different resources, and the right pane reveals the resource histogram over time. Mary's overallocations are also gone, as a result of this resource leveling.

6. *View the Gantt chart again.* Click the **Gantt Chart** icon on the View bar and move the split bar to review the Gantt chart, if necessary. The project completion date has been moved back to 9/26 as a result of the resource leveling.

7. *View the Leveling Gantt chart.* Click **View** on the menu bar, click **More Views**, and double-click **Leveling Gantt**. Click **Zoom Out**, if necessary, to adjust the timescale to show quarters and months, and use the horizontal scroll bar to reveal all the symbols on the Leveling Gantt chart, until your screen resembles Figure A-56. Project 2002 adds a green bar to the Gantt chart to show leveled tasks.

Figure A-56. Leveling Gantt Chart View

8. *Save the file as level.mpp.* Click **File** on the menu bar, click **Save As**, and name the file **level.mpp**. Close the file.

If you want to undo the leveling, immediately select Undo from the Standard toolbar. Alternatively, return to the Resource Leveling dialog box, and click the Clear Leveling button.

If you want to check your work, a copy of level.mpp is available on the companion Web site for this text or from your instructor.

Consult the Project 2002 Help feature and use the key word "level" for more information on resource leveling. Also, be careful when setting options for this feature, so that the software adjusts resources only when it should. For example, the end date for the Project Tracking Database project was pushed back because you set the leveling options to allow the dates to slip. In this case, the project manager might prefer to ask her team to work a little overtime to remain on schedule.

Now that you have learned how to change resource calendars, view resource histograms, and level resources, you are ready to learn how to use Project 2002 to assist in project communications management.

PROJECT COMMUNICATIONS MANAGEMENT

Project 2002 can help you generate, collect, disseminate, store, and report project information. There are many different tables, views, and reports to aid in project communications, as you have seen in the previous sections. This section highlights some common reports and views. It also describes how to use templates and insert hyperlinks from Project 2002 into other project documents, how to save Project 2002 files as Web pages, and how to use Project 2002 in a workgroup setting.

Common Reports and Views

To use Project 2002 to enhance project communications, it is important to know when to use the many different ways to collect, view, and display project information. Table A-5 provides a brief summary of Project 2002 features and their functions, which will help you understand when to use which feature. Examples of most of these features are provided as figures in this appendix.

You can see from Table A-5 that many different reports are available in Project 2002. The overview reports provide summary information that top management might want to see, such as a project summary or a report of milestone tasks. Current activities reports help project managers stay abreast of and control project activities. The reports of unstarted tasks and slipping tasks alert project managers to problem areas. Cost reports provide information related to the cash flow for the project, budget information, overbudget items, and earned value management. Assignment reports help the entire project team by providing different views of who is doing what on a project. You can see who is overallocated by running the Overallocated Resources report or by viewing the two workload reports. You can also create custom reports based on any project information you have entered into Project 2002.

Table A-5: Functions of Project 2002 Features

FEATURE	FUNCTION
Gantt Chart view, Entry table	Enter basic task information
Network Diagram view	View task dependencies and critical path graphically
Schedule table	View schedule information in a tabular form
Cost table	Enter fixed costs or view cost information
Resource Sheet view	Enter resource information
Resource Information and Gantt Chart split view	Assign resources to tasks
Save Baseline	Save project baseline plan
Tracking toolbar	Enter actual information
Earned Value table	View earned value information
Resource Graph	View resource allocation
Resource Usage	View detailed resource usage
Resource Management	View resource usage and Gantt chart to find toolbar overallocation problems
Resource Leveling	Level resources
Overview Reports	Project summary, Top-Level Tasks, Critical Tasks, Milestones, Working Days
Current Activities Reports	Unstarted Tasks, Tasks Starting Soon, Tasks In Progress, Completed Tasks, Should Have Started Tasks, Slipping Tasks
Cost Reports	Cash Flow, Budget, Overbudget Tasks, Overbudget Resources, Earned Value
Assignments Reports	Who Does What, Who Does What When, To-Do List, Overallocated Resources
Workload Reports	Task Usage, Resource Usage
Custom Reports	Allows customization of each type of report
Save As Web page	Save files as HTML documents
Insert Hyperlink	Inserts hyperlinks to other files or Web sites

Using Templates and Inserting Hyperlinks and Comments

Chapter 10, Project Communications Management, and several other chapters provide examples of templates that you can use to improve project communications. Because it is often difficult to create good project plans, some organizations, such as Northwest Airlines, Microsoft, and various government agencies, keep a repository of sample Project 2002 files. Project 2002 includes several templates.

To access the Project 2002 templates:

1. *Open the General templates dialog box.* Open **level.mpp**, if necessary, click **File** on the menu bar, and then click **New** to display the New Project Task pane, as shown in Figure A-57.

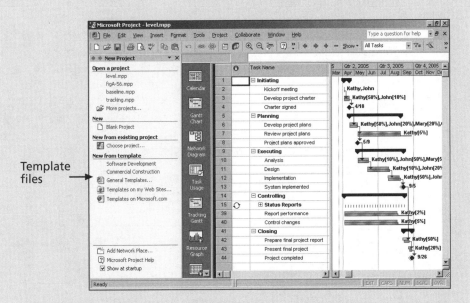

Figure A-57. Accessing Template Files

2. *Open the Templates dialog box.* Click the **General Templates** link below the **New from Template** section in the New Project Task pane. The Templates dialog box is displayed. Click the **Project Templates** tab, as shown in Figure A-58. As you can see, there are many templates available. You can also access additional templates from the **Templates on Microsoft.com** link from the New Project Task pane, from other Web sites, or from your own organization. See Appendix D of this text for information on templates.

Figure A-58. Project Templates

3. *Open the Microsoft Office Templates Gallery.* Click **Cancel** in the Templates dialog box, click the **Templates on Microsoft.com** link from the New Project Task pane, and then click **United States** (or the appropriate country for you) on Microsoft's Tools on the Web page to go to the Template Gallery, as shown in Figure A-59. Your screen may vary based on the templates Microsoft is providing now.

Figure A-59. Microsoft Office Template Gallery

4. *Close your browser and the level.mpp file.*

Using templates can help you prepare your project files, but be careful to address the unique needs of your project and organization. For example, even though the Home Move template file can provide guidance in preparing a Project 2002 file for a specific home move project, you need to tailor the file for your particular situation and schedule. You can also create your own Project 2002 templates by saving your file as a template using the Save As option from the File menu. Simply change the Save As Type option to Template. Project 2002 then gives the file an .mpt extension to identify the file as a template (Microsoft Project Template).

In addition to using templates for Project 2002 files, it is helpful to use templates for other project documents and insert hyperlinks to them from Project 2002. For example, your organization might have templates for meeting agendas, project charters, status reports, and project plans. See Appendix D and the companion Web site for examples of other template files. Next, you will create hyperlinks to some template files created in other applications.

To insert a hyperlink within a Project 2002 file:

1. *View the Entry table.* Open the **baseline.mpp** file. Right-click the **Select All** button and click **Entry** to return to the Entry table view.

2. *Select the task in which you want to insert a hyperlink.* Click the Task Name for Task 2, **Kickoff meeting**.

3. *Open the Insert Hyperlink dialog box.* Click **Insert** on the menu bar, and then click **Hyperlink**. Alternatively, you can right-click the task name and select **Hyperlink...**. The Insert Hyperlink dialog box opens, as shown in Figure A-60. You may have different folders visible based on your computer's directory structure.

Figure A-60. Insert Hyperlink Dialog Box

4. *Enter the filename of the hyperlink file.* Select the **Browse for File** icon 📂 to browse for the kickoffmeeting.doc file you downloaded from the companion Web site for this text, as shown in Figure A-60. Double-click on the **kickoffmeeting.doc** file, and then click **OK**.

5. *Reveal the Indicators column.* Click **Insert** on the menu bar, and then click **Column**. Click the **Field Name** drop-down list, click **Indicators**, and then click **OK**. A Hyperlink icon 🔗 should appear in the Indicators column to the left of the Task Name for Task 2. Move your mouse over the hyperlink icon until the mouse pointer changes to the Hand symbol 🖑 to reveal the name of the hyperlinked file, as shown in Figure A-61.

Figure A-61. Hyperlink Indicator

Clicking the Hyperlink icon in the Indicators column will automatically open the hyperlinked file. Using hyperlinks is a good way to keep all project documents organized.

It is also a good idea to insert notes or comments into Project 2002 files to provide more information on specific tasks.

To insert a note for Task 3:

1. *Open the Task Information dialog box for Task 3.* Open **baseline.mpp**, if necessary, and then double-click the Task Name for Task 3, **Develop Project Charter**. Click the **Notes** tab.

2. *Enter your note text.* Type "The charter was developed as a joint effort." in the Notes text box, as shown in Figure A-62.

Figure A-62. Task Information Dialog Box Notes Tab

3. *See the resulting Notes icon.* Click **OK** to enter the note, and notice the Notes icon 🗒 in the Indicators column next to Task 3.

4. *Open the note.* Double-click the **Notes** icon in the Indicators column for Task 3 to view the note, and then click **OK**.

5. *Close the file without saving it.* Click **File** on the menu bar, and then click **Close**. Click **No** when you are prompted to save the file.

Saving Files as Web Pages

Many organizations use the Web to disseminate information to employees and to communicate with customers. Project 2002 includes a feature that enables you to save project information as a Web page. As with most Microsoft applications, this feature in Project 2002 is available by selecting Save As Web Page from the File menu. You can create Web pages in several standard formats, including Compare to Baseline, Cost Data By Task, Earned Value Information, Top Level Tasks List, a "Who Does What" Report, and more. You can also design your own formats for information to be saved as Web pages.

To save project information as a Web page:

1. *Open the Save As dialog box for Web pages.* Open **baseline.mpp**. Click **File** on the menu bar, and then click **Save As Web Page**.

2. *Name the Web page file.* Type **taskinfo** in the File name text box as the new filename, and then click **Save** and save the file to your Data Disk or hard drive. The Export Wizard dialog box appears. Click **Next** to go to the next step of the Export Wizard, click the option to **Use existing map**, and then click **Next**, so your screen resembles Figure A-63.

Figure A-63. Export Wizard — Map Selection Dialog Box

3. *Select the Import/export map.* Click **Default Task Information** from the Import/export map list, and then click **Finish**. Project 2002 saves an HTML version of task information from baseline.mpp. The name of the new Web page is taskinfo.html.

In order to see the new Web page, you must open the new HTML file you created.

To open the taskinfo.html file:

1. *Open Windows Explorer.* Click **Start** on the Windows taskbar, select **Programs** (or **All Programs** in Windows XP), then **Accessories**, and then click **Windows Explorer**. Navigate your folders to find the taskinfo.html file. You can also use other methods to open Windows Explorer or use My Computer to find the desired file. Alternatively, start your Web browser, and use the Open Page feature from the File menu to open the file.

2. *Open the taskinfo.html file in your Web browser.* Double-click **taskinfo.html** to open it in your Web browser. The file extension for the file may not appear, depending on your settings. Figure A-64 shows part of the taskinfo.html file in Microsoft Internet Explorer.

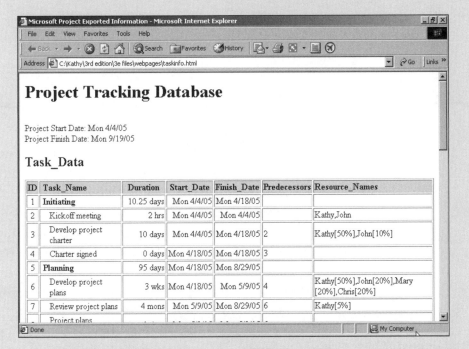

Figure A-64. Default Task Information HTML File

3. *Close the taskinfo.html file and exit Project 2002.* Click **Close** to close your browser, and then go back to your Project 2002 window and exit Project 2002 without saving the baseline file.

There are several ways to change or customize the standard HTML documents in Project 2002. The Export Wizard allows you to change formats for existing maps, and you can also create your own map. Consult the Project 2002 Help topic "HTML" to find more information on HTML features of Project 2002 and saving your files as Web pages.

Saving project information as Web pages includes text and numerical information, but not graphic information. You can create Web displays of graphic views, such as Gantt charts or network diagrams, but the process requires knowledge of how to include images in Web pages. For detailed information on working with graphic views, see the Project 2002 online Help topic "Copy a Microsoft Project picture into another program or a Web page." If Web users have Project 2002 loaded on their local PCs, you can share Project 2002 files over the Web by creating a hyperlink to the Project 2002 file stored on a Web server.

Recall that this appendix uses Project 2002 Standard. The Microsoft Solution for Enterprise Project Management includes many features to make it easy to share project information over the Internet. Consult Microsoft's Web site on Project 2002 for more information.

As you can see, Project 2002 is a very powerful tool. If used properly, it can greatly help users successfully manage projects.

CLOSING CASE

Terry Dunlay was amazed at how much you could do with Project 2002. She could see that it was important to understand several project management concepts before using the software. For example, she had to look up information on creating WBSs, critical path analysis, resource leveling, and earned value management to understand what she was doing with Project 2002. Kathy Nelson was pleased to see how quickly Terry learned to use the software. She asked Terry if she would like to help her on her next project, and Terry was thrilled to have the opportunity to learn even more.

DISCUSSION QUESTIONS

1. What are some of the main differences between Project 2002 and previous versions of Microsoft Project?
2. How do you use Project 2002 to create a WBS hierarchy?

3. Summarize how you use Project 2002 to assist in time management. How do you enter durations, link tasks, and view critical path information?

4. How can Project 2002 assist you in project cost management? How do you enter fixed costs? How do you enter resources and assign them to tasks? How can you view earned value information?

5. Briefly describe how to change resource calendars, view resource histograms, and level resources.

6. Summarize different ways to communicate information with Project 2002. How do you link to other files from within your Project 2002 file? How can you use templates and share information on the Web?

SUGGESTED READINGS

1. Bunin, Rachel, Lisa Friedrichsen, and Kathy Schwalbe. *New Perspectives on Project 2002, Introductory*. Boston: Course Technology (2003).

   This book walks you through step-by-step tutorials on how to use Project 2002 on one continuous running case. It includes six tutorials on Planning a Project, Creating a Project Schedule, Communicating Project Information, Assigning Resources and Costs, Tracking Progress and Closing a Project, and Sharing Information Across Projects, Applications, and the World Wide Web. There are review activities and several cases for each chapter. There are also supporting files and other information on the publisher's Web site at www.course.com.

2. Marmel, Elaine. *Microsoft Project 2002 Bible*, Hungry Minds, Inc. (2002).

   This comprehensive text (754 pages) includes detailed information on how to use Project 2002. It also includes a CD-ROM with links to Microsoft Project partners, including several sample files and demo products.

3. Pyron, Tim et al. *Special Edition: Using Microsoft Project 2002*, Que Publishing (2002).

   This text includes detailed coverage of Project 2002 Standard, Professional, Server, and Web Access. It includes 1200 pages of text and additional information on Que's Web site. Tim Pyron is a Microsoft Project consultant and trainer and editor of Woody's Project Watch, a free newsletter for Project users (www.woodyswatch.com).

4. Stover, Teresa S. *Microsoft Project 2002 Inside Out*. Microsoft Press (2002).

   This text includes 963 pages of text and a CD-ROM with an HTML interface, a collection of Microsoft add-ins and utilities, and a complete eBook for easy browsing and printing.

EXERCISES

The more you practice using the different features of Project 2002, the more quickly you will master the application and be able to use it to manage projects. This section includes exercises based on three examples of information technology projects that could benefit from using Project 2002 to assist in project scope, time, cost, human resource, and communications management. It also includes an exercise you can use to apply Project 2002 on a real project. There are simpler exercises that use Project 2002 at the end of several other chapters in this text. You can also re-create several screen shots in this text to practice your skills, or you can open and then make slight changes to template files or the files that come with this text.

Exercise A-1: Web Site Development

A nonprofit organization would like you to lead a Web site development project for them. The organization has Internet access that includes space on a Web server, but has no experience developing Web pages or Web sites. In addition to creating their Web site, they would like you to train two people on their staff to do simple Web page updates. The organization wants their Web site to include the following basic information, as a minimum: description of the organization (mission, history, and recent events), list of services, and contact information. They want the Web site to include graphics (photographs and other images) and have an attractive, easy-to-use layout.

1. Project Scope Management

   Create a WBS for this project and enter the tasks in Project 2002. Create milestones and summary tasks. Assume that some of the project management tasks you need to do are similar to tasks from the Project Tracking Database example. Some of the specific analysis, design, and implementation tasks will be to:

   a. Collect information on the organization in hardcopy and digital form (brochures, reports, organization charts, photographs, and so on).

   b. Research Web sites of similar organizations.

   c. Collect detailed information about the customer's design preferences and access to space on a Web server.

   d. Develop a template for the customer to review (background color for all pages, position of navigation buttons, layout of text and images, typography, including basic text font and display type, and so on).

   e. Create a site map or hierarchy chart showing the flow of Web pages.

f. Digitize the photographs and find other images for the Web pages; digitize hard-copy text.

g. Create the individual Web pages for the site.

h. Test the pages and the site.

i. Implement the Web site on the customer's Web server.

j. Get customer feedback.

k. Incorporate changes.

l. Create training materials for the customer on how to update the Web pages.

m. Train the customer's staff on updating the Web pages.

2. Project Time Management

a. Enter realistic durations for each task, and then link the tasks as appropriate. Be sure that all tasks are linked (in some fashion) to the start and end of the project. Assume that you have four months to complete the entire project. (*Hint*: Use the Project Tracking Database as an example.)

b. Print the Gantt Chart view and Network Diagram view for the project.

c. Print the Schedule table to see key dates and slack times for each task.

3. Project Cost Management

a. Assume that you have three people working on the project, and each of them would charge $20 per hour. Enter this information in the Resource Sheet.

b. Estimate that each person will spend an average of about five hours per week for the four-month period. Assign resources to the tasks, and try to make the final cost in line with this estimate.

c. Print the budget report for your project.

4. Project Human Resource Management

a. Assume that one project team member will be unavailable (due to vacation) for two weeks in the middle of the project. Make adjustments to accommodate this vacation so that the schedule does not slip and the costs do not change. Document the changes from the original plan and the new plan.

b. Use the Resource Usage view to see each person's work each month. Print a copy of the Resource Usage view.

5. Project Communications Management

a. Print a Gantt chart for this project. Use a timescale that enables the chart to fit on one page.

b. Print a "To-do List" report for each team member.

c. Create a "Who Does What Report" HTML document. Print it from your Web browser.

Exercise A-2: Software Training Program

ABC Company has 50,000 employees in its headquarters. The company wants to increase employee productivity by setting up an internal software applications training program for its employees. The training program will teach employees how to use software programs such as Windows, Word, Excel, PowerPoint, Access, and Project 2002. Courses will be offered in the evenings and on Saturdays and will be taught by qualified volunteer employees. The instructors will be paid $40 per hour. In the past, various departments sent employees to courses offered by local vendors during company time. In contrast to local vendors' programs, this internal training program should save the company money on training as well as make its people more productive. The Human Resources department will manage the program, and any employee can take the courses. Employees will receive a certificate for completing courses, and a copy of the certificate will be put in their personnel files. The company has decided to use off-the-shelf training materials, but is not sure which vendor's materials to use. The company needs to set up a training classroom, survey employees on what courses they want to take, find qualified volunteer instructors, and start offering courses. The company wants to offer the first courses within six months. One person from Human Resources is assigned full-time to manage this project, and top management has pledged its support of the project.

1. Project Scope Management

   Create a WBS for this project and enter the tasks in Project 2002. Create milestones and summary tasks. Assume that some of the project management tasks you need to do are similar to tasks from the Project Tracking Database example. Some of the tasks specific to this project will be to:

   a. Review off-the-shelf training materials from three major vendors and decide which materials to use.

   b. Negotiate a contract with the selected vendor for their materials.

   c. Develop communications information about this new training program. Disseminate the information via department meetings, e-mail, the company's intranet, and fliers to all employees.

   d. Create a survey to determine the number and type of courses needed and employees' preferred times for taking courses.

   e. Administer the survey.

   f. Solicit qualified volunteers to teach the courses.

   g. Review resumes, interview candidates for teaching the courses, and develop a list of preferred instructors.

h. Coordinate with the facilities department to build two classrooms with twenty personal computers in each, a teacher station, and an overhead projection system (assume that the facilities department will manage this part of the project).

i. Schedule courses.

j. Develop a fair system for signing up for classes.

k. Develop a course evaluation form to assess the usefulness of each course and the instructor's teaching ability.

l. Offer classes.

2. Project Time Management

a. Enter realistic durations for each task and then link appropriate tasks. Be sure that all tasks are linked in some fashion to the start and end of the project. Use the Project Tracking Database as an example. Assume that you have six months to complete the entire project.

b. Print the Gantt Chart view and Network Diagram view for the project.

c. Print the Schedule table to see key dates and slack times for each task.

3. Project Cost Management

a. Assume that you have four people from various departments available part-time to support the full-time Human Resources person, Terry, on the project. Assume that Terry's hourly rate is $40. Two people from the Information Technology department will each spend up to 25 percent of their time supporting the project. Their hourly rate is $50. One person from the Marketing department is available 25 percent of the time at $40 per hour, and one person from Corporate is available 30 percent of the time at $35 per hour. Enter this information about time and hourly wages into the Resource Sheet. Assume that the cost to build the two classrooms will be $100,000, and enter it as a fixed cost.

b. Using your best judgment, assign resources to the tasks.

c. View the Resource Graphs for each person. If anyone is overallocated, make adjustments.

d. Print the budget report for the project.

4. Project Human Resource Management

a. Assume that the marketing person will be unavailable for one week, two months into the project, and for another week, four months into the project. Make adjustments to accommodate this unavailability, so the schedule does not slip and costs do not change. Document the changes from the original plan and the new plan.

b. Add to each resource a 5 percent raise that starts three months into the project. Print a new budget report.

c. Use the Resource Usage view to see each person's work each month. Print a copy.

5. Project Communications Management

    a. Print a Gantt chart for this project. Use a timescale that enables the chart to fit on one page.

    b. Print a "To-do List" report for each team member.

    c. Create an HTML document using the "Default task information" map. Print it from your Web browser.

Exercise A-3: Project Tracking Database

Expand the Project Tracking Database example. Assume that XYZ Company wants to create a history of project information, and the Project Tracking Database example is the best format for this type of history. The company wants to track information on twenty past and current projects and wants the database to be able to handle 100 projects total. It wants to track the following project information:

- Project name
- Sponsor name
- Sponsor department
- Type of project
- Project description
- Project manager
- Team members
- Date project was proposed
- Date project was approved or denied
- Initial cost estimate
- Initial time estimate
- Dates of milestones (for example, project approval, funding approval, project completion)
- Actual cost
- Actual time
- Location of project files

1. Project Scope Management

Open scope.mpp and add more detail to Executing tasks, using the information in Table A-6.

Table A-6: Company XYZ Project Tracking Database Executing Tasks

ANALYSIS TASKS	DESIGN TASKS	IMPLEMENTATION TASKS
Collect list of 20 projects	Gather detailed requirements for desired outputs from the database	Enter project data
Gather information on projects	Create fully attributed, normalized data model	Test database
Define draft requirements	Create list of edit rules for fields, queries, and default values, and format reports about projects masks for data	Make adjustments, as needed
Create entity relationship diagram for database	Develop list of queries required for database	Conduct user testing
Create sample entry screen	Review design information with customer	Make adjustments based on user testing
Create sample report	Make adjustments to design based on customer feedback	Create online Help, user manual, and other documentation
Develop simple prototype of database	Create full table structure in database prototype	Train users on the system
Review prototype with customer	Create data input screens	
Make adjustments based on customer feedback	Write queries	
	Create reports	
	Create main screen	
	Review new prototype with customer	

2. Project Time Management

   a. Enter realistic durations for each task, and then link appropriate tasks. Make the additional tasks fit the time estimate: 20 days for analysis tasks, 30 days for design tasks, and 20 days for implementation tasks. Assume that all of the executing tasks have a total duration of 70 days. Do not overlap analysis, design, and implementation tasks.

   b. Print the Gantt Chart view and Network Diagram view for the project.

   c. Print the Schedule table to see key dates and slack times for each task.

3. Project Cost Management

   a. Use the resource and cost information provided in resource.mpp.

   b. Assign resources to the new tasks. Try to make the final cost about the same as that shown in the Project Tracking Database example: about $50,000.

   c. Print the budget report for the project.

4. Project Human Resources Management

   a. Two months after the project begins, give everyone on the team a 10 percent raise. Document the increase in costs that these raises cause.

   b. Use the Resource Usage view to see each person's work each month. Print a copy of the Resource Usage view.

5. Project Communications Management

   a. Print a Gantt chart for this project. Use a timescale that enables the chart to fit on one page.

   b. Print a "Top-Level Tasks" report.

   c. Create a "Cost Data by Task" HTML document. Print it from your Web browser.

Exercise A-4: Real Project Application

If you are doing a group project as part of your class or for a project at work, use Project 2002 to create a detailed file describing the work you plan to do. Enter a sufficient WBS, estimate task durations, link tasks, enter resources and costs, assign resources, and so on. Then save your file as a baseline and track your progress. View earned value information when you are about halfway done with the project or course. Export the earned value information into Excel and create an Earned Value Chart. Continue tracking your progress until the project or course is finished. Print your Gantt chart, Resource Sheet, Project Summary report, and relevant information. In addition, write a two- to three-page paper describing your experience. What did you learn about Project 2002 from this exercise? What did you learn about managing a project? How do you think Project 2002 helps in managing a project? You may also want to interview people who use Project 2002 and ask them about their experiences and suggestions.

RUNNING CASES

Review the running cases provided in Appendix C. Several questions involve using Project 2002.

Appendix B

Advice for the Project Management Professional (PMP) Exam and Related Certifications

OPENING CASE

Lew French had worked in a variety of positions at Fortune 500 companies for the last twenty-eight years. He received many promotions and made a few lateral moves, too. He recently moved into a position where he was team leader on a Six Sigma project. He had completed several training classes on Six Sigma and had more than twenty years of project management experience. However, he was also very eager to learn more about PMI and the PMP certification program. Two other colleagues in his department already had their Project Management Professional (PMP) designations. His company provided PMP certification preparation classes, and there was one coming up soon.

Lew talked to his boss, who agreed that it would be a good idea for Lew to pursue the PMP certification. The company was managing more and more projects every year, and by having the PMP certification, Lew would demonstrate the expertise he had developed over the years as well as his desire to be recognized as a professional project manager.

INTRODUCTION TO PROJECT MANAGEMENT CERTIFICATION PROGRAMS

This appendix provides information on Project Management Professional (PMP), IT Project+, and related certifications and offers some advice for taking the PMP and IT Project+ exams. It briefly describes various certification programs, requirements for earning and maintaining PMP and IT Project+ certification, and the structure and content of these exams. It also provides suggestions on preparing for the exams, tips for taking the exams, sample questions, and final advice on PMP and IT Project+ certification and project management in general.

What Is PMP Certification?

The Project Management Institute (PMI) offers certification as a Project Management Professional (PMP). As mentioned in Chapter 1, the number of people earning PMP certification grew rapidly in the past ten years, and there were 52,443 PMPs at the end of 2002. There are PMPs in 120 countries throughout the world. A major milestone occurred in 1999 when PMI's Certification Program Department became the first professional certification program department in the world to attain ISO 9001 recognition. Detailed information about PMP certification, the PMP Certification Handbook, and an online application are available from PMI's Web site (*www.pmi.org*) under Professional Development & Careers. The following information is quoted from PMI's Web site:

> The Project Management Institute (PMI®) is project management's leading global professional association, and as such, it administers a globally accepted and recognized, rigorous, examination-based, professional certification program of the highest caliber. The PMI Certification Program maintains ISO 9001 certification in Quality Management Systems as evidence of its commitment to professional excellence.
>
> If you enjoy the prestige that comes from being the best in your field, then you will appreciate the professional advantages derived from becoming a PMP. PMP certification is the profession's most globally recognized and respected certification credential. The PMP designation following your name tells current and potential employers that you have a solid foundation of project management knowledge that you can readily apply in the workplace.[1]

---

[1] Project Management Institute, "PMI® Certifications," (*www.pmi.org/info/PDC_CertificationsOverview.asp?nav=0401*) (March 2003).

Many companies and organizations are recommending or even requiring PMP certification for their project managers. A February 2003 newsletter reported that Microsoft chose PMI's PMP Certification Program as the certification of choice for its Microsoft Services Operation. Microsoft chose the PMP certification because of its global recognition and proven record in professional development for project managers.[2]

Certification Magazine published its annual review of how certification affects salaries of information technology professionals. This industry-wide study uses real-world numbers to show how education and experience affect a person's bottom-line salary. During a down market, people might ask why they should seek additional technical certifications. According to Gary Gabelhouse of *Certification Magazine*, "Perhaps it is best expressed in two words: job security. In boom times, one constantly reviews the rate of growth in salary as a key personal-success measurement. However, in down times, job security is paramount."[3]

What Are the Requirements for Earning and Maintaining PMP Certification?

You can now apply to take the PMP exam online. You can also fill out the application forms and mail them in with a check, if desired. Before you apply to take the exam, you must meet the following four requirements:

1. Have experience working in the field of project management. When you apply to take the exam, you enter the role you played on one or several projects and how many hours you worked in each of the five project management process groups for each project. Roles include:

   ■ project contributor
   ■ supervisor
   ■ manager
   ■ project leader
   ■ project manager
   ■ educator
   ■ consultant
   ■ administrator
   ■ other

[2] Project Management Institute Information Systems Specific Interest Group (ISSIG), *ISSIG Bits* (*www.pmi-issig.org*) (February 2003).

[3] Global Knowledge, "2003 Certification Salary by Certmag," *Global Knowledge E-Newsletter*, Issue #56 (March 2003).

Note that you do *not* have to have experience as a project manager to take the PMP exam; any of these roles will suffice. PMP certification requires all applicants with a bachelor's degree to have at least 4500 hours of project management experience within the six-year period prior to application. Applicants without a bachelor's degree are required to have 7500 hours of experience within an eight-year period prior to application. You must fill out a simple form listing the title of a project or projects you worked on, the start and end date for when you worked on the project(s), your role on the project, the number of hours you spent in initiating, planning, executing, controlling, and closing processes, and a list of the deliverables that you managed on the project. PMI staff will review your qualifications and let you know if you are qualified to take the PMP exam. You cannot take the exam without this experience qualification.

2. Document at least 35 contact hours of project management education. There is no time frame for this requirement. A university or college, a training company or consultant, a PMI chapter, a PMI Registered Education Provider (REP), a company-sponsored program, or a distance-learning company can provide the education. The hours must include content on project quality, scope, time, cost, human resources, communications, risk, procurement, and integration management. You must list the course title, institution, date, and number of hours for each course. For example, if Lew French from the opening case took a principles of management course at a university twenty-two years ago, he could list that on the education form along with a one- or two-day PMP exam preparation class sponsored by his company, a local PMI chapter or REP, a private training company or consultant, or a distance-learning program.

3. Agree to the PMP certificant and candidate agreement and release statement. This form certifies that application information is accurate and complete and that candidates will conduct themselves in accordance with the PMP Code of Professional Conduct, professional development requirements, and other PMI certification program policies and procedures. You can simply check a box saying you agree to this information when applying online.

4. Pay the appropriate exam fee. As of March 2003, the PMP certification fee was $405 for PMI members and $555 for non-members. The re-examination fee was $275 for PMI members and $375 for non-members. The individual PMI membership fee was $129 as of March 2003. Note that students, or anyone enrolled in a degree-granting program at an accredited, or globally equivalent, college or university, can join PMI at the student member rate of only $40. If you want to earn your PMP certification, it makes sense to join PMI, not only for the financial savings, but also for other benefits. Consult PMI's Web site for membership information.

After PMI sends you an eligibility letter to take the PMP exam, you can sign up to take it at several different Prometric (formerly Sylvan Technology) testing sites. You must take the exam within six months of receiving your eligibility letter. The eligibility letter will include complete details for scheduling your exam.

The last step in earning PMP certification is passing the exam! The PMP exam consists of 200 four-option multiple-choice questions. Although you cannot use any study aids during the exam, you can bring a non-programmable calculator to assist in performing calculations required to answer some of the questions. You are also given two blank pieces of paper, so you can write down formulas or other information when you enter the exam room, but you cannot bring in any notes or other materials. The questions on each test are randomly selected from a large test bank, so each person taking the exam receives different questions. Test takers have 4.5 hours to take this computerized exam, and a passing score is 68.5%, or at least 137 correct answers out of 200. PMI reviews and revises the exam annually.

Starting in January 1999, PMI instituted a new professional development program for maintaining the PMP certification. To maintain your PMP status, you must earn at least sixty Professional Development Units within three years and agree to continue to adhere to PMI's PMP Code of Professional Conduct. PMI will mail you its Professional Development Program Handbook after you earn PMP certification. This handbook provides more details on maintaining your PMP status.

What Is IT Project+ Certification?

More than 250,000 people worldwide have earned Computing Technology Industry Association (CompTIA) certifications in PC service, networking, document imaging, Internet, and PC Server technologies. In April 2001, CompTIA started offering its IT Project+ certification, which was purchased from Prometric-Thomson Learning, and recognized as the Gartner Institute Certification Program.

The following information is quoted from CompTIA's Web site:

The CompTIA IT Project+ certification validates your knowledge and abilities as an IT project manager.

IT Project+ certification is a vendor-neutral credential recognized worldwide. Having an IT Project+ certification is proof that you have the critical knowledge of business practices, interpersonal skills and project management processes, in addition to the ability to effectively plan, implement and complete IT projects on time and within your budget.

The demand for skilled project managers is significant and many companies either preferentially hire or require the IT Project+ certification, including State Farm,

Comark, CompuCom, Cable Express Technologies, Emtec and Care Call, Inc. IT Project+ is also endorsed by the National Skill Standards Board.[4]

What Are the Requirements for Earning and Maintaining IT Project+ Certification?

Unlike the PMP exam, there are very few requirements you need to meet to take the IT Project+ exam. CompTIA does not require that you have any work experience or formal education in project management before you can take the IT Project+ exam. You can register to take the IT Project+ exam online from ComTIA's Web site. Testing sites include Prometric and Pearson VUE. The main requirements include paying a fee and passing the exam, as summarized below:

1. As of March 2003, the cost for taking the IT Project+ exam was $155 for CompTIA members and $207 for non-members.
2. To pass this two-hour, 84-question exam, you must score at least 63%.
3. You do not need to renew your IT Project+ certification.
4. IT Project+ certification is one of the prerequisites or equivalents to Novell's Certified Novell Engineer (CNE), Certified Novell Administrator (CNA), and Certified Novell Instructor (CNI) certifications. CompTIA's IT Project+ certification is also a Continuing Certification Requirement (CCR) to maintaining a Master CNE certification.

What Other Certification Exams Related to Project Management Are Available?

In recent years, PMI and other organizations have been developing more certifications related to information technology project management. Below are brief descriptions of other existing certifications and some currently in development.

■ PMI's Certificates of Added Qualification (CAQ): The PMP exam is not industry-specific, so you will not be tested on information unique to a particular industry, such as information technology. PMI developed additional exams called Certificates of Added Qualification (CAQ) to test industry-specific knowledge. PMPs with experience in certain fields could take CAQ exams. The first PMI CAQ exam, given in August 2000, was the Automotive Product Development (APD) CAQ. In 2002, PMI released a CAQ for people in the information technology field. However, PMI

---

[4] CompTIA Web site (*www.comptia.org/certification/itproject*) (March 2003).

suspended the CAQ exams in 2003. Consult PMI's Web site (*www.pmi.org*) for further information on CAQ exams.

- Microsoft provides certification as a Microsoft Office Specialist (MOS) to recognize the ability to demonstrate proficiency in using its software products. You can earn certification for Microsoft Project at the core or comprehensive level. Consult Microsoft's Web site for detailed information on the MOS certification.

- The International Project Management Association (IPMA) offers a four-level certification program. The main requirements for each level are derived from typical activities, responsibilities, and requirements from practice. The IPMA four-level certification system, in descending order, includes the Certified Project Director, Certified Project Manager, Certified Project Management Professional, and Certified Project Management Practitioner. See IPMA's Web site (*www.ipma.org*) for more details.

- Certified IT Project Manager (CITPM): In 1998 the Singapore Computer Society collaborated with the Infocomm Development Authority of Singapore to establish an Information Technology Project Management Certification Program. By the fall of 2000, about 400 participants had joined the program, with 248 professionals currently having the title of Certified IT Project Manager (CITPM). PMI recently signed a Memorandum of Understanding with the Singapore Computer Society to support and advance the global credentialing of project management and information technology expertise.[5]

WHAT IS THE STRUCTURE AND CONTENT OF THE PMP EXAM?

The PMP exam is based on information from the entire project management body of knowledge (PMBOK®) as well as the area of professional responsibility. Essentially, the exam reviews concepts and terminology in PMI's PMBOK® Guide 2000, and texts such as this one will help to reinforce your understanding of key topics in project management. Table B-1 shows the approximate breakdown of questions on the PMP exam by process groups as of early 2003. Candidates should review updated exam information on PMI's Web site to make sure they are prepared for their PMP exams. For example, PMI provides sample PMP questions and answers that you can download for free from their Web site.

---

[5] Schwalbe, Kathy, "Certification News," *ISSIG Review* (Fourth Quarter 2000).

Table B-1: Breakdown of Questions on the PMP by Process Groups

PROCESS GROUP	PERCENTAGE OF QUESTIONS ON PMP EXAM
Initiating	8.5
Planning	23.5
Executing	23.5
Controlling	23
Closing	7
Professional Responsibility	14.5

Study Table 3-1 from Chapter 3 of this text to understand the relationships among project management process groups, activities, and knowledge areas. Table 3-1 briefly outlines which activities are performed during each of the project management process groups and what is involved in each of the knowledge areas. It is also important to understand what each of the project management activities includes. Several questions on the certification exam require an understanding of this framework for project management, and many questions require an understanding of the various tools and techniques described in the PMBOK® Guide 2000 and this text.

The PMP exam includes conceptual questions, application questions, and evaluative questions. Conceptual questions test your understanding of key terms and concepts in project management. For example, you should know basic definitions such as what a project is, what project management is, and what key activities are included in project scope management. Application questions test your ability to apply techniques to specific problems. For example, a question might provide information for constructing a network diagram and ask you to find the critical path. You might be given cost and schedule information and be asked to find the schedule or cost performance index. Evaluative questions provide situations for you to analyze, and your response will indicate how you would handle them. For example, a project might have many problems. A question might ask what you would do in that situation, given the information provided. Remember that all questions are multiple choice, so you must select the best answer from the options provided.

HOW SHOULD YOU PREPARE FOR THE PMP EXAM?

To prepare for the PMP exam, it is important to understand your own learning and testing style, and to use whatever resources and study techniques work best for you. Several companies offer sample tests, CD-ROMs, and courses designed to help people pass the exam. Several chapters of PMI offer PMP exam review courses. Many people form study groups to provide support in preparing for the exam. Having some "peer pressure" might help to ensure that you actually take and pass the exam. If you do not pass the exam, you have to pay to take it again, so you want to make sure you are ready the first time.

Even if you think you know about project management, studying the material in the PMBOK® Guide 2000 and this book before taking the exam will help you. If you are comfortable with the concepts in the PMBOK® Guide 2000 and this text and have some experience working on projects, you should do fine.

Recall the triple constraint of project management—scope, time, and cost management. Focus on these three knowledge areas when studying. Be familiar with project charters, WBSs, network diagrams, critical path analysis, cost estimates, earned value, and so on. Many questions on the exam are related to project time management, so study that area closely. In addition, be sure to memorize earned value formulas.

To pass the exam, you need to answer only 68.5 percent of the questions correctly. Your PMP certificate will not display your final score, so it does not matter if you get 70, 80, 90, or even 100 percent correct.

TEN TIPS FOR TAKING THE PMP EXAM

1. The PMP exam is computer based and begins with a short tutorial on how to use the testing software. The software makes it easy to mark questions you want to review later, so learning how to mark questions is helpful. Using this feature can give you a feel for how well you are doing on the test. It is a good idea to go through every question somewhat quickly and mark those questions on which you want to spend more time. If you mark fewer then 60 out of 200 questions before you finish, odds are that you will get at least 68.5 percent correct and pass the exam. You can miss 63 questions and still pass.

2. The time allotted for the exam is 4.5 hours, and each multiple-choice question has four alternatives. You should have plenty of time to complete the exam. Use the first two to three hours to go through all of the questions. Do not spend too much time on any one question. As you work, mark each question that you would like to return to for further consideration. Then use the remaining time to check the questions you are not sure of.

3. Some people believe it is better to change answers you were originally unsure of. If you think that a different answer is better, after reading the question again, then change your answer. If you are still unsure, leave the question marked so you can see a final total of unmarked questions.

4. Do not try to read more into the questions than what is stated. There are no trick questions. Most of the questions are relatively short, and there are only four alternatives from which to choose the answer. Few of the alternatives include "all of the above" or "none of the above." Some questions might include the alternative "There is not enough information."

5. To increase your chances of getting the right answer, first eliminate obviously wrong alternatives, and then choose among the remaining alternatives.

6. Some questions require doing calculations such as earned value management. It is worthwhile to memorize the earned value formulas to make answering these questions easier. You may use a non-programmable calculator while taking the exam, so be sure to bring one to make performing calculations easier.

7. You will be given a couple of pieces of blank paper to use during the exam. Before starting the test, you might want to write down important equations so that you do not have to rely on your memory. When you come to a question involving calculations, write the calculations down so you can check your work for errors.

8. Read all questions carefully. A few sections of the test require that you answer three to four questions about a scenario. These questions can be difficult; it can seem as if two of the alternatives could be correct, although you can choose only one. Read the directions for these types of questions several times to be sure you know exactly what you are supposed to do. Also, remember important concepts such as using a WBS, emphasizing teamwork, and practicing professional integrity.

9. If you do not know an answer and need to guess, be wary of alternatives that include words such as always, never, only, must, and completely. These extreme words usually indicate incorrect alternatives because there are many exceptions to rules.

10. After about two hours, take a short break to clear your mind. You might want to bring a snack to have during your break.

SAMPLE PMP EXAM QUESTIONS

A few sample questions similar to those you will find on the PMP exam are provided below. You can check your answers at the end of this appendix. If you miss fewer than six out of these twenty questions, you are probably ready to take the PMP exam.

1. A document that formally recognizes the existence of a project is a

   A. Gantt chart

   B. WBS

   C. Project charter

   D. Scope statement

2. Decomposition is used in developing
   A. the management plan
   B. the communications plan
   C. the earned value
   D. the WBS

3. The critical path on a project represents
   A. the shortest path through a network diagram
   B. the longest path through a network diagram
   C. the most important tasks on a project
   D. the highest-risk tasks on a project

4. If the EV for a project is $30,000, the AC is $33,000, and the PV is $25,000, what is the cost variance?
   A. $3,000
   B. –$3,000
   C. $5,000
   D. –$5,000

5. If the EV for a project is $30,000, the AC is $33,000, and the PV is $25,000, how is the project performing?
   A. The project is over budget and ahead of schedule.
   B. The project is over budget and behind schedule.
   C. The project is under budget and ahead of schedule.
   D. The project is under budget and behind schedule.

6. What is the target goal for defects per million opportunities using Six Sigma?
   A. 1
   B. 3.4
   C. 34
   D. 100

7. Project human resource management does not include which of the following processes?
   A. organizational planning
   B. staff acquisition
   C. team development
   D. resource planning

8. If a project team goes from three to five people, how many more communications channels are there?
   A. 7
   B. 6
   C. 5
   D. 4

9. Your project team has identified several risks related to your project. You decide to take actions to reduce the impact of a particular risk event by reducing the probability of its occurrence. What risk response strategy are you using?

   A. risk avoidance

   B. risk acceptance

   C. risk mitigation

   D. contingency planning

10. Which type of contract provides the least amount of risk for the buyer?

    A. firm fixed price

    B. fixed price incentive

    C. cost plus incentive fee

    D. cost plus fixed fee

11. Suppose you have a project with four tasks as follows:

    ■ Task 1 can start immediately and has an estimated duration of 1.

    ■ Task 2 can start after Task 1 is completed and has an estimated duration of 4.

    ■ Task 3 can start after Task 2 is completed and has an estimated duration of 5.

    ■ Task 4 can start after Task 1 is completed and must be completed when Task 3 is completed. Its estimated duration is 8.

    What is the length of the critical path for this project?

    A. 9

    B. 10

    C. 11

    D. 12

12. In which of the following project management process groups are the most time and money typically spent?

    A. initiating

    B. planning

    C. executing

    D. controlling

13. Creating a probability/impact matrix is part of which risk management process?

    A. risk management planning

    B. risk identification

    C. qualitative risk analysis

    D. quantitative risk analysis

14. It is crucial that your project team finish your project on time. Your team is using a technique to account for limited resources. You have also added a project buffer before the end date and feeding buffers before each critical task. What technique are you using?

    A. critical path analysis

    B. PERT

    C. critical chain

    D. earned value

15. One of your senior technical specialists informs you a major design flaw exists in a systems development project you are managing. You are already testing the system and planned to roll it out to more than 5,000 users in a month. You know that changing the design now will cause several cost and schedule overruns. As project manager, what should you do first?

    A. Issue a stop work order until you understand the extent of the project.

    B. Notify your project sponsor immediately to see if there are additional funds available to work on this problem.

    C. Notify your senior management and let them decide what to do.

    D. Hold a meeting as soon as possible with key members of your project team to discuss possible solutions to the problem.

16. You are a member of a large government project. You know that the contract insists that all equipment be manufactured in the United States. You see a senior member of your team replacing a company etching on a piece of equipment that was made in a foreign country. You confront this person, and he says he is following the project manager's orders. What should you do?

    A. Nothing; the project manager made the decision.

    B. Immediately report the violation to the government.

    C. Update your resume and look for another job.

    D. Talk to the project manager about the situation, and then decide what to do.

17. Which of the following is not an output of the overall change control process?

    A. project plan updates

    B. corrective action

    C. change requests

    D. lessons learned

18. Which type of dependency is inherent in the nature of the work being done?

    A. mandatory dependency

    B. discretionary dependency

    C. external dependency

    D. internal dependency

19. Obtaining quotes, bids, offers, or proposals is part of which project procurement management process?

   A. procurement planning

   B. solicitation planning

   C. source selection

   D. solicitation

20. Your boss believes that all of your project team members avoid work as much as possible. He or she often uses threats and various control schemes to make sure people are doing their jobs. Which approach to managing people does your boss follow?

   A. Maslow's hierarchy of needs

   B. Theory X

   C. Theory Y

   D. Herzberg's motivation and hygiene factors

The Project Management Institute started providing free sample PMP exam questions on its Web site in 1999. You can download several PMP sample exam questions from *www.pmi.org* under the certification section. PMI's Web site also provides answers to frequently asked questions about the PMP certification application and PMI's professional development program. Several individuals and companies also provide books, CD-ROMs, online practice tests, courses, and so on to help individuals prepare for the PMP exam. You can search for PMP exam resources on any Web search engine or online bookstores. As of March 2003, performing a search using the keyword "PMP" at *www.amazon.com* resulted in 35 different books and related products. Performing a search using the keywords "PMP exam" at *www.google.com* resulted in 41,000 hits in March 2003. See the Suggested Readings section for more information.

ADDITIONAL INFORMATION ON THE IT PROJECT+ EXAM

Because there are no experience or education requirements for taking the IT Project+ exam, you might want to take it early in your career. Once you have enough experience and education to take the PMP exam, you might want to earn and maintain PMP certification.

As stated earlier, the IT Project+ exam consists of 84 questions. Table B-2 shows the approximate breakdown of questions on the IT Project+ exam by four domain areas. Candidates should review updated exam information on CompTIA's Web site to make sure they are prepared for the exams. For example, CompTIA provides a detailed list of objectives you should understand before taking the IT Project+ exam. Studying information in this book will also help you prepare for the IT Project+ exam. You might also want to purchase an exam guide to get specific information and access to more sample questions.

Table B-2: Breakdown of Questions on the IT Project+ Exam by Domain Areas

DOMAIN AREA	PERCENTAGE OF QUESTIONS ON IT PROJECT+ EXAM
Scope Definition	27
Preliminary Planning/Project Planning	39
Project Execution	29
Closure	5

Much of the advice for taking the PMP exam also applies to taking the IT Project+ exam. Many people find the questions to be similar on both exams, although the PMP exam is longer and more comprehensive. Below are some of the main differences between the content and types of questions you will find on the IT Project+ exam:

- The IT Project+ exam includes scenarios and content specific to the information technology industry. For example, you should understand the various systems development life cycles and issues that often occur on information technology projects.
- You should understand the various roles of people on information technology projects, such as business analysts, database analysts, programmers, and so on.
- Although many of the questions are multiple-choice like the PMP exam questions, several questions involve choosing two or more correct answers. Several questions also involve matching or putting items in order, as shown in the next section.
- The IT Project+ exam is not based on the PMBOK® Guide 2000, so you do not need to know the processes involved in the various knowledge areas. However, much of the terminology, concepts, tools, and techniques are the same on both exams.

SAMPLE IT PROJECT+ EXAM QUESTIONS

A few sample questions similar to those you will find on the IT Project+ exam follow. You can check your answers at the end of this appendix.

1. Two software developers on your project disagree on how to design an important part of a system. There are several technologies and methodologies they could use. What should be the primary driver in deciding how to proceed?
   A. following corporate standards
   B. following industry standards
   C. meeting business needs
   D. using the lowest-cost approach

2. Match the following items to their descriptions:

Stakeholder	A. Acts as a liaison between the business area and developers
Project manager	B. Writes software code
Business analyst	C. Person involved in or affected by project activities
Programmer	D. Responsible for managing project activities

3. For a project to be successful, the project manager should strive to understand and meet certain goals. What are the three main project goals to meet? Select three answers.

   A. scope or performance goals

   B. time goals

   C. political goals

   D. cost goals

   E. stock price goals

4. You have received an incomplete project scope definition. Put the following actions in order of how you should proceed to complete them.

   A. Incorporate changes to the scope definition document.

   B. Review the draft scope definition document with your project team.

   C. Get signatures on the completed scope definition document.

   D. Rewrite the draft scope definition document with your users and project team.

5. What term is used to describe the process of reaching agreement on a collective decision?

   A. collaboration

   B. cooperation

   C. coordination

   D. concensus

6. What is a variance?

   A. A buffer in a duration estimate

   B. A small amount of money set aside for contingencies

   C. A form of risk management

   D. A deviation from the project plan

7. Which of the following would be legitimate reasons for a vendor to request a delay in delivering a product? Select two answers.

   A. The vendor may have underestimated the amount of time required to produce and deliver the product.

   B. The project contact from the vendor's organization may be going on vacation.

   C. The vendor might be able to provide a better product by delivering the product late.

   D. The vendor might lose money by delivering the product late.

8. When should you involve stakeholders in the change control process on information technology projects?

   A. before a change is submitted

   B. after a change is submitted

   C. when a change is submitted

   D. throughout the life of a project

9. Which of the following techniques can be used to help management requirements? Select three answers.

   A. prototyping

   B. use case modeling

   C. JAD

   D. worst-case modeling

10. Match the following items to their descriptions:

Lessons learned	A. Leave a clear and complete history of a project
Project audits	B. Review project progress and results
Project archives	C. Document what went right or wrong on a project

FINAL ADVICE ON CERTIFICATION AND PROJECT MANAGEMENT IN GENERAL

Now that you have read this text and discussed project management with others, I hope that you have grown to see project management as a valuable skill, especially in the information technology field. Project management certification can be your first step toward advancing your career in the fast-paced world of technology, because no matter how much things change, the need for projects and project managers is a constant.

The knowledge and experience I have gained working on and managing projects continues to help me in my career and my personal life. I was fortunate to step into a project management role very early in my career as an Air Force officer. My first real job at the age of twenty-two was as a project manager. I have held several job titles since then—systems analyst, engineer, technical specialist, information technology management consultant, independent consultant, college professor, and now author. All of these jobs included working on or managing projects. As a mother of three, I can also attest to the fact that project management skills help in planning and executing social activities (weddings, birthday parties, fundraisers, and so on) and in dealing with the joys and challenges of everyday life.

CLOSING CASE

Lew French took a two-day PMP exam preparation class offered on-site through his company. He was surprised to see several people in their mid-twenties in the class, but they had enough education and experience to take the exam. Lew learned a lot in the class and met some great people. He passed the PMP exam on his first try and is very proud of his accomplishment.

DISCUSSION QUESTIONS

1. What is PMP certification, and why do you think the number of people earning it has grown so much in the past ten years?
2. What do you need to do before you can take the PMP exam? What is the exam itself like? What do you need to do to maintain PMP certification? What do you need to do to take the IT Project+ exam? How does the IT Project+ exam differ from the PMP exam? Do you need to renew IT Project+ certification?
3. Briefly describe project management certification programs other than the PMP or IT Project+ certifications.
4. What is the difference between conceptual, application, and evaluative questions? Which project management process groups have the most questions on the PMP exam? What are the four domain areas tested on the IT Project+ exam?
5. If you plan to take the IT Project+ or PMP exam soon, what should you do to prepare?
6. Which tips for taking the PMP exam do you think would be most helpful for you?

EXERCISES

1. Go to PMI's Web site and review the information about taking the PMP exam. Write a one- to two-page paper summarizing what you found.
2. Go to CompTIA's Web site and review the information about taking the IT Project+ exam. Write a one- to two-page paper summarizing what you found.
3. Download the sample PMP questions and answers from PMI's Web site. Take the sample test, and then score your results. Summarize how you did and note the areas you would need to study before you could take the PMP exam.
4. Interview someone who has PMP or IT Project+ certification. Ask him or her why he or she earned the certification and how it has affected his or her career. Write your findings in a one- to two-page paper.

5. Do an Internet search on earning PMP or IT Project+ certification. Be sure to search for Yahoo Groups related to these topics. What are some of the options you found to help people prepare for either exam? If you were to take one of the exams, what do you think you would do to help study for it? Do you think you would need additional information beyond what is in this text to help you pass? Write a one- to two-page paper describing your findings and opinions.

6. Read a recent issue of PMI's *PM Network* magazine. You can order a copy from PMI's Web site, or contact someone from a local PMI chapter to obtain a copy. Summarize all of the ads you find in the magazine related to earning PMP certification in a one- to two-page paper. Also, include your opinion on which ad, course, books, CD-ROMs, or other media appeals to you the most.

SUGGESTED READINGS

1. Amazon.com, search all products for "PMP" or "IT Project+" (*www.amazon.com*) (2003).

   There are many books, CD-ROMS, and audiotapes available for sale from Amazon.com or similar sites, with new offerings coming out every year. The "Dummy" series even published a book called "PMP Certification for Dummies" in 2003. Read the descriptions and decide which resources best meet your needs and budget.

2. Boson Software, CompTIA, and PMI Practice Tests (*www.boson.com*) (2003).

   Boson Software is well known for their sample tests for various certification exams. They also provide a money-back guarantee. As of March 2003, they provided four different sample tests for the PMP exam (one in Spanish) and two for the IT Project+ exam. You can download a free demo and then purchase a sample test with 204 questions, explanations, and references.

3. Cheetah Learning, "Cheetah Accelerator Toolkit™ for the PMP®" (*www.cheetahlearning.com*) (January, 2003).

   You can download this free 28-page booklet with several good suggestions to help you prepare for the PMP exam. Cheetah Learning also offers online and instructor-led courses related to project management and the PMP exam.

4. Project Management Institute (PMI) Standards Committee, *A Guide to the Project Management Body of Knowledge (PMBOK®)*, 2000.

   You get this guide on CD-ROM as part of your PMI membership. Be sure to study the PMBOK® Guide 2000 since it provides the basis for what is on the PMP exam. Also, download PMI's latest set of sample questions and answers from www.pmi.org. PMI also provides links to several sites that provide project management and PMP-related materials and courses.

5. Yahoo Groups. (*www.yahoo.group.com*) (2003).

> *There are several Yahoo Groups related to PMP and IT Project+ certification. You can also start your own group at no cost, if you like. One of the most popular PMP groups (over 2100 members as of March 2003) was at* http://groups.yahoo.com/group/PMPCert.

ANSWERS TO SAMPLE PMP QUESTIONS

1. C
2. D
3. B
4. B
5. A
6. B
7. D
8. A
9. C
10. A
11. B
12. C
13. C
14. C
15. D
16. D
17. C
18. A
19. D
20. B

ANSWERS TO SAMPLE IT PROJECT+ QUESTIONS

1. C
2. C, D, A, B
3. A, B, D
4. B, D, A, C
5. D
6. D
7. A, C
8. D
9. A, B, C
10. C, B, A

Appendix C
Running Cases

INTRODUCTION

The purpose of this appendix is to provide some running cases that you can use to practice your project management skills. Four running cases are included in this appendix, and several more are on the companion Web site for this text (*www.course.com/mis/schwalbe*). Each running case includes five parts—initiating, planning, executing, controlling, and closing—with scenario-based information and several tasks to complete under each part. Several of the questions involve using templates provided in Appendix D. See your instructor for suggested solutions.

RUNNING CASE 1: RECREATION AND WELLNESS INTRANET PROJECT

This running case continues the Minicase scenarios from Chapters 2 and 3. It most closely resembles the JWD Consulting case study presented in Chapter 3.

Part 1: Initiating

Scenario

Manage Your Health, Inc. (MYH) is a Fortune 500 company that provides a variety of health care services across the globe. MYH has more than 20,000 full-time employees and more than 5,000 part-time employees. MYH recently updated its strategic plan, and key goals include reducing internal costs, increasing cross selling of products, and exploiting new Web-based technologies to help employees, customers, and suppliers work together to improve the development and delivery of its health care products and services. Below are some ideas the information technology department has developed for supporting these strategic goals:

1. Recreation and Wellness Intranet Project: Provide an application on the current Intranet to help employees improve their health. A recent study

found that MYH, Inc. pays 20 percent more than the industry average for employee health care premiums, primarily due to the poor health of its employees. You believe that this application will help improve employee health within one year of its rollout so that you can negotiate lower premiums, saving about $30/employee/year for full-time employees over the next four years. This application would include the following capabilities:

■ Allow employees to register for company-sponsored recreational programs, such as soccer, softball, bowling, jogging, walking, and other sports

■ Allow employees to register for company-sponsored classes and programs to help them manage their weight, reduce stress, stop smoking, and manage other health-related issues

■ Track data on employee involvement in these recreational and health-management programs

■ Offer incentives for people to join the programs and do well in them (i.e., incentives for achieving weight goals, winning sports team competitions, etc.)

2. Health Coverage Costs Business Model: Develop an application to track employee health care expenses and company health care costs. Health care premiums continue to increase, and the company has changed insurance carriers several times in the past ten years. This application should allow business modeling of various scenarios as well as tracking and analyzing current and past employee health care expenses and company health care costs. This application must be secure and run on the current intranet so several managers and analysts could access it and download selective data for further analysis. The new application must also import data from the current systems that track employee expenses submitted to the company and the company's costs to our insurance provider. You believe that having this data will help you revise policies concerning employee contributions to health care premiums and help you negotiate for lower premiums with insurance companies. You estimate that this application would save your company about $20/employee/year for full-time employees over the next four years and cost about $100,000 to develop.

3. Cross-Selling System: Develop an application to improve cross-selling to current customers. The current sales management system has separate sections for major product/service categories and different sales representatives based on those products and services. You see great opportunities to increase sales to current customers by providing them discounts when they purchase multiple products/services. You estimate that this system would increase profits by $1 million each year for the next three years and cost about $800,000 each year for development and maintenance.

4. Web-Enhanced Communications System: Develop a Web-based application to improve development and delivery of products and services. There

are currently several incompatible systems related to the development and delivery of products and services to customers. This application would allow customers and suppliers to provide suggestions, enter orders, view the status and history of orders, and use electronic commerce capabilities to purchase and sell their products. You estimate that this system would save your company about $2 million each year for three years after implementation. You estimate it will take one year and $3 million to develop and require 20 percent of development costs each year to maintain.

Tasks

1. Summarize each of the above-proposed projects in a simple table format suitable for presentation to top management. Include the name for each project, identify how each one supports business strategies, assess the potential financial benefits and other benefits of each project, and provide your initial assessment of the value of each project. Write your results in a one- to two-page memo to top management, including appropriate back-up information and calculations.

2. Prepare a weighted decision matrix using the template provided in Appendix D to evaluate these four projects. Develop at least four criteria, assign weights to each criterion, assign scores, and then calculate the weighted scores. Print the spreadsheet and bar chart with the results. Also write a one-page paper describing this weighted decision matrix and what the results show.

3. Assume that you have been assigned to work on the Recreation and Wellness Intranet Project described above. You are part of a team to help select the project manager for this project. Write a one- to two-page job description for this person, including a description of job duties, required experience and education, and desired skills.

4. Prepare a business case for the Recreation and Wellness Intranet Project. Assume the project will take six months to complete and cost about $200,000. Use the business case template provided in Appendix D to help calculate the NPV, ROI, and year in which payback occurs.

5. Prepare a project charter for the Recreation and Wellness Intranet Project. Assume the project will take six months to complete and cost about $200,000. Use the project charter template provided in Appendix D.

Part 2: Planning

Scenario

Managers at MYH selected Tony Prince as the project manager for the Recreation and Wellness Intranet Project. Tony had previous project management and systems analysis experience within the company, and he was an avid sports participant. Tony was starting to put the project team together. Team members so far included you, a programmer/analyst and aspiring project manager, Patrick, a network specialist, Nancy, a business analyst, and Bonnie, another programmer/analyst. Tony also had a few people from other departments supporting the project, such as James, from human resources, and Cassandra from finance. Tony knew he would have to develop a survey to solicit input from all employees about this new system and make sure it was very user-friendly.

Tasks

1. Develop a team contract for this project. Use the template provided in Appendix D as a guide. Also, look at the examples of project charters in Chapters 3 and 5.

2. Prepare a stakeholder analysis for this project, using the template provided in Appendix D and example in Chapter 4 as guides. Be creative in making up information about stakeholders.

3. Develop a scope statement for the project. Use the template provided in Appendix D and the example in Chapter 3 as guides. Be as specific as possible in describing product characteristics and requirements, as well as product deliverables.

4. Develop a work breakdown structure (WBS) for the project. Break down the work to level 2 or level 3, as appropriate. Use the template in Appendix D and samples in Chapters 3 and 5 as guides. Print the WBS in list form as a Word file. Be sure the WBS is based on the project charter, scope statement, and other relevant information.

5. Use the WBS you developed in number 4 above (enter the WBS items in the Task Name column) to create a Gantt chart and network diagram in Project 2002 for the project. Estimate task durations and enter dependencies, as appropriate. Also include several milestones. Remember that your schedule goal for the project is six months. Print the Gantt chart and network diagram.

6. Assign resources to the tasks in the Gantt chart using Project 2002. See Appendix A for instructions on setting up a resource sheet and assigning resources. Assume that Tony works an average of twenty hours per week on the project and earns $100 per hour and the other team members and support staff earn $60 per hour. Assume no one is paid overtime. Assume team members can work up to full-time on the project and support staff can work up to ten hours per week. Remember that your budget for the six-month project is $200,000. Assume all costs are for internal labor. Print the cash flow report showing costs each month. Also, print the resource usage report showing the hours each person is assigned to work on each task each week.

7. Create a probability/impact matrix and list of prioritized risks for the project. Include at least ten risks. Use the templates in Appendix D and examples in Chapters 3 and 11.

Part 3: Executing

Scenario

Tony Prince and the project team have run into a few challenges in executing the Recreation and Wellness Intranet Project. It has been two months since the project started. The company recently fired its Chief Financial Officer (CFO), and the new CFO likes to keep a very close eye on all projects and expenditures that she considers mission-critical, including your project. She questions the potential savings projected from your project. In addition, James, the person supporting your project from the human resources area, has not been very helpful. Your project team thought there were a lot of acronyms and terms unique to the information technology area, but you're having a hard time understanding several of James's suggestions because he uses a lot of human resources and wellness terminology. The same is true for Cassandra, your support person from the finance area. She also voiced her frustration at the project team meetings, which seemed to focus excessively on technical details with little or no discussion of financial topics. You are also having some difficulties working with Bonnie, the other main programmer on this project. She insists that you document and study everything very closely before writing any code. You prefer to develop screens and other ideas quickly to get user feedback. You also did not receive much response from the survey you sent to all employees asking about their interest in this new system and suggestions for what features it should have. The project team's mentality is to plow ahead to make the schedule, but you wonder if the system will be of value if you do not get more support for it.

Tasks

1. Develop a few prototype screens for the Recreation and Wellness Intranet. Include a home page and at least two other screens. Use any software to create the prototype.

2. Develop a project Web site for the Recreation and Wellness Intranet Project. Use the sample template from Chapter 10, and include links to all project documents and the prototype screens created in Step 1 above. Post the pages on the Web site.

3. Your project manager, Tony, has to give a short presentation at a management review meeting with the new CFO, the Chief Information Officer (CIO), and several other top managers at MYH. Tony asked you to help him prepare the presentation. He needed to make it five to ten minutes long. Since the CFO did not know much about this or other projects, the main intent of the meeting was to inform everyone about the project and its progress. Prepare five to ten slides that Tony could use as part of his presentation.

4. The CFO has asked all project teams to submit weekly milestone reports directly to her. Prepare a milestone report for the project based on information in your Gantt chart. Assume you are two months into your project. Add additional milestones, if needed, to make the report more meaningful. Use the milestone report template in Appendix D and the sample from Chapter 3.

5. Develop an agenda for a project team meeting to work through some of the challenges you are facing in executing the project so far. Write a one- to two-page paper with several alternatives for addressing these challenges.

6. After the management review meeting described in Step 1 above, management has decided to change the scope of the project. In addition to providing information for employees related to recreational activities and wellness programs sponsored by your company, the system should also include information about similar programs offered through a health club chain MYH recently acquired and all similar community-sponsored programs within twenty miles of MYH. You estimate this will increase the number of sports teams, wellness programs, and so on tremendously. You originally planned for the system to handle about fifty recreational activities and wellness programs offered by your company, but now you estimate there are at least two hundred items you could include that are offered by the health club chain, and another three hundred offered by local community programs. Management has asked your project team to provide new schedule and cost estimates. They are willing to spend more money and suggest outsourcing some of the work, and they would like to see the system operational within nine months from now. Document all of the planning information that you will need to change based on this information. Estimate how many hours it will take to update all the plans.

7. Develop a detailed list of assumptions and questions that the project team has related to the new scope of the project. Document the information in a one- to two-page memo to management.

Part 4: Controlling

Scenario

The project team was pretty much on schedule until management increased the scope of the project. You are working on updating all of your project plans. You decided to outsource much of the work related to user requirements, user interface design, and collecting information about similar programs offered through the health club chain MYH recently acquired and all similar community-sponsored programs within twenty miles from the company. The CFO strongly recommended a consulting firm she had worked with in the past, and she personally helped write the contract for its work. She is also monitoring the project team's project very closely.

Tasks

1. Review the Gantt chart you created in Part 2, and then update it based on the new scope of the project. Assume that everything was going as planned for the first two months of the project, and then you created a new baseline after management changed the project scope.

2. Update the project charter and scope statement. Briefly describe other plans you have created so far that you think you should update.

3. Assume that the contract and statement of work (SOW) for the outsourced work is complete except for information on controlling—specifically, how the supplier will report its progress information. Write a few paragraphs describing what information you want to see to make sure the supplier is doing a good job.

4. Review the templates in Appendix D and on the companion Web site for this text (*www.course.com/mis/schwalbe*) related to controlling. Decide which ones you want the supplier to use for its work and include instructions for using them.

5. Update your list of prioritized risks. Write a one- to two-page paper explaining the new risk information.

Part 5: Closing

Scenario

It has been almost nine months since the project scope changed. You did roll out the Recreation and Wellness Intranet last week, and so far the user response has been very positive. The CIO wants to know what will be involved in maintaining the system after the project is over. Your final project presentation to top management is next week. You did meet your schedule goal, and the final cost for the whole system was about $500,000. Top management is pleased with the results, even though it cost about $100,000 more than planned.

Tasks

1. Prepare a short presentation with ten to fifteen slides that Tony could use at the final project presentation to summarize the results of the project. Assume that Tony has fifteen to twenty minutes to give the presentation. Include slides describing each phase of the project, from initiating through closing. Highlight project changes and how they were managed. Be creative in describing the results of the project by including user feedback and other positive aspects of the project.

2. Prepare a lessons-learned report for the entire project. Include input from all stakeholders in summarizing the lessons learned. Use the template provided in Appendix D, and be creative in your response.

3. Prepare a final project report, using information provided in the template for final project documentation in Appendix D.

4. Write a one- to two-page transition report describing what will be involved in maintaining the new system. Look for templates for transition reports on the companion Web site or samples of actual transition reports available on the Internet.

5. Update the project Web site so that it includes all final project and product deliverables.

RUNNING CASE 2: SIX SIGMA PROJECT

This running case will help you learn more about Six Sigma as well as project management. It also includes the use of many templates provided in Appendix D.

Part 1: Initiating

Scenario

Global Computers, Inc. provides a variety of computer-related goods and services to customers throughout the world. The corporate headquarters has more than 50,000 full-time employees, and another 30,000 full- and part-time employees are located throughout the world. Your new CEO, John, came from General Electric, and he and his management team are ready to revamp the entire organization by implementing Six Sigma principles. The initial focus will be on training employees in Six Sigma and finding high-payoff projects that will improve operations and decrease costs. You are a project manager in the Information Technology department, and you have been with the company for more than ten years. You have met with other managers in your area, and you have suggested the following potential projects:

1. Online Six Sigma Training Program: Since the CEO wants to train many employees in Six Sigma very quickly, your team thinks it makes sense to provide much of the training in an online format. You have researched several suppliers and you believe you could outsource much of this project.

2. Coding Performance Improvement Project: Even though Global Computers is in the computer industry, the information technology department does not think it uses computing power economically

because many programs are not coded for efficiency. The company spends hundreds of millions of dollars for supercomputer and mainframe computing power, and you believe that this project can reduce the number of large computers the company needs.

3. Web Site Redesign Project: The company's corporate Web site has grown tremendously in the last few years, and you are not sure that the current design is attracting new customers or meeting current customer needs very well.

4. Wireless Technology Project: There have been great advancements in wireless technologies, and you believe that the company is not taking advantage of developing or using various wireless products.

Tasks

1. Summarize each of the above-proposed projects in a simple table format suitable for presentation to management. Include the name for each project, identify how each one would be suitable as a Six Sigma project, how each could improve operations and decrease costs, and use your best judgment to rank each project. Write your results in a one- to two-page memo to top management, including appropriate back-up information and calculations.

2. Prepare a weighted decision matrix using the template provided in Appendix D to evaluate these four projects. Develop at least four criteria, assign weights to each criterion, assign scores, and then calculate the weighted scores. Print the spreadsheet and bar chart with the results. Write a one-page paper describing this weighted decision matrix and what the results show.

3. Assume that you have been assigned to work on the Coding Performance Improvement Project described above. You are part of a team to define skill requirements for the project manager for this project. You would actually like to get the project manager position. Write a one- to two-page job description for this person, including a description of job duties, required experience, education, and desired skills.

4. Research information on the Internet related to similar projects done to improve coding efficiency and reduce the amount of processing power needed by large companies. Write a one- to two-page paper with your findings. Include at least two references.

5. Prepare a business case for the Coding Performance Improvement Project. Assume this Six Sigma project will take six months to complete and cost about $300,000. After it is completed, there would be another project to implement the new coding processes, which will take about nine months and $700,000 to do. Estimated benefits are more than $4 million the first year after implementation and $1 million per year for the next three

years. Use the business case template provided in Appendix D to help calculate the NPV, ROI, and year in which payback occurs.

6. Prepare a project charter for the Coding Performance Improvement Project. Assume the project will take six months to complete and cost about $300,000. Use the project charter template provided in Appendix D.

Part 2: Planning

Scenario

Congratulations! You have been selected as the project manager for the Coding Performance Improvement Project. The company's CIO, Mary, is the project sponsor. Now you need to put together your project team and get to work on this high-visibility project. Top management has told you that you can hand-pick your team. In addition, you need to attend a five-day Six Sigma class before you start working on this project, and you will have a senior manager at your company and Six Sigma Master Black Belt, Kevin, acting as your mentor throughout the project. Kevin will also oversee your team's progress.

Tasks

1. Write a one- to two-page paper describing your "dream team" for this project. What type of people would you want on your team? Make up names, backgrounds, and personality traits for your dream team. Assume you are considering having four full-time people and four people working 25 percent of their time on your team.

2. Develop a team contract for this project using the members of your team defined in Step 1 above. Use the template provided in Appendix D as a guide. Look at the examples of project charters in Chapters 3 and 5.

3. Prepare a stakeholder analysis for this project, using the template provided in Appendix D and example in Chapter 4 as guides. Be creative in making up information about stakeholders.

4. Develop a scope statement for the project. Use the template provided in Appendix D and the example in Chapter 3 as guides. Be as specific as possible in describing product characteristics, requirements, and deliverables.

5. Develop a work breakdown structure (WBS) for the project. Break down the work to level 2 or level 3, as appropriate. Use the Six Sigma DMAIC steps described in Chapter 8 of this text as part of your WBS. Include two days of Six Sigma training for all team members as part of your WBS. Use the template in Appendix D and samples in Chapters 3 and 5 as guides. Print the WBS in list form as a Word file. Be sure to base your WBS on the project charter, scope statement, and other relevant information.

6. Use the WBS you developed in number 4 above (enter the WBS items in the Task Name column) to create a Gantt chart and network diagram in Project 2002 for the project. Estimate task durations and enter dependencies, as appropriate, and include several milestones. Remember that your schedule goal for the project is six months. Print the Gantt chart and network diagram.

7. Create a spreadsheet to estimate project costs, clearly labeling all information. List each team member's name, how many hours he or she works each week on average, the hourly labor rate, and the total labor cost per person and for the entire project. Assume the six-month project is twenty-six weeks. Also, assume that you work an average of thirty hours per week on the project and earn $80 per hour. Assume that Kevin, your mentor, works ten hours per week on the project and earns $120 per hour, and the other team members earn $50 per hour. Assume no one is paid overtime. Include Six Sigma training costs for all full-time team members at $1000 each. Print your spreadsheet with your cost estimate based on this data. Remember that your budget for the six-month project is $300,000. Propose what changes you would make to meet this budget goal.

8. Create a probability/impact matrix and list of prioritized risks for the project. Include at least ten risks. Use the templates provided in Appendix D and examples in Chapters 3 and 11.

Part 3: Executing

Scenario

You and your project team have run into a few challenges in executing the Coding Performance Improvement Project. Three months have passed since the project started. You selected four full-time people for your dream team and four people working quarter-time. Unfortunately, you selected full-time team members who were all pretty much like you—all drivers and natural leaders. Everyone kept trying to say how to do things, but no one wanted to pay attention to the details and get the actual work done. In addition, no one on your team really understood how Six Sigma projects worked, and the training was not very helpful. No one realized how many process diagrams and how much data you had to collect. No one understood the statistics you had to run as part of the project, and the programmers on your team think all of the paperwork and number crunching is a waste of time. They think all you need to do is hold a few meetings with all of the programmers and have them follow a few basic guidelines to make their code more efficient. Kevin, your mentor and a Master Black Belt, is trying to help out, but he is ready to reassign at least half of the people on your project to his own dream team.

Tasks

1. Research the DMAIC process. Write a two- to three-page paper summarizing the steps involved and how it would apply to this project. Include at least three references, including at least one article from the *www.isixsigma.com* Web site. Discuss whether or not you think some of the team members have legitimate complaints and why.

2. Create a Pareto diagram related to the Coding Performance Improvement Project using the template provided in Appendix D. Assume you have been tracking the common coding inefficiencies programmers make by examining random samples of code from several programs. Assume there are six categories of inefficiencies with the following frequency counts:

   ■ Category 1: 24
   ■ Category 2: 72
   ■ Category 3: 28
   ■ Category 4: 15
   ■ Category 5: 57
   ■ Category 6: 4

3. Develop a project Web site for the Coding Performance Improvement Project. See the sample template provided in Chapter 10. Include links to all project documents you have created so far, as well the items you prepared for Steps 1 and 2 above. Include links to helpful Web sites about Six Sigma. Post the pages on the Internet.

4. Prepare five to ten slides for a short presentation you need to make to the Six Sigma Steering Committee, a group of senior managers in your company, about the status of the project so far. You have been instructed to be very honest in your presentation and raise all important issues. You know that this committee also likes to see suggested solutions to problems.

5. Prepare a milestone report for the project based on information in your Gantt chart. Assume you have completed the first three months of your project. Add additional milestones, if needed, to make the report more meaningful. Use the milestone report template in Appendix D and the sample provided in Chapter 3.

6. Develop an agenda for a project team meeting to work through some of the challenges you are facing in executing the project so far. Write a one- to two-page paper with several alternatives for addressing these challenges.

Part 4: Controlling

Scenario

After the Six Sigma Steering Committee meeting halfway though your project, you have been demoted to team member and Kevin, your mentor, is taking over the role as project manager. Kevin kept all four part-time team members and one of the full-time team members, but he replaced the other three people with his own staff. You had the option to quit the project, but you decided to stay on since you knew much of the history, and you really did want to see how someone else would run things. You respected Kevin's work so far and wanted to learn more from him. Kevin still plans to finish the project in three more months. He has been given authority to go a bit over budget as long as a new, more conservative financial analysis shows that this project and the following one are good investments.

Tasks

1. Review the Gantt chart you created in Part 2. Enter actuals for the first three months, assuming the project was about 20 percent behind schedule and 20 percent over budget. Print the earned value table. Import the data into Excel and create and print an earned value chart.

2. Update the cost estimate you created in Part 2. Assume that for the first 13 weeks of the project, you only used two people working 25 percent of the time for 10 hours each week, you cut your hours to 20 hours per week instead of 30, and you cut Kevin's hours to 6.5 hours per week instead of 10 hours per week. The four full-time team members did charge 40 hours per week and attended Six Sigma training that cost $1000 for each of them. Create a new spreadsheet showing these actual costs to date. Then provide an estimate of what it will cost to complete the project in the next three months. Assume Kevin works 30 hours per week, you work 30 hours per week, the two part-time people work 20 hours per week, and all of Kevin's new team members work an average of 120 hours per week total at $100 per hour.

3. Create an updated financial analysis for this project. Use the cost estimate you created in Step 2 for the project's cost, and include the cost of the next nine-month project at $700,000. Assume ongoing costs of $100,000 each year for the following four years. Assume the estimated benefits are $2 million the first year after implementation and $1 million per year for the next two years. Use an 8 percent discount rate. Determine the NPV, ROI, and payback period for both of these projects using this information. Use the financial analysis for business case template in Appendix D or the examples in Chapter 5.

4. Kevin has asked you to meet personally with the four project team members that he decided to pull off the project. Write a one- to two-page paper describing what you would say to these members of your former dream team. Be careful to put yourself in their shoes and make the meetings good learning experiences for everyone involved.

Part 5: Closing

Scenario

Kevin did a great job managing the rest of the Coding Performance Improvement Project. You were very impressed with his ability to pull everyone together to meet project goals. Your final project presentation to top management is next week. You did meet your schedule goal, and the final cost for the whole system was about $400,000. Top management is pleased with the results, even though it cost about 25 percent more than planned. They appreciated Kevin's and your honesty in redoing the financial estimates, which still made the project look like an excellent investment.

Tasks

1. Kevin asked you to help him prepare and present the final project presentation. Prepare a short presentation to summarize results of the project with ten to fifteen slides that you would give in a fifteen- to twenty-minute presentation to the Six Sigma Steering Committee. Include slides describing each phase of the project, from initiating through closing. Highlight project changes and how they were managed. Be sure to emphasize the financial value of this project for this group.

2. Prepare a lessons-learned report for what you personally learned from this project. Use the template provided in Appendix D, and be creative in your response.

3. Prepare a final project report, using information provided in the template for final project documentation in Appendix D.

4. The Six Sigma Steering Committee asked you to prepare more detailed information on how the financial benefits of this Coding Performance Improvement Project and the follow-up implementation project were determined. They also want you to meet with a senior member of the finance department to help them audit the results each year. Write a one- to two-page report describing how the benefits could be calculated and how they could be automated. Assume that all of the projected savings come from a decrease in the number of mainframe computers and supercomputers the company will need to run since the code is more efficient.

5. Update the project Web site to make sure it includes all final project and product deliverables.

RUNNING CASE 3: WEDDING PROJECT

This running case does not directly involve an information technology project, although you will need to use several software applications to complete the running case. These scenarios may be easier for you to relate to, and you can easily change parts of it to a similar scenario, such as planning a large reunion or similar event. This running case also involves using many of the templates provided in this text.

Part 1: Initiating

Scenario

Your best friend, Chelsea, just got engaged, and she's terrible at planning anything. She is highly emotional, and she relies on you, her down-to-earth, rational friend, in many situations. The couple has asked you to be in their wedding and to help with a lot of the wedding planning. The wedding is next summer, so you have one year to plan everything. Chelsea's parents gave them a very limited budget for the wedding, $5,000, and her fiancé's family has agreed to pay for all of the beverages that will be served at the wedding reception. They will also host the rehearsal dinner. Both the bride and groom have many friends and relatives, and their initial estimate is that 200 people would attend. They really want to have a nice wedding, with a full dinner, large bridal parties, a live band, a honeymoon in Hawaii, and so on. Neither Chelsea nor her fiancé, Chris, have much money of their own to contribute to wedding expenses, but they think they can come up with another $5,000 at the most if they have to.

Tasks

1. You know that Chelsea has grand plans for her wedding and you have to help define the scope to meet her budget. Many bridal magazines and Web sites have checklists to help identify everything that needs to be done and bought for a wedding. Research at least three different sources, and then put together a checklist and list of questions you want Chelsea and Chris to answer to help in planning the wedding.

2. Research information about the costs of weddings and determine the major cost drivers. Add some questions to the list you created in Step 1 to help prioritize the most important elements of their wedding where you might be able to cut costs. You know that Chelsea's aunt is a great seamstress, and Chris's mother runs a flower shop.

3. Prepare a project charter for Chelsea's wedding. Assume that Chelsea and Chris are the project sponsors and you are the project manager. The budget is $10,000, and you have no more than twelve months to plan the wedding. Include both sets of parents as key stakeholders on the project charter.

Part 2: Planning

Scenario

After researching several checklists and having many discussions with their parents and you, Chelsea and Chris are beginning to firm up some of the wedding plans. They realize they will have to be extremely budget-conscious, but they think that everyone can enjoy their wedding without spending a fortune. Key cost drivers will be the cost of the meal, the number of people who come for the meal, the cost to rent a hotel function room or hall (or provide tables and chairs if they decide to have the reception at one of their parents' homes), and the cost for the honeymoon. Chelsea's aunt has agreed to make her dress and veil as well as the bridesmaids' gowns, but Chelsea still needs to pay for the materials. Chris's mom said that her shop would donate all of the flowers, as long as the flowers are not too lavish. Chris has some friends in a small band who have agreed to play at the reception for a very small fee, and the couple decided that they could put off going to Hawaii for a few years. As long as they can spend at least three nights in a nice inn or bed and breakfast by a lake or beach, they will be happy.

Tasks

1. Develop a scope statement for the project. Use the template provided in Appendix D and the example in Chapter 3 as guides. Be as specific as possible in describing product characteristics and requirements as well as deliverables. Be creative, and pretend that you are the bride and groom coming up with these requirements.

2. Develop a work breakdown structure (WBS) for the project. Use some of the wedding planning checklists to help decide on the WBS categories. For example, food will be a major category, as will clothing. Break down the work to Level 2 or Level 3, as appropriate. Use the template in Appendix D and samples in Chapters 3 and 5 as guides. Print the WBS in list form as a Word file. Be sure to base your WBS on the project charter, scope statement, and other relevant information.

3. Use the WBS you developed in Step 2 (enter the WBS items in the Task Name column) to create a Gantt chart and network diagram in Project 2002 for the project. Estimate task durations and enter dependencies, as appropriate, and include several milestones. For example, you know that you have to reserve the church as soon as possible since many items depend on having a firm date. Chelsea's aunt would like to have at least four months to make all of the dresses, and an additional month after the fitting to make any adjustments. Print the Gantt chart and network diagram.

4. Create a spreadsheet to model the costs for the entire wedding and honeymoon. Clearly label all of the items, quantities, assumptions, and so on. Include a notes column to keep track of special notes or comments. Make the input variables very clear, such as number of people attending, number of bridesmaids and groomsmen, number of nights in a hotel for the honeymoon, number of pictures they plan to order, and so on. Enter what you think are realistic numbers and print your results. Write a one- to two-page paper explaining the cost model. If the final estimate is above or below $10,000, describe suggestions for bringing the costs closer to the budget.

5. Create a probability/impact matrix and list of prioritized risks for the wedding. For example, you know that Chris has friends with a band who said they would play for a minimal fee, but if they get a job that pays them more money that night, they might back out. Include at least ten risks in your list. Use the templates provided in Appendix D and examples in Chapters 3 and 11.

Part 3: Executing

Scenario

It is just two weeks before the wedding. Chelsea is trying on her dress, and she looks fabulous. However, you are running into some problems. Chris's friends with the band just called and said they cannot make it. So, you are scrambling to find another band or possibly a deejay at this late notice. Chelsea's mom told you in confidence that her flower shop is not doing very well. She knows that Chelsea really wants to have a lot of flowers in the church and orchids in her bouquet, but it would be much cheaper to use flowers that are in season, such as the lilacs that are just starting to bloom. She asked you if you would talk to Chelsea about changing her plans for flowers. You made what you thought were final arrangements with the food caterer and planned to have 200 people, but you are estimating that only 180 guests will attend, at most. You want to see if the caterer will not charge for those twenty people, even though your contract with them was based on serving 200 people.

Tasks

1. Develop a project Web site for Chris and Chelsea's wedding. See the sample template provided in Chapter 10, but feel free to make it more personal. Include links to all project documents you have created so far. Also, include links to helpful sites about weddings, including online gift registries, and the couple's unique interests. Be creative in developing the Web site and in finding interesting links to include. Post the pages on the Internet.

2. As part of your surprise wedding gift to Chris and Chelsea, you plan to take digital photos and short videos and put them on their own Web site. You bought them their own domain name, and you have scanned several photos of Chris and Chelsea that you got from their parents. Develop a prototype of this Web site using some of your own personal photos. Post the pages on the Internet. Include a page that explains how you purchased the domain name, how you scanned in photos, how you created short videos that run on the Web, and how you created the site. Consider using a tool such as FrontPage or Dreamweaver to create the site.

3. Chris and Chelsea's parents really appreciate all the help you have given their children in planning their wedding. They want to know how well you are meeting the budget goals for the wedding, and they wonder if you would talk to their children about controlling their money once they are married. Write a one- to two-page paper summarizing where you have spent the wedding money so far and projecting future expenditures. Include advice on how to control costs after they are married, too. Assume that Chris and Chelsea will need to spend $5,000 of their own money. Their parents know they only had $3,000 and will need a loan for the rest of the money. They also know that their children have credit card debt, and they would be better off taking out a different type of loan. Include a sheet in your spreadsheet called loans to help the couple figure out what their payments will be for a loan with varying interest rates and terms.

4. Chelsea's older sister, Alexis, has not been very cooperative throughout the wedding planning process. She just burst into the room and told Chelsea how much she hated the bridesmaid dress she had to wear and told her she is crazy to get married to the first guy she fell for. Alexis vowed not to get married until she is over thirty, and Chelsea is only twenty-three. Chelsea started to cry, and you know you have to say something. Write a short paper describing what you might say at that instant and how you can help Chelsea to deal with her older, very opinionated sister in the future.

Write a one- to two-page paper describing your suggestions for handling the conflicts related to the band, the flowers, and the caterer. Be specific in describing your options and what you think would work best.

Part 4: Controlling

Scenario

The wedding ceremony was beautiful, and people are just lining up outside the reception hall to greet the bride and groom when they come in. You thought you could just enjoy the wedding day, especially after you worked so hard to plan everything just right. Well, the caterer recognizes you from a prior meeting, and tells you there are only enough of the china serving plates for 150 people instead of 180 because someone misread the contract. There are some paper plates in the kitchen. Can they use those? The photographer just realized that he forgot to bring his flash. He asks you if someone can take pictures outside for him for about thirty minutes while he runs home to get the flash. You hear the band warming up in the background, and they sound awful! You suddenly awaken from your dream and realize you still have one more week before the wedding.

Tasks

1. Instead of making yourself available to take care of last-minute problems on the day of the wedding, you decide to do a little more verification work this week and develop contingency plans. You also plan to ask your sister, who is also very organized and level-headed, to take over for you on the day of the wedding. Write a one- to two-page paper listing important items you want to verify before the wedding day and describing some contingency plans to have in place to avoid any problems. Include a list of issues in your paper that you need to address.

2. Create a checklist for all of the items you want to verify this week. Include extra items you might keep in your sister's car trunk, such as plates, a good camera, and so on.

3. Update the project charter, scope statement, and Gantt chart to reflect what actually happened with the project planning and execution. Set a baseline on your Gantt chart, and then enter actual schedule information. Assume that some things took longer than planned (like deciding on the location and menu), while others took less time (like getting the dresses made and finalizing the honeymoon plans). Print a tracking Gantt chart to show the results.

4. Update your list of prioritized risks. Write a one- to two-page paper explaining the new risk information.

Part 5: Closing

Scenario

Chelsea and Chris just returned from their honeymoon. They had a wonderful time, and they keep thanking you for all of the help you provided. They love the Web site you made for them, and they are even meeting with someone to talk about a new loan, as you suggested. Chris also announced that his older brother just got engaged, and he wondered if you would be willing to share some advice and maybe think about being a paid wedding planner for their wedding. Chelsea's mom also mentioned that some of her customers at the flower shop have asked her if she could refer any wedding planners and if you would be interested in preparing a business card she could hand out.

Tasks

1. Prepare a short presentation with ten to fifteen slides that you could use as a final project presentation. Assume that you have fifteen to twenty minutes to give the presentation, and make the presentation interesting for your audience—the newlyweds and their parents. Also, assume that you would like to use some of the information to help you find future jobs as a wedding planner.

2. Prepare a lessons-learned report for the entire wedding project. Include input from all stakeholders in summarizing the lessons learned. Use the template provided in Appendix D, and be creative in your response.

3. Prepare a final project report, using information provided in the template for final project documentation in Appendix D.

4. You did enjoy planning Chelsea and Chris's wedding. Research the possibility of working part-time as a wedding planner, and find at least two local companies or individuals that provide wedding planning services. Estimate how much you could charge for your services and how much work you think you could handle while keeping your full-time job. Write a one- to two-page paper describing your findings and if you think it is worth pursuing a part-time or potential full-time job as a wedding planner.

5. Update the project Web site to make sure it includes all final project and product deliverables.

RUNNING CASE 4: REAL GROUP PROJECT

This running case assumes that you will work in groups and complete a real project for a real project sponsor. It uses several of the templates provided in Appendix D.

Part 1: Initiating

Scenario

It is important to initiate projects that add value to an organization. It is also important to follow good project management practices by officially recognizing projects with project charters, assigning competent project managers and other team members, and preparing business cases, if appropriate.

Tasks

1. Based on your experience and personal contacts, propose a potential project that you could work on as part of a team. Complete the Potential Project Proposal Form Template found in Appendix D (see Figure D-36). The instructor or a student should summarize each of the proposed projects in a simple spreadsheet. List the name of the student who proposed the project, the project name, and the project sponsor, as a minimum.

2. Write a one- to two-page paper proposing how to select projects and groups.

3. After projects and groups have been selected, each group should select a project manager. Write a one-page paper (one per group) describing how you decided who would be the project manager. Discuss the key roles of other group members, referring to the project proposal for more information.

4. If applicable, prepare a business case for your group project. Assume the project will last the entire class term, and the only costs will be the number of hours your group puts into the project. Use the business case template provided in Appendix D to help calculate the NPV, ROI, and year in which payback occurs.

5. Prepare a project charter for your group project. Be sure that your sponsor and all group members sign it. Send out a draft before requesting signatures. Use the project charter template provided in Appendix D.

Part 2: Planning

Scenario

Your project team should communicate with each other, your sponsor, and other key stakeholders to develop the following plans. If you can meet in person, it is often helpful. If the project is virtual, be as clear as possible in your communications.

Tasks

1. Develop a team contract for your project. Use the template provided in Appendix D as a guide. Refer to the examples of a team contract provided in Chapter 3.

2. Prepare a stakeholder analysis for your project, using the template provided in Appendix D and the example in Chapter 4 as guides. Be creative in making up information about stakeholders.

3. Prepare a communications management plan to address how your team will communicate with each other and key stakeholders on this project.

Review Chapter 10 for information on this topic.

4. Develop a scope statement for the project. Use the template provided in Appendix D and the example in Chapter 3 as guides. Be as specific as possible in describing product characteristics, requirements, and deliverables.

5. Develop a work breakdown structure (WBS) for the project. Break down the work to level 2 or level 3, as appropriate. Use the template in Appendix D and samples in Chapters 3 and 5 as guides. Print the WBS in list form as a Word file. Be sure to base your WBS on the project charter, scope statement, and other relevant information.

6. Use the WBS you developed in Step 4 above (enter the WBS items in the Task Name column) to create a Gantt chart and network diagram in Project 2002 for the project. Estimate task durations and enter dependencies, as appropriate. Include several milestones, such as dates for presenting status reports and giving your final project presentation. Print the Gantt chart and network diagram.

7. Assign resources to the tasks in your Gantt chart using Project 2002. See Appendix A for instructions on setting up a resource sheet and assigning resources. Decide as a team how many hours each person would be expected to work on the project in total and roughly how many hours each person might work each week. To simplify, enter a labor rate of $10 for each team member. If there are any fixed costs, enter them also. Print out appropriate cost and human resource management reports.

8. Create a probability/impact matrix and list of prioritized risks for the project. Include at least ten risks. Use the templates provided in Appendix D and examples in Chapters 3 and 11.

Part 3: Executing

Scenario

Every project will have different challenges during execution. Each project should also have some real product-related deliverables, such as creating a Web site, developing and administering a survey, preparing a research study, and so on.

Tasks

1. Develop a project Web site for your project. Use the sample template provided in Chapter 10, if desired. Include links to all project documents and the products created as part of the project, if possible. Also, include information on each team member to personalize the site. Post the pages on the Internet.

2. Create the products you described in your scope statement. Send a note to your instructor and sponsor when key products are completed. If possible, create links to those products on your project Web site. For example, if you create a survey, include a link to the survey.

3. Prepare a milestone report for your project based on information in your Gantt chart. Assume you are halfway through your term, but your project sponsor and instructor are not sure if you are getting much work accomplished. Add additional milestones, if needed, to make the report more meaningful. See Appendix D for a milestone report template and the sample provided in Chapter 3.

4. Develop an agenda for a project team meeting to work through some of the challenges you are facing in executing the project so far. Write a one- to two-page paper with several alternatives for addressing these challenges.

Part 4: Controlling

Scenario

Your instructor and sponsor should let you know what their expectations are for overseeing your project. Each group's project manager may also have different expectations. You should have included controlling-related items in your WBS and Gantt chart. If not, go back, add them, and complete the steps below.

Tasks

1. Prepare an initial progress report and presentation. Use your project Web site when giving the presentation. Review of all of your planning documents and a summary progress report form similar to the one shown in Appendix D. Be sure to list any issues you have already had or foresee with the project.

2. Prepare at least one more progress report and presentation later in the term. Follow the format used in Step 1 above. Show the results of the work you have completed for each presentation. For example, if you are building a Web site for a real project sponsor, show the progress on the Web site during the progress presentations.

3. Review the Gantt chart you created in Part 2, and save it as a baseline. Enter the actual start and end dates for each task as it is completed. Enter actual hours and costs, if applicable. Print a tracking Gantt chart at least twice during the term of your project, once about halfway through the project and once at the end of the project.

4. Update the project charter and scope statement as needed. Update your list of prioritized risks and any other plans that need to be updated based on the actual project's execution. Include a summary of these updates in your progress reports and post them on your project Web site.

Part 5: Closing

Scenario

Use each project's closing as a learning experience for each student, group, and the entire class.

Tasks:

1. Prepare a twenty- to thirty-minute final presentation for your project. Be as professional as possible in your presentation, and be sure it is a group effort. Show your final project Web site as part of the presentation, and include a link to this final presentation on the Web site. Emphasize what went right and what went wrong on your project.

2. Prepare a lessons-learned report describing what you personally learned by doing the group project. Include lessons learned by classmates if they affected you, too. Use the template provided in Appendix D, and be honest in your response.

3. Prepare a final project notebook, including a final project report, using information provided in the template for final project documentation in Appendix D. Submit the notebook to your instructor and your project sponsor. Include copies of all deliverables. Be sure to have the project name, team member names, project Web site URL, and a date on the front page. Include a table of contents and clearly number or label each part of the report.

4. Update the project Web site to make sure it includes all final project and product deliverables.

Appendix D

Templates

INTRODUCTION

As mentioned throughout this text, many people benefit from using templates to help them prepare various project management documents, charts, and other files. This appendix provides hard copies of the templates used in this text as well as information on where to find additional templates. It also discusses how you can create your own templates to meet your specific project needs. The companion Web site for this text (*www.course.com/mis/schwalbe*) includes links to the application files used to create the templates shown in this appendix as well as links to many other Web sites that provide additional templates.

TEMPLATES FROM THIS TEXT

This section provides blank templates for all of the templates mentioned in this text. Table D-1 provides a summary of the template name, figure number, chapter number, project management process group(s) where you normally use the template, application software used to create it, and the filename for the template. Note that you could use other software to create some of the templates. For example, the stakeholder analysis template was created using Word, but you could also create it using Excel. The templates are listed in the order in which you might use them on a real project. You can download the files for all of these templates from the companion Web site for this text. Be careful to change inputs for files created in Excel or Project 2002, and feel free to modify the templates to meet your particular project needs.

Table D-1: Template Matrix

Template Name	Figure Number	Process Group	Chapter(s) Where Used	Application Software	Filename on Companion Web Site
Kickoff Meeting	D-1	Initiating	Appendix A	Word	kickoff_meeting.doc
Business Case	D-2	Initiating	3	Word	business_case.doc
Business Case Financial Analysis	D-3	Initiating	3	Excel	business_case_financials.xls
Weighted Scoring Model	D-4	Initiating	5	Excel	wtd_scoring.xls
Project Charter	D-5	Initiating	3, 5	Word	charter.doc
Team Contract	D-6	Planning	3	Word	team_contract.doc
Scope Statement	D-7	Planning	3	Word	scope_statement.doc
Statement of Work	D-8	Planning	12	Word	statement_of_work.doc
Request for Proposal Outline	D-9	Planning	12	Word	rfp_outline.doc
Stakeholder Analysis	D-10	Planning	4	Word	stakeholder_analysis.doc
Software Project Management Plan	D-11	Planning	4	Word	sw_project_mgt_plan.doc
Work Breakdown Structure	D-12	Planning	3, 5	Word	wbs.doc
Gantt Chart	D-13	Planning, Executing	3, 6	Project 2002	Gantt_chart.mpp
Network Diagram	D-14	Planning, Executing	3, 6	Project 2002	network_diagram.mpp
Project Cost Estimate Overview	D-15	Planning	7	Word	cost_estimate_overview.doc
Quality Assurance Plan	D-16	Planning	8	Word	quality_assurance_plan.doc
Pareto Diagram	D-17	Planning, Executing	8	Excel	pareto_diagram.xls

Table D-1: Template Matrix (continued)

TEMPLATE NAME	FIGURE NUMBER	PROCESS GROUP	CHAPTER(S) WHERE USED	APPLICATION SOFTWARE	FILENAME ON COMPANION WEB SITE
Project Organizational Chart	D-18	Planning, Executing	9	PowerPoint	project_org_chart.ppt
Responsibility Assignment Matrix	D-19	Planning, Executing	9	Excel	ram.xls
Resource Histogram	D-20	Planning	9	Excel	resource_histogram.xls
Stakeholder Analysis for Project Communications	D-21	Planning	10	Word	stakeholder_anal_comm.doc
Project Description (text)	D-22	Planning	10	Word	project_description_text.doc
Project Description (Gantt chart)	D-23	Planning	10	Project 2002	project_description_Gantt.mpp
Milestone Report	D-24	Executing	3	Word	milestone_report.doc
Weekly Status Report	D-25	Controlling	3	Word	weekly_status_report.doc
Monthly Progress Report	D-26	Controlling	10	Word	monthly_progress_report.doc
Probability/Impact Matrix	D-27	Planning, Executing, Controlling	11	PowerPoint	prob_impact_matrix.ppt
List of Prioritized Risks	D-28	Planning, Executing, Controlling	3, 11	Word	list_of_risks.doc
Lessons-Learned Report	D-29	Closing	3, 10	Word	lessons_learned.doc
Final Project Documentation	D-30	Closing	3, 10	Word	final_documentation.doc

**Project Name
Kickoff Meeting
Date**

Meeting Objective:

Agenda:
- Introductions of attendees
- Background of project
- Review of project-related documents (i.e. business case, project request, contract)
- Discussion of project organizational structure
- Discussion of project scope, time, and cost goals
- Discussion of other important topics
- List of action items from meeting

Action Item	Assigned To	Due Date

Date and time of next meeting:

Figure D-1. Kickoff Meeting Template

Business Case for Project Name
Date:
Prepared by:

1.0 Introduction/ Background
2.0 Business Objective
3.0 Current Situation and Problem/Opportunity Statement
4.0 Critical Assumption and Constraints
5.0 Analysis of Option and Recommendation
6.0 Preliminary Project Requirements
7.0 Budget Estimate and Financial Analysis
8.0 Schedule Estimate
9.0 Potential Risks
10.0 Exhibits Exhibit A: Financial Analysis

Figure D-2. Business Case Template

Financial Analysis for Project Name

Created by: **Date:**

Note: Change the inputs, such as interest rate, number of years, costs, and benefits. Formulas are entered in the Excel file used to create this template. Be sure to double-check the

Discount rate   8%

Assume the project is completed in Year 0

	0	1	2	3	Total
Costs	100	100	100	100	
Discount factor	1.00	0.93	0.86	0.79	
Discounted costs	**100**	**93**	**86**	**79**	**358**
Benefits	0	200	200	200	
Discount factor	1.00	0.93	0.86	0.79	
Discounted benefits	**0**	**185**	**171**	**159**	**515**
Discounted benefits - costs	(100)	93	86	79	**158** ◄— NPV
Cumulative benefits - costs	(100)	(7)	78	158	

ROI ─────────────────────► **44%**
                            Payback before Year X

Assumptions
Enter assumptions here

Figure D-3. Business Case Financial Analysis Template

Weighted Scoring Model for Project Name

Created by: Date:

Criteria	Weight	Project 1	Project 2	Project 3	Project 4
A	25%	90	90	50	20
B	15%	70	90	50	20
C	15%	50	90	50	20
D	10%	25	90	50	70
E	5%	20	20	50	90
F	20%	50	70	50	50
G	10%	20	50	50	90
Weighted Project Scores	**100%**	**56**	**78.5**	**50**	**41.5**

Figure D-4. Weighted Scoring Model Template

Project Charter

Project Title:
Project Start Date: **Projected Finish Date:**
Budget Information:

Project Manager: Name, phone, e-mail
Project Objectives:

Approach:

Roles and Responsibilities

Role	Name	Organization/ Position	Contact Information

Sign-off: (Signatures of all above stakeholders; can sign by their names in table above)

Comments: (Handwritten or typed comments from above stakeholders, if applicable)

Figure D-5. Project Charter Template

Team Contract

Project Name:

Project Team Members Names and Sign-off:

Name	Sign-off on Team Contract

Code of Conduct: As a project team, we will:

-
-
-

Participation: We will:

-
-
-

Communication: We will:

-
-
-

Problem Solving: We will:

-
-
-

Meeting Guidelines: We will:

-
-
-

Figure D-6. Team Contract Template

Scope Statement

Project Title:
Date: Prepared by:
Project Justification:

Product Characteristics and Requirements:
    1.
    2.
    3.
    4.

Summary of Project Deliverables
Project management-related deliverables: Business case, charter, team contract, scope statement, WBS, schedule, cost baseline, status reports, final project presentation, final project report, lessons-learned report, and any other documents required to manage the project.
Product-related deliverables: Research reports, design documents, software code, hardware, etc.
    1.
    2.
    3.
    4.

Project Success Criteria:

Figure D-7. Scope Statement Template

Statement of Work for Project Name

Prepared by: **Date:**

I.   Scope of Work:
Describe the work to be done in detail. Specify the hardware and software involved and
the exact nature of the work.

II.   Location of Work:
Describe where the work must be performed. Specify the location of hardware and
software and where the people must perform the work.

III.   Period of Performance:
Specify when the work is expected to start and end, working hours, number of hours that
can be billed per week, where the work must be performed, and related schedule
information.

IV.   Deliverables Schedule:
List specific deliverables, describe them in detail, and specify when they are due.

V.   Applicable Standards:
Specify any company or industry-specific standards that are relevant to performing the
work.

VI.   Acceptance Criteria:
Describe how the buyer organization will determine if the work is acceptable.

VII.   Special Requirements:
Specify any special requirements such as hardware or software certifications, minimum
degree or experience level of personnel, travel requirements, and so on.

Figure D-8. Statement of Work Template

Request for Proposal (RFP) Outline
for Project Name

Prepared by: **Date:**

I.   Purpose of RFP

II.   Organization's Background

III.   Basic Requirements

IV.   Hardware and Software Environment

V.   Description of RFP Process

VI.   Statement of Work and Schedule Information

VII.   Possible Appendices

        A. Current System Overview
        B. System Requirements
        C. Volume and Size Data
        D. Required Contents of Vendor's Response to RFP
        E. Model Contract

Figure D-9. Request for Proposal Outline Template

Stakeholder Analysis for Project Name

Prepared by: **Date:**

	Stakeholder 1	Stakeholder 2	Stakeholder 3
Organization			
Role on project			
Unique facts about stakeholder			
Level of interest			
Level of influence			
Suggestions on managing relationships			

Figure D-10. Stakeholder Analysis Template

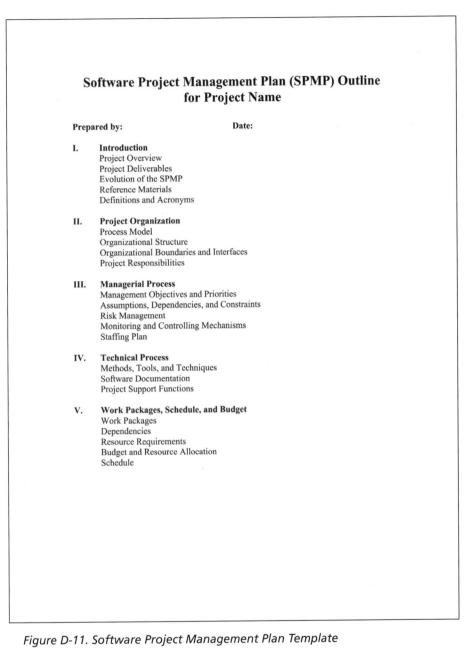

Software Project Management Plan (SPMP) Outline for Project Name

Prepared by: Date:

I.   **Introduction**
      Project Overview
      Project Deliverables
      Evolution of the SPMP
      Reference Materials
      Definitions and Acronyms

II.   **Project Organization**
      Process Model
      Organizational Structure
      Organizational Boundaries and Interfaces
      Project Responsibilities

III. **Managerial Process**
      Management Objectives and Priorities
      Assumptions, Dependencies, and Constraints
      Risk Management
      Monitoring and Controlling Mechanisms
      Staffing Plan

IV.   **Technical Process**
      Methods, Tools, and Techniques
      Software Documentation
      Project Support Functions

V.   **Work Packages, Schedule, and Budget**
      Work Packages
      Dependencies
      Resource Requirements
      Budget and Resource Allocation
      Schedule

Figure D-11. Software Project Management Plan Template

Work Breakdown Structure Template for Project Name

Prepared by: **Date:**

1.0 Main category 1
    1.1 Subcategory
    1.2 Subcategory
        1.2.1 Sub-subcategory
        1.2.2 Sub-subcategory
    1.3 Subcategory
    1.4 Subcategory

2.0 Main category 2
    2.1 Subcategory
    2.2 Subcategory
        2.2.1 Sub-subcategory
        2.2.2 Sub-subcategory
    2.3 Subcategory
    2.4 Subcategory

3.0 Main category 3
    3.1 Subcategory
    3.2 Subcategory
        3.2.1 Sub-subcategory
        3.2.2 Sub-subcategory
    3.3 Subcategory
    3.4 Subcategory

4.0 Main category 4
    4.1 Subcategory
    4.2 Subcategory
        4.2.1 Sub-subcategory
        4.2.2 Sub-subcategory
    4.3 Subcategory
    4.4 Subcategory

Figure D-12. Work Breakdown Structure Template

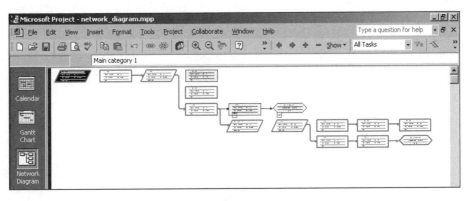

Figure D-13. Gantt Chart Template

Figure D-14. Network Diagram Template

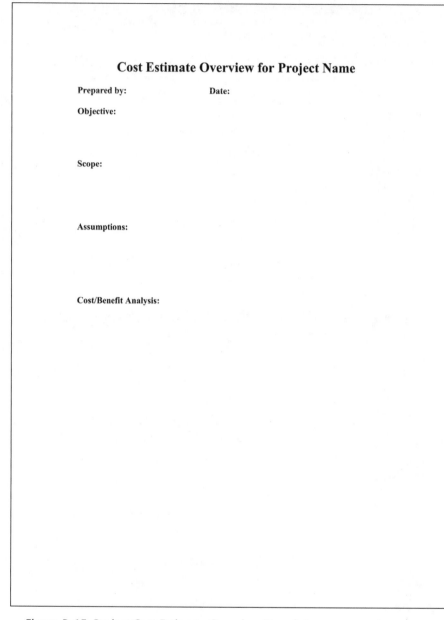

Cost Estimate Overview for Project Name

Prepared by: **Date:**

Objective:

Scope:

Assumptions:

Cost/Benefit Analysis:

Figure D-15. Project Cost Estimate Overview Template

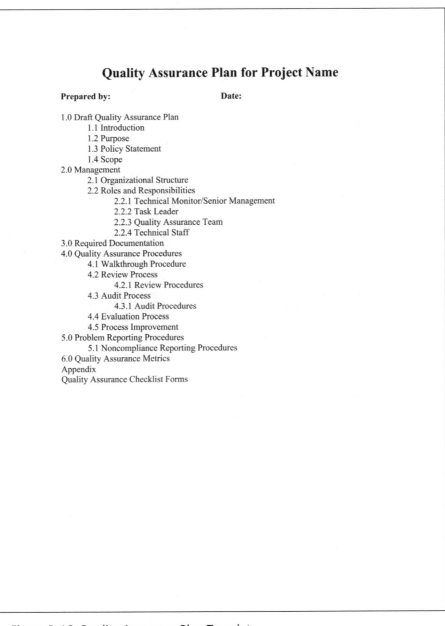

Quality Assurance Plan for Project Name

Prepared by: **Date:**

1.0 Draft Quality Assurance Plan
       1.1 Introduction
       1.2 Purpose
       1.3 Policy Statement
       1.4 Scope
2.0 Management
       2.1 Organizational Structure
       2.2 Roles and Responsibilities
              2.2.1 Technical Monitor/Senior Management
              2.2.2 Task Leader
              2.2.3 Quality Assurance Team
              2.2.4 Technical Staff
3.0 Required Documentation
4.0 Quality Assurance Procedures
       4.1 Walkthrough Procedure
       4.2 Review Process
               4.2.1 Review Procedures
       4.3 Audit Process
               4.3.1 Audit Procedures
       4.4 Evaluation Process
       4.5 Process Improvement
5.0 Problem Reporting Procedures
       5.1 Noncompliance Reporting Procedures
6.0 Quality Assurance Metrics
Appendix
Quality Assurance Checklist Forms

Figure D-16. Quality Assurance Plan Template

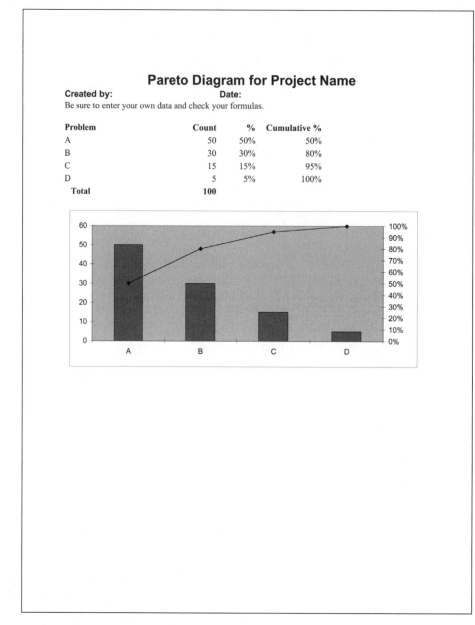

Figure D-17. Pareto Diagram Template

Organizational Chart for Project Name
Prepared by:
Date:

Figure D-18. Project Organizational Chart Template

Responsibility Assignment Matrix for Project Name

Prepared by: Date:

WBS Activities → OBS Units	1.1.1	1.1.2	1.1.3	1.2.1	1.2.2	1.2.3	2.1.1	2.1.2	2.1.3
A									
B									
C									
D									
E									
F									

Key:

Figure D-19. Responsibility Assignment Matrix Template

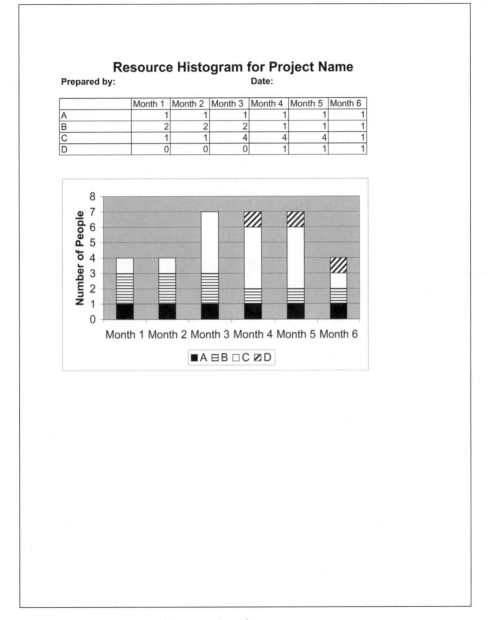

Resource Histogram for Project Name

Prepared by: Date:

	Month 1	Month 2	Month 3	Month 4	Month 5	Month 6
A	1	1	1	1	1	1
B	2	2	2	1	1	1
C	1	1	4	4	4	1
D	0	0	0	1	1	1

Figure D-20. Resource Histogram Template

Stakeholder Analysis for Project Communications

Project Name:
Date:
Created by:

Stakeholder Name	Document Name	Document Format	Contact Person	Due Date

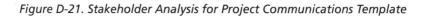

Figure D-21. Stakeholder Analysis for Project Communications Template

Project Name Description

Prepared by: **Date:**

Objective:

Scope:

Assumptions:

Cost:

Schedule:

Figure D-22. Project Description Template (text)

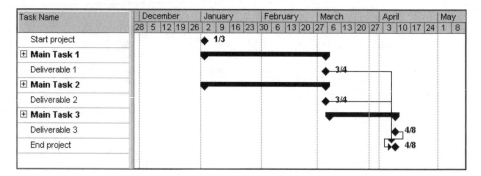

Figure D-23. Project Description Template (Gantt Chart)

Milestone Report for Project Name

Prepared by: **Date:**

Milestone	Date	Status	Responsible	Issues/Comments
Initiating				
Planning				
Executing				
Controlling				
Closing				

Figure D-24. Milestone Report Template

Weekly Status Report

Project Name: **Team Member Name:** **Date:**
Work completed this week:
Work to complete next week:
What's going well and why:
What's not going well and why:
Suggestions/Issues:
Project changes:

Figure D-25. Weekly Status Report Template

Monthly Progress Report for Project Name

Prepared by: **Date:**

I.   **Accomplishments for Month of January (or appropriate month):**

II.   **Plans for February (or following month):**

III.   **Issues:**

IV.   **Project Changes (Date and Description):**

Figure D-26. Monthly Progress Report Template

Figure D-27. Probability/Impact Matrix Template

List of Prioritized Risks for Project Name

Prepared by: **Date:**

Ranking	Potential Risk
1	
2	
3	
4	
5	
6	
7	
8	
9	
10	

Figure D-28. List of Prioritized Risks Template

Lessons Learned Report

Prepared by:	Date:
Project Name:	
Project Sponsor:	
Project Manager:	
Project Dates:	
Final Budget:	

1. Did the project meet scope, time, and cost goals?

2. What was the success criteria listed in the project scope statement?

3. Reflect on whether or not you met the project success criteria.

4. In terms of managing the project, what were the main lessons your team learned?

5. Describe one example of what went right on this project.

6. Describe one example of what went wrong on this project.

7. What will you do differently on the next project, based on your experience working on this project?

Figure D-29. Lessons Learned Report Template

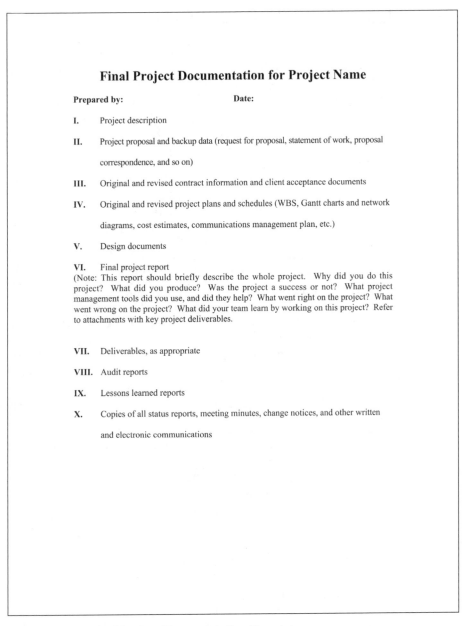

Final Project Documentation for Project Name

Prepared by: **Date:**

I.   Project description

II.   Project proposal and backup data (request for proposal, statement of work, proposal

        correspondence, and so on)

III.   Original and revised contract information and client acceptance documents

IV.   Original and revised project plans and schedules (WBS, Gantt charts and network

        diagrams, cost estimates, communications management plan, etc.)

V.   Design documents

VI.   Final project report
(Note: This report should briefly describe the whole project. Why did you do this
project? What did you produce? Was the project a success or not? What project
management tools did you use, and did they help? What went right on the project? What
went wrong on the project? What did your team learn by working on this project? Refer
to attachments with key project deliverables.

VII.   Deliverables, as appropriate

VIII.   Audit reports

IX.   Lessons learned reports

X.   Copies of all status reports, meeting minutes, change notices, and other written

        and electronic communications

Figure D-30. Final Project Documentation Template

TEMPLATES FROM OTHER SOURCES

In the spring of 2003, four students in a project management course at the University of Minnesota did a group project to help find and evaluate templates. Bill Hunt was the project manager, and his team members included Heather Fox, Kinwai Chan, and Roy Musselman. All four students took the class in a virtual environment. I was the sponsor for the project and their instructor for the course, which met one night a week from mid-January through early May.

In their preliminary project report, one of the first product-related deliverables for the project, the students reported finding 54 unique Web sites that offered project management templates. Thirty-five of these sites offered free templates with no required registration or membership, 11 sites required a registration or free membership to access the templates, and 18 sites required a fee for either membership or for direct purchase of the templates. The students used several different search engines and entered keywords such as "free project management templates" and similar variations.

Table D-2 provides a count of the templates the students found by knowledge area and process group. Note that the students found 333 templates. Figure D-31 provides a bar chart showing the number of templates found by knowledge area. Figure D-32 shows the percentage of templates found by progress group in a pie chart. Note that most of the templates—41 percent—fit under the planning process group, followed by the controlling process group with 32 percent. The majority of the templates fell under the communications and scope management knowledge areas, with 73 (22 percent) and 70 (21 percent) each, respectively. These results make sense given the nature of the documentation involved in managing projects.

Table D-2: Number of Templates Found by Knowledge Area and Process Group

KNOWLEDGE AREA	PROJECT PROCESS GROUPS					TOTAL
	INITIATING	PLANNING	EXECUTING	CONTROLLING	CLOSING	
Integration		15	7	30		52
Scope	35	23		12		70
Time		10		8		18
Cost		14		4		18
Quality		11	14	5		30
Human Resource		13	4			17
Communications		12	11	35	15	73
Risk		38		12		50
Procurement		4	1		0	5
Total	**35**	**140**	**37**	**106**	**15**	**333**

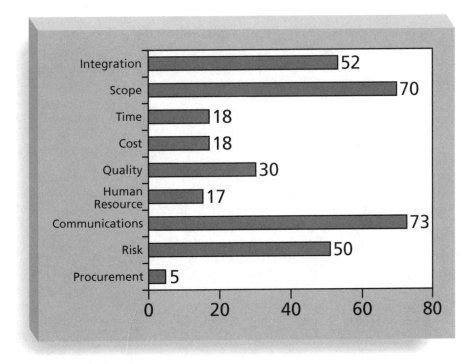

Figure D-31. Number of Templates Found by Knowledge Area

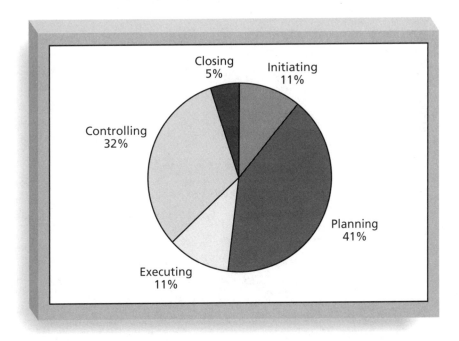

Figure D-32. Percentage of Templates Found by Project Management
Process Group

As you can see, there are plenty of templates available to assist in most areas of project management. This student project was by no means an exhaustive search. For example, typing "contract templates" into *google.com* on March 2003 produced several sites that provided free templates related to contract administration, and typing "contract closeout templates" produced several sites with free templates related to contract closeout.

Many software packages also include template files. You learned how to access several template files to use with Project 2002 in Appendix A. Word, Excel, and PowerPoint also include several template files, and you can save your own files as template files as well.

Table D-3 lists the Web site names for the sites found that provided twenty or more free templates. To help avoid information overload, I asked the students to evaluate all of the sites they found and recommend which ones they thought were most valuable. Table D-4 lists the students' ranking of the "top ten" Web sites for accessing free templates. The students evaluated the Web sites based on the overall quality of the free templates as well as the quantity of templates available. Note that most of the Web sites that provide free templates are created by government agencies, which do not profit from selling project management services. They may make their templates available as part of their public service. The companion Web site for this text provides links for accessing all of these sites. You can determine for yourself which templates are of most value in meeting your particular needs.

Table D-3: Web Sites with Twenty or More Free Templates

NUMBER OF FREE TEMPLATES	WEB SITE NAME
50	allPM.com
44	TenStep.com
40	California Department of Finance Statewide Information Technology Web Site
40	University of Waterloo Information Systems and Technology—Project Templates
30	Minnesota Office of Technology
23	New York State Office for Technology
22	Michigan Department of Information Technology
20	smartdraw.com
20	U.S. Department of Energy Office of the Chief Information Officer
20	Process Impact
20	Army Training Information Management Program
20	State of Texas Department of Information Resources

Table D-3: Web Sites with Twenty or More Free Templates (continued)

NUMBER OF FREE TEMPLATES	WEB SITE NAME
20	Project Management—Tasmanian State Government
20	ProjectWare

Table D-4: Students' "Top Ten" Web Sites with Free Templates

WEB SITE NAME
Michigan Department of Information Technology
Minnesota Office of Technology
New York State Office for Technology
Project Management—Tasmanian State Government
U.S. Department of Energy Office of the Chief Information Officer
University of Waterloo Information Systems and Technology—Project Templates
Method 123
ProjectWare
State of Texas Department of Information Resources
U.S. Defense Logistics Agency

CREATING YOUR OWN TEMPLATES

If you cannot find an existing template that meets your needs, you can create your own. It often helps to review other templates to get ideas from them and then prepare several draft templates for people in your organization to review. You can save files as template files in most software applications, or you can simply save them as regular files and make them easily available in your organization. Consult the Help facility of your software application to learn how to save files as templates. Be sure to find out if you need formal approval to create or use new templates in your organization and work within the proper channels. Remember that the main goal of using templates is to make it easier for people to manage projects.

In addition, be sure to follow copyright laws. All of the templates created for this text and shown in Figures D-1 through D-30 and Figures D33 through D35 have no copyright restrictions. You can use them freely. Most of the free templates listed on the companion Web site also have no copyright restrictions. If there is a copyright statement on a template, you must first ask for permission to use the form from that source.

The following figures provide a few additional templates that I have created using Microsoft Word. For example, since scope creep is a major problem on many information technology projects, it is important to have a good process for managing changes. Using a template for submitting change requests is one way to help manage changes. It is also helpful to include some type of client acceptance or project completion form to officially close out projects. After reviewing several change request and client acceptance/project completion templates, I put together the ones shown in Figure D-33 and D-34.

Change Request Form

Project Name:
Date Request Submitted:
Title of Change Request
Change Order Number:
Submitted by: (name and contact information)
Change Category: Scope   Schedule   Cost   Technology   Other

Description of change requested:

Events that made this change necessary or desirable:

Justification for the change/why it is needed/desired to continue/complete the project:

Impact of the proposed change on:
Scope:
Schedule:
Cost:
Staffing:
Risk:
Other:

Suggested implementation if the change request is approved:

Required approvals:

Name/Title	Date	Approve/Reject

Figure D-33. Change Request Form Template

Client Acceptance/Project Completion Form

Project Name: _____
Project Manager: _____

I (We), the undersigned, acknowledge and accept delivery of the work completed for this project on behalf of our organization. My (Our) signature(s) attest to my (our) agreement that this project has been completed. No further work should be done on this project.

Name	Title	Signature	Date

1. Was this project completed to your satisfaction?   Yes   No

2. Please provide the main reasons for your satisfaction or dissatisfaction with this project.

3. Please provide suggestions on how our organization could improve its project delivery capability in the future.

Thank you for your inputs.

Figure D-34. Client Acceptance/Project Completion Form Template

I also created two templates to help me teach my project management classes. Figure D-35 provides a template for a student survey I have students fill out on the first day of class. Note that the template includes important contact information, asks students about their strengths, weaknesses, and what they want to learn in the class, and provides an easy means for getting signatures to meet academic honesty policies. Figure D-36 provides a template that students use as part of their first homework assignment, in which they propose potential projects to do as part of the course. Since the project teams normally include three to five students, we get more than enough good projects for students to work on.

These two examples illustrate how you can create templates to meet your specific needs. If you want to get to know students better and get them involved in a class early, it helps to use a survey. If you want your class to do real projects, it helps to provide a template for students to submit their own project ideas. Similarly, if you want to get to know stakeholders on a project, you could design your own survey template to collect information about them. If you want people to propose project ideas, you also could develop a template for that purpose.

Project Management Student Survey

Student Name:_____ Date: _____
E-mail: _____
Major: _____ Minor: _____
Daytime Phone: _____ Employer: _____
Brief job description, if currently working full- or part-time:

1. In this class, all students are resources. What are some of your abilities/strengths as they might relate to this class?

2. What are some of your weaknesses or areas you would like to improve?

3. What are you most interested in learning from this class?

4. Are you currently working on a project or have you recently completed one? Please describe the project below.

5. For the group project, team members will need skills in several areas. Please rank yourself on a scale of 1 to 10 (10 being best) on each of the skills listed below.
   - ☐ Real project experience
   - ☐ Customer interaction
   - ☐ Research skills
   - ☐ Writing skills
   - ☐ Presentation skills
   - ☐ Software skills (project management software)
   - ☐ Web site creation
   - ☐ Leadership
   - ☐ Teamwork facilitation
6. List any questions or concerns on the back of this page. Thanks!

ACADEMIC HONESTY:
I have read and understand the policies regarding academic honesty. I understand how they apply to this course, and I pledge myself to abide by the policies and work to create an atmosphere of academic integrity on the campus.
Signature: _____ Date: _____

Figure D-35. Project Management Student Survey Template

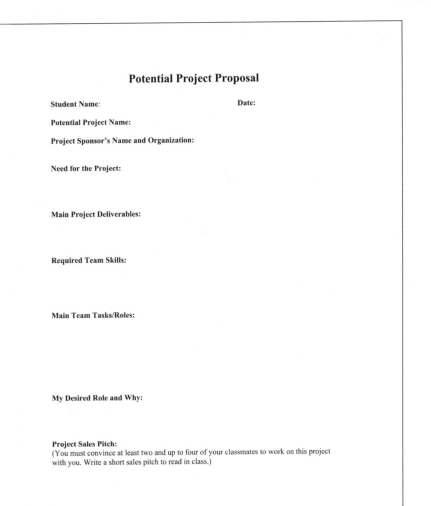

Figure D-36. Potential Project Proposal Template

As you can see, templates can assist project managers, team members, and instructors in doing their jobs. There are many templates available from this text, and many more available from numerous Web sites. It is also relatively easy to create your own templates based on your unique needs. Consult the companion Web site to download templates included in this text and to access templates on other Web sites.

Glossary

A

acceptance decisions — decisions that determine if the products or services produced as part of the project will be accepted or rejected

activity — an element of work, normally found on the WBS, that has an expected duration, cost, and resource requirements; also called **task**

activity definition — identifying the specific activities that the project team members and stakeholders must perform to produce the project deliverables

activity duration estimating — estimating the number of work periods that are needed to complete individual activities

activity sequencing — identifying and documenting the relationships between project activities

activity-on-arrow (AOA) or **arrow diagramming method (ADM)** — a network diagramming technique in which activities are represented by arrows and connected at points called nodes to illustrate the sequence of activities

actual cost (AC) — the total of direct and indirect costs incurred in accomplishing work on an activity during a given period, formerly called the actual cost of work performed (ACWP)

Adaptive Software Development (ASD) — a SDLC model that assumes that systems development follows an adaptive or non-prescriptive approach because the scope of the project cannot be clearly articulated and the schedule and cost cannot be accurately predicted

administrative closure — generating, gathering, and disseminating information to formalize phase or project completion

analogous estimates — a cost estimating technique that uses the actual cost of a previous, similar project as the basis for estimating the cost of the current project, also called **top-down estimates**

analogy approach — creating a WBS by using a similar project's WBS as a starting point

appraisal cost — the cost of evaluating processes and their outputs to ensure that a project is error-free or within an acceptable error range

B

backward pass — a project network diagramming technique that determines the late start and late finish dates for each activity in a similar fashion

balanced scorecard — a methodology that converts an organization's value drivers, such as customer service, innovation, operational efficiency, and financial performance, to a series of defined metrics

baseline — the original project plan plus approved changes

baseline dates — the planned schedule dates for activities in a Gantt chart

benchmarking — a technique used to generate ideas for quality improvements by comparing specific project practices or product characteristics to those of other projects or products within or outside the performing organization

bottom-up approach — creating a WBS by having team members identify as many specific tasks related to the project as possible and then grouping them into higher-level categories

bottom-up estimates — a cost estimating technique based on estimating individual work items and summing them to get a project total

brainstorming — a technique by which a group attempts to generate ideas or find a solution for a specific problem by amassing ideas spontaneously and without judgment

budget at completion (BAC) — the original total budget for a project

budgetary estimate — a cost estimate used to allocate money into an organization's budget

buffer — additional time to complete a task, added to an estimate to account for various factors

burst — when a single node is followed by two or more activities on a network diagram

C

Capability Maturity Model (CMM) — a five-level model laying out a generic path to process improvement for software development in organizations

cash flow — benefits minus costs or income minus expenses

cash flow analysis — a method for determining the estimated annual costs and benefits for a project

Change Control Board (CCB) — a formal group of people responsible for approving or rejecting changes on a project

change control system — a formal, documented process that describes when and how official project documents may be changed

closing processes — formalizing acceptance of the project or phase and bringing it to an orderly end

COCOMO II — a newer, computerized cost-estimating tool based on Boehm's original model that allows one to estimate the cost, effort, and schedule when planning a new software development activity

coercive power — using punishment, threats, or other negative approaches to get people to do things they do not want to do

communications infrastructure — a set of tools, techniques, and principles that provide a foundation for the effective transfer of information among people

communications management plan — a document that guides project communications

communications planning — determining the information and communications needs of the stakeholders: who needs what information, when will they need it, and how will the information be given to them

compromise mode — using a give-and-take approach to resolving conflicts; bargaining and searching for solutions that bring some degree of satisfaction to all the parties in a dispute

computerized tools — cost-estimating tools that use computer software, such as spreadsheets and project management software

configuration management — a process that ensures that the descriptions of the project's products are correct and complete

conformance — delivering products that meet requirements and fitness for use

conformance to requirements — the project processes and products meet written specifications

confrontation mode — directly facing a conflict using a problem-solving approach that allows affected parties to work through their disagreements

constructive change orders — oral or written acts or omissions by someone with actual or apparent authority that can be construed to have the same effect as a written change order

Constructive Cost Model (COCOMO) — a parametric model developed by Barry Boehm for estimating software development costs

contingency plans — predefined actions that the project team will take if an identified risk event occurs

contingency reserves — dollars included in a cost estimate to allow for future situations that may be partially planned for (sometimes called **known unknowns**) and are included in the project cost baseline

contract — a mutually binding agreement that obligates the supplier to provide the specified products or services, and obligates the buyer to pay for them

contract administration — managing the relationship with the supplier

contract close-out — completion and settlement of the contract, including resolution of any open items

control chart — a graphic display of data that illustrates the results of a process over time

controlling processes — actions to ensure that project objectives are met

cost baseline — a time-phased budget that project managers use to measure and monitor cost performance

cost budgeting — allocating the overall cost estimate to individual work items to establish a baseline for measuring performance

cost control — controlling changes to the project budget

cost estimating — developing an approximation or estimate of the costs of the resources needed to complete the project

cost management plan — a document that describes how cost variances will be managed on the project

cost of capital — the return available by investing the capital elsewhere

cost of nonconformance — taking responsibility for failures or not meeting quality expectations

cost of quality — the cost of conformance plus the cost of nonconformance

cost performance index (CPI) — the ratio of earned value to actual cost; can be used to estimate the projected cost to complete the project

cost plus fixed fee (CPFF) contract — a contract in which the buyer pays the supplier for allowable performance costs plus a fixed fee payment usually based on a percentage of estimated costs

cost plus incentive fee (CPIF) contract — a contract in which the buyer pays the supplier for allowable performance costs along with a predetermined fee and an incentive bonus

cost plus percentage of costs (CPPC) contract — a contract in which the buyer pays the supplier for allowable performance costs along with a predetermined percentage based on total costs

cost-reimbursable contracts — contracts involving payment to the supplier for direct and indirect actual costs

cost variance (CV) — the earned value minus the actual cost

crashing — a technique for making cost and schedule tradeoffs to obtain the greatest amount of schedule compression for the least incremental cost

critical chain scheduling — a method of scheduling that takes limited resources into account when creating a project schedule and includes buffers to protect the project completion date

critical path — the longest path through a network diagram that determines the earliest completion of a project

critical path method (CPM) or **critical path analysis** — a project network analysis technique used to predict total project duration

D

defect — any instance where the product or service fails to meet customer requirements

Define, Measure, Analyze, Improve, Control (DMAIC) — a systematic, closed-loop process for continued improvement that is scientific and fact based

definitive estimate — a cost estimate that provides an accurate estimate of project costs

deliverable — a product, such as a report or segment of software code, produced as part of a project

Delphi technique — an approach used to derive a consensus among a panel of experts, to make predictions about future developments

dependency — the sequencing of project activities or tasks; also called a **relationship**

deputy project managers — people who fill in for project managers in their absence and assist them as needed, similar to the role of a vice president

design of experiments — a quality technique that helps identify which variables have the most influence on the overall outcome of a process

direct costs — costs that are related to a project and can be traced back in a cost-effective way

directives — new requirements imposed by management, government, or some external influence

discount factor — a multiplier for each year based on the discount rate and year

discount rate — the minimum acceptable rate of return on an investment; also called the required rate of return, hurdle rate, or opportunity cost of capital, shown in a percentage format, such as 8 percent

discretionary dependencies — sequencing of project activities or tasks defined by the project team and used with care since they may limit later scheduling

dummy activities — activities with no duration and no resources used to show a logical relationship between two activities in the arrow diagramming method of project network diagrams

duration — the actual amount of time worked on an activity *plus* elapsed time

E

early finish date — the earliest possible time an activity can finish based on the project network logic

early start date — the earliest possible time an activity can start based on the project network logic

earned value (EV) — the percentage of work actually completed multiplied by the planned cost, formerly called the **budgeted cost of work performed (BCWP)**

earned value management (EVM) — a project performance measurement technique that integrates scope, time, and cost data

effort — the number of workdays or work hours required to complete a task

empathic listening — listening with the intent to understand

enterprise project management software — software that integrates information from multiple projects to show the status of active, approved, and future projects across an entire organization

estimate at completion (EAC) — an estimate of what it will cost to complete the project based on performance to date

executing processes — coordinating people and other resources to carry out the project plans and produce the products or deliverables of the project

executive steering committee — a group of senior executives from various parts of the organization who regularly review important corporate projects and issues

expected monetary value (EMV) — the product of the risk event probability and the risk event's monetary value

expert power — using one's personal knowledge and expertise to get people to change their behavior

external dependencies — sequencing of project activities or tasks that involve relationships between project and non-project activities

external failure cost — a cost related to all errors not detected and corrected before delivery to the customer

Extreme Programming (XP) — an approach to programming for developing software in rapidly changing environments

extrinsic motivation — causes people to do something for a reward or to avoid a penalty

F

fallback plans — plans developed for risks that have a high impact on meeting project objectives, to be implemented if attempts to reduce the risk are not effective

fast tracking — a schedule compression technique in which you do activities in parallel that you would normally do in sequence

features — the special characteristics that appeal to users

feeding buffers — additional time added before tasks on the critical path that are preceded by non-critical-path tasks

finish-to-finish dependency — a relationship on a network diagram where the "from" activity must be finished before the "to" activity can be finished

finish-to-start dependency — a relationship on a project network diagram where the "from" activity must be finished before the "to" activity can be started

fishbone diagrams — diagrams that trace complaints about quality problems back to the responsible production operations; sometimes called **Ishikawa diagrams**

fitness for use — a product can be used as it was intended

fixed-price or **lump-sum contracts** — contracts with a fixed total price for a well-defined product or service

flowcharts — diagrams that show how various elements of a system relate to each other

forcing mode — using a win-lose approach to conflict resolution to get one's way

formal acceptance — documentation that the project's sponsor or customer signs to show they have accepted the products of the project

forward pass — a network diagramming technique that determines the early start and early finish dates for each activity

free slack (free float) — the amount of time an activity can be delayed without delaying the early start of any immediately following activities

function points — technology independent assessments of the functions involved in developing a system

functional organizational structure — an organizational structure that groups people by functional areas such as information technology, manufacturing, engineering, and accounting

functionality — the degree to which a system performs its intended function

G

Gantt chart — a standard format for displaying project schedule information by listing project activities and their corresponding start and finish dates in a calendar format

groupthink — conformance to the values or ethical standards of a group

H

hierarchy of needs — a pyramid structure illustrating Maslow's theory that people's behaviors are guided or motivated by a sequence of needs

human resources frame — focuses on producing harmony between the needs of the organization and the needs of people

I

indirect costs — costs that are related to the project but cannot be traced back in a cost-effective way

influence diagrams — diagrams that represent decision problems by displaying essential elements, including decisions, uncertainties, and objectives, and how they influence each other

information distribution — making needed information available to project stakeholders in a timely manner

initiating processes — actions to commit to begin or end projects and project phases

initiation — committing the organization to begin a project or continue to the next phase of a project

intangible costs or benefits — costs or benefits that are difficult to measure in monetary terms

integrated change control — coordinating changes across the entire project

integration testing — testing that occurs between unit and system testing to test functionally grouped components to ensure a subset(s) of the entire system works together

interface management — identifying and managing the points of interaction between various elements of a project

internal failure cost — a cost incurred to correct an identified defect before the customer receives the product

internal rate of return (IRR) — the discount rate that makes the net present value equal to zero; also called **time-adjusted rate of return**

interviewing — a fact-finding technique that is normally done face-to-face or via telephone, but can also occur through e-mail or instant messaging

intrinsic motivation — causes people to participate in an activity for our own enjoyment

ISO 9000 — a quality system standard developed by the International Organization for Standardization (ISO) that includes a three-part, continuous cycle of planning, controlling, and documenting quality in an organization

J

Joint Application Design (JAD) — using highly organized and intensive workshops to bring together project stakeholders—the sponsor, users, business analysts, programmers, and so on—to jointly define and design information systems

K

kickoff meeting — a meeting held at the beginning of a project or project phase where all major project stakeholders discuss project objectives, plans, and so on

L

late finish date — the latest possible time an activity can be completed without delaying the project finish date

late start date — the latest possible time an activity may begin without delaying the project finish date

learning curve theory — a theory that states that when many items are produced repetitively, the unit cost of those items normally decreases in a regular pattern as more units are produced

legacy systems — older information systems that usually ran on an old mainframe computer

legitimate power — getting people to do things based on a position of authority

lessons learned — reflective statements written by project managers and their team members

life cycle costing — considers the total cost of ownership, or development plus support costs, for a project

M

maintainability — the ease of performing maintenance on a product

make-or-buy decision — when an organization decides if it is in its best interests to make certain products or perform certain services inside the organization, or if it is better to buy them from an outside organization

Malcolm Baldrige National Quality Award — an award started in 1987 to recognize companies that have achieved a level of world-class competition through quality management

management reserves — dollars included in a cost estimate to allow for future situations that are unpredictable (sometimes called **unknown unknowns**)

mandatory dependencies — sequencing of project activities or tasks that are inherent in the nature of the work being done on the project

matrix organizational structure — an organizational structure in which employees are assigned to both functional and project managers

maturity model — a framework for helping organizations improve their processes and systems

mean — the average value of a population

measurement and test equipment cost — the capital cost of equipment used to perform prevention and appraisal activities

merge — when two or more nodes precede a single node on a network diagram

milestone — a significant event on a project with zero duration

mind mapping — a technique that can be used to develop WBSs by using branches radiating from a central core idea to structure thoughts and ideas

mirroring — the matching of certain behaviors of the other person

Monte Carlo analysis — a risk quantification technique that simulates a model's outcome many times, to provide a statistical distribution of the calculated results

multitasking — when a resource works on more than one task at a time

Murphy's Law — if something can go wrong, it will

Myers-Briggs Type Indicator (MBTI) — a popular tool for determining personality preferences

N

net present value (NPV) analysis — a method of calculating the expected net monetary gain or loss from a project by discounting all expected future cash inflows and outflows to the present point in time

network diagram — a schematic display of the logical relationships or sequencing of project activities

node — the starting and ending point of an activity on an activity-on-arrow diagram

normal distribution — a bell-shaped curve that is symmetrical about the mean of the population

O

opportunities — chances to improve the organization

organizational breakdown structure (OBS) — a specific type of organizational chart that shows which organizational units are responsible for which work items

organizational planning — identifying, assigning, and documenting project roles, responsibilities, and reporting relationships

overallocation — when more resources than are available are assigned to perform work at a given time

overrun — the additional percentage or dollar amount by which actual costs exceed estimates

P

parametric modeling — a cost-estimating technique that uses project characteristics (parameters) in a mathematical model to estimate project costs

Pareto analysis — identifying the vital few contributors that account for most quality problems in a system

Pareto diagrams — histograms that help identify and prioritize problem areas

Parkinson's Law — work expands to fill the time allowed

payback period — the amount of time it will take to recoup, in the form of net cash inflows, the net dollars invested in a project

performance — how well a product or service performs the customer's intended use

performance reporting — collecting and disseminating performance information, which includes status reports, progress measurement, and forecasting

PERT weighted average =

$$\frac{\text{optimistic time} + 4 \times \text{most likely time} + \text{pessimistic time}}{6}$$

phase exit or **kill point** — management review that should occur after each project phase to determine if projects should be continued, redirected, or terminated

planned value (PV) — that portion of the approved total cost estimate planned to be spent on an activity during a given period, formerly called the budgeted cost of work scheduled (BCWS)

planning processes — devising and maintaining a workable scheme to accomplish the business need that the project was undertaken to address

political frame — addresses organizational and personal politics

politics — competition between groups or individuals for power and leadership

power — the potential ability to influence behavior to get people to do things they would not otherwise do

Precedence Diagramming Method (PDM) — a network diagramming technique in which boxes represent activities

predictive life cycle — a SDLC model in which the scope of the project can be clearly articulated and the schedule and cost can be accurately predicted

prevention cost — the cost of planning and executing a project so that it is error-free or within an acceptable error range

probabilistic time estimates — duration estimates based on using optimistic, most likely, and pessimistic estimates of activity durations instead of using one specific or discrete estimate

probability/impact matrix or chart — a matrix or chart that lists the relative probability of a risk occurring on one side of a matrix or axis on a chart and the relative impact of the risk occurring on the other

problems — undesirable situations that prevent the organization from achieving its goals

process — a series of actions directed toward a particular result

process adjustments — adjustments made to correct or prevent further quality problems based on quality control measurements

procurement — acquiring goods and/or services from an outside source

procurement planning — determining what to procure and when

profit margin — the ratio between revenues and profits

profits — revenues minus expenses

program — a group of projects managed in a coordinated way

Program Evaluation and Review Technique (PERT) — a project network analysis technique used to estimate project duration when there is a high degree of uncertainty with the individual activity duration estimates

progress reports — reports that describe what the project team has accomplished during a certain period of time

project — a temporary endeavor undertaken to create a unique product or service

project acquisition — the last two phases of a project life cycle, implementation and close-out

project archives — a complete set of organized project records that provide an accurate history of the project

project buffer — additional time added before the project's due date

project charter — a document that formally recognizes the existence of a project and provides direction on the project's objectives and management

project cost management — the processes required to ensure that the project is completed within the approved budget

project feasibility — the first two phases of a project life cycle, concept and development

project forecasting — predicting future project status and progress based on past information and trends

project integration management — includes the processes involved in coordinating all of the other project management knowledge areas throughout a project's life cycle. The main processes involved in project integration management include project plan development, project plan execution, and integrated change control

project life cycle — the collection of project phases—concept, development, implementation, and close-out

project management — the application of knowledge, skills, tools, and techniques to project activities in order to meet project requirements

Project Management Institute (PMI) — international professional society for project managers

project management knowledge areas — project integration management, scope, time, cost, quality, human resource, communications, risk, and procurement management

Project Management Office (PMO) — an organizational group responsible for coordinating the project management functions throughout an organization

project management process groups — the progression of project activities from initiating to planning, executing, controlling, and closing

Project Management Professional (PMP) — certification provided by PMI that requires documenting project experience and education, agreeing to follow the PMI code of ethics, and passing a comprehensive examination

project management software — software specifically designed for project management

project management tools and techniques — methods available to assist project managers and their teams.

project manager — the person responsible for working with the project sponsor, the project team, and the other people involved in a project to meet project goals

project organizational structure — an organizational structure that groups people by major projects, such as specific aircraft programs

project plan — a document used to coordinate all project planning documents and guide project execution and control. Key parts of a project plan include an introduction or overview of the project, a description of how the project is organized, the management and technical processes used on the project, and sections describing the work to be done, schedule information, and budget information

project plan development — taking the results of other planning processes and putting them into a consistent, coherent document—the project plan

project plan execution — carrying out the project plan by performing the activities it includes

project portfolio management — when organizations group and manage projects as a portfolio of investments that contribute to the entire enterprise's success

project procurement management — the processes required to acquire goods and services for a project from outside the performing organization

project quality management — ensuring that a project will satisfy the needs for which it was undertaken

project scope management — the processes involved in defining and controlling what is or is not included in a project

project sponsor — the person who provides the direction and funding for a project

project time management — the processes required to ensure timely completion of a project

prototyping — developing a working replica of the system or some aspect of the system to help define user requirements

Q

qualitative risk analysis — qualitatively analyzing risks and prioritizing their effects on project objectives

quality — the totality of characteristics of an entity that bear on its ability to satisfy stated or implied needs

quality assurance — periodically evaluating overall project performance to ensure the project will satisfy the relevant quality standards

quality audits — structured reviews of specific quality management activities that help identify lessons learned and can improve performance on current or future projects

quality circles — groups of nonsupervisors and work leaders in a single company department who volunteer to conduct group studies on how to improve the effectiveness of work in their department

quality control — monitoring specific project results to ensure that they comply with the relevant quality standards and identifying ways to improve overall quality

quality planning — identifying which quality standards are relevant to the project and how to satisfy them

quantitative risk analysis — measuring the probability and consequences of risks and estimating their effects on project objectives

R

RACI charts — charts that show Responsibility, Accountability, Consultation, and Informed roles for project stakeholders

rapport — a relation of harmony, conformity, accord, or affinity

referent power — getting people to do things based on an individual's personal charisma

reliability — the ability of a product or service to perform as expected under normal conditions without unacceptable failures

Request for Proposal (RFP) — a document used to solicit proposals from prospective suppliers

Request for Quote (RFQ) — a document used to solicit quotes or bids from prospective suppliers

required rate of return — the minimum acceptable rate of return on an investment

reserves — dollars included in a cost estimate to mitigate cost risk by allowing for future situations that are difficult to predict

residual risks — risks that remain after all of the response strategies have been implemented

resource histogram — a column chart that shows the number of resources assigned to a project over time

resource leveling — a technique for resolving resource conflicts by delaying tasks

resource loading — the amount of individual resources an existing schedule requires during specific time periods

resource planning — determining what resources (people, equipment, and materials) and what quantities of each resource should be used to perform project activities

responsibility assignment matrix (RAM) — a matrix that maps the work of the project as described in the WBS to the people responsible for performing the work as described in the OBS

return on investment (ROI) — (benefits minus costs) divided by costs

reward power — using incentives to induce people to do things

rework — action taken to bring rejected items into compliance with product requirements or specifications or other stakeholder expectations

risk — the possibility of loss or injury

risk acceptance — accepting the consequences should a risk occur

risk avoidance — eliminating a specific threat or risk, usually by eliminating its causes

risk events — specific circumstances that may occur to the detriment of the project

risk factors — numbers that represent overall risk of specific events, given their probability of occurring and the consequence to the project if they do

risk identification — determining which risks are likely to affect a project and documenting the characteristics of each

risk management plan — a plan that documents the procedures for managing risk throughout the project

risk management planning — deciding how to approach and plan the risk management activities for a project, by reviewing the project charter, WBS, roles and responsibilities, stakeholder risk tolerances, and the organization's risk management policies and plan templates

risk mitigation — reducing the impact of a risk event by reducing the probability of its occurrence

risk monitoring and control — monitoring known risks, identifying new risks, reducing risks, and evaluating the effectiveness of risk reduction throughout the life of the project

risk response planning — taking steps to enhance opportunities and reduce threats to meeting project objectives

risk symptoms or **triggers** — indications for actual risk events

risk transference — shifting the consequence of a risk and responsibility for its management to a third party

risk utility or **risk tolerance** — the amount of satisfaction or pleasure received from a potential payoff

risk-averse — having a low tolerance for risk

risk-neutral — a balance between risk and payoff

risk-seeking — having a high tolerance for risk

Robust Design methods — methods that focus on eliminating defects by substituting scientific inquiry for trial-and-error methods

rough order of magnitude (ROM) estimate — a cost estimate prepared very early in the life of a project to provide a rough idea of what a project will cost

runaway projects — projects that have significant cost or schedule overruns

S

schedule control — controlling and managing changes to the project schedule

schedule development — analyzing activity sequences, activity duration estimates, and resource requirements to create the project schedule

schedule performance index (SPI) — the ratio of earned value to planned value; can be used to estimate the projected time to complete a project

schedule variance (SV) — the earned value minus the planned value

scope — all the work involved in creating the products of the project and the processes used to create them

scope change control — controlling changes to project scope

scope creep — the tendency for project scope to keep getting bigger and longer

scope definition — subdividing the major project deliverables into smaller, more manageable components

scope planning — developing documents to provide the basis for future project decisions, including the criteria for determining if a project or phase has been completed successfully

scope statement — a document used to develop and confirm a common understanding of the project scope

scope verification — formalizing acceptance of the project scope, sometimes by customer sign-off

Scrum — an iterative development process for software where the iterations are referred to as sprints, which normally last thirty days

secondary risks — risks that are a direct result of implementing a risk response

seven run rule — if seven data points in a row on a quality control chart are all below the mean, above the mean, or are all increasing or decreasing, then the process needs to be examined for nonrandom problems

six 9s of quality — a measure of quality control equal to 1 fault in 1 million opportunities

Six Sigma — a comprehensive and flexible system for achieving, sustaining, and maximizing business success that is uniquely driven by close understanding of customer needs, disciplined use of facts, data, statistical analysis, and diligent attention to managing, improving, and reinventing business processes

slack — the amount of time a project activity may be delayed without delaying a succeeding activity or the project finish date; also called **float**

slipped milestone — a milestone activity that is completed later than planned

SMART criteria — guidelines to help define milestones that are **s**pecific, **m**easurable, **a**ssignable, **r**ealistic, and **t**ime-framed

smoothing mode — deemphasizing or avoiding areas of differences and emphasizing areas of agreements

solicitation — obtaining quotes, bids, offers, or proposals as appropriate

solicitation planning — documenting product requirements and identifying potential sources

source selection — choosing from among potential suppliers

staff acquisition — getting the needed personnel assigned to and working on the project

staffing management plan — a document that describes when and how people will be added to and taken off the project team

stakeholder analysis — an analysis of information such as key stakeholders' names and organizations, their roles on the project, unique facts about each stakeholder, their level of interest in the project, their influence on the project, and suggestions for managing relationships with each stakeholder

stakeholders — people involved in or affected by project activities

standard deviation — a measure of how much variation exists in a distribution of data

start-to-finish dependency — a relationship on a project network diagram where the "from" activity cannot start before the "to" activity is finished

start-to-start dependency — a relationship in which the "from" activity cannot start until the "to" activity starts

statement of work (SOW) — a description of the work required for the procurement

statistical sampling — choosing part of a population of interest for inspection

status reports — reports that describe where the project stands at a specific point in time

status review meetings — regularly scheduled meetings used to exchange project information

strategic planning — determining long-term objectives by analyzing the strengths and weaknesses of an organization, studying opportunities and threats in the business environment, predicting future trends, and projecting the need for new products and services

structural frame — deals with how the organization is structured (usually depicted in an organizational chart) and focuses on different groups' roles and responsibilities to meet the goals and policies set by top management

subproject managers — people responsible for managing the subprojects that a large project might be broken into

sunk cost — money that has been spent in the past

symbolic frame — focuses on the symbols, meanings, and culture of an organization

synergy — an approach where the whole is greater than the sum of the parts

system outputs — the screens and reports the system generates

system testing — testing the entire system as one entity to ensure it is working properly

systems — sets of interacting components working within an environment to fulfill some purpose

systems analysis — a problem-solving approach that requires defining the scope of the system to be studied, and then dividing it into its component parts for identifying and evaluating its problems, opportunities, constraints, and needs

systems approach — a holistic and analytical approach to solving complex problems that includes using a systems philosophy, systems analysis, and systems management

systems development life cycle (SDLC) — a framework for describing the phases involved in developing and maintaining information systems

systems management — addressing the business, technological, and organizational issues associated with making a change to a system

systems philosophy — an overall model for thinking about things as systems

systems thinking — taking a holistic view of an organization to effectively handle complex situations

T

tangible costs or **benefits** — costs or benefits that can be easily measured in dollars

team development — building individual and group skills to enhance project performance

termination clause — a contract clause that allows the buyer or supplier to end the contract

Theory of constraints (TOC) — a management philosophy that states that any complex system at any point in time often has only one aspect or constraint that is limiting its ability to achieve more of its goal

time and material contracts — a hybrid of both fixed-price and cost-reimbursable contracts

Top Ten Risk Item Tracking — a qualitative risk analysis tool for identifying risks and maintaining an awareness of risks throughout the life of a project

top-down approach — creating a WBS by starting with the largest items of the project and breaking them into their subordinate items

total slack (total float) — the amount of time an activity may be delayed from its early start without delaying the planned project finish date

tracking Gantt chart — a Gantt chart that compares planned and actual project schedule information

triggers — indications for actual risk events

triple constraint — balancing scope, time, and cost goals

U

unit price contract — a contract where the buyer pays the supplier a predetermined amount per unit of service, and the total value of the contract is a function of the quantities needed to complete the work

unit test — a test of each individual component (often a program) to ensure it is as defect-free as possible

use case modeling — a process for identifying and modeling business events, who initiated them, and how the system should respond to them

user acceptance testing — an independent test performed by end users prior to accepting the delivered system

W

weighted scoring model — a technique that provides a systematic process for basing project selection on numerous criteria

withdrawal mode — retreating or withdrawing from an actual or potential disagreement

work authorization system — a method for ensuring that qualified people do the work at the right time and in the proper sequence

work breakdown structure (WBS) — a deliverable-oriented grouping of the work involved in a project that defines the total scope of the project

work package — a task at the lowest level of the WBS

workarounds — unplanned responses to risk events that project teams use when there is no contingency plan in place

Y

yield — represents the number of units handled correctly through the development process

Index